CAMBRIDGE MC

ON MATHEMATI

MW01421527

General Editors: P.V. Landshoff, D.W. Sciama, S. Weinberg

GAUGE FIELD THEORIES

GAUGE FIELD THEORIES

STEFAN POKORSKI

Professor of Theoretical Physics, University of Warsaw

The right of the University of Cambridge to print and sell all manner of books was granted by Henry VIII in 1534. The University has printed and published continuously since 1584.

CAMBRIDGE UNIVERSITY PRESS

Cambridge

New York New Rochelle

Melbourne Sydney

Published by the Press Syndicate of the University of Cambridge
The Pitt Building, Trumpington Street, Cambridge CB2 1RP
32 East 57th Street, New York, NY 10022, USA
10 Stamford Road, Oakleigh, Melbourne 3166, Australia

First published 1987
First paperback edition 1989

Printed in Great Britain at University Press, Cambridge

British Library Cataloguing in Publication Data

Pokorski, Stefan
Gauge field theories – (Cambridge monographs
on mathematical physics)

1. Gauge fields (Physics) 2. Particles
(Nuclear physics)
I. Title
539.7'21 QC793.3.F5

Library of Congress Cataloguing in Publication Data

Pokorski, Stefan, 1942–
Gauge field theories.

(Cambridge monographs on mathematical physics)
Bibliography:
Includes index.
1. Gauge fields (Physics) 2. Quantum field theory.
3. Quantum chromodynamics. 4. Symmetry (Physics)
I. Title. II. Series.
QC793.3.F5P65 1987 530.1'43 86–18831

ISBN 0 521 26537 1 hard covers
ISBN 0 521 36846 4 paperback

TM

In memory of Osterns – my mother's family

Contents

Preface

This book has its origin in a long series of lectures given at the Institute for Theoretical Physics, Warsaw University. It is addressed to graduate students and to young research workers in theoretical physics who have some knowledge of quantum field theory in its canonical formulation, for instance at the level of two volumes by Bjorken & Drell (1964, 1965). The book is intended to be a relatively concise reference to some of the field theoretical tools used in contemporary research in the theory of fundamental interactions. It is a technical book and not easy reading. Physical problems are discussed only as illustrations of certain theoretical ideas and of computational methods. No attempt has been made to review systematically the present status of the theory of fundamental interactions.

I am grateful to Wojciech Królikowski, Maurice Jacob and Peter Landshoff for their interest in this work and strong encouragement. My warm thanks go to Antonio Bassetto, Wilfried Buchmüller, Wojciech Królikowski, Heinrich Leutwyler, Peter Minkowski, Olivier Piguet, Jacek Prentki, Marco Roncadelli, Henri Ruegg and Wojtek Zakrzewski for reading various chapters of this book and for many useful comments, and especially to Peter Landshoff for reading most of the preliminary manuscript.

I am also grateful to several of my younger colleagues at the Institute for Theoretical Physics in Warsaw for their stimulating interactions. My thanks go to Andrzej Czechowski for his collaboration at the early stage of this project and for numerous useful discussions. I am grateful to Wojciech Dębski, Marek Olechowski, Jacek Pawełczyk, Andrzej Turski, Robert Budzynski, Krzysztof Meissner and Michał Spalinski, and particularly to Paweł Krawczyk for checking a large part of calculations contained in this book.

Finally my thanks go to Zofia Ziółkowska for her contribution to the preparation of the manuscript.

Warsaw, 1985 Stefan Pokorski

1
Introduction

1.1 Gauge invariance

It seems appropriate to begin this book by quoting the following experimental information (Review of Particle Properties 1984):

electron life-time	$> 2 \times 10^{22}$ years
neutron life-time for the electric charge nonconserving decays ($n \to p +$ neutrals)	$\gtrsim 10^{19}$ years
proton life-time	$> 10^{32}$ years
photon mass	$< 6 \times 10^{-22}$ MeV
neutrino (ν_e) mass	$< 46 \times 10^{-6}$ MeV

From the first three lines of this Table we see that the conservation of the electric charge is proved experimentally much worse than the conservation of the baryonic charge. Nevertheless, nobody is seriously contesting electric charge conservation whereas experiments searching for proton decay belong to the present frontiers in physics. The reason lies in the general conviction that the theory of electromagnetism has gauge symmetry† whereas no gauge invariance principle can be invoked to protect baryonic charge conservation. Exact gauge invariance protects the conservation of the electric charge. It also implies the masslessness of the photon and as seen from the Table the present experimental limit on the photon mass is indeed many orders of magnitude better than for the other 'massless' particle: the electron neutrino.

It should be stressed at this point that $U(1)$ gauge invariance implies global $U(1)$ invariance but, of course, the opposite is not true. A tiny mass for the photon would destroy the gauge invariance of electrodynamics but leave unaffected all its Earth-bound effects, including the quantum ones, as long as the electric charge

† The reader who is not familiar with notions of global symmetry and gauge symmetry is advised to read Section 1.3 first.

was conserved. In particular, such a theory is also renormalizable (Matthews 1949, Boulware 1970, Salam & Strathdee 1970) since longitudinally polarized photons decouple from the conserved current.

Thus one may ask what are the virtues of gauge invariance? We trust electric charge conservation, because we expect gauge invariance is behind it, and we doubt baryon charge conservation, though this has been much better proved experimentally than the former one. We do not trust global symmetries as candidates for underlying first principles! A global symmetry gives us a freedom of convention: choice of a reference frame (the phase of the electron wave function for the $U(1)$ symmetry of electrodynamics). It can be redefined freely, provided that all observers in the universe redefine it in exactly the same way. This sounds unphysical and we are led to propose that this freedom of convention is present independently at every space-time point or is not present at all as an exact law of nature. (Approximate global symmetries may, nevertheless, be and are very useful in describing the fundamental interactions.) This aesthetical argument may not convince everybody. Those who remain sceptical should then remember that gauge theories give an economical description of the laws of nature based on well-defined underlying principles which has been phenomenologically successful. As we know at present, this statement accounts not only for electrodynamics with its $U(1)$ abelian gauge symmetry, but also for weak and strong interactions successfully described by gauge field theories with non-abelian symmetry groups: $SU(2) \times U(1)$ and $SU(3)$, respectively. And non-abelian gauge symmetries are more restrictive and more profound than the $U(1)$ symmetry. In particular, non-abelian gauge bosons carry the group charges and their mass terms in the lagrangian would in general, unless introduced by spontaneous symmetry breaking, destroy not only the gauge symmetry but also the current conservation and therefore the renormalizability of the theory. The standard experimental evidence for gauge theories of weak and strong interactions is briefly summarized in the next Section.

We end this Section with a short historical 'footnote' (Pauli 1933). The terminology 'gauge invariance' can be traced back to Weyl's studies (Weyl 1919) of invariance under space-time-dependent changes of gauge (scale) in an attempt to unify gravity and electromagnetism. This attempt proved, however, unsuccessful. In 1926, Fock observed (Fock 1926) that one could base quantum electrodynamics (QED) of scalar particles on the operator

$$-\mathrm{i}\hbar\frac{\partial}{\partial x^{\mu}} - \frac{e}{c}A_{\mu}$$

where A_μ is the electromagnetic four-potential and that the equations were invariant under the transformation

$$A_\mu \to A_\mu + \frac{\partial f(x)}{\partial x^\mu}, \qquad \Phi \to \Phi \exp\left[\mathrm{i}ef(x)/\hbar c\right]$$

which he called gradient transformation. London (1927) pointed out the similarity of Fock's to Weyl's earlier work: instead of Weyl's scale change a local phase change was considered by Fock. In 1929, Weyl studied invariance under this phase change but he kept unchanged his earlier terminology 'gauge invariance' (Weyl 1929). The concept of gauge transformations was generalized to non-abelian gauge groups by Yang & Mills (1954). Similar ideas were also proposed much earlier by Klein (1939) and by Shaw (1955).

1.2 Reasons for gauge theories of strong and electroweak interactions

We summarize very briefly the standard arguments in favour of gauge theories in elementary particle physics. Both quantum chromodynamics (QCD) and the Glashow–Salam–Weinberg theory are syntheses of our understanding of fundamental interactions progressing over many past years.

QCD

QCD emerged as a development of the Gell-Mann–Zweig quark model for hadrons (Gell-Mann 1964, Zweig 1964). The latter was postulated as a rationale for the successful $SU(3)$ classification of hadrons (today one should say flavour $SU(3)$). Assigning quarks q to the fundamental representation of $SU(3)$, not realized by any known hadrons, and giving them spin one-half one obtains the phenomenologically successful $SU(3)$ and $SU(6)$ schemes. $SU(6)$ is obtained by adjoining the group $SU(2)$ of spin rotation to the internal symmetry group $SU(3)$ for baryons (qqq) and mesons ($q\bar{q}$). In particular, the known hadrons indeed realize only those representations of $SU(3)$ which are given by the composite model. The quark model for hadrons, successful as it was, appeared, however, to have difficulties in reconciling the Fermi statistics for quarks with the most natural assumption that in the lowest-lying hadronic states all the relative angular momenta among constituent quarks vanish (s-wave states). Thus, baryon wave functions should be antisymmetric in spin and flavour degrees of freedom. This is not the case in the original quark model as can be immediately seen from inspection of the $\Delta^{++}(\frac{3}{2}^+)$ wave function which must be $u\uparrow u\uparrow u\uparrow$; u denotes the quark with electric charge $Q=\frac{3}{2}$, the arrow denotes spin $S_z=\frac{1}{2}$ for each quark.

The difficulty can be resolved by postulating a new internal quantum number for quarks which has been called colour (Greenberg 1964, Han & Nambu 1965, Nambu 1966 and Bardeen, Fritzsch & Gell-Mann 1973). If a quark of each flavour has three, otherwise indistinguishable, colour states, Fermi statistics is saved by using a totally antisymmetric colour wave function $\varepsilon_{abc}u_a\uparrow u_b\uparrow u_c\uparrow$. Assuming furthermore that (i) strong interactions are invariant under global $SU(3)_{\text{colour}}$ transformations (the states may then be classified by their $SU(3)_{\text{colour}}$ representation) (ii) physical hadrons are colourless i.e. they are singlets under $SU(3)_{\text{colour}}$ (quark confinement) we can understand why only qqq and $q\bar{q}$ states, and not qq or qqqq etc., exist in nature: the singlet representation appears only in the $3 \times 3 \times 3$ and $3 \times \bar{3}$ products.

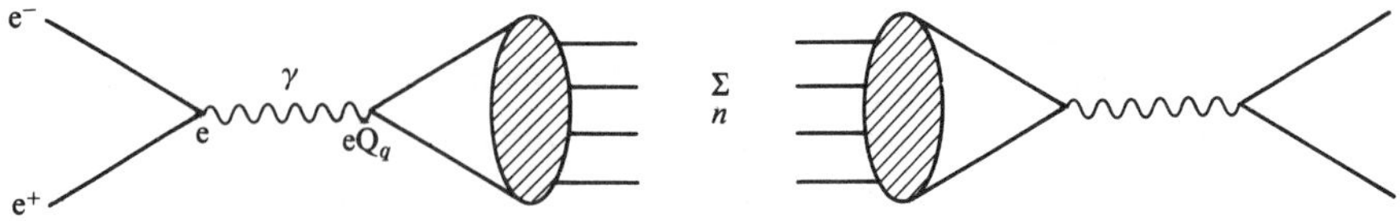

Fig. 1.1 The process $e^+e^- \to \gamma \to$ hadrons in the parton model. The sum is taken over all hadronic states in the reaction $e^+e^- \to \gamma \to \bar{q}q \to$ hadrons.

The concept of colour is supported also by at least two other, strong arguments. One is based on the parton model (Feynman 1972) approach to the reaction $e^+e^- \to$ hadrons. The total cross section for this process is then given by the diagram in Fig. 1.1 and the ratio

$$R = \frac{\sigma(e^+e^- \to \text{hadrons})}{\sigma(e^+e^- \to \mu^+\mu^-)} \tag{1.1}$$

is predicted to be

$$R = \frac{e^2 \sum_q Q_q^2}{e^2} = \sum_q Q_q^2 = 3 \times (\tfrac{4}{9} + \tfrac{1}{9} + \tfrac{4}{9} + \tfrac{1}{9} + \tfrac{1}{9} + \cdots)$$
$$= \tfrac{11}{3} \text{ (including quarks up to b)} \tag{1.2}$$

The experimental value of R is in good agreement with this prediction and in poor agreement with the colourless prediction $\frac{11}{9}$.

Yet another reason for colour is provided by the decay $\pi^0 \to 2\gamma$. Here again the number of quark states matters in explaining the width of $\pi^0 \to 2\gamma$. This problem will be discussed in more detail in Chapter 12.

The concept of colour certainly underlies what we believe to be the true theory of strong interactions, namely QCD. However the theory also has several other basic features which are partly suggested by experimental observations and partly required by theoretical consistency. Firstly, it is assumed that strong interactions act on the colour quantum numbers and only on them. Experimentally there is no evidence for any flavour dependence of strong forces; all flavour-dependent effects can be explained by quark mass differences and the origin of the quark masses, though not satisfactorily understood yet, is expected to be outside of QCD. In addition, only colour symmetry can be assumed to be an exact symmetry (flavour symmetry is evidently broken) and this, combined with the assumption that it is a gauge symmetry (Han & Nambu 1965, Fritzsch, Gell-Mann & Leutwyler 1973), has profound implications: asymptotic freedom (Gross & Wilczek 1973, Politzer 1973) and presumably, though not proven, confinement of quarks. Both are welcome features. Asymptotic freedom means that the forces become negligible at short distances and consequently the interaction between quarks by exchange of non-abelian gauge fields (gluons) is consistent with the successful, as the first approximation, description of the deep inelastic scattering in the frame-work of the parton model. It has been shown that only non-abelian gauge theories are asymptotically free (Coleman & Gross 1973). Confinement of the colour quantum numbers, i.e. of quarks and gluons, has not yet been proved to follow from QCD but it is likely to be true, reflecting strongly singular structure of the non-abelian gauge theory in the IR region. Once we assume colourful quarks as elementary objects in hadrons, confinement of colour is desirable in view of the so far unsuccessful experimental search for free quarks and to avoid a proliferation of unwanted states.

An important line of argument in favour of gluons as vector bosons begins with approximate chiral symmetry of strong interactions (Chapter 9). Coupling of fermions with vector and axial-vector fields, but not with scalars or pseudoscalars, is chirally invariant. A theory with axial-vector gluons based on the $SU(3)$ group cannot be consistently renormalized because of anomalies (Chapter 12). Thus we arrive at vector gluon interaction.

In recent years there has been a lot of research in QCD perturbation theory neglecting the unsolved confinement problem. Of course this cannot be fully satisfactory since experimenters collide hadrons and not quarks and gluons. One can nevertheless argue that it is a justifiable approximation at short distances. Thus, at its present stage the theory provides us with calculable corrections to the free-field behaviour of quarks and gluons in the parton model and can be tested in the deep inelastic region. Given the accuracy of the calculations and of the experimental data one cannot claim yet to have strongly positive experimental verification of perturbative QCD predictions. However, experimental results are certainly consistent with QCD (in particular jet physics provides us already with good evidence for the vector nature of gluons) and in view of its elegance and self-consistency there are few sceptical of its chance of being the theory of strong interactions.

Electroweak theory

There is, at present, impressive experimental evidence for the electroweak gauge theory with the gauge symmetry spontaneously broken. To introduce the Glashow–Salam–Weinberg theory we recall first that the effective Fermi lagrangian for the charged-current weak interactions, valid at low energies, has conventionally been taken to be (for a systematic account of the weak interactions phenomenology see e.g. Gasiorowicz 1966 and more recently Abers & Lee 1973 and Taylor 1976)

$$\mathscr{L}_{\text{eff}}(x) = \sqrt{2}G_{\text{F}} j^{\dagger}_{\mu}(x) j^{\mu}(x) \tag{1.3}$$

where the Fermi β-decay constant $G_{\text{F}} = 1.165 \times 10^{-5}\,\text{GeV}^{-2}$ ($\hbar = c = 1$) and the charged current $j_{\mu}(x)$ is composed of several pieces, each with V–A structure. In terms of the lepton and quark fields it can be written as follows

$$j^{\mu} = \sum_{i} \bar{\Psi}^{i}_{\text{L}} \gamma^{\mu} T^{-} \Psi^{i}_{\text{L}} \tag{1.4}$$

with

$$T^{\pm} = \tfrac{1}{2}(\tau^{1} \pm \text{i}\tau^{2}) = T^{1} \pm \text{i}T^{2}$$

where τ^i are Pauli matrices and for weak interactions, as known in 1960s,

$$\Psi_L^i = \tfrac{1}{2}(1-\gamma_5)\left\{\begin{pmatrix} \nu_e \\ e^- \end{pmatrix}, \begin{pmatrix} \nu_\mu \\ \mu^- \end{pmatrix}, \begin{pmatrix} u \\ d' \end{pmatrix}\right\} \tag{1.5}$$

The subscript L stands for the left-handed fermions. The prime superscript indicates the existence of mixing of the quark fields observed in strong interactions (mass eigenstates):

$$d' = d\cos\Theta_C + s\sin\Theta_C$$

where the angle Θ_C is known as the Cabibbo angle and has been measured in weak decays of strange particles.

We now extend the Fermi–Cabibbo theory by several additional assumptions. Firstly, we postulate the existence of a symmetry group, global for the time being, for the weak interactions. Having the current (1.4) it is natural to postulate $SU(2)$ symmetry and consequently the existence of the neutral current corresponding to the third generator of the $SU(2)$

$$[T^+, T^-] = 2T^3 \tag{1.6}$$

which would induce transitions, such as, for instance, those in Fig. 1.2 occurring with a similar strength to the charged-current reactions. Neutral-current weak transitions with the expected strength have been discovered at CERN (Hasert *et al.* 1973). However, they (i) are not of purely V–A character as expected from the $SU(2)$ model, and (ii) always conserve strangeness to a very good accuracy. According to the existing experimental limits the strangeness non-conserving neutral-current transitions (like $K_L^0 \to \mu^+\mu^-$ or $K^0 \leftrightarrow \bar{K}^0$) are suppressed by many orders of magnitude as compared to the standard weak processes. Both factors call for further invention in searching for a realistic theory of weak interactions. Glashow, Iliopoulos & Maiani (1970) have discovered that the problem of the strangeness non-conserving neutral current is solved if the set of fermionic doublets Ψ_L^i is completed with a fourth one

$$\begin{pmatrix} c \\ s' \end{pmatrix}, \quad s' = -d\sin\Theta_C + s\cos\Theta_C.$$

One can immediately check that with the s' orthogonal to d' the neutral current is diagonal in flavour. Thus, they have predicted the existence of the charm quark discovered later at Slac (Aubert *et al.* 1974, Augustin *et al.* 1974). Also the doublet classification of the left-handed fermions with the equal number of lepton and quark doublets, now further confirmed by the experimental discovery of the

$$\begin{pmatrix} \nu_\tau \\ \tau \end{pmatrix} \text{ and } \begin{pmatrix} t \\ b \end{pmatrix}$$

doublets†, has emerged as an important property of weak interactions. This has

† For the t quark the experimental situation is still unclear.

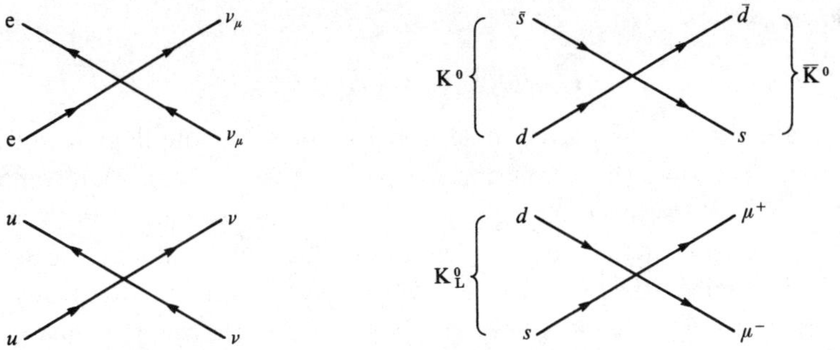

Fig. 1.2 Some neutral-current weak transitions.

profound implications for a successful extension of the effective model into a non-abelian local gauge field theory: it assures the cancellation of the chiral anomaly and consequently the renormalizability of the theory. A highly consistent scheme begins to be expected.

The next step towards the final form of the Glashow–Salam–Weinberg theory is a 'unification' of weak and electromagnetic interactions (Schwinger 1957, Glashow 1961, Salam & Ward 1964). Thus, we want the electric charge Q also to be the generator of the symmetry group of our theory. To achieve this in the most economical way we notice that for the left-handed doublets of fermions we can define a new quantum number Y (weak hypercharge)

$$\tfrac{1}{2}Y = Q - T^3 \tag{1.7}$$

such that each doublet of the left-handed fermions is an eigenvector of the operator Y (e.g. $Y_{\nu_L} = Y_{e_L} = -1$). Therefore Y commutes with the generators of $SU(2)$ and including the right-handed fermions by the prescription $\frac{1}{2}Y = Q$ (they are singlets with respects to $SU(2)$) we arrive at the $SU(2) \times U(1)$ symmetry group for the electroweak interactions. The $U(1)$ current reads

$$2j^\mu_Y = \sum_i \bar{\Psi}^i_L \gamma^\mu Y \Psi^i_L + \sum_i \bar{l}^i_R \gamma^\mu Y l^i_R + \sum_i \bar{q}^i_R \gamma^\mu Y q^i_R \tag{1.8}$$

where $l^i_R = \frac{1}{2}(1+\gamma_5)l^i$ and l^i and q^i are lepton and quark fields, respectively. According to (1.7), the electromagnetic current can be written as follows

$$\begin{aligned} j^\mu_{em} &= j^\mu_Y + \sum_i \bar{\Psi}^i_L \gamma^\mu T^3 \Psi^i_L \\ &= \sum_i \bar{l}^i Q \gamma^\mu l^i + \sum_i \bar{q}^i Q \gamma^\mu q^i \end{aligned} \tag{1.9}$$

The $SU(2) \times U(1)$ group is the minimal one which contains the electromagnetic and weak currents. With electromagnetism being described by a gauge field A_μ our minimal model 'unifying' electromagnetic and weak interactions requires the Yang–Mills gauge fields W^α_μ and B_μ to couple to the $SU(2) \times U(1)$ currents giving the interaction

$$igj^{\alpha}_{\mu}W^{\alpha\mu}+ig'j^{\mu}_{Y}B_{\mu} \tag{1.10}$$

Since we are dealing with a direct group product the couplings g and g' are independent. Therefore the significance of the electroweak unification comes from the choice of a minimal group structure which takes account of both currents. Even with two independent coupling constants, this 'unification' has a predictive power: it determines the structure of the parity violation in the neutral weak current (see (1.15) below) and gives the W^3–γ mixing in terms of g' and g, which as we will see, reduces the number of free parameters of the theory.

The charged gauge bosons

$$W^{\pm}_{\mu}=\frac{1}{\sqrt{2}}(W^1_{\mu}\pm iW^2_{\mu}) \tag{1.11}$$

couple directly to the experimentally observed weak currents. As follows from the formula

$$Q=\frac{1}{g}(gT^3)+\frac{1}{g'}(g'\tfrac{1}{2}Y) \tag{1.12}$$

the photon field A_{μ} (determined by the fact that it couples to the electromagnetic current) must be a combination of the W^3 and B bosons

$$A_{\mu}=\left(\frac{1}{g^2}+\frac{1}{g'^2}\right)^{-1/2}\left(\frac{1}{g}W^3_{\mu}+\frac{1}{g'}B_{\mu}\right) \tag{1.13}$$

The orthogonal combination

$$Z_{\mu}=\left(\frac{1}{g^2}+\frac{1}{g'^2}\right)^{-1/2}\left(\frac{1}{g'}W^3_{\mu}-\frac{1}{g}B_{\mu}\right) \tag{1.14}$$

couples to the neutral weak current j^{nc}_{μ}

$$j^{\mathrm{nc}}_{\mu}=\frac{g}{\cos\Theta_{\mathrm{W}}}(j^3_{\mu}-\sin^2\Theta_{\mathrm{W}}j_{\mu\mathrm{em}}) \tag{1.15}$$

where j^{μ}_{em} is given by (1.9) and we have introduced the standard notation in terms of the so-called Weinberg angle (Glashow 1961, Weinberg 1967*b*)

$$\sin^2\Theta_{\mathrm{W}}=\frac{g'^2}{g^2+g'^2}\quad\text{or}\quad tg\,\Theta_{\mathrm{W}}=\frac{g'}{g}\quad\text{or}\quad e=g\sin\Theta_{\mathrm{W}}=g'\cos\Theta_{\mathrm{W}}. \tag{1.16}$$

(the last relation follows from the fact that the electromagnetic coupling constant is electric charge e).

If initially all vector fields are massless, we must now introduce masses for intermediate vector boson fields $W^{\pm}$ and Z to account for the weakness of the weak interactions as compared to electromagnetism. One knows that the massive Yang–Mills theory, although less divergent than a theory with arbitrary couplings, is not renormalizable unless the vector boson masses are introduced by means of the Higgs mechanism which breaks the non-abelian gauge invariance spontaneously ('t Hooft 1971*a*, *b*). In the minimal model this is achieved (Weinberg 1967*b*, Salam 1968) with one complex $SU(2)$ doublet of scalar fields whose neutral component has a non-vanishing vacuum expectation value v. Then the boson masses are (at the tree level)

$$m_W = \tfrac{1}{2}vg, \quad m_Z = \tfrac{1}{2}v(g'^2 + g^2)^{1/2}$$

so that

$$m_W/m_Z \cos\Theta_W = 1 \tag{1.17}$$

Thus, apart from the fermion mass matrix and the scalar self-coupling we have a three parameter (g, g', v) electroweak theory. For instance, knowing the electromagnetic coupling $\alpha = e^2/4\pi$, the Fermi constant G_F and the Weinberg angle (parity non-conservation in all neutral-current transitions is consistent with the form (1.15) and with $\sin^2\Theta_W \approx 0.21$) one predicts, at the tree level,

$$m_W = m_Z \cos\Theta_W = \left(\frac{\pi\alpha}{\sqrt{2}G_F}\right)^{1/2} \frac{1}{\sin\Theta_W} \approx 80\,\text{GeV} \tag{1.18}$$

Vector bosons with the expected masses have been discovered at CERN in 1982! Though this triumph of the spontaneously broken Yang–Mills theory has still to be confirmed by higher accuracy data compared with higher order calculations (and by the discovery of the Higgs particle!) we would like to stress the role of gauge symmetry and the unification idea, apparently coupled to each other, in this so far most economical and so far very successful description of the electroweak interactions.

Non-abelian gauge theories are considered at present the most promising theoretical framework for fundamental interactions. They are also extensively studied in the hope of providing the unified description of all interactions (grand unification, supergravity) going beyond the so-called standard $SU(3) \times SU(2) \times U(1)$ model summarized in this Section.

1.3 Non-abelian gauge field lagrangian

$U(1)$ gauge symmetry

Gauge invariance in electrodynamics is often introduced as follows: consider a free-field theory of n Dirac particles with the lagrangian density

$$\mathscr{L} = \sum_{i=1}^{n} (\bar{\Psi}_i \mathrm{i}\gamma^\mu \partial_\mu \Psi_i - m\bar{\Psi}_i \Psi_i) \tag{1.19}$$

Then define an $U(1)$ group of transformations on the fields by

$$\Psi'_i(x) = \exp(-\mathrm{i}q_i\Theta)\Psi_i(x) \tag{1.20}$$

where the parameter q_i is an eigenvalue of the generator Q of $U(1)$ and numbers the representation to which the field Ψ_i belongs. The lagrangian (1.19) is invariant under that group of transformations.

Symmetries of the lagrangian imply conservation laws (Noether's theorem). For a system described by a lagrangian

$$L = \int \mathrm{d}^3x \, \mathscr{L}(\Phi_i(x), \partial_\mu \Phi_i(x))$$

with the classical equations of motion

$$\partial_\mu \frac{\delta \mathscr{L}}{\delta(\partial_\mu \Phi_i)} - \frac{\delta \mathscr{L}}{\delta \Phi_i} = 0 \tag{1.21}$$

any continuous symmetry transformation which leaves the action $S = \int L \mathrm{d}t$ invariant implies the existence of a conserved current

$$\partial^\mu j_\mu(x) = 0$$

and the charge

$$Q(t) = \int \mathrm{d}^3x \, j_0(x) \tag{1.22}$$

which is a constant of motion

$$\mathrm{d}Q/\mathrm{d}t = 0$$

provided the current falls off sufficiently rapidly at spatial infinity. If the lagrangian density $\mathscr{L}$ is invariant under some global internal symmetry group G, i.e. under the infinitesimal transformation

$$\Phi_i(x) \to \Phi_i'(x) = \Phi_i(x) + \delta\Phi_i(x) \tag{1.23}$$

where

$$\delta\Phi_i(x) = -\mathrm{i}\Theta^\alpha T^\alpha_{ij} \Phi_j(x)$$

Θ^α are x-independent and the T^α are a set of matrices satisfying the Lie algebra of the group G

$$[T^\alpha, T^\beta] = \mathrm{i}c^{\alpha\beta\gamma} T^\gamma, \quad \mathrm{Tr}[T^\alpha T^\beta] = \tfrac{1}{2}\delta^{\alpha\beta} \tag{1.24}$$

then the conserved currents read

$$j^\alpha_\mu = -\mathrm{i} \frac{\delta \mathscr{L}}{\delta(\partial^\mu \Phi_i)} T^\alpha_{ij} \Phi_j \tag{1.25}$$

By Noether's theorem the $U(1)$ symmetry of the lagrangian (1.19) implies the existence of the conserved current

$$j_\mu(x) = \sum_i q_i \bar{\Psi}_i \gamma_\mu \Psi_i \tag{1.26}$$

and therefore conservation of the corresponding charge (1.22). We now consider gauge transformations (local phase transformations in which Θ is now allowed to vary with x)

$$\Psi_i'(x) = \exp[-iq_i\Theta(x)]\Psi_i(x) \tag{1.27}$$

It is straightforward to verify that lagrangian (1.19) is not invariant under gauge transformations because transformation of the derivatives of fields gives extra terms proportional to $\partial_\mu\Theta(x)$. To make the lagrangian invariant one must introduce a new term which can compensate for the extra terms. Equivalently, one should find a modified derivative $D_\mu\Psi_i(x)$ which transforms like $\Psi_i(x)$

$$[D_\mu\Psi_i(x)]' = \exp[-iq_i\Theta(x)]D_\mu\Psi_i(x) \tag{1.28}$$

and replace ∂_μ by D_μ in the lagrangian (1.19).

The derivative D_μ is called a covariant derivative. The covariant derivative is constructed by introducing a vector (gauge) field $A_\mu(x)$ and defining

$$D_\mu\Psi_i(x) = [\partial_\mu + iq_i eA_\mu(x)]\Psi_i(x) \tag{1.29}$$

The transformation rule (1.28) is ensured if the gauge field $A_\mu(x)$ transforms as follows

$$A_\mu'(x) = A_\mu(x) + \frac{1}{e}\partial_\mu\Theta(x) \tag{1.30}$$

Covariant derivatives play an important role in gauge theories. In particular one may construct new covariant objects by repeated application of covariant derivatives. For the antisymmetric product of two derivatives

$$[D_\mu, D_\nu]\Psi_i = D_\mu(D_\nu\Psi_i) - D_\nu(D_\mu\Psi_i)$$

one gets

$$[D_\mu, D_\nu]\Psi_i = iq_i e[\partial_\mu A_\nu(x) - \partial_\nu A_\mu(x)]\Psi_i \tag{1.31}$$

By comparing the gauge transformation properties of both sides in (1.31) we conclude that

$$F_{\mu\nu}(x) = \partial_\mu A_\nu(x) - \partial_\nu A_\mu(x) \tag{1.32}$$

is gauge invariant. The field strength tensor $F_{\mu\nu}$ can be used to complete the lagrangian with the gauge-invariant kinetic energy term for the gauge field itself (assume one fermion field for definiteness)

$$\mathscr{L} = \bar{\Psi}(i\not{D} - m)\Psi - \tfrac{1}{4}F_{\mu\nu}^2 \tag{1.33}$$

The Euler–Lagrange equations of motion are now

$$\partial_\mu F^{\mu\nu} = eq\bar{\Psi}\gamma^\nu\Psi$$
$$(i\not{\partial} - m)\Psi = eq\not{A}\Psi$$

or

$$(i\not{D} - m)\Psi = 0$$

and identifying e with electric charge we recognize in our theory the Maxwell–Dirac electrodynamics.

Of course, $U(1)$ gauge invariance implies global $U(1)$ invariance and the conservation of current (1.26) and charge (1.22). It also implies absence of the gauge field mass term $m^2 A_\mu A^\mu$. But such a term does not break the global $U(1)$ symmetry.

Non-abelian gauge symmetry

To construct a non-abelian gauge field lagrangian we repeat the same steps. We start with the free-field lagrangian for Dirac fields Ψ which transform according to a representation of some non-abelian Lie group

$$\Psi'(x) = \exp(-\mathrm{i}\Theta^\alpha T^\alpha)\Psi(x) \tag{1.34}$$

where the T^α are the matrix representations of the generators of the group, appropriate for the fields Ψ and satisfying relations (1.24).

The free-field lagrangian is invariant under global group transformations (1.34). Let us now consider the extension of the group G to a group of local gauge transformations. Generalizing the $U(1)$ case we seek a covariant derivative such that

$$[D_\mu \Psi(x)]' = \exp[-\mathrm{i}\Theta^\alpha(x) T^\alpha] D_\mu \Psi(x) = U(x) D_\mu \Psi(x) \tag{1.35}$$

By analogy with (1.29) we expect it to be given by a combination of the normal derivative and a transformation on the fields Ψ

$$D_\mu \Psi(x) = [\partial_\mu + A_\mu(x)]\Psi(x) \tag{1.36}$$

where A_μ is an element of the Lie algebra

$$A_\mu(x) = -\mathrm{i} g A_\mu^\alpha(x) T^\alpha \tag{1.37}$$

Thus we need gauge fields in the number given by the number of generators of the group. The constant g is arbitrary. In classical physics, g is a dimensionful parameter and therefore can always be scaled to one. At the quantum level g is relevant since quantum theory contains the Planck constant $\hbar$ and g is dimensionless in units $\hbar = 1$. Equivalently, at the quantum level the normalization of the field is fixed by the normalization of the single particle state. With the form (1.36) the condition (1.35) is satisfied if the gauge transformation rule for $A_\mu(x)$ is as follows:

$$A'_\mu(x) = U(x) A_\mu(x) U^{-1}(x) - [\partial_\mu U(x)] U^{-1}(x) \tag{1.38}$$

or for an infinitesimal transformation

$$\delta A_\mu(x) = A'_\mu(x) - A_\mu(x) = -\partial_\mu \Theta(x) + [\Theta(x), A_\mu(x)] \tag{1.39}$$

where

$$\Theta(x) = -\mathrm{i}\Theta^\alpha(x) T^\alpha \tag{1.40}$$

is an infinitesimal gauge parameter in the matrix notation. The corresponding transformation for the gauge fields $A^\alpha_\mu(x)$ reads

$$\delta A^\alpha_\mu(x) = -(1/g)\partial_\mu \Theta^\alpha(x) + c_{\alpha\beta\gamma}\Theta^\beta(x)A^\gamma_\mu(x) \tag{1.41}$$

It can be checked that transformations (1.39) and (1.41) form a group. It is also seen from (1.41) that under global transformations ($\partial_\mu \Theta = 0$) the gauge fields transform according to the adjoint representation of the group with $(T^\alpha)_{\beta\gamma} = -\mathrm{i}c_{\alpha\beta\gamma}$.

Similarly, as before, we can construct new covariant quantities by repeated application of the covariant derivative. We get, for instance,

$$[D_\mu, D_\nu]\Psi(x) = (\partial_\mu A_\nu(x) - \partial_\nu A_\mu(x) + [A_\mu(x), A_\nu(x)])\Psi(x) \tag{1.42}$$

Thus, the antisymmetric tensor (the field strength)

$$G_{\mu\nu}(x) = \partial_\mu A_\nu(x) - \partial_\nu A_\mu(x) + [A_\mu(x), A_\nu(x)] \tag{1.43}$$

is a covariant quantity transforming under gauge transformations as follows

$$G'_{\mu\nu}(x) = U(x)G_{\mu\nu}(x)U^{-1}(x) \tag{1.44}$$

For an infinitesimal transformation we get

$$\delta G_{\mu\nu}(x) = [\Theta(x), G_{\mu\nu}(x)] \tag{1.45}$$

It is obvious from (1.43) that the tensor $G_{\mu\nu}$ can be decomposed in terms of group generators

$$G_{\mu\nu} = -\mathrm{i}gG^\alpha_{\mu\nu}T^\alpha \tag{1.46}$$

where

$$G^\alpha_{\mu\nu} = \partial_\mu A^\alpha_\nu - \partial_\nu A^\alpha_\mu + gc^{\alpha\beta\gamma}A^\beta_\mu A^\gamma_\nu \tag{1.47}$$

The transformation rule for $G^\alpha_{\mu\nu}(x)$ follows from (1.44) and (1.45).

Finally, we can generalize the definition of the covariant derivative to apply it to any Lie algebra element $\xi(x) = -\mathrm{i}\xi^\alpha(x)T^\alpha$. By applying the covariant derivative $D_\rho = \partial_\rho + A_\rho(x)$ to the object $\xi\Psi$ and insisting on the Leibnitz rule

$$D_\rho(\xi\Psi) = (D_\rho\xi)\Psi + \xi(D_\rho\Psi)$$

we get

$$D_\mu\xi = \partial_\mu\xi + [A_\mu, \xi] \tag{1.48}$$

Then the gauge transformation (1.39) can be written simply as

$$\delta A_\mu(x) = -D_\mu\Theta(x) \tag{1.49}$$

We are ready to write down the gauge invariant lagrangian for a non-abelian gauge field theory with fermions. It reads

$$\mathscr{L} = +\frac{1}{2g^2}\mathrm{Tr}[G_{\mu\nu}G^{\mu\nu}] + \bar{\Psi}(\mathrm{i}\not{D} - m)\Psi \tag{1.50}$$

($\mathrm{Tr}[G_{\mu\nu}G^{\mu\nu}]$ is gauge invariant because $\mathrm{Tr}[UG_{\mu\nu}G^{\mu\nu}U^{-1}] = \mathrm{Tr}[G_{\mu\nu}G^{\mu\nu}]$). Terms of higher order in fields are not allowed for quantum field theory if it is to be renormalizable (Chapter 4). The term $G_\mu{}^\mu$ is zero and a possibility of a parity non-conserving term $\varepsilon^{\mu\nu\rho\delta}G_{\mu\nu}G_{\rho\delta}$ in the lagrangian will be discussed later on. The pure gauge field term must be present in the lagrangian because we need the kinetic energy term quadratic in the gauge fields. Its gauge-invariant form implies then the presence of two gauge self-interaction terms of the order g and g^2, respectively

$$\frac{1}{2g^2}\mathrm{Tr}[G_{\mu\nu}G^{\mu\nu}] = -\tfrac{1}{4}(\partial_\mu A_\nu^\alpha - \partial_\nu A_\mu^\alpha)^2 - gc^{\alpha\beta\gamma}A_\mu^\beta A_\nu^\gamma \partial^\mu A^{\nu\alpha} - \tfrac{1}{4}g^2 c^{\alpha\beta\gamma}c^{\alpha\sigma\delta}A_\mu^\beta A_\nu^\gamma A^{\mu\sigma}A^{\nu\delta} \tag{1.51}$$

(we recall our normalization $\mathrm{Tr}[T_\alpha T_\beta] = \frac{1}{2}\delta_{\alpha\beta}$).

We end this section with a derivation of the Euler–Lagrange equations for gauge fields. One requires the action to be stationary with respect to small variations of the gauge fields

$$A_\mu \to A_\mu + \delta A_\mu \tag{1.52}$$

and its derivative

$$\delta(\partial_\nu A_\mu) = \partial_\nu(\delta A_\mu)$$

where we formally treat δA_μ as a covariant quantity transforming in an adjoint representation of the group

$$\delta A'_\mu(x) = U(x)\delta A_\mu(x)U^{-1}(x) \tag{1.53}$$

The change of the field strength (1.43) under the variation (1.52) can be expressed as follows

$$G_{\mu\nu} \to G_{\mu\nu} + D_\mu(\delta A_\nu) - D_\nu(\delta A_\mu) \tag{1.54}$$

Here we have used definition (1.48) of a covariant derivative acting on a covariant object δA_μ. The variation of the gauge field action is

$$\delta\int \mathrm{d}^4x \frac{1}{2g^2}\mathrm{Tr}[G_{\mu\nu}G^{\mu\nu}] = \frac{2}{g^2}\int \mathrm{d}^4x\,\mathrm{Tr}[G^{\mu\nu}(D_\mu \delta A_\nu)] \tag{1.55}$$

(antisymmetry of $G_{\mu\nu}$ in μ and ν has been used). Using the relation

$$D_\mu \mathrm{Tr}[(G^{\mu\nu}\delta A_\nu)] = \mathrm{Tr}[(D_\mu G^{\mu\nu})\delta A_\nu] + \mathrm{Tr}[G^{\mu\nu}(D_\mu \delta A_\nu)] \tag{1.56}$$

and the invariance of $\mathrm{Tr}[G_{\mu\nu}\delta A^\nu]$ under gauge transformations (therefore, on the l.h.s. of (1.56) we may replace the covariant derivative by the ordinary derivative) we conclude, by integrating (1.56) and neglecting the surface terms at infinity on the l.h.s., that

$$\delta\int \mathrm{d}^4x \frac{1}{2g^2}\mathrm{Tr}[G_{\mu\nu}G^{\mu\nu}] = -\frac{2}{g^2}\int \mathrm{d}^4x\mathrm{Tr}[(D_\mu G^{\mu\nu})\delta A_\nu] \tag{1.57}$$

The remaining term

$$\delta \int d^4x \bar{\Psi}(i\not{D} - m)\Psi$$

can be written as

$$\delta \int d^4x \bar{\Psi}(i\not{D} - m)\Psi = -2 \int d^4x \mathrm{Tr}[j_\mu \delta A^\mu] \tag{1.58}$$

where the current j_μ defined by the variation (1.58) reads

$$j_\mu(x) = -i j^\alpha_\mu(x) T^\alpha$$

with

$$j^\alpha_\mu = \bar{\Psi}\gamma_\mu T^\alpha \Psi \tag{1.59}$$

Combining (1.57) and (1.58) we get the equation of motion

$$(1/g^2) D_\mu G^{\mu\nu} + j^\nu = 0 \tag{1.60}$$

Applying the covariant derivative to (1.60) one derives the covariant divergence equation for the current (1.59)

$$D_\mu j^\mu(x) = 0 \tag{1.61}$$

Thus, gauge fields can only couple consistently to currents which are covariantly conserved. We also see from (1.61) that the non-abelian charge associated with the fermionic current is not a constant of motion; since gauge fields are not neutral their contribution must be included to find conserved charges.

Problems

1.1 Show that for an abelian gauge theory the Noether current corresponding to the gauge symmetry transformation (1.27) and (1.30) is

$$S^\mu = \partial_\rho (F^{\rho\mu} \Theta(x))$$

and that the associated conserved charge

$$Q_S = \int d^3x S^0(\mathbf{x}, t)$$

either vanishes or reduces to the electric charge associated with global $U(1)$ invariance provided Θ approaches an angle-independent limit at spatial infinity. Thus gauge invariance leads to no new conservation laws as compared to global invariance.

1.2 In abelian gauge theory the dual field strength tensor $\tilde{F}^{\mu\nu}$ is defined as

$$\tilde{F}^{\mu\nu} = \tfrac{1}{2}\varepsilon^{\mu\nu\rho\sigma} F_{\rho\sigma}, \quad \varepsilon^{0123} = -1$$

Show that $F_{\mu\nu}\tilde{F}^{\mu\nu}$ is a total divergence

$$F^{\mu\nu}\tilde{F}_{\mu\nu} = \partial_\mu K^\mu$$

where

$$K^\mu = \varepsilon^{\mu\nu\rho\sigma} A_\nu F_{\rho\sigma}$$

In a non-abelian case a matrix-valued dual field strength is

$$\tilde{G}^{\mu\nu} = \tfrac{1}{2}\varepsilon^{\mu\nu\rho\sigma} G_{\rho\sigma}$$

and

$$\mathrm{Tr}[G_{\mu\nu}\tilde{G}^{\mu\nu}] = \partial_\mu K^\mu$$

where

$$K^\rho = \varepsilon^{\rho\sigma\mu\nu}\mathrm{Tr}[G_{\sigma\mu}A_\nu + \tfrac{2}{3}A_\sigma A_\mu A_\nu]$$

1.3 Check the Jacobi identity for the covariant derivatives

$$[D_\mu,[D_\nu,D_\rho]] + [D_\nu,[D_\rho,D_\mu]] + [D_\rho,[D_\mu,D_\nu]] = 0$$

and the Bianchi identity

$$D_\mu G_{\nu\sigma} + D_\nu G_{\sigma\mu} + D_\sigma G_{\mu\nu} = 0$$

or

$$D_\mu \tilde{G}^{\mu\nu} = 0$$

1.4 In pure Yang–Mills theory the Noether current for global symmetry transformations is

$$^{\mathrm{Y}}j^a_\mu = c_{abc} G^{\mu\nu}_b A_{\nu c}$$

and for gauge transformations

$$^{\mathrm{Y}}j^\mu_\Theta = \frac{2}{g^2}\mathrm{Tr}[G^{\nu\mu}D_\nu\Theta]$$

Using field equations $D_\mu G^{\mu\nu} = 0$ show that both currents are conserved. Show that the charge

$$Q^a = \int \mathrm{d}^3x\, {}^{\mathrm{Y}}j^a_0(\mathbf{x}, t)$$

is time independent, provided ${}^{\mathrm{Y}}\mathbf{j}^a$ falls off sufficiently rapidly at large $|\mathbf{x}|$. Since ${}^{\mathrm{Y}}\mathbf{j}^a$ is not gauge invariant, the fall-off requirement restricts the large $|\mathbf{x}|$ behaviour of gauge transformation. Show that Q^a is gauge covariant against gauge transformations which approach a definite angle-independent limit as $|\mathbf{x}| \to \infty$. Show that the current ${}^{\mathrm{Y}}j^\mu_\Theta$ generates no new charges provided Θ approaches an angle-independent limit at spatial infinity.

Include fermion fields. Using Noether's theorem derive the conserved symmetry current:

$$J^a_\mu = j^a_\mu + c_{abc} G^{\mu\nu}_b A_{\nu c}$$

where j^a_μ is given by (1.59). Show that the ordinary conservation of J^a_μ is equivalent to the covariant conservation of j^a_μ and the gauge field equation (1.60). Show that

$$\partial_\mu G^{\mu\nu}_a = J^\nu_a$$

2

Path integral formulation of quantum field theory

2.1 Path integrals in quantum mechanics

Our first aim is to show that the matrix elements of quantum mechanical operators can be written as functional (path) integrals over all trajectories, with the integrand dependent on the action integral. Intuitively, the need for integration is obvious: the position operator does not commute with the momentum or the hamiltonian. In consequence, time evolution changes the position eigenstate into one in which position is not determined. The quantum system has no definite trajectory and it is necessary to take a sum over all possible ones, according to the superposition principle.

Transition matrix elements as path integrals

Consider a quantum mechanical system with one degree of freedom. The *eigenstates* of the position operator are introduced as follows

$$X_H(t)|x,t\rangle = x|x,t\rangle \qquad \text{Heisenberg picture}$$
$$X_S|x\rangle = x|x\rangle \qquad \text{Schrödinger picture}$$

with the relation

$$|x\rangle = \exp[-(\mathrm{i}/\hbar)Ht]|x,t\rangle$$

where H denotes the hamiltonian of the system. The matrix element

$$\langle x',t'|x,t\rangle = \langle x'|\exp[-(\mathrm{i}/\hbar)H(t'-t)]|x\rangle \tag{2.1}$$

corresponds to the transition from the eigenstate $|x\rangle$ at the moment of time t to the state $|x'\rangle$ at the time t', and is a Green's function: define $|t\rangle$ by $H|t\rangle = \mathrm{i}\hbar(\partial/\partial t)|t\rangle$, then

$$\langle x|t'\rangle = \int \mathrm{d}x' < x|\exp[-(\mathrm{i}/\hbar)H(t'-t)]|x'\rangle\langle x'|t\rangle$$

The matrix element (2.1) we shall first represent as a multiple integral which shall then be used to define the functional integral by a limiting procedure. First we divide the time interval $(t'-t)$ into $(n+1)$ equal parts of length ε

$$t' = (n+1)\varepsilon + t$$
$$t_j = j\varepsilon + t \qquad (j = 1, \ldots, n)$$

Next, we use the completeness relation at each of the times t_j:

$$\int \mathrm{d}x_j |x_j, t_j\rangle\langle x_j, t_j| = 1 \tag{2.2}$$

together with

$$\langle x_j, t_j | x_{j-1}, t_{j-1}\rangle = \left\langle x_j \middle| \exp\left(-\frac{\mathrm{i}}{\hbar}\varepsilon H\right) \middle| x_{j-1}\right\rangle = \langle x_j | x_{j-1}\rangle - \frac{\mathrm{i}\varepsilon}{\hbar}\langle x_j | H | x_{j-1}\rangle + O(\varepsilon^2) \tag{2.3}$$

where $x_0, x_{n+1}, t_0, t_{n+1}$ are to be understood as x, x', t, t' respectively. Choosing the hamiltonian $H = H(P, X)$ to be of the form $H = f(P) + g(X)$ we can write

$$\begin{aligned}\langle x_j | H | x_{j-1}\rangle &= \int \mathrm{d}p_j \langle x_j | p_j\rangle\langle p_j | H | x_{j-1}\rangle \\ &= \int \frac{\mathrm{d}p_j}{2\pi\hbar} \exp\left[\frac{\mathrm{i}}{\hbar} p_j(x_j - x_{j-1})\right] H(p_j, x_{j-1})\end{aligned} \tag{2.4}$$

where $H(p, x)$ is now the classical c-number hamiltonian. Using (2.4), (2.3) becomes

$$\begin{aligned}\langle x_j, t_j | x_{j-1}, t_{j-1}\rangle &= \int \frac{\mathrm{d}p_j}{2\pi\hbar} \exp\left[\frac{\mathrm{i}}{\hbar} p_j(x_j - x_{j-1})\right]\left[1 - \frac{\mathrm{i}}{\hbar}\varepsilon H(p_j, x_{j-1})\right] + O(\varepsilon^2) \\ &= \int \frac{\mathrm{d}p_j}{2\pi\hbar} \exp\left[\frac{\mathrm{i}}{\hbar} p_j(x_j - x_{j-1}) - \frac{\mathrm{i}}{\hbar}\varepsilon H(p_j, x_{j-1})\right] + O(\varepsilon^2)\end{aligned} \tag{2.5}$$

and we obtain the following expression for the matrix element (2.1):

$$\langle x', t' | x, t\rangle = \lim_{n\to\infty} \int \prod_{j=1}^{n} \mathrm{d}x_j \int \prod_{j=1}^{n+1} \frac{\mathrm{d}p_j}{2\pi\hbar} \exp\left\{\frac{\mathrm{i}}{\hbar}\sum_{j=1}^{n+1} [p_j(x_j - x_{j-1}) - H(p_j, x_{j-1})(t_j - t_{j-1})]\right\} \tag{2.6}$$

where the limit $n \to \infty$ $(\varepsilon \to 0)$ has been taken and the $O(\varepsilon^2)$ terms neglected. This result we shall write in the compact form

$$\langle x', t' | x, t\rangle = \int \frac{\mathscr{D}x\mathscr{D}p}{2\pi\hbar} \exp\left\{\frac{\mathrm{i}}{\hbar}\int_t^{t'} [p\dot{x} - H(p, x)]\,\mathrm{d}\tau\right\} \tag{2.7}$$

$\int(\mathscr{D}x\mathscr{D}p/2\pi\hbar) \equiv \int\prod_\tau(\mathrm{d}x(\tau)\,\mathrm{d}p(\tau)/2\pi\hbar)$ is called a functional integration over all phase space, with the boundary conditions $x(t) = x$, $x(t') = x'$ implied in this case. Equation (2.7) is the promised path integral representation of $\langle x', t' | x, t \rangle$.

If the hamiltonian is of a simple form

$$H = (1/2m)P^2 + V(X) \tag{2.8}$$

it is convenient to perform the momentum integrations in (2.6). Shifting the integration variables: $p_j \to p_j - m(\Delta x_j/\varepsilon)$ we obtain

$$\int \frac{\mathrm{d}p_j}{2\pi\hbar} \exp\left[\frac{\mathrm{i}}{\hbar}\left(p_j \Delta x_j - \frac{p_j^2}{2m}\varepsilon\right)\right] = \frac{1}{N_j}\frac{1}{2\pi\hbar}\exp\left[\frac{\mathrm{i}}{\hbar}\varepsilon\frac{m}{2}\left(\frac{\Delta x_j}{\varepsilon}\right)^2\right] \tag{2.9}$$

where $\Delta x_j = x_j - x_{j-1}$ and

$$\frac{1}{N_j} = \int \mathrm{d}p_j \exp\left(-\frac{\mathrm{i}}{\hbar}\frac{p_j^2}{2m}\varepsilon\right)$$

The final result has the form of a functional integral over configuration space:

$$\langle x', t' | x, t \rangle = \frac{1}{N}\int \frac{\mathscr{D}x}{2\pi\hbar} \exp\left\{\frac{\mathrm{i}}{\hbar} S[x]\right\} \tag{2.10}$$

Here $S[x] = \int_t^{t'} L(x, \dot{x})\mathrm{d}\tau$ is the action integral over the trajectory $x(\tau)$ where $L(x, \dot{x}) = \frac{1}{2}m\dot{x}^2 - V(x)$ is the Lagrange function and the normalization factor N is given by

$$\frac{1}{N} = \int \mathscr{D}p \exp\left(-\frac{\mathrm{i}}{\hbar}\int_t^{t'} \frac{p^2}{2m}\mathrm{d}\tau\right)$$

Starting with the canonically quantized theory described by hamiltonian (2.8) we have derived path integral representation (2.10). We can use another approach, namely, to define the quantum theory by functional integral (2.10) i.e. we can choose the path integral formulation as the quantization prescription for a system with the classical hamiltonian in the form (2.8). Then our derivation proves the equivalence of path integral and canonical quantization methods for systems described by hamiltonian (2.8). There are, however, systems for which canonical quantization is ambiguous due to non-commutativity of operators P and X. Here belong, for instance, theories where the lagrangian is

$$L = \tfrac{1}{2}f(x)\dot{x}^2 + g(x)\dot{x} - V(x) \tag{2.11}$$

On the other hand quantum theory, which in the classical limit gives the theory (2.11), is unambiguously defined by the path integral *Ansatz* in configuration space, insensitive to the ordering of P and X. It can be shown (Cheng 1972) that the appropriately generalized path integral reads

$$\langle x', t' | x, t \rangle \sim \lim_{j\to\infty} \int \prod_{t_j} \frac{\mathrm{d}x(t_j)}{2\pi\hbar} f^{1/2}(x(t_j)) \exp\left\{\frac{\mathrm{i}}{\hbar} S[x(t_j)]\right\} \tag{2.12}$$

Equivalently, one can postulate (2.7) with the c-number hamiltonian corresponding to lagrangian (2.11), supplemented by the rule that, whenever there is an ambiguity, the integrals over p_j are to be performed before the x-integration. Integrating over p_j one derives (2.12) (see Problem 2.1). The modification of the functional measure $\mathscr{D}x$ into

$$\mathscr{D}x = \lim_{i\to\infty} \int \prod_i \mathrm{d}x_i f^{1/2}(x_i)$$

comes from the integral

$$\frac{1}{N_j} = \int \mathrm{d}p_j \exp\left[-\frac{\mathrm{i}}{2\hbar} f^{-1}(x_j) p_j^2 \varepsilon \right]$$

which now depends on x_j. The functional integral (2.12) defines the same dynamics as the Schrödinger equation with one particular ordering of the operators P and X, namely the symmetric ordering, e.g. $[f(x)p]_S = \frac{1}{2}[f(x)p + pf(x)]$.

Theories like (2.11) are of physical interest. A lagrangian of this kind appears, for instance, for a system of two interacting particles

$$L = \frac{1}{2m}\dot{x}^2 + \frac{1}{2m}\dot{y}^2 - V(x, y) \tag{2.13}$$

subject to a constraint equation $y = c(x)$. Therefore

$$L = \frac{1}{2m}\left[1 + \left(\frac{\mathrm{d}c}{\mathrm{d}x}\right)^2 \dot{x}^2\right] - V(x, c(x)) \tag{2.14}$$

We shall encounter a similar situation in Chapter 3 when quantizing gauge field theories.

In summary, in the following we shall consider quantum theories defined by the path integral formulation. Their correspondence to canonically quantized theories, if of interest, must be verified case by case. For field theories considered in the book this, in fact, can be done at least in the framework of perturbation theory.

If $S[x(\tau)] \gg \hbar$ we can evaluate (2.10) by use of the saddle point approximation. The integral is then dominated by trajectories close to the classical one $x_c(t)$, which satisfies $\delta S[x_c(\tau)] = 0$. That is, the path integral formulation allows a relatively simple understanding of the classical and semi-classical limit.

Matrix elements of position operators

The matrix element $\langle x', t' | x, t \rangle$ determines all transition probabilities between quantum mechanical states. In view of further applications of functional formalism to quantum field theories it is also important to know the path integral representation of the matrix elements of the position operators, corresponding to the field operator of quantum field theory. For the time-ordered product of N such operators the following expression can be shown to hold

$$\langle x',t'|TX(t_1)\ldots X(t_N)|x,t\rangle = \int \frac{\mathscr{D}x\mathscr{D}p}{2\pi\hbar} x(t_1)\ldots x(t_N)\exp\left\{\frac{i}{\hbar}\int_t^{t'}[p\dot{x} - H(p,x)]\,d\tau\right\} \tag{2.15}$$

Let us check (2.15) for the product of two operators: $X(\tau_1)X(\tau_2)$ at $\tau_1 > \tau_2$. Again we divide the time axis into small intervals, choosing $t_1\ldots t_n$ in such a way that

$$\tau_1 = t_{i_1}, \qquad \tau_2 = t_{i_2}$$

and apply the completeness relation at each t_i. We obtain

$$\begin{aligned}\langle x',t'|X(\tau_1)X(\tau_2)|x,t\rangle &= \int \prod_i dx_i \langle x',t'|x_n,t_n\rangle \cdots \langle x_{i_1},t_{i_1}|X(\tau_1)|x_{i_1-1},t_{i_1-1}\rangle \cdots \\ &\quad * \langle x_{i_2},t_{i_2}|X(\tau_2)|x_{i_2-1},t_{i_2-1}\rangle \cdots \langle x_1,t_1|x,t\rangle \\ &= \int \prod_i dx_i x_{i_1} x_{i_2} \langle x',t'|x_n,t_n\rangle \cdots \langle x_1,t_1|x,t\rangle \end{aligned} \tag{2.16}$$

Proceeding exactly as when deriving (2.6) we obtain (2.15). Note that (2.16) is true for $\tau_1 > \tau_2$. When $\tau_1 < \tau_2$, the r.h.s. of (2.16) corresponds to the matrix element $\langle x',t'|X(\tau_2)X(\tau_1)|x,t\rangle$. Therefore the path integral, like (2.15), defines the matrix element of the time-ordered product of the position operators

$$\int \frac{\mathscr{D}x\mathscr{D}p}{2\pi\hbar} x(t_1)x(t_2)\exp\left\{\frac{i}{\hbar}\int_t^{t'}[p\dot{x} - H]\,d\tau\right\} = \begin{cases}\langle x',t'|X(t_1)X(t_2)|x,t\rangle, & t_1 > t_2 \\ \langle x',t'|X(t_2)X(t_1)|x,t\rangle, & t_1 < t_2\end{cases} \tag{2.17}$$

As before, it is also possible to change from phase space path integrals to path integrals over configuration space.

Also, let us note that the transition amplitude in presence of an external source $J(\tau)$

$$\langle x',t'|x,t\rangle^J = \int \frac{\mathscr{D}x\mathscr{D}p}{2\pi\hbar} \exp\left\{\frac{i}{\hbar}\int_t^{t'}[p\dot{x} - H(p,x) + \hbar J(\tau)x(\tau)]\,d\tau\right\} \tag{2.18}$$

corresponding to the usual transition amplitude with the hamiltonian modified by a source term: $H \to H - \hbar Jx$, can be used as a generating functional of the matrix elements of the position operators, which are then given by its functional derivatives with respect to $J(\tau)$

$$\langle x',t'|TX(t_1)\ldots X(t_N)|x,t\rangle = \left(\frac{1}{i}\right)^N \frac{\delta^N}{\delta J(t_1)\ldots\delta J(t_N)}\bigg|_{J=0} \langle x',t'|x,t\rangle^J \tag{2.19}$$

Instead of formal definition of the functional derivative, for our purpose it is sufficient to know that, in the case of the functional $F[J]$ defined as

$$F[J] = \int dq_1 \ldots \int dq_n f(q_1\ldots q_n)J(q_1)\ldots J(q_n) \tag{2.20}$$

the functional derivative with respect to J is simply (f can be chosen to be symmetric in all variables)

$$\frac{\delta F[J]}{\delta J(q)} = \int \mathrm{d}q_1 \ldots \mathrm{d}q_{n-1} J(q_1) \ldots J(q_{n-1}) n f(q_1 \ldots q_{n-1} q) \tag{2.21}$$

This corresponds to the usual rule of differentiating the monomials. If the functional is defined instead by the series

$$\Phi[J] = \sum_{n=1}^{\infty} \frac{1}{n!} \int \mathrm{d}q_1 \ldots \mathrm{d}q_n \phi_n(q_1 \ldots q_n) J(q_1) \ldots J(q_n) \tag{2.22}$$

then

$$\phi_n(q_1 \ldots q_n) = \left. \frac{\delta^n \Phi[J]}{\delta J(q_1) \ldots \delta J(q_n)} \right|_{J=0} \tag{2.23}$$

which compares with the Taylor expansion of the usual functions.

2.2 Vacuum-to-vacuum transitions and the imaginary time formalism

General discussion

In field theoretical applications we shall mainly deal with the matrix elements of the field operators taken between the vacuum states: the Green's functions. Let us first consider the analogous problem in quantum mechanics. Assume that the lagrangian L of the system is time independent. The energy eigenstates correspond to the wave functions $\Phi_n(x) = \langle x|n\rangle$. In particular, the ground state or *the vacuum* is described by the function $\Phi_0(x) = \langle x|0\rangle$. It will be convenient to use $\Phi_0(x,t)$ defined as

$$\Phi_0(x,t) = \exp[-(\mathrm{i}/\hbar)E_0 t]\langle x|0\rangle = \langle x|\exp[-(\mathrm{i}/\hbar)Ht]|0\rangle = \langle x,t|0\rangle \tag{2.24}$$

We are interested in the matrix element $\langle 0|TX(t_1)\ldots X(t_n)|0\rangle$. It reads

$$\langle 0|TX(t_1)\ldots X(t_n)|0\rangle = \int \mathrm{d}x'\,\mathrm{d}x\,\Phi_0^*(x',t')\langle x',t'|TX(t_1)\ldots X(t_n)|x,t\rangle \Phi_0(x,t) \tag{2.25}$$

and for the matrix element $\langle x',t'|TX(t_1)\ldots X(t_n)|x,t\rangle$ the functional form (2.15) can be used. The considered vacuum expectation value can also be obtained from a generating functional

$$\langle 0|TX(t_1)\ldots X(t_N)|0\rangle = \left(\frac{1}{\mathrm{i}}\right)^N \left. \frac{\delta^N}{\delta J(t_1)\ldots\delta J(t_N)} \right|_{J=0} W[J] \tag{2.26}$$

where

$$W[J] = \langle 0|0\rangle^J = \int \mathrm{d}x'\,\mathrm{d}x\,\Phi_0^*(x',t')\langle x',t'|x,t\rangle^J \Phi_0(x,t) \tag{2.27}$$

with $\langle x',t'|x,t\rangle^J$ given by (2.18). It is very important that the generating functional $W[J]$ can be also derived in another way. We shall show that

$$W[J] = \lim_{\substack{T_1 \to +i\infty \\ T_2 \to -i\infty}} \frac{\exp[(i/\hbar)E_0(T_2 - T_1)]}{\Phi_0^*(x_1)\Phi_0(x_2)} \langle x_2, T_2 | x_1, T_1 \rangle^J \tag{2.28}$$

This implies that $W[J]$ is, in fact, determined by the transition amplitude $\langle x_2, T_2 | x_1, T_1 \rangle^J$ at any given x_1, x_2 (for instance $x_1 = x_2 = 0$) provided that the analytic continuation to the imaginary values of T_1 and T_2 is performed. To derive (2.28) let us choose that the source $J(t)$ vanishes outside the time interval (t, t') with $T_2 > t' > t > T_1$. Then we can write

$$\langle x_2, T_2 | x_1, T_1 \rangle^J = \int dx' \, dx \, \langle x_2, T_2 | x', t' \rangle \langle x', t' | x, t \rangle^J \langle x, t | x_1, T_1 \rangle \tag{2.29}$$

where

$$\begin{aligned} \langle x, t | x_1, T_1 \rangle &= \langle x | \exp[-(i/\hbar)H(t - T_1)] | x_1 \rangle \\ &= \sum_n \Phi_n(x)\Phi_n^*(x_1) \exp[-(i/\hbar)E_n(t - T_1)] \end{aligned}$$

and similarly for $\langle x_2, T_2 | x', t' \rangle$. The only T-dependent terms are now the factors $\exp[-(i/\hbar)E_n(t - T_1)]$ and we can continue to $T_1 \to +i\infty$ explicitly. Recalling that E_0 is the lowest energy eigenvalue we get

$$\begin{aligned} \lim_{T_2 \to -i\infty} \exp[(i/\hbar)E_0 T_2] \langle x, t | x_1, T_1 \rangle &= \Phi_0(x) \exp[-(i/\hbar)E_0 t] \Phi_0^*(x_1) \\ &= \Phi_0(x, t)\Phi_0^*(x_1) \end{aligned}$$

and, in the same way

$$\lim_{T_2 \to -i\infty} \exp[(i/\hbar)E_0 T_2] \langle x_2, T_2 | x', t' \rangle = \Phi_0^*(x', t')\Phi_0(x_2)$$

After employing (2.27) and (2.29), (2.28) follows. Notice also that (2.28) and (2.26) imply

$$\begin{aligned} \langle 0 | TX(t_1) \ldots X(t_N) | 0 \rangle = \lim_{\substack{T_1 \to +i\infty \\ T_2 \to -i\infty}} & \frac{\exp[(i/\hbar)E_0(T_2 - T_1)]}{\Phi_0^*(x_1)\Phi_0(x_2)} \\ & * \langle x_2, T_2 | TX(t_1) \ldots X(t_N) | x_1, T_1 \rangle \end{aligned} \tag{2.30}$$

We conclude that the vacuum matrix elements can be calculated by taking functional derivatives of the generating functional $W[J]$, given by (2.28). The J-independent factors are irrelevant, because we can always consider quantities like

$$\frac{1}{\langle 0 | 0 \rangle} \langle 0 | TX(t_1) \ldots X(t_N) | 0 \rangle = \left(\frac{1}{i}\right)^N \frac{1}{W[0]} \frac{\delta^N}{\delta J(t_1) \ldots \delta J(t_N)}\bigg|_{J \equiv 0} W[J] \tag{2.31}$$

Consequently, instead of (2.28) we can write

$$W[J] = \lim_{\substack{T_1 \to +i\infty \\ T_2 \to -i\infty}} \int_{\substack{x(T_1) = x_1 \\ x(T_2) = x_2}} \mathscr{D}x \exp\left\{(i/\hbar) \int_{T_1}^{T_2} [L(x, \dot{x}) + \hbar J x] \, dt\right\} \tag{2.32}$$

where x_1, x_2, are arbitrary.

Harmonic oscillator

As an example of particular interest in view of further field theoretical applications we shall consider the case of the harmonic oscillator

$$L=\tfrac{1}{2}\dot{x}^2-\tfrac{1}{2}\omega^2x^2=\tfrac{1}{2}(\mathrm{d}/\mathrm{d}t)(x\dot{x})-\tfrac{1}{2}x\ddot{x}-\tfrac{1}{2}\omega^2x^2$$

The action integral, including the source term, can be written as follows

$$S^J[x]=\int_{T_1}^{T_2}[L+\hbar Jx]\,\mathrm{d}t=\tfrac{1}{2}x\dot{x}|_{T_1}^{T_2}+\tfrac{1}{2}(x,Ax)+(\hbar J,x) \tag{2.33}$$

where (f,g) denotes the 'scalar product'

$$(f,g)=\int_{T_1}^{T_2}\mathrm{d}t\,f(t)g(t)$$

and the operator A is defined by the equation

$$\int_{T_1}^{T_2}\mathrm{d}t\,\mathrm{d}t'\,x(t)A(t,t')x(t')=\int_{T_1}^{T_2}(-x\ddot{x}-\omega^2x^2)\,\mathrm{d}t \tag{2.34}$$

so that

$$A(t,t')=-(\mathrm{d}^2/\mathrm{d}t'^2+\omega^2)\delta(t'-t) \tag{2.35}$$

and

$$Ax\equiv\int_{T_1}^{T_2}\mathrm{d}t'\,A(t,t')x(t')$$

Let us now change the functional integration variable in (2.32) by introducing $z(t)$

$$x(t)=x_c(t)+z(t)$$

where $x_c(t)$ is a solution of the classical equation of motion (in the presence of the source)

$$\ddot{x}+\omega^2x=\hbar J$$

Requiring that $x_c(t)$ satisfies the boundary conditions $x_c(T_1)=x_1, x_c(T_2)=x_2$, we have $z(T_1)=z(T_2)=0$. The classical trajectory $x_c(t)$ can be written as a superposition of a solution $x_0(t)$ of the homogeneous equation, and the special solution of the complete equation

$$x_c(t)=x_0(t)+\hbar\int\mathrm{d}t'G(t-t')J(t') \tag{2.36}$$

Here $G(t-t')$ is the Green's function of the classical equation of motion

$$(\mathrm{d}^2/\mathrm{d}t^2+\omega^2)G(t-t')=\delta(t-t') \tag{2.37}$$

so that $G=-A^{-1}$.† Reexpressing the action integral (2.33) in terms of the classical trajectory and the new integration variable $z(t)$ we get

$$S^J[x_c+z]=\tfrac{1}{2}x_c\dot{x}_c|_{T_1}^{T_2}+\tfrac{1}{2}(z,Az)+\tfrac{1}{2}\hbar(J,x_0)+\tfrac{1}{2}\hbar^2(J,GJ) \tag{2.38}$$

† In the space of functions $x(t)$ with zero modes $x_0(t)$ of A excluded.

where we have used the relations

$$Ax_c = -\hbar J, \quad (x_c, Az) = (Ax_c, z) - \dot{z}x_c|_{T_1}^{T_2}$$

together with (2.36).

The resulting $W[J]$ is of the following form

$$W[J] = \lim_{\substack{T_1 \to +i\infty \\ T_2 \to -i\infty}} \exp[(i/2\hbar)\hbar^2(J, GJ)] \exp\{(i/2\hbar[x_c\dot{x}_c|_{T_1}^{T_2} + \hbar(J, x_0)]\}$$

$$* \int_{\substack{z(T_1)=0 \\ z(T_2)=0}} \mathscr{D}z \exp\left[(i/\hbar)\int_{T_1}^{T_2} L(z)\,dt\right] \tag{2.39}$$

The important observation is that if the Green's function (2.37) is chosen to satisfy the conditions

$$G(t) \xrightarrow[t \to \pm i\infty]{} 0, \qquad dG(t)/dt \xrightarrow[t \to \pm i\infty]{} 0 \tag{2.40}$$

the only functional dependence of $W[J]$ on J is given by the factor

$$W[J] \sim \exp[(i/2\hbar)\hbar^2(J, GJ)] \tag{2.41}$$

To prove this assertion we should show that $x_c\dot{x}_c|_{T_1}^{T_2}$ and (J, x_0) are J independent whenever the boundary conditions (2.40) are satisfied. In fact (2.36) and (2.40) imply

$$x_c\dot{x}_c|_{T_1}^{T_2} \xrightarrow[\substack{T_1 \to +i\infty \\ T_2 \to -i\infty}]{} x_0\dot{x}_0|_{T_1}^{T_2}$$

with x_0 obviously J independent. For (J, x_0) the argument is somewhat more involved. Consider the solution of the homogeneous equation

$$x_0(t) = A\exp(i\omega t) + B\exp(-i\omega t)$$

with A, B chosen to satisfy the boundary conditions $x_0(T_1) = x_1, x_0(T_2) = x_2$

$$A = \frac{x_2\exp(-i\omega T_1) - x_1\exp(-i\omega T_2)}{\exp[i\omega(T_2 - T_1)] - \exp[-i\omega(T_2 - T_1)]},$$

$$B = \frac{x_1\exp(i\omega T_2) - x_2\exp(i\omega T_1)}{\exp[i\omega(T_2 - T_1)] - \exp[-i\omega(T_2 - T_1)]}$$

In the limit $T_1 \to +i\infty$, $T_2 \to -i\infty$ we have

$$A \approx x_2\exp(-\omega|T_2|), \qquad B \approx x_1\exp(-\omega|T_1|)$$

implying that (J, x_0) also vanishes in this limit.

Note that in (2.41) the limit $T_1 \to +i\infty$, $T_2 \to -i\infty$ has in fact been omitted. This could have been done because the integral

$$\int_{T_1}^{T_2} dt \int_{T_1}^{T_2} dt'\, J(t)G(t - t')J(t') \equiv (J, GJ)$$

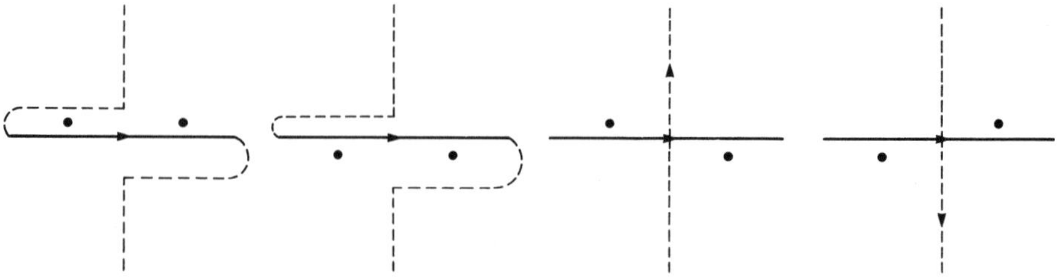

Fig. 2.1 Possible prescriptions for bypassing the singularity at $\nu = \pm\omega$ in the integral (2.42) in the complex ν plane.

is independent of the limits of integration due to the requirement that $J(t)$ vanishes outside some finite time interval (t_1, t_2) such that $T_2 > t_2 > t_1 > T_1$. The dependence on T_1 and T_2 can now occur only in the factors which are J independent because of the boundary conditions (2.40), and therefore irrelevant for our discussion.

To complete the case of the harmonic oscillator we must find the Green's function $G(t - t')$ consistent with the boundary conditions specified by (2.40). Introducing the Fourier transform $\tilde{G}(\nu)$

$$G(t) = \int_{-\infty}^{\infty} \frac{d\nu}{2\pi} \tilde{G}(\nu) \exp(-i\nu t)$$

we obtain the formal solution of (2.37)

$$G(t) = -\int_{-\infty}^{\infty} \frac{d\nu}{2\pi} \frac{1}{\nu^2 - \omega^2} \exp(-i\nu t) \tag{2.42}$$

This result represents, in fact, four distinct solutions depending on the prescription for bypassing the singularity at $\nu = \pm\omega$, which may be one of those shown in Fig. 2.1. Cases (*a*) and (*b*) are easily seen to be in conflict with the boundary conditions (2.40). It suffices to perform the analytic continuation of (2.42) by deforming the contour as shown in the picture and then to integrate for imaginary t. The two remaining solutions agree with (2.40): we again continue analytically to imaginary t by deforming the contour of the ν integration from the real to the imaginary axis. They correspond to adding to the denominator of the integrand in (2.42) $(+i\varepsilon)$ and $(-i\varepsilon)$, respectively

$$G(t) = -\int_{-\infty}^{\infty} \frac{d\nu}{2\pi} \frac{1}{\nu^2 - \omega^2 \pm i\varepsilon} \exp(-i\nu t) \tag{2.43}$$

Adding $(\pm i\varepsilon)$ to the Green's function denominator is equivalent to introducing the extra term $(\pm i\varepsilon x^2)$ to the lagrangian function. Only for $(+i\varepsilon x^2)$ is the functional integral well defined, so that the final solution for the Green's function is specified uniquely

$$G(t) = -\int_{-\infty}^{\infty} \frac{d\nu}{2\pi} \frac{1}{\nu^2 - \omega^2 + i\varepsilon} \exp(-i\nu t) \tag{2.44}$$

After the contour integration this becomes (for real t)

$$G(t) = \frac{i}{2\omega} [\exp(-i\omega t)\Theta(t) + \exp(i\omega t)\Theta(-t)] \tag{2.45}$$

The physical interpretation of the $(+i\varepsilon)$ prescription is thus made clear: it corresponds to the positive frequency solutions propagating into the future.

Euclidean Green's functions

Now let us observe that the same Green's function can also be derived in another way, that is by the analytic continuation from the imaginary time region. To this end we first define the 'Euclidean' (that is, imaginary time) Green's function $G_E(\tau)$, which obeys the equation

$$(-(d^2/d\tau^2) + \omega^2)G_E(\tau) = \delta(\tau) \tag{2.46}$$

Consequently

$$G_E(\tau) = \int_{-\infty}^{\infty} \frac{d\nu}{2\pi} \frac{1}{\nu^2 + \omega^2} \exp(-i\nu t) \tag{2.47}$$

As the singularities of the integrand appear at $\nu = \pm i\omega$, away from the integration contour, $G_E(\tau)$ is uniquely specified to be

$$G_E(\tau) = -\left(\frac{1}{2\omega}\right) \exp(-\omega|\tau|) \tag{2.48}$$

in agreement with the boundary conditions specified for $G(t)$. We shall now show that the real time Green's function $G(t)$ is given by the analytic continuation of $G_E(\tau)$ to $\tau = it$

$$G(t) = -iG_E(it) \tag{2.49}$$

In order to perform this continuation the integration contour in (2.47) should be deformed to keep the integral convergent during the $\tau = it$; such a rotation is illustrated in Fig. 2.2. The result is

$$G(t) = -\int_{-\infty}^{\infty} \frac{d\nu}{2\pi} \frac{1}{\nu^2 - \omega^2 + i\varepsilon} \exp(-i\nu t)$$

which coincides with (2.44).

It is also possible to continue to $\tau = -it$. In this case, however, the resulting real time Green's function appears with the opposite, $(-i\varepsilon)$ prescription. We conclude that the analytic continuation is correct for $\tau = it$. One way to see the difference is to make the substitution $t = \mp i\tau$ under the functional integral. The

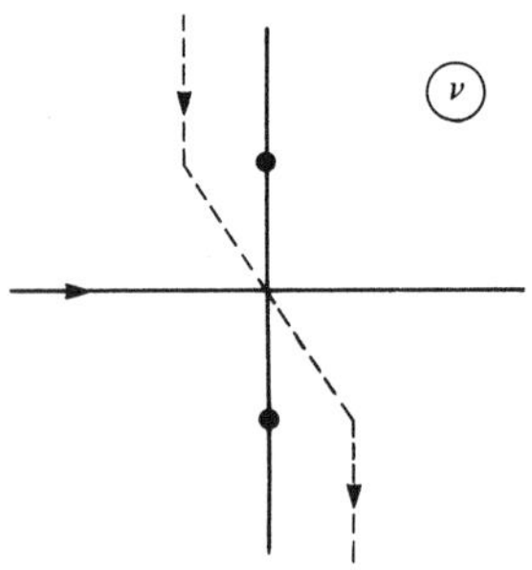

Fig. 2.2 Rotation of the integration contour from the real to the imaginary axis in the complex ν-plane which defines the analytic continuation of $G_E(\tau)$ to $\tau = it$.

action changes as follows:

$$\frac{i}{\hbar}S[x] = \frac{i}{\hbar}\int dt\left[\frac{1}{2}\left(\frac{dx}{dt}\right)^2 - V(x)\right] \xrightarrow[t=\pm i\tau]{}$$

$$* \frac{i}{\hbar}(\pm i)\int d\tau\left[-\frac{1}{2}\left(\frac{dx}{d\tau}\right)^2 - V(x)\right] = \pm\frac{1}{\hbar}S_E[x] \tag{2.50}$$

where $S_E[x]$ is the 'Euclidean' action. The resulting functional integral is of the form

$$\int \mathscr{D}x \exp\{\pm(1/\hbar)S_E[x]\} \tag{2.51}$$

and only for the $(-)$ sign is it well defined.

For the generating functional $W_E[J]$ one gets

$$W_E[J] = \int \mathscr{D}x \exp\left\{-(1/\hbar)S_E[x] + \int d\tau J(\tau)x(\tau)\right\} \sim \exp\left[\tfrac{1}{2}\hbar(J, G_E J)\right]$$

where $G_E(\tau)$ is defined by (2.46).

2.3 Path integral formulation of quantum field theory

Green's functions as path integrals

The results of Sections 2.1 and 2.2 are easily generalized to the case of more than one degrees of freedom. If the number of degrees of freedom is to be n, the coordinate x should be replaced by an n-component vector. The functional integral would now correspond to the sum over all trajectories in the n-dimensional configuration space, satisfying appropriate boundary conditions.

In the field theory, the trajectory $x(t)$ is replaced by a field function $\Phi(\mathbf{x}, t)$. The degrees of freedom are now labelled by the continuous index $\mathbf{x}$; the number of degrees of freedom is obviously infinite. To define the appropriate path integral one can start from a multiple integral on a discrete and, for a beginning, finite lattice of space-time points. This amounts to defining the quantum field theory as a limit of a theory with only a finite number of degrees of freedom.

The limit of an infinite lattice, related to the thermodynamical limit of statistical mechanics, already defines a theory with infinite number of degrees of freedom. However, this lattice theory has not enough space-time invariance and a continuous theory must be defined. The latter limit is accompanied by infinities, the 'UV divergences' of quantum field theory. The definition of the functional integral in quantum field theory is thus more ambiguous than in the case of quantum mechanics. Nevertheless, the functional formalism in quantum field theory is of great heuristic value. It is a very convenient tool for studying perturbation theory and allows a natural description of some non-perturbative phenomena.

The quantum field theory is usually formulated in terms of the vacuum expectation values of the chronologically ordered products of the field operators, the Green's functions:

$$G^{(n)}(x_1,\ldots,x_n) = \langle 0|\,T\Phi(x_1)\ldots\Phi(x_n)|0\rangle \tag{2.52}$$

Using our experience from the previous Section we shall write down the path integral representation for them. In particular, it is important to remember the role played by (2.28) and (2.30) in getting rid of the vacuum wave functions originally present in (2.25) as the boundary conditions. Thus, by analogy with the results of the previous Section, we postulate the following path integral representation

$$G^{(n)}(x_1,\ldots,x_n) \sim \int \mathscr{D}\Phi\Phi(x_1)\ldots\Phi(x_n)\exp\left[(\mathrm{i}/\hbar)\int \mathrm{d}^4x\mathscr{L}\right] \tag{2.53}$$

($\mathscr{D}\Phi$ denotes integration over all functions $\Phi(\mathbf{x},t)$ of space and time, because, for each value of $\mathbf{x}$, $\Phi(\mathbf{x},t)$ corresponds to a separate degree of freedom; $\mathscr{L}$ is the lagrangian density). Equation (2.30) suggests that any boundary conditions are, in fact, irrelevant here, provided that we take the imaginary time limit. In particular, the trajectories may be not constrained at all.

In analogy with (2.30), the rotation of the time contour away from the real axis should start somewhere at large absolute values of time, so that the arguments of the Green's functions: $t_1\ldots t_n$ stay real. If instead the whole time axis is rotated, $t_i = -\mathrm{i}\tau_i$, the result is an Euclidean Green's function (see the end of Section 2.2). The latter has a particularly convenient path integral representation, because the weight factor in the integrand: $\exp(-S_{\mathrm{E}}/\hbar)$ is then non-negative. This Euclidean path integral formalism can be used to define the Minkowski space Green's functions by an analytic continuation of the Euclidean ones. An expression like (2.53) would then be understood as an analytic continuation in the variables $t_1,\ldots,t_n$ of the analogous Euclidean formula. Equivalently one can work in Minkowski space and use the $(+\mathrm{i}\varepsilon)$ prescription (see (2.62)).

Equation (2.53) should be regarded as the formulation of the theory. What is the relation between the path integral and the usual (canonical) operator formulation of quantum field theory (e.g. Bjorken & Drell 1965) based on the same lagrangian $\mathscr{L}$? For our purposes these are equivalent if they imply the same perturbation theory Feynman rules. This has to be checked in each case of interest. However, the derivation of the perturbation theory rules is much simpler in the functional framework, particularly in the case of gauge field theories. In the following we shall often use the path integral formulation, referring to the operator formalism only if it helps to make the presentation more concise, like in the case of the scattering operator discussed in Section 2.7.

It is convenient to normalize the Green's functions by factorizing out the vacuum amplitude

$$G^{(n)}(x_1,\ldots,x_n) = \langle 0|T\Phi(x_1)\ldots\Phi(x_n)|0\rangle/\langle 0|0\rangle$$
$$= N\int \mathscr{D}\Phi\Phi(x_1)\ldots\Phi(x_n)\exp\left[(\mathrm{i}/\hbar)\int \mathrm{d}^4x\mathscr{L}\right] \tag{2.54}$$

where

$$1/N = \int \mathscr{D}\Phi\exp\left[(\mathrm{i}/\hbar)\int \mathrm{d}^4x\mathscr{L}\right] \sim \langle 0|0\rangle \tag{2.55}$$

This removes the extra factors like those appearing in (2.30). The T-product defined by (2.54) is a covariant quantity (see Section 10.1 on ambiguities of T-products). The derivatives of the Green's functions are functional integrals of derivatives of fields, e.g.

$$\Box_x G^{(2)}(x,y) \sim \int \mathscr{D}\Phi\Box_x\Phi(x)\Phi(y)\exp\left[(\mathrm{i}/\hbar)\int \mathrm{d}^4x\mathscr{L}\right] \tag{2.56}$$

The Green's functions (2.54) satisfy the standard equations of motion which can be derived using the invariance of the functional integral under a shift of variables $\Phi(x)\to\Phi(x)+\varepsilon f(x)$ (see Problem 2.2): e.g. in a scalar massless field theory defined by the lagrangian

$$\mathscr{L} = \tfrac{1}{2}\partial_\mu\Phi\partial^\mu\Phi - V(\Phi)$$

we have

$$\Box_x G^{(2)}(x,y) = -\langle 0|TV'(\Phi(x))\Phi(y)|0\rangle - \mathrm{i}\delta(x-y) \tag{2.56a}$$

We observe also that given a theory formulated in terms of the Green's functions (2.54) quantities like $\langle 0|T\Box_x\Phi(x)\Phi(y)|0\rangle$ remain *a priori* undefined. We can define them by requiring that they satisfy certain constraints like e.g. for a scalar massless field theory

$$\Box_x\langle 0|T\Phi(x)\Phi(y)|0\rangle = \langle 0|T\Box_x\Phi(x)\Phi(y)|0\rangle - \mathrm{i}\delta(x-y)$$

where the l.h.s. is given by (2.56a), which amounts to specifying a regularization prescription for them.

The Green's functions (2.54) are given by the functional derivatives of the functional $W[J]$ equivalent to the vacuum transition amplitude in presence of the external source $J(x)$

$$W[J]=N\int \mathscr{D}\Phi \exp\left\{(\mathrm{i}/\hbar)\int \mathrm{d}^4x[\mathscr{L}+\hbar J(x)\Phi(x)]\right\} \tag{2.57}$$

Expanding in powers of J and using (2.54) we can rewrite $W[J]$ as follows

$$W[J]=\sum_{n=0}^{\infty}\frac{\mathrm{i}^n}{n!}\int \mathrm{d}x_1\ldots\mathrm{d}x_n G^{(n)}(x_1,\ldots,x_n)J(x_1)\ldots J(x_n)$$

Consequently

$$G^{(n)}(x_1,\ldots,x_n)=\left(\frac{1}{\mathrm{i}}\right)^n \frac{\delta^n}{\delta J(x_1)\ldots\delta J(x_n)}\bigg|_{J\equiv 0} W[J] \tag{2.58}$$

The Green's functions can also be considered as the analytic continuation of those obtained from the generating functional defined in the Euclidean space with $x_0=-\mathrm{i}\bar{x}_0$ where $\bar{x}_0$ is real

$$W_{\mathrm{E}}[J]=N\int \mathscr{D}\Phi \exp\left\{-(1/\hbar)S_{\mathrm{E}}[\Phi(\bar{x})]+\int \mathrm{d}^3x\,\mathrm{d}\bar{x}_0 J\Phi\right\}$$

where e.g. for a free scalar field theory

$$S_{\mathrm{E}}[\Phi(\bar{x})]=\tfrac{1}{2}\int \mathrm{d}^3x\mathrm{d}\bar{x}_0\left[\left(\frac{\partial\Phi}{\partial\bar{x}_0}\right)^2+(\boldsymbol{\nabla}\Phi)^2+m^2\Phi^2\right].$$

The Euclidean Green's functions are given by

$$G_{\mathrm{E}}^{(n)}(\bar{x}_1,\ldots,\bar{x}_n)=\frac{\delta}{\delta J(\bar{x}_1)}\cdots\frac{\delta}{\delta J(\bar{x}_n)}\bigg|_{J\equiv 0} W_{\mathrm{E}}[J]$$

Using path integral formalism one can derive equations of motion for them (Problem 2.2), e.g. in free scalar field theory

$$\left(-\frac{\partial^2}{\partial\bar{x}_0^2}-\boldsymbol{\nabla}+m^2\right)G_{\mathrm{E}}^{(2)}(\bar{x},\bar{y})=\delta(\bar{x}_0-\bar{y}_0)\delta(\mathbf{x}-\mathbf{y})$$

analogous to those in Minkowski space.

The Minkowski space Green's functions are given by analytic continuation

$$G^{(n)}(x_1,\ldots,x_n)=(\mathrm{i})^n G_{\mathrm{E}}^{(n)}(\bar{x}_1,\ldots,\bar{x}_n)$$

where

$$\mathbf{x}=\bar{\mathbf{x}}_n,\quad x_0=-\mathrm{i}\bar{x}_0$$

Since in Euclidean space the exponent in the generating functional $W_{\mathrm{E}}[J]$ is bounded from above we can evaluate the functional integral by the saddle point method (see Problem 2.8). We recall that the saddle point approximation rests on the fact that the integral

$$I = \int dx \exp[-a(x)]$$

can be successfully approximated by

$$I \cong \exp[-a(x_0)] \int dx \exp[-\tfrac{1}{2}(x-x_0)^2 a''(x_0)]$$

where x_0 satisfies $a'(x_0)=0$, if $a''(x_0)>0$ and if the points away from the minimum do not contribute much.

We remark also that we can define the field operator evolution for imaginary time

$$\Phi(x) = \exp(iHx_0)\Phi(0,\mathbf{x})\exp(-iHx_0) = \exp(H\bar{x}_0)\Phi(0,\mathbf{x})\exp(-H\bar{x}_0)$$

and also the operator ordering $\bar{T}$ with respect to $\bar{x}_0$ analogously to the T-product definition, e.g.

$$\bar{T}\Phi(\bar{x})\Phi(\bar{y}) = \Theta(\bar{x}_0 - \bar{y}_0)\Phi(\bar{x})\Phi(\bar{y}) + \Theta(\bar{y}_0 - \bar{x}_0)\Phi(\bar{y})\Phi(\bar{x})$$

whose vacuum expectation value satisfies then the same equation of motion as $G_E^{(2)}(\bar{x},\bar{y})$ defined by the generating functional in Euclidean space.

Action quadratic in fields

In the case when the classical action is quadratic in the field variable $\Phi(x)$

$$S = \frac{1}{2}\int d^4x d^4y \Phi(x) A(x,y) \Phi(y) \tag{2.59}$$

the generating functional $W_0[J]$ (index 0 corresponds to the specific form (2.59) of the action) is easily obtained in a closed form. Repeating the steps that led us to the derivation of the formula (2.41) we obtain

$$W_0[J] = \exp\left[\left(\frac{i}{2\hbar}\right)\hbar^2 \int d^4x d^4y J(x) G(x,y) J(y)\right] \tag{2.60}$$

where $G(x,y)$ is the Green's function of the classical field equation

$$\int A(x,y)\Phi(y) d^4y = -\hbar J(x) \tag{2.61}$$

satisfying conditions (2.40).

One example of a theory with quadratic action is the theory of a free field, with the lagrangian density

$$\mathscr{L} = \tfrac{1}{2}(\partial_\mu\Phi\partial^\mu\Phi - m^2\Phi^2) + \tfrac{1}{2}i\varepsilon\Phi^2 \tag{2.62}$$

The extra term $\frac{1}{2}i\varepsilon\Phi^2$ has been added in accordance with the discussion following (2.43): it makes the functional integral (2.57) well-defined and simultaneously assures the correct boundary conditions for the Green's function $G(x, y)$. As

$$\partial_\mu \Phi \partial^\mu \Phi = \partial_\mu(\Phi \partial^\mu \Phi) - \Phi \partial_\mu \partial^\mu \Phi \tag{2.63}$$

and neglecting in the action the surface term at infinity, the action can be written in the form

$$S = \frac{1}{2} \int d^4x d^4y \Phi(x)[-\partial_\mu \partial^\mu - m^2 + i\varepsilon]\delta(x-y)\Phi(y) \tag{2.64}$$

That is

$$A(x, y) = -[\partial_\mu \partial^\mu + m^2 - i\varepsilon]\delta(x-y) \tag{2.65}$$

The classical field equation

$$(\partial_\mu \partial^\mu + m^2 - i\varepsilon)\Phi(x) = \hbar J(x) \tag{2.66}$$

is then solved by

$$\Phi(x) = \hbar \int d^4y G(x-y) J(y) \tag{2.67}$$

with the Green's function $G(x-y)$ satisfying

$$[\partial_\mu \partial^\mu_{(x)} + m^2 - i\varepsilon] G(x-y) = \delta(x-y) \tag{2.68}$$

Introducing the Fourier transform $\tilde{G}(k)$

$$G(x-y) = \frac{1}{(2\pi)^4} \int d^4k \tilde{G}(k) \exp[-ik(x-y)] \tag{2.69}$$

we obtain from (2.68)

$$\tilde{G}(k) = -\frac{1}{k^2 - m^2 + i\varepsilon} \tag{2.70}$$

and therefore

$$G(x-y) = -\frac{1}{(2\pi)^4} \int d^4k \frac{1}{k^2 - m^2 + i\varepsilon} \exp[-ik(x-y)] \tag{2.71}$$

Again, the Green's function agrees with the analytic continuation from the Euclidean region.

Gaussian integration

We end this Section with a comment on gaussian integration. Having defined functional integrals as an appropriate limit of multiple integrals one encounters integrals of the type

$$I = \int_{-\infty}^{\infty} dx \exp(-b\varepsilon x^2 + ibx^2) = 2 \int_0^{\infty} dx \exp[-bx^2(\varepsilon - i)] \tag{2.72}$$

with $\varepsilon \to 0$ where the ε-term corresponds to the $(+\mathrm{i}\varepsilon)$ prescription. Rotating the contour of integration into $x' = x\exp(\mathrm{i}\varphi)$ such that $\exp(2\mathrm{i}\varphi)(\varepsilon - \mathrm{i}) = 1$ one easily finds

$$I = \exp[\mathrm{i}(\tfrac{1}{4}\pi - \tfrac{1}{2}\varepsilon)]\left(\frac{\pi}{b}\right)^{1/2} \underset{\varepsilon\to 0}{\longrightarrow} (\mathrm{i}\pi/b)^{1/2} \tag{2.73}$$

which is the analytic continuation of the gaussian integral

$$\int_{-\infty}^{\infty} \mathrm{d}x \exp(-ax^2) = (\pi/a)^{1/2} \tag{2.74}$$

for complex a (and $\mathrm{Re}\, a > 0$). A similar result holds for integration over the complex variable $z = x + \mathrm{i}y$

$$\int \mathrm{d}z^* \mathrm{d}z \exp(-az^*z) \equiv 2\int_{-\infty}^{\infty} \mathrm{d}x \int_{-\infty}^{\infty} \mathrm{d}y \exp[-a(x^2+y^2)] = 2\pi/a \tag{2.75}$$

for complex a (and $\mathrm{Re}\, a > 0$). We write $\int \mathrm{d}z^* \mathrm{d}z$ for $2\int \mathrm{d}\,\mathrm{Re}\, z \int \mathrm{d}\,\mathrm{Im}\, z$ by matter of convention. For many degrees of freedom $z_1, \ldots, z_n$ we can define complex vector z

$$z = \begin{pmatrix} z_1 \\ \vdots \\ z_n \end{pmatrix}$$

and denote by (z^*, Az) the scalar product of vectors z and Az, where A is a $n \times n$-dimensional complex matrix. Then we have the following generalization of relation (2.75)

$$\int \mathrm{d}z_1^* \mathrm{d}z_1 \ldots \int \mathrm{d}z_n^* \mathrm{d}z_n \exp[\mathrm{i}(z^*, Az)] = (2\pi)^n \mathrm{i}^n / \det A \tag{2.76}$$

for any positive-definite matrix A which can be diagonalized by unitary transformations. The result is obvious for a diagonal matrix A (it follows from multiple use of (2.75)). For the general case one has to show that the integration measure is invariant under the unitary transformation diagonalizing A (the scalar product (z^*, Az) is obviously invariant). Denote the unitary transformation by $U = A + \mathrm{i}B$ with A and B real matrices. From the unitarity of U: $UU^\dagger = 1$ it follows that

$$AA^{\mathrm{T}} + BB^{\mathrm{T}} = 1, \quad BA^{\mathrm{T}} - AB^{\mathrm{T}} = 0 \tag{2.77}$$

The change of integration variables $z' = Uz$ can be written as

$$\begin{pmatrix} x' \\ y' \end{pmatrix} = \begin{pmatrix} A & -B \\ B & A \end{pmatrix} \begin{pmatrix} x \\ y \end{pmatrix} = M \begin{pmatrix} x \\ y \end{pmatrix} \tag{2.78}$$

and using (2.77) we conclude that the transformation matrix M is orthogonal ($MM^{\mathrm{T}} = 1$). Therefore the integration measure is indeed invariant under the unitary transformation U. For integration over n real variables $x \equiv (x_1, \ldots, x_n)$ one has

$$\int dx_1 \ldots dx_n \exp[i(x, Ax)] = \pi^{n/2} i^{n/2} / [\det A]^{1/2} \tag{2.79}$$

In the limit $n \to \infty$ integral (2.79) is the J-independent path integral in (2.39). Although we do not have to evaluate it explicitly, it is worth remembering its compact form (2.79). Another point is that in addition to results (2.76) and (2.79) we can also easily calculate integrals in which the product (z^*, Az) is replaced by an expression

$$(z^*, Az) + (b^*, z) + (z^*, b) \equiv F(z) \tag{2.80}$$

where b is a constant vector, and matrix A is now hermitean. Indeed, expression (2.80) can be written as

$$F(z) = (\omega^*, A\omega) - (b^*, A^{-1}b) \tag{2.81}$$

where $\omega = z - z_0$ and $z_0 = -A^{-1}b$ is the minimum of $F(z)$. Actually, (2.81) has been used in the derivation of (2.41) and (2.60) for the generating functional $W_0[J]$.

Finally we recall that for any matrix L which can be diagonalized by a unitary transformation

$$\det(1 - L) = \exp \operatorname{Tr} \ln(1 - L)$$

where

$$\operatorname{Tr} \ln(1 - L) = -\operatorname{Tr}[L + \tfrac{1}{2}L^2 + \tfrac{1}{3}L^3 + \cdots]$$

2.4 Introduction to perturbation theory

Perturbation theory and the generating functional

We shall discuss first a simple case of a scalar field theory described by the lagrangian

$$\mathscr{L} = \tfrac{1}{2}[\partial_\mu \Phi(x) \partial^\mu \Phi(x) - m^2 \Phi^2(x)] - \frac{\lambda}{4!} \Phi^4(x) \tag{2.82}$$

where x denotes four coordinates in Minkowski space. We seek a method for calculating an arbitrary Green's function in such a theory. For that purpose it is convenient to use the functional formulation of the theory in which the Green's functions are given by functional derivatives of the generating functional $W[J]$

$$G^{(n)}(x_1, \ldots, x_n) = \left(\frac{1}{i}\right)^n \frac{\delta}{\delta J(x_1)} \cdots \frac{\delta}{\delta J(x_n)}\bigg|_{J \equiv 0} W[J] \tag{2.83}$$

where

$$W[J] = N \int \mathscr{D}\Phi \exp\left\{(i/\hbar) \int d^4x [\mathscr{L} + \hbar J(x) \Phi(x)]\right\} \tag{2.84}$$

and factor N is defined by (2.55).

We recall at this point that, so far, we have been able to calculate exactly the generating functional $W_0[J]$ for a theory of non-interacting scalar fields with the action S_0 given by

$$S_0 = \int d^4x \mathscr{L}_{\text{free}} = \int d^4x \tfrac{1}{2}(\partial_\mu \Phi \partial^\mu \Phi - m^2\Phi^2) \tag{2.85}$$

or, more generally, containing terms at most quadratic in fields. The analogous, exact, solution is not known for the full theory (2.82) and the method to be used for calculating the Green's functions (2.83) is a perturbative expansion in terms of powers of S_I defined as follows

$$S_\text{I} = S - S_0 \tag{2.86}$$

where

$$S = \int d^4x \mathscr{L}$$

Such an expansion should be understood as an expansion in a neighbourhood of the vacuum state for which $\Phi(x) = 0$ so that the action S_I can be regarded as a small parameter. It is convenient to note the following functional identity

$$\int \mathscr{D}\Phi \exp\left(\frac{i}{\hbar}\left\{S_0[\Phi] + S_\text{I}[\Phi] + \hbar \int J(x)\Phi(x) d^4x\right\}\right)$$

$$= \exp\left\{\frac{i}{\hbar} S_\text{I}\left[\frac{1}{i}\frac{\delta}{\delta J}\right]\right\} \int \mathscr{D}\Phi \exp\left(\frac{i}{\hbar}\left\{S_0[\Phi] + \hbar \int J(x)\Phi(x) d^4x\right\}\right) \tag{2.87}$$

so that (2.84) can be rewritten as

$$W[J] = \frac{\exp\left\{\frac{i}{\hbar} S_\text{I}\left[\frac{1}{i}\frac{\delta}{\delta J}\right]\right\} W_0[J]}{\exp\left\{\frac{i}{\hbar} S_\text{I}\left[\frac{1}{i}\frac{\delta}{\delta J}\right]\right\} W_0[J]\Big|_{J\equiv 0}} = N \exp\left\{\frac{i}{\hbar} S_\text{I}\left[\frac{1}{i}\frac{\delta}{\delta J}\right]\right\} W_0[J] \tag{2.88}$$

The perturbation series is generated by expanding the exponential factor $\exp\{(i/\hbar)S_\text{I}[(1/i)(\delta/\delta J)]\}$ in powers of S_I and performing the functional differentiations as indicated. This is equivalent to expanding $\exp\{(i/\hbar)S_\text{I}[\Phi]\}$ under the path integral. In perturbation theory one gets, therefore, the following general formula for the Green's functions (2.83)

$$G^{(n)}(x_1, \ldots, x_n) = \frac{\int \mathscr{D}\Phi \Phi(x_1)\ldots\Phi(x_n)\left[\sum_{N=0}^{\infty} \frac{1}{N!}\left(\frac{i}{\hbar}S_\text{I}\right)^N\right] \exp\left(\frac{i}{\hbar}S_0\right)}{\int \mathscr{D}\Phi \left[\sum_{N=0}^{\infty} \frac{1}{N!}\left(\frac{i}{\hbar}S_\text{I}\right)^N\right] \exp\left(\frac{i}{\hbar}S_0\right)} \tag{2.89}$$

If S_I can be written as an integral over polynomials in fields (as in our example) any Green's function can be calculated term by term using (2.89). Each term can be written as a respective derivative of the functional $W_0[J]$.

Wick's theorem

As an introduction to the detailed exposition of perturbative rules we consider again the case of non-interacting scalar fields and prove Wick's theorem. The n-point Green's function $G_0^{(n)}(x_1, \ldots, x_n)$ is given by (2.83) with $W[J]$ replaced by $W_0[J]$. From the previous considerations we know that

$$W_0[J] = \exp\left[\tfrac{1}{2}\mathrm{i}\hbar \int \mathrm{d}^4x\, \mathrm{d}^4y J(x) G(x-y) J(y)\right] \tag{2.90}$$

up to the overall J-independent normalization factor which cancels out in the definition of the Green's functions, where $G(x-y)$ is the Green's function for the free classical Klein–Gordon equation

$$G(x-y) = -\frac{1}{(2\pi)^4}\int \mathrm{d}^4k \frac{\exp[-\mathrm{i}k(x-y)]}{k^2 - m^2 + \mathrm{i}\varepsilon} \tag{2.91}$$

Referring again to (2.83) one observes that the n-point Green's function is given by the $(\frac{1}{2}n)$th term of the Taylor expansion of exponent (2.90). For the free propagator one immediately gets

$$G_0^{(2)}(x_1, x_2) = -\mathrm{i}\hbar G(x_1 - x_2) \tag{2.92}$$

(the symmetry $G(x_1 - x_2) = G(x_2 - x_1)$ has been taken into account). The case $n = 4$ is somewhat more complicated. (Notice that all $G^{(n)}$-functions with n odd equal zero.) It is convenient to convert the integrals in (2.90) into sums over discrete points and write

$$\mathrm{i}^4 G_0^{(4)}(x_1, \ldots, x_4) = \left.\frac{\delta^4}{\delta J_1 \delta J_2 \delta J_3 \delta J_4}\right|_{J=0} \left(\frac{\mathrm{i}}{2}\hbar\right)^2 \frac{1}{2!} \sum_{i,j,k,l} J_i J_j J_k J_l G_{ij} G_{kl} \tag{2.93}$$

where

$$J_i = J(x_i), \qquad G_{ij} = G(x_i - y_j)$$

The structure of (2.93) can be represented diagrammatically by the 'prototype' diagram

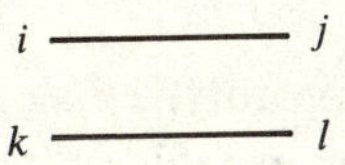

$$\equiv (-\mathrm{i}\hbar G_{ij}) \times (-\mathrm{i}\hbar G_{kl})$$

and the differentiation with respect to $J_1 \ldots J_4$ leads to all possible assignments of subscripts 1–4 to subscripts i, j, k, l. One gets

$$G_0^{(4)}(x_1,\ldots,x_4) = \tfrac{1}{8}\left(\begin{array}{c}\underset{3\quad\quad 4}{\overset{1\quad\quad 2}{\rule{2cm}{0.4pt}}} + \underset{3\quad\quad 4}{\overset{2\quad\quad 1}{\rule{2cm}{0.4pt}}} + \text{6 other permutations}\end{array}\right)$$

$$+ \tfrac{1}{8}\left(\begin{array}{c}\underset{2\quad\quad 4}{\overset{1\quad\quad 3}{\rule{2cm}{0.4pt}}} + \text{7 other permutations}\end{array}\right)$$

$$+ \tfrac{1}{8}\left(\begin{array}{c}\underset{2\quad\quad 3}{\overset{1\quad\quad 4}{\rule{2cm}{0.4pt}}} + \text{7 other permutations}\end{array}\right)$$

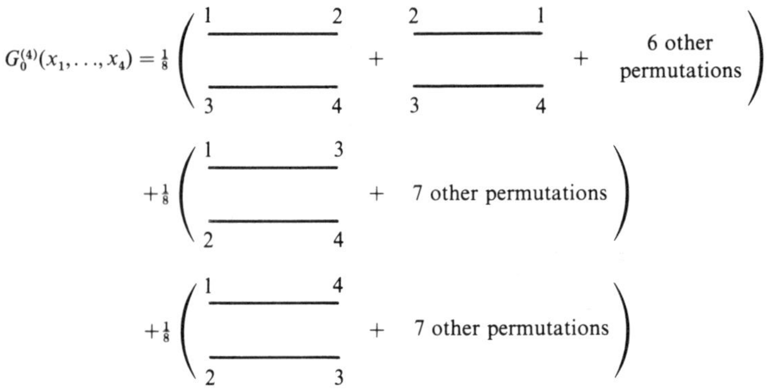

So finally

$$G_0^{(4)}(x_1,\ldots,x_4) = G_0^{(2)}(x_1,x_2)G_0^{(2)}(x_3,x_4) + G_0^{(2)}(x_1,x_3)G_0^{(2)}(x_2,x_4) \\ + G_0^{(2)}(x_1,x_4)G_0^{(2)}(x_2,x_3) \tag{2.94}$$

This is the content of Wick's theorem (see, e.g. Bjorken & Drell 1965) which states that any n-point free Green's function can be written as a sum over all possible products of $\tfrac{1}{2}n$ two-point Green's function $G_0^{(2)}(x_i,x_j)$.

An example: four-point Green's function in $\lambda\Phi^4$

Coming back to our main problem we shall now discuss several examples in order to derive general rules for perturbative calculations in the theory of interacting fields. Let us consider the four-point Green's function and discuss the numerator of (2.89) in the first order in S_I which reads

$$\int \mathscr{D}\Phi\,\Phi(x_1)\ldots\Phi(x_4)(-i\lambda)\frac{1}{4!\hbar}\int d^4y\,\Phi^4(y)\exp\left(\frac{i}{\hbar}S_0\right) \tag{2.95}$$

or, using the generating functional $W_0[J]$,

$$(-i\lambda)\frac{1}{4!\hbar}\int d^4y\,\frac{\delta}{\delta J(x_1)}\cdots\frac{\delta}{\delta J(x_4)}\frac{\delta^4}{\delta J(y)}W_0[J]\bigg|_{J\equiv 0} \tag{2.96}$$

It is clear that the non-zero contribution to (2.96) comes from the term in the Taylor expansion of (2.90) which contains four propagators

$$(\tfrac{1}{2}i\hbar)^4 \frac{1}{4!}\left[\int d^4x\,d^4y\,J(x)G(x-y)J(y)\right]^4 \tag{2.97}$$

and can be represented graphically by the prototype diagram in Fig. 2.3. To make the combinatorics transparent we have again replaced the integrations in (2.97) by discrete sums. The differentiations of (2.97) now provide all possible assignments of points x_1,x_2,x_3,x_4 and four points y (it is convenient to split them into y_1,y_2,y_3,y_4, remembering that the limit $y_1=y_2=y_3=y_4$ should be understood) to the points i,j,k,l,m,n,o,p. It is easy to discuss them in a systematic fashion. Let us

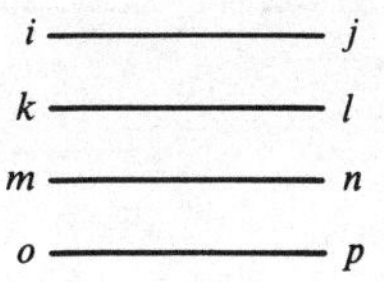

Fig. 2.3

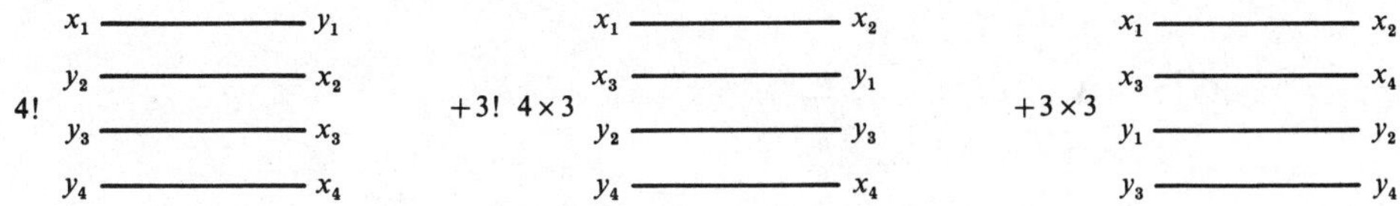

Fig. 2.4

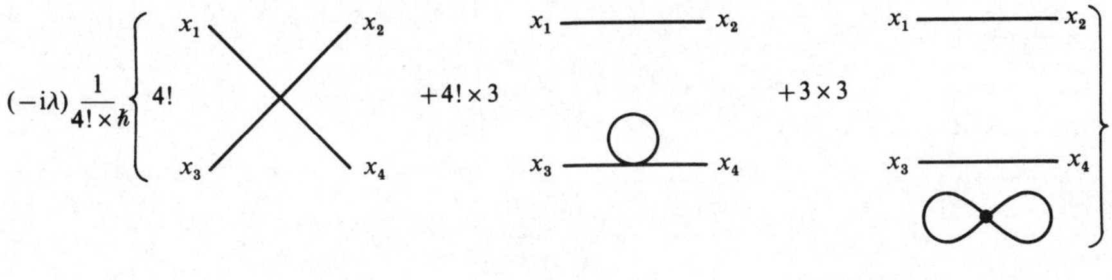

Fig. 2.5

first assume that pairs of points joined by the propagators are specified, e.g. (x_1, y_1), (x_2, y_2), (x_3, y_3), (x_4, y_4). Then, we have an additional combinatorial factor following from the 'horizontal' symmetry (e.g. $x_1 = i$, $y_1 = j$ and $x_1 = j$, $y_1 = i$) which is 2^n for n pairs, and another factor from the 'vertical' symmetry ((x_1, y_1) can be assigned to (i, j) or (k, l), etc.) which is $n!$ (again for n pairs). These factors cancel with the factor $1/(2^n n!)$ present in (2.97) and this happens for any term of expansion (2.89). So we have to remember about $1/N!$ present in (2.89) ($N = 1$ in our examples) and about $1/4!$ present in the definition of the coupling constant and, on the other hand, we have to take account of all possible choices of pairs of points x_i and y_i joined by propagators. The result is as shown in Fig. 2.4 with self-evident classification of different possibilities. In the limit $y_1 = y_2 = y_3 = y_4 = y$ we get the diagrams shown in Fig. 2.5 where the dot represents the integral of a product of two propagators taken at the same space-time points y (notice that the combinatorial factors in front of the second and the third diagrams include different permutations of x_1, x_2, x_3, x_4). The analytic expression for the first diagram reads

$$(-\mathrm{i}\lambda/\hbar) \int \mathrm{d}^4 y G_0^{(2)}(x_1, y) G_0^{(2)}(x_2, y) G_0^{(2)}(x_3, y) G_0^{(2)}(x_4, y) \tag{2.98}$$

It remains to extend our considerations to terms S_{I}^N, $N > 1$ and to discuss the role of the denominator in (2.89). Both problems are related since $G_N^{(n)}$ involves in principle a quotient of terms of arbitrary order. We observe that e.g. the S_{I}^2

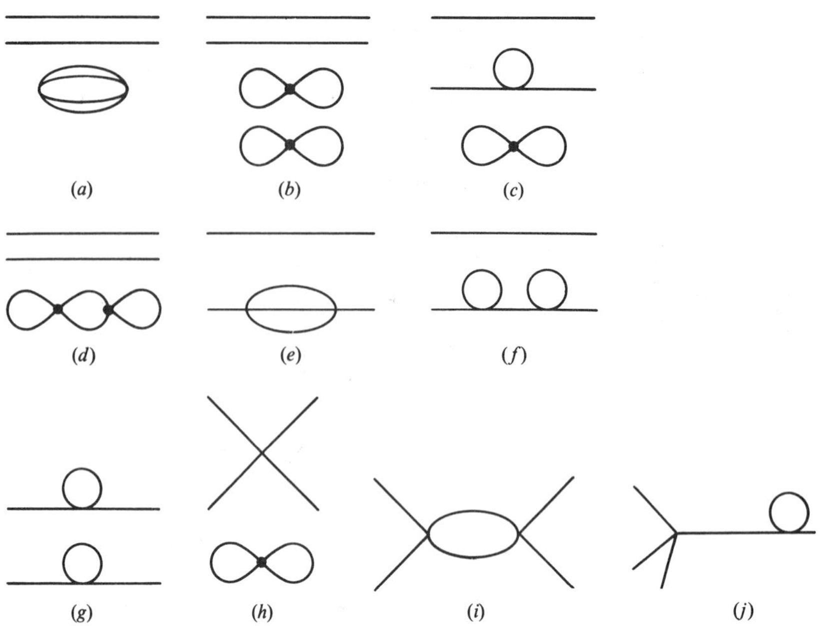

Fig. 2.6

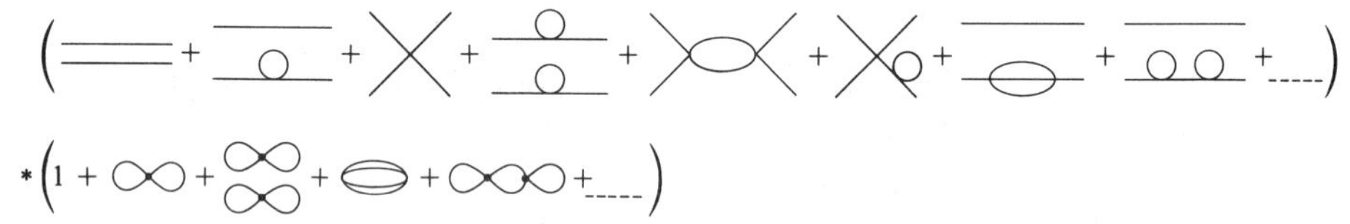

Fig. 2.7

term in the numerator is given by the diagrams (*a*)–(*j*) in Fig. 2.6. In general, one can convince oneself that the numerator of (2.89) can be represented by a product of two infinite series of amplitudes as shown in Fig. 2.7. The second bracket contains the sum of the so-called vacuum amplitudes which do not depend on the Green's function considered and, most important, it happens to be just the denominator of (2.89). (The proof of this again requires some attention to the combinatorial factors.) So we may ignore the denominator of (2.89) and simultaneously all the diagrams containing vacuum subgraphs. Actually, for a given Green's function we shall be interested only in connected diagrams like those in Fig. 2.8 which give non-trivial *S*-matrix elements for $G^{(4)}(x_1, \ldots, x_4)$. They define the connected Green's functions $G^{(n)}_{\text{conn}}$. The amplitude for e.g. the central diagram can be easily read off from the numerator of (2.89). In the following we shall understand the central diagram to be the one in which the pairs of the external points, which are joined by propagators to each of the interaction points, have been specified. In momentum space this corresponds to assigning to each

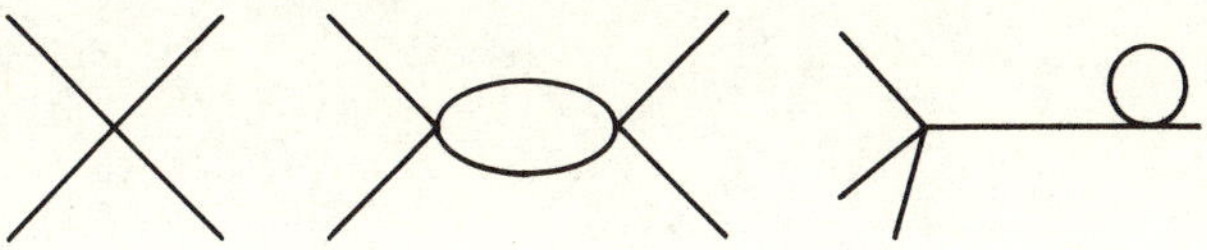

Fig. 2.8

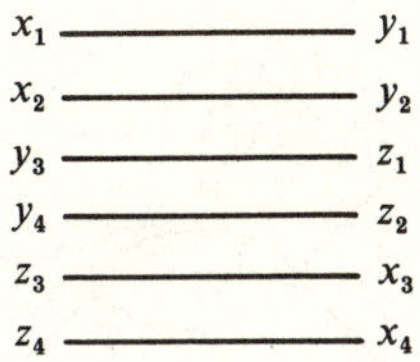

Fig. 2.9

vertex definite external momenta. In this sense we have three different central diagrams, each of which can be discussed as follows. There are twelve fields under the functional integral, so the 'prototype' diagram with six propagators is relevant for combinatorics. No two external points can be joined by a propagator. Two external points must be joined with the interaction point at y and the other two with the interaction point at z. In addition two propagators must connect y and z. We can have therefore the situation depicted in Fig. 2.9 and, in addition y_1 can be replaced by any y_i (factor 4), y_2 by any $y_j \neq y_i$ (factor 3), the same for z_4 and z_3 (factor 4×3), there are two combinations for the remaining y_l joining z_k (factor 2) and finally y can be interchanged with z in connections with external points (factor 2). We get $4! \times 4!$ to cancel $1/4! \times 4!$ from the definition of the coupling constant. (The factors from the 'vertical' and 'horizontal' symmetries cancel as usual.) Since $N = 2$ there is an extra factor $\frac{1}{2}$ (we will call it the symmetry factor) and the amplitude reads

$$\tfrac{1}{2}(-\mathrm{i}\lambda/\hbar)^2 \int \mathrm{d}^4y\,\mathrm{d}^4z\, G_0^{(2)}(x_1, y)G_0^{(2)}(x_2, y)G_0^{(2)}(y, z)G_0^{(2)}(y, z)G_0^{(2)}(x_3, z)G_0^{(2)}(x_4, z) \tag{2.99}$$

Our examples can be generalized to the following rules. To calculate $G^{(n)}_{\mathrm{conn}}(x_1, \ldots, x_n)$ to order N in S_{I} draw all connected diagrams consistent with the structure of the numerator in (2.89). For each diagram calculate the combinatorial (symmetry) factor. Attach $(-\mathrm{i}\lambda/\hbar)$ to each vertex and $G_0^{(2)}$ to each line.

Momentum space

One is usually interested in Green's functions in momentum space defined as Fourier transforms of the respective Green's functions in configuration space. We define $\tilde{G}^{(n)}(p_1, \ldots, p_n)$ by the relation

$$G^{(n)}(x_1, \ldots, x_n) = \int \frac{\mathrm{d}^4p_1}{(2\pi)^4} \cdots \frac{\mathrm{d}^4p_n}{(2\pi)^4} \exp(-\mathrm{i}p_1x_1) \ldots \exp(-\mathrm{i}p_nx_n)\tilde{G}^{(n)}(p_1, \ldots, p_n) \tag{2.100}$$

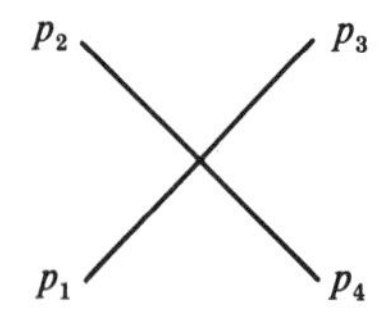

Fig. 2.10

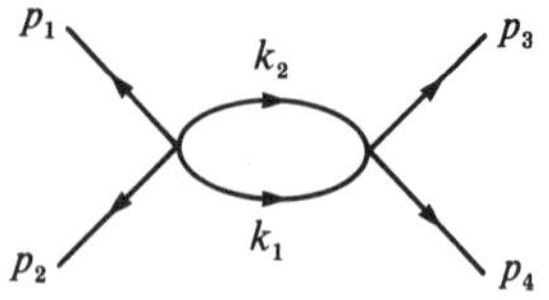

Fig. 2.11

so that all momenta are treated as outgoing (due to $(+\mathrm{i}\varepsilon)$) in propagators, positive frequencies propagate into future $x_n^0 \to +\infty$ (see (2.45)). In actual physical transitions

$$x_1^0 \ldots x_k^0 \to -\infty$$
$$x_{k+1}^0 \ldots x_n^0 \to +\infty$$

and therefore the momenta $p_1, \ldots, p_n$ defined by (2.100) are related to the physical momenta $\tilde{p}_i$ ($\tilde{p}_i^0 > 0$) of the initial and final particles as follows

$$\begin{aligned} p_i &= -\tilde{p}_i \qquad & i &= 1, \ldots, k \\ p_i &= \tilde{p}_i \qquad & i &= k+1, \ldots, n \end{aligned}$$

In momentum space we get from (2.92) and (2.91)

$$\tilde{G}_0^{(2)}(p_1, p_2) = (2\pi)^4 \delta(p_1 + p_2) \hbar \frac{\mathrm{i}}{p_1^2 - m^2 + \mathrm{i}\varepsilon} \tag{2.101}$$

For the diagram shown in Fig. 2.10, represented by expression (2.98) the momentum space amplitude reads

$$(2\pi)^4 \delta\left(\sum_{i=1}^{4} p_i \right) (-\mathrm{i}\lambda/\hbar) \prod_{i=1}^{4} \hbar \frac{\mathrm{i}}{p_i^2 - m^2 + \mathrm{i}\varepsilon} \tag{2.102}$$

and for the diagram shown in Fig. 2.11 we get, from (2.99)

$$\tfrac{1}{2}(-\mathrm{i}\lambda/\hbar)^2 \prod_{i=1}^{4} \hbar \frac{\mathrm{i}}{p_i^2 - m^2 + \mathrm{i}\varepsilon} \int \frac{\mathrm{d}^4 k_1}{(2\pi)^4} \frac{\mathrm{d}^4 k_2}{(2\pi)^4} (2\pi)^4 \delta(p_1 + p_2 + k_1 + k_2)(2\pi)^4$$

$$* \delta(p_3 + p_4 - k_1 - k_2) \hbar \frac{\mathrm{i}}{k_1^2 - m^2 + \mathrm{i}\varepsilon} \hbar \frac{\mathrm{i}}{k_2^2 - m^2 + \mathrm{i}\varepsilon} \tag{2.103}$$

The Feynman rules in momentum space are now obvious. Usually, the diagrams

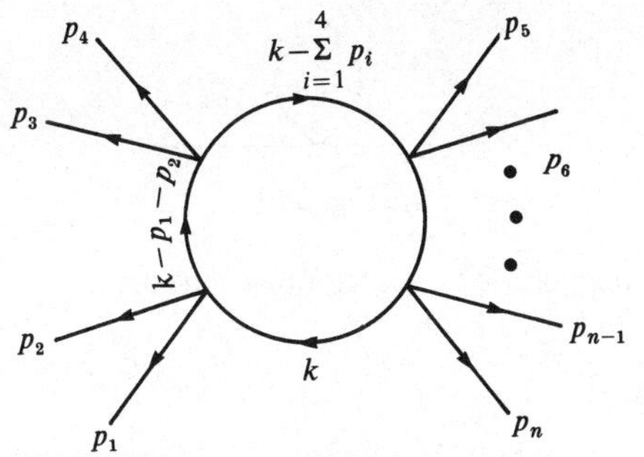

Fig. 2.12

as above are assumed to represent amplitudes without the propagators on external lines (Feynman diagrams). Observe that each vertex contributes a factor $\hbar^{-1}$ and each internal line gives a factor $\hbar$. Thus a Feynman diagram with L loops is of the order $O(\hbar^{L-1})$, because $L=$ (number of internal lines − number of vertices + 1).

It is easy to extend these considerations to a loop with N vertices. Again, assume that the pairs of external points which are joined to each vertex as well as the ordering of vertices along the loop are specified; this corresponds to a well-defined situation in momentum space. We have the factor $(1/N!)(1/4!)^N$ from the general formula (2.89) and an extension of our discussion of the $N=2$ case gives $(3\times 4)^N \times 2^{N-1}\times N!$. So we again get the symmetry factor $\frac{1}{2}$ for a Feynman diagram in momentum space, with momenta specified along all the internal lines, e.g. as in Fig. 2.12. For N vertices there are $(2N-1)(2N-3)\ldots 1*(N-1)!$ different graphs in momentum space; the factor $(2N-1)\ldots 1$ gives the number of different groupings of $2N$ momenta in N pairs and $(N-1)!$ gives the number of their different orderings along the loop.

An important modification of the lagrangian (2.82) follows if we specify that all products of the field operators in it are normal-ordered (Bjorken & Drell 1965, p. 91)

$$\mathscr{L}=\tfrac{1}{2}{:}\partial_\mu\Phi(x)\partial^\mu\Phi(x){:}-\tfrac{1}{2}m^2{:}\Phi^2(x){:}-(\lambda/4!){:}\Phi^4(x){:} \tag{2.104}$$

This can be rewritten in a form similar to (2.82), but with a number of counterterms; remember that under the path integral the fields are c-numbers and the only way of specifying their order is by introducing appropriate counterterms. From Wick's theorem it follows that (see Section 2.7):

$$\Phi^2(x)={:}\Phi^2(x){:}+\text{const.} \tag{2.105}$$

$$\Phi^4(x)={:}\Phi^4(x){:}+6G_0^{(2)}(x,x){:}\Phi^2(x){:}+\text{const.} \tag{2.106}$$

so that

$${:}\Phi^4(x){:}=\Phi^4(x)-6G_0^{(2)}(x,x)\Phi^2(x)-\text{const.} \tag{2.107}$$

and the other counterterms are constants. In the resulting perturbation expansion the diagram 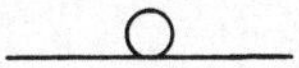 is now cancelled by the $-6G_0^{(2)}(x,x)\Phi^2(x)$ counterterm. Note, however, that the higher order corrections to this diagram, like that shown in

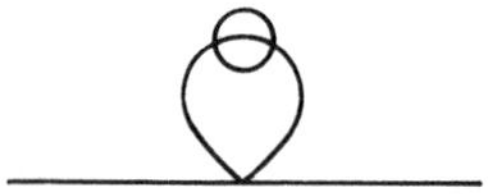

Fig. 2.13

Fig. 2.13 are not cancelled. The constant counterterms cancel some vacuum diagrams.

2.5 Path integrals for fermions; Grassmann algebra

Anticommuting c-numbers

The path integral quantization method involves classical fields only and the theory is formulated directly in terms of Green's functions given by path integrals like (2.53). Since for Fermi fields we must have, e.g.

$$G^{(2)}(x_1, x_2) = -G^{(2)}(x_2, x_1) \tag{2.108}$$

the fermion classical fields in the path integral must be taken to be anticommuting quantities. The anticommuting c-number variables, or functions, are said to form Grassmann algebra.

There follows a brief introduction to the subject of Grassmann algebra. Let us take η to be a Grassmann variable,

$$\{\eta, \eta\} = 0 \quad \text{or} \quad \eta^2 = 0 \tag{2.109}$$

Any function $f(\eta)$ can be written as

$$f(\eta) = f_0 + \eta f_1 \tag{2.110}$$

For instance if $f(\eta)$ is taken to be an ordinary number, then f_0 and f_1 are ordinary and Grassmann numbers, respectively. One defines the left and right derivatives by the equation

$$\frac{\mathrm{d}}{\mathrm{d}\eta}\eta = \eta\frac{\overleftarrow{\mathrm{d}}}{\mathrm{d}\eta} = 1 \tag{2.111}$$

Therefore, when f_1 is a Grassmann number,

$$\frac{\mathrm{d}}{\mathrm{d}\eta}f(\eta) = f_1 = -f(\eta)\frac{\overleftarrow{\mathrm{d}}}{\mathrm{d}\eta} \tag{2.112}$$

Furthermore, we define the integration by the requirement of linearity and the following relations

$$\int \mathrm{d}\eta 1 = 0, \qquad \int \mathrm{d}\eta\, \eta = 1 \tag{2.113}$$

The first relation follows from the requirement that the property of a convergent integral over commuting numbers

$$\int_{-\infty}^{\infty} \mathrm{d}x\, f(x) = \int_{-\infty}^{\infty} \mathrm{d}x\, f(x+a)$$

valid for any finite a, holds for an integral over anticommuting numbers as well. The second relation is the normalization convention. We also add a specification that rule (2.113) applies when $\mathrm{d}\eta$ and η are next to each other. Comparing (2.113) with (2.111) we see that the integrals and the left derivatives are identical

$$\int \mathrm{d}\eta\, f(\eta) = \frac{\mathrm{d}}{\mathrm{d}\eta} f(\eta) = f_1 \tag{2.114}$$

The integral of the derivative vanishes

$$\int \mathrm{d}\eta \frac{\mathrm{d}}{\mathrm{d}\eta} f(\eta) = \frac{\mathrm{d}^2}{\mathrm{d}\eta^2} f(\eta) = 0 \tag{2.115}$$

Consider now the change of integration variable $\eta \to \eta' = a + b\eta$. One gets

$$\int \mathrm{d}\eta\, f(\eta) = \int \mathrm{d}\eta' \left(\frac{\mathrm{d}\eta}{\mathrm{d}\eta'}\right)^{-1} f(\eta(\eta')) \tag{2.116}$$

i.e. the standard Jacobian appears inverted. All the rules can be easily generalized to the case of n real Grassmann variables. For a complex Grassmann variable the real and imaginary parts can be replaced by η and η^* as independent generators of Grassmann algebra.

In the following Chapter the Gaussian integral for complex Grassmann variables will be shown to be very useful. It can be shown that instead of (2.76) we now have

$$\int \mathrm{d}\eta_1\, \mathrm{d}\eta_1^* \ldots \mathrm{d}\eta_n\, \mathrm{d}\eta_n^* \exp(\eta^*, A\eta) = \det A \tag{2.117}$$

where

$$(\eta^*, A\eta) = \sum_{i,j} \eta_i^* A_{ij} \eta_j$$

Equation (2.117) is valid for an arbitrary A. For a diagonal A the proof of (2.117) follows immediately from the integration rules for Grassmann variables. The corresponding formula also holds for an integral over real variables (see Problem 2.4).

Gaussian integrals can be used to define the determinant of a matrix acting in superspace. Coordinates in superspace are commuting and anticommuting variables, z and η respectively. We can define a linear transformation in superspace

$$\begin{pmatrix} z' \\ \eta' \end{pmatrix} = \begin{bmatrix} A & D \\ C & B \end{bmatrix} \begin{pmatrix} z \\ \eta \end{pmatrix} \equiv M \begin{pmatrix} z \\ \eta \end{pmatrix}$$

where matrices A and B are commuting, and C and D are anticommuting. Vectors in superspace are decomposed in terms of the variables z and η.

The superdeterminant of a superspace matrix M can be constructed by calculating a generalized Gaussian integral

$$(\det M)^{-1} = \int \frac{\mathrm{d}z\,\mathrm{d}z^*}{2\pi}\,\mathrm{d}\eta^*\,\mathrm{d}\eta \exp\left[-(z^*, Az) - (z^*, D\eta) - (\eta^*, Cz) - (\eta^*, B\eta)\right]$$

To evaluate the integral we make a shift in integration variables

$$\begin{aligned} \eta &= \eta' - B^{-1}Cz \\ \eta^* &= \eta^{*\prime} - z^*DB^{-1} \end{aligned}$$

and using (2.76) and (2.119) we get

$$\det M = \frac{\det(A - DB^{-1}C)}{\det B}$$

If we shift z and z^* we get

$$\det M = \frac{\det A}{\det(B - CA^{-1}D)}$$

These results for the superdeterminant are not surprising if we write

$$M = \begin{bmatrix} A & 0 \\ C & 1 \end{bmatrix} \begin{bmatrix} 1 & A^{-1}D \\ 0 & B - CA^{-1}D \end{bmatrix}$$

or

$$M = \begin{bmatrix} 1 & D \\ 0 & B \end{bmatrix} \begin{bmatrix} A - DB^{-1}C & 0 \\ B^{-1}C & 1 \end{bmatrix}$$

Several other properties of the superdeterminant are listed in Problem 2.9.

Fermion propagator

The classical fermion fields $\Psi(x)$ and $\bar{\Psi}(x)$ are taken to be elements of an infinite-dimensional Grassmann algebra. The generating functional $W_0^{\Psi}[\alpha]$ for a free-fermion field described by the lagrangian

$$\mathscr{L}_{\Psi} = \bar{\Psi}(\mathrm{i}\partial\!\!\!/ - m)\Psi$$

is

$$W_0^{\Psi}[\alpha, \bar{\alpha}] = N \int \mathscr{D}\Psi\mathscr{D}\bar{\Psi} \exp\left\{(\mathrm{i}/\hbar) \int \mathrm{d}^4x[\mathscr{L}_{\Psi}(x) + \hbar\bar{\alpha}(x)\Psi(x) + \hbar\bar{\Psi}(x)\alpha(x)]\right\} \tag{2.18}$$

where

$$N^{-1} = \langle 0|0\rangle = W_0^{\Psi}[0, 0]$$

and $\alpha(x)$ and $\bar{\alpha}(x)$ are sources of fermion fields. Using a formula analogous to (2.81) but for Grassmann fields

$$\int \mathscr{D}\eta\mathscr{D}\eta^* \exp[\mathrm{i}(\eta^*, A\eta) + \mathrm{i}(\beta^*, \eta) + \mathrm{i}(\eta^*, \beta)] = \det(\mathrm{i}A) \exp[\mathrm{i}(\beta^*, -A^{-1}\beta)] \tag{2.119}$$

we can rewrite $W_0^\Psi[\alpha, \bar{\alpha}]$ as follows

$$W_0^\Psi[\alpha,\bar{\alpha}] = \exp\left[i\hbar \int d^4x d^4y \bar{\alpha}(x) S(x-y) \alpha(y) \right] \tag{2.120}$$

where

$$(i\partial\!\!\!/ - m) S(x-y) = -\mathbb{1}\delta(x-y) \tag{2.121}$$

Therefore

$$S(x-y) = -\int \frac{d^4k}{(2\pi)^4} \mathbb{1} \frac{k\!\!\!/ + m}{k^2 - m^2 + i\varepsilon} \exp[-ik(x-y)] \tag{2.122}$$

The Feynman propagator defined as

$$\begin{aligned} S_F(x-y) &= \langle 0 | T\Psi(x)\bar{\Psi}(y) | 0 \rangle / \langle 0 | 0 \rangle = \int \mathscr{D}\Psi \mathscr{D}\bar{\Psi} \Psi(x) \bar{\Psi}(y) \\ &* \exp\left[(i/\hbar) \int d^4x \mathscr{L}_\Psi(x) \right] \Big/ \int \mathscr{D}\Psi \mathscr{D}\bar{\Psi} \exp\left[(i/\hbar) \int d^4x \mathscr{L}_\Psi(x) \right] \end{aligned} \tag{2.123}$$

reads

$$S_F(x-y) = \left(\frac{1}{i}\right)^2 \frac{\delta}{\delta\bar{\alpha}(x)} W_0^\Psi[\alpha,\bar{\alpha}] \frac{\overleftarrow{\delta}}{\delta\alpha(y)} \bigg|_{\alpha=\bar{\alpha}=0} \tag{2.124}$$

and using (2.120) one gets

$$S_F(x-y) = -i\hbar S(x-y) \tag{2.125}$$

Imagine now a theory of interacting fermions with the action

$$\begin{aligned} S &= \int d^4x \mathscr{L}_\Psi(x) + \int d^4x \bar{\Psi}(x) V(x) \Psi(x) \\ &= S_0 + S_I \end{aligned} \tag{2.126}$$

Following Section 2.4 we can develop perturbation theory with the interaction term as the perturbation. In the expansion in powers of S_I of the e.g. vacuum-to-vacuum amplitude (the denominator in (2.89)) we find under the path integral the following products of the Ψ and $\bar{\Psi}$ fields

$$\int dx_1 \ldots dx_k \bar{\Psi}(x_1) V(x_1) \Psi(x_1) \bar{\Psi}(x_2) V(x_2) \Psi(x_2) \ldots \bar{\Psi}(x_k) V(x_k) \Psi(x_k) \tag{2.127}$$

which correspond to fermion loops like that shown in Fig. 2.14. The path integral with insertion (2.127) can be calculated explicitly by differentiating $W_0^\Psi[\alpha, \bar{\alpha}]$ given by (2.120) with respect to α and $\bar{\alpha}$. Before making the substitution

$$\Psi(x_i) \to \frac{1}{i} \frac{\overrightarrow{\delta}}{\delta\bar{\alpha}(x_i)}, \qquad \bar{\Psi}(x_i) \to \frac{1}{i} \frac{\overleftarrow{\delta}}{\delta\alpha(x_i)} \tag{2.128}$$

we must however, rearrange the fields as follows

$$V(x_1)\Psi(x_1)\bar{\Psi}(x_2)V(x_2)\Psi(x_2)\bar{\Psi}(x_3) \cdots V(x_k)\Psi(x_k)\bar{\Psi}(x_1)$$

This reordering introduces the familiar minus sign for a closed fermion loop. From

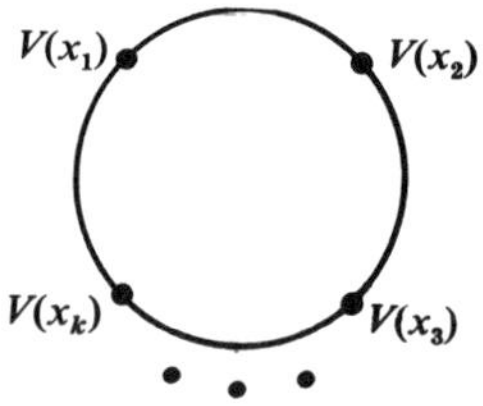

Fig. 2.14

(2.120) and (2.128) we obtain for it the following result

$$(-)S_F(x_1-x_2)V(x_2)S_F(x_2-x_3)\ldots V(x_k)S_F(x_k-x_1) \tag{2.129}$$

2.6 Generating functionals for Green's functions and proper vertices; effective potential

Classification of Green's functions and generating functionals

There are three basic sorts of Green's functions. The full Green's functions are defined by (2.54)

$$G^{(n)}(x_1,\ldots,x_n) = \langle 0|T\Phi(x_1)\ldots\Phi(x_n)|0\rangle/\langle 0|0\rangle$$
$$= \int \mathscr{D}\Phi\Phi(x_1)\ldots\Phi(x_n)\exp\{(\mathrm{i}/\hbar)S[\Phi]\}\bigg/\int\mathscr{D}\Phi\exp\{(\mathrm{i}/\hbar)S[\Phi]\} \tag{2.54}$$

with appropriate generalization to the case of other fields; for simplicity, in this Section we shall always consider the case of a single field. In perturbation theory the full Green's functions $G^{(n)}$ are given by the sum of all diagrams with n external legs, including disconnected ones, with the exception of the vacuum diagrams which are cancelled by the normalization factor.

The connected Green's functions $G^{(n)}_{\text{conn}}(x_1,\ldots,x_n)$ are defined as the connected part of $G^{(n)}$. A connected Green's function is obtained by disregarding all terms which factorize into two or more functions with no overlapping arguments. For a Fourier transform this implies that no subset of n external momenta is conserved separately. The Feynman diagrams contributing to $G^{(n)}_{\text{conn}}$ are all connected.

The connected proper vertex functions $\Gamma^{(n)}(x_1,\ldots,x_n)$, also called the one-particle-irreducible (1PI) Green's functions, are given by the Feynman diagrams which are one-particle irreducible; that is, they remain connected after an arbitrary internal line is cut. In the lowest order (no loops) the connected proper vertex functions coincide when appropriately normalized with the vertices of the original lagrangian.

To each type of the Green's functions we assign the corresponding generating functional. These are defined as follows

$$W[J] = \sum_{n=0}^{\infty}\frac{\mathrm{i}^n}{n!}\int \mathrm{d}^4x_1\ldots\mathrm{d}^4x_n G^{(n)}(x_1,\ldots,x_n)J(x_1)\ldots J(x_n) \tag{2.130}$$

$$Z[J] = \hbar \sum_{n=1}^{\infty} \frac{i^{n-1}}{n!} \int d^4x_1 \ldots d^4x_n G^{(n)}_{\text{conn}}(x_1, \ldots, x_n) J(x_1) \ldots J(x_n) \tag{2.131}$$

and

$$\Gamma[\Phi] = \sum_{n=1}^{\infty} \frac{1}{n!} \int d^4x_1 \ldots d^4x_n \Gamma^{(n)}(x_1, \ldots, x_n) \Phi(x_1) \ldots \Phi(x_n) \tag{2.132}$$

Observe that in the generating functional $\Gamma[\Phi]$ we denote the arguments as $\Phi(x)$, not $J(x)$: this is convenient because of the relationship between $\Gamma[\Phi]$ and the action integral $S[\Phi]$ which we shall discuss below. It follows from definitions (2.130)–(2.132) that

$$G^{(n)}(x_1, \ldots, x_n) = \left(\frac{1}{i}\right)^n \frac{\delta^n W[J]}{\delta J(x_1) \ldots \delta J(x_n)}\bigg|_{J \equiv 0} \tag{2.133}$$

$$G^{(n)}_{\text{conn}}(x_1, \ldots, x_n) = \frac{1}{\hbar}\left(\frac{1}{i}\right)^{n-1} \frac{\delta^n Z[J]}{\delta J(x_1) \ldots \delta J(x_n)}\bigg|_{J \equiv 0} \tag{2.134}$$

and

$$\Gamma^{(n)}(x_1, \ldots, x_n) = \frac{\delta^n \Gamma[\Phi]}{\delta \Phi(x_1) \ldots \delta \Phi(x_n)}\bigg|_{\Phi \equiv 0} \tag{2.135}$$

The generating functional of the full Green's functions has been studied in the previous Sections. It is given by the path integral formula

$$W[J] = \int \mathscr{D}\Phi \exp\left[(i/\hbar)\left\{S[\Phi] + \hbar \int d^4x J(x)\Phi(x)\right\}\right] \Big/ \int \mathscr{D}\Phi \exp\left\{(i/\hbar)S[\Phi]\right\} \tag{2.136}$$

We want to show that the functionals $Z[J]$ and $\Gamma[\Phi]$ are given by the following relations:

$$W[J] = \exp\{(i/\hbar)Z[J]\} \tag{2.137}$$

and

$$\Gamma[\Phi_{\text{CL}}] = Z[J] - \hbar \int d^4x J(x) \Phi_{\text{CL}}(x) \tag{2.138}$$

where the 'classical' field $\Phi_{\text{CL}}(x)$ is defined as the field which minimizes the combination

$$\Gamma[\Phi] + \hbar \int d^4x J(x) \Phi(x)$$

or, by the equation

$$\frac{\delta \Gamma[\Phi]}{\delta \Phi(x)}\bigg|_{\Phi = \Phi_{\text{CL}}} = -\hbar J(x) \tag{2.139}$$

By differentiating (2.138) with respect to $J(x)$ we also find

$$\frac{\delta Z[J]}{\delta J(x)} = \int d^4y \frac{\delta \Gamma[\Phi_{\text{CL}}]}{\delta \Phi_{\text{CL}}(y)} \frac{\delta \Phi_{\text{CL}}(y)}{\delta J(x)} + \hbar \Phi_{\text{CL}}(x) + \hbar \int d^4y \frac{\delta \Phi_{\text{CL}}(y)}{\delta J(x)} J(y) \tag{2.140}$$

or, with (2.139),

$$\Phi_{\mathrm{CL}}(x) = \frac{1}{\hbar}\frac{\delta Z[J]}{\delta J(x)} = \frac{\langle 0|\Phi(x)|0\rangle_J}{\langle 0|0\rangle_J} \tag{2.141}$$

Equation (2.138) is the Legendre transform. If $Z[J]$ is known we can use (2.141) to determine $J(x)$ in terms of $\Phi_{\mathrm{CL}}(x)$ so that the r.h.s. of (2.138) can be written as a functional of $\Phi_{\mathrm{CL}}(x)$ which determines $\Gamma[\Phi_{\mathrm{CL}}]$. The value of (2.141) when the external source is turned off ($J(x) \equiv 0$) is the vacuum expectation value of the field $\Phi(x)$ which, for the time being, we assume to be zero

$$\left.\frac{\delta Z[J]}{\delta J(x)}\right|_{J\equiv 0} = 0 \tag{2.142}$$

The case of a non-zero field vacuum expectation value reflecting spontaneous symmetry breaking will be discussed later on in this Section.

Equation (2.139) expresses $J(x)$ in terms of $\Phi_{\mathrm{CL}}(x)$ and in this sense it is the inverse of (2.141). In particular, together with (2.141) and (2.142), it implies that

$$\left.\frac{\delta\Gamma[\Phi_{\mathrm{CL}}]}{\delta\Phi_{\mathrm{CL}}(x)}\right|_{\Phi_{\mathrm{CL}}\equiv 0} = 0 \tag{2.143}$$

because when $J \equiv 0$, Φ_{CL} takes the value 0 and vice versa.

It remains to prove that (2.137) and (2.138) hold for the generating functionals defined by (2.130)–(2.132). This can be checked by induction (see e.g. Abers & Lee 1973). Here we shall use another argument which allows a simple understanding of the relationship between $\Gamma[\Phi]$ and $S[\Phi]$.

Effective action

Consider the classical field equation (we take the case of the $\lambda\Phi^4$ theory for definiteness)

$$(\partial_\mu\partial^\mu + m^2)\Phi_{\mathrm{c}} = -(\lambda/3!)\Phi_{\mathrm{c}}^3 + \hbar J(x), \quad \lambda > 0 \tag{2.144}$$

Equation (2.144) follows from the action principle

$$\left.\frac{\delta}{\delta\Phi}\right|_{\Phi=\Phi_{\mathrm{c}}}\left\{S[\Phi] + \hbar\int \mathrm{d}^4x J(x)\Phi(x)\right\} = 0 \tag{2.145}$$

(do not confuse Φ_{c} and Φ_{CL}) where

$$S[\Phi] = \int \mathrm{d}^4x[-\tfrac{1}{2}\Phi(\partial_\mu\partial^\mu + m^2)\Phi - (\lambda/4!)\Phi^4]$$

Equation (2.144) can be solved perturbatively. In the zeroth order in λ we obtain

$$\Phi_{\mathrm{c}}^{(0)}(x) = \int \mathrm{d}^4y[-\mathrm{i}\hbar G(x-y)]\mathrm{i}J(y) \tag{2.146}$$

$\Phi_c^{(1)}(x) =$ 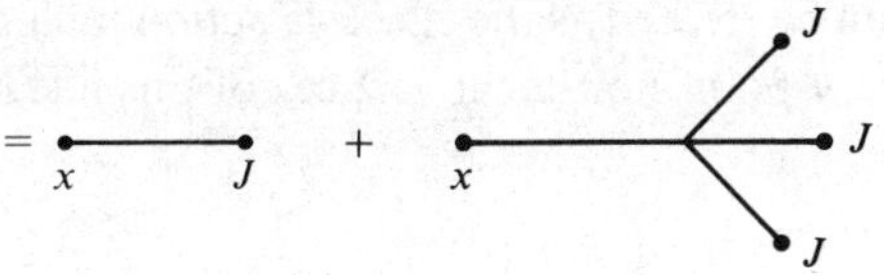

Fig. 2.15

where $G(x-y)$ is defined by (2.68)

$$(\partial_\mu \partial^\mu + m^2)G(x-y) = \delta(x-y) \tag{2.68}$$

(the iε term should always be remembered). To obtain the first order correction we insert $\Phi_c^{(0)}$ in the r.h.s. of (2.144). The result reads

$$\Phi_c^{(1)}(x) = \Phi_c^{(0)}(x) + \int d^4y[-i\hbar G(x-y)](-i\lambda/\hbar 3!)\left\{\int d^4z[-iG(y-z)]iJ(z)\right\}^3 \tag{2.147}$$

which can be represented diagrammatically as shown in Fig. 2.15. Repeating the iterations we obtain the perturbation series for the classical field $\Phi_c(x)$. It is easy to see that only tree diagrams appear and that they are all connected.

Consider now the generating functional $W[J]$, given by (2.136), in the classical approximation. This corresponds to replacing $\Phi(x)$ in the integrand of (2.136) by the classical solution $\Phi_c(x)$. Only the numerator then contributes because $J \equiv 0$ in the denominator, and we get

$$W[J] = \exp\left[(i/\hbar)\left\{S[\Phi_c] + \hbar\int d^4x J(x)\Phi_c(x)\right\}\right] \tag{2.148}$$

Thus, if we assume (2.137), in the classical approximation $Z_c[J]$ reads

$$Z_c[J] = S[\Phi_c] + \hbar\int d^4x J(x)\Phi_c(x) \tag{2.149}$$

and it is indeed given by the connected tree diagrams. This can be seen by calculating $\delta^n Z/\delta J(x_1)\ldots\delta J(x_n)|_{J\equiv 0}$ using (2.149) and the expansion (2.147) for the $\Phi_c(x)$.

Moreover, from (2.138) and (2.141), in the classical approximation $\Gamma[\Phi_{CL}] = S[\Phi_c]$ so in this approximation $\Gamma[\Phi_c]$ as given by (2.138) is indeed the generating function for the 1PI Green's functions (because $\delta^n S[\Phi_c]/\delta\Phi_c(x_1)\ldots\delta\Phi_c(x_n)|_{\Phi_c=0}$ generates proper vertex functions in the tree approximation).

When the loop corrections are taken into account the connected Green's functions as well as the generating functional $Z[J]$ can still be understood as given by the tree diagrams but with the elementary vertices replaced by 1PI Green's functions. If in (2.136) we replace the action $S[\Phi]$ by the functional $\Gamma[\Phi]$ and instead of doing the path integration exactly, again make in the integrand the 'classical' approximation $\Phi(x) \to \Phi_{CL}(x)$ with $\Phi_{CL}(x)$ defined by (2.139), we shall obtain the exact result. This is because all loop corrections are already included in $\Gamma[\Phi]$. In this sense $\Gamma[\Phi]$ can be treated as the effective action including the loop corrections. $\Gamma^{(n)}(x_1,\ldots,x_n)$s are n-point non-local vertices of this effective action. We conclude that to all orders

$$W[J] = \exp\left[(\mathrm{i}/\hbar)\left\{\Gamma[\Phi_{\mathrm{CL}}] + \hbar\int \mathrm{d}^4x J(x)\Phi_{\mathrm{CL}}(x)\right\}\right] \tag{2.150}$$

and therefore, as $\Gamma[\Phi_{\mathrm{CL}}]$ and Φ_{CL} are both given by connected diagrams, $Z[J]$ given by (2.137) is indeed connected (the connected diagrams exponentiate). Also (2.138) holds to all orders.

It is worth observing explicitly that differentiating (2.141) with respect to $\Phi_{\mathrm{CL}}(y)$, using (2.139) and setting $\Phi_{\mathrm{CL}}(x) = 0$ one gets

$$\int \mathrm{d}^4z\Gamma^{(2)}(x-z)G^{(2)}_{\mathrm{conn}}(z-x) = \mathrm{i}\hbar\delta(x-y) \tag{2.151}$$

or, in momentum space

$$\tilde{\Gamma}^{(2)}(p)\tilde{G}^{(2)}(p) = \mathrm{i}\hbar \tag{2.152}$$

(we draw attention to the factor i). For higher n, strictly speaking, $\mathrm{i}\Gamma^{(n)}$s are the 1PI functions given directly in terms of Feynman diagrams.

Spontaneous symmetry breaking and effective action

The formalism of the effective action can be extended to the case of a non-vanishing field vacuum expectation value which corresponds to spontaneous symmetry breaking (the latter is discussed in detail in Chapter 9). The functionals $W[J]$ and $Z[J]$ are taken as given by (2.136) and (2.137), respectively. However, we now suppose that (2.142) is replaced by

$$\frac{\langle 0|\Phi(x)|0\rangle}{\langle 0|0\rangle} = \frac{1}{\hbar}\frac{\delta Z[J]}{\delta J(x)}\bigg|_{J\equiv 0} = v \tag{2.153}$$

where $v = \text{const.}$ from translational invariance. It can be proved, e.g. by induction, that higher derivatives of $Z[J]$ at $J\equiv 0$ are connected Green's functions of the field $\bar{\Phi} = \Phi - v$ whose vacuum expectation value vanishes:

$$\frac{1}{\hbar}\left(\frac{1}{\mathrm{i}}\right)^{n-1}\frac{\delta^n Z[J]}{\delta J(x_1)\ldots\delta J(x_n)}\bigg|_{J\equiv 0} = \langle 0|T\bar{\Phi}(x_1)\ldots\bar{\Phi}(x_n)|0\rangle \tag{2.154}$$

The effective action is defined by (2.138) and (2.139), so that

$$\frac{\delta\Gamma[\Phi_{\mathrm{CL}}]}{\delta\Phi_{\mathrm{CL}}(x)}\bigg|_{\Phi_{\mathrm{CL}}=v} = 0 \tag{2.155}$$

This follows from (2.139), (2.141) and (2.153) because when $J\equiv 0$, Φ_{CL} takes the value v and vice versa. Also, we can prove, e.g. by induction, that $\Gamma[\Phi_{\mathrm{CL}}]$ given by (2.138) generates 1PI vertices for the field $\bar{\Phi} = \Phi - v$

$$\left.\frac{\delta^n\Gamma[\Phi_{CL}]}{\delta\Phi_{CL}(x_1)\ldots\delta\Phi_{CL}(x_n)}\right|_{\Phi_{CL}=v} = \Gamma^{(n)}_{\bar{\Phi}=\Phi-v}(x_1,\ldots,x_n) \tag{2.156}$$

and

$$\Gamma[\Phi_{CL}] = \sum_{n=1}^{\infty}\frac{1}{n!}\int d^4x_1\ldots d^4x_n\Gamma^{(n)}_{\bar{\Phi}=\Phi-v}(x_1,\ldots,x_n)(\Phi_{CL}(x_1)-v)\ldots(\Phi_{CL}(x_n)-v) \tag{2.157}$$

It is important to observe that even for $\langle 0|\Phi(x)|0\rangle = v \neq 0$ the effective action $\Gamma[\Phi_{CL}]$ can also be expanded in terms of the original fields $\Phi(x)$, see (2.132). The coefficients of such an expansion

$$\left.\frac{\delta^n\Gamma[\Phi_{CL}]}{\delta\Phi_{CL}(x_1)\ldots\delta\Phi_{CL}(x_n)}\right|_{\Phi_{CL}=0} = \Gamma^{(n)}_{\Phi}(x_1,\ldots,x_n) \tag{2.158}$$

are the 1PI Green's functions for the original fields $\Phi(x)$. This can be proved e.g. by repeating the inductive arguments leading to (2.156) but for $\Phi_{CL}=0$ and correspondingly for $J(x) = J_0$ where

$$\left.\frac{\delta Z[J]}{\delta J(x)}\right|_{J\equiv J_0} = 0 \tag{2.159}$$

One can first check by induction that higher derivatives of $Z[J]$ taken for $J \equiv J_0$ give connected Green's functions $\langle 0|T\Phi\ldots\Phi|0\rangle_{\text{conn}}$. Then we differentiate the appropriate number of times the relation $\delta Z[J]/\delta J(x) = \Phi_{CL}(x)$ with respect to Φ_{CL} and express the derivatives $\delta^n J(x)/\delta\Phi_{CL}(x_1)\ldots\delta\Phi_{CL}(x_n)$ in terms of $\Gamma[\Phi_{CL}]$ defined by (2.138) and (2.139). In the limit $J \equiv J_0$ and $\Phi_{CL} \equiv 0$ we get the desired result. The relation between the Green's functions $\Gamma^{(n)}_{\bar{\Phi}=\Phi-v}$ and $\Gamma^{(n)}_{\Phi}$ is given by the analogue of the relation

$$\left(\frac{d}{dx}\right)^n f(x)|_{x=a} = \sum_{m=0}^{\infty}\frac{a^m}{m!}\left(\frac{d}{dx}\right)^{n+m} f(x)|_{x=0}$$

Thus we have

$$\Gamma^{(n)}_{\bar{\Phi}=\Phi-v}(x_1,\ldots,x_n) = \sum_{m=0}^{\infty}(1/m!)v^m\int d^4x_{n+1}\ldots d^4x_{n+m}\Gamma^{(n+m)}_{\Phi}(x_1,\ldots,x_n,\ldots,x_{n+m}) \tag{2.160}$$

or, in momentum space

$$\tilde{\Gamma}^{(n)}_{\bar{\Phi}}(p_1,\ldots,p_n) = \sum_{m=0}^{\infty}(1/m!)v^m\tilde{\Gamma}^{(n+m)}_{\Phi}(p_1,\ldots,p_n,\underbrace{0,\ldots,0}_{m}) \tag{2.161}$$

where $\tilde{\Gamma}^{(n)}(p_1,\ldots,p_n)$ is the Fourier transform of the proper vertex $\Gamma^{(n)}(x_1,\ldots,x_n)$

$$\tilde{\Gamma}^{(n)}(p_1,\ldots,p_n)(2\pi)^4\delta(p_1+\cdots+p_n) = \int\prod_{i=1}^{n} d^4x\exp(ip_ix_i)\Gamma^{(n)}(x_1,\ldots,x_n)$$

Equation (2.161) expresses the 1PI Green's functions in a theory with spontaneously broken symmetry in terms of the 1PI Green's functions calculated as in the symmetric mode.

Effective potential

Equation (2.155) has very important implications in studies of spontaneous symmetry breaking: it tells us that the field vacuum expectation value is the value of $\Phi_{CL}(x)$ which extremizes $\Gamma[\Phi_{CL}]$. These implications can be discussed more conveniently if we introduce the so-called effective potential $V(\varphi)$. It is defined by the equation

$$\Gamma[\Phi(x)=\varphi=\text{const.}]=-(2\pi)^4\delta(0)V(\varphi) \tag{2.162}$$

($V(\varphi)$ is a function, not a functional). The condition (2.155) translates now into

$$\left.\frac{\mathrm{d}V}{\mathrm{d}\varphi}\right|_{\varphi=v}=0 \tag{2.163}$$

which is more useful than the original one because the effective potential $V(\varphi)$ can be calculated perturbatively in a systematic way. Using representation (2.132) for the $\Gamma[\Phi]$ we get

$$V(\varphi)=-\sum_{n=2}^{\infty}(1/n!)\tilde{\Gamma}_{\Phi}^{(n)}(0,0,\ldots,0)\varphi\ldots\varphi \tag{2.164}$$

so that the nth derivative of V at $\varphi=0$ is the sum of all 1PI diagrams for the *original* fields of the lagrangian, with n vanishing external momenta. (In the tree approximation V is just the ordinary potential, the negative sum of all non-derivative terms in the lagrangian density.) Alternatively, one can rewrite the lagrangian of the theory in terms of shifted fields $\bar{\Phi}=\Phi-v$ considering v as an extra free parameter, develop the appropriate Feynman rules and fix v directly from (2.155) and (2.156) i.e. by demanding that all tadpole diagrams sum to zero. This method of calculation is equivalent to computing directly the derivative of V and demanding that it vanishes, without computing V first, but it makes use of the formulation of the theory in which the underlying spontaneously broken symmetry is not explicit.

Knowledge of the effective potential means that one knows the structure of spontaneous symmetry breaking. In principle, using (2.164), we can calculate the effective potential in perturbation theory but the calculation involves a double sum: over all 1PI Green's functions and for each 1PI Green's function there is the expansion in powers of the coupling constant. A sensible way of organizing this double series is the loop expansion. It has been seen at the end of Section 2.4 that loop expansion is an expansion in powers of Planck's constant $\hbar$: a Feynman diagram with L loops is $O(\hbar^{L-1})$. Thus, the loop expansion preserves any symmetry of the lagrangian and it is unaffected by shifts of fields and by the redefinition of the division of the lagrangian into free and interacting parts associated with such shifts. Indeed expansion in powers of $\hbar$ is an expansion in a parameter that multiplies the total Lagrange density; $(1/\hbar)\mathscr{L}\to\mathscr{L}$ corresponds to change of units into $\hbar=1$.

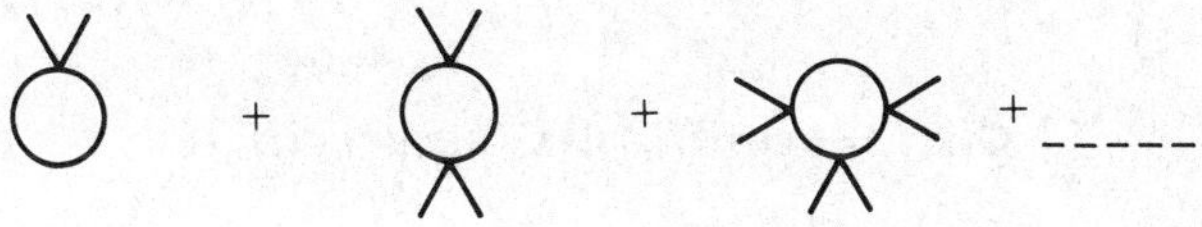

Fig. 2.16

The calculation of the effective potential for the $\lambda\Phi^4$ theory in the one-loop approximation is simple. To the lowest order (tree approximation) the only non-vanishing 1PI Green's functions are $\tilde{\Gamma}^{(2)}(p,p) = p^2 - m^2$ and $\tilde{\Gamma}^{(4)} = -\lambda$, and we get

$$V_0(\varphi) = +\tfrac{1}{2}m^2\varphi^2 + (\lambda/4!)\varphi^4 \tag{2.165}$$

To the next order (one-loop approximation) we have the infinite series of diagrams shown in Fig. 2.16 which gives the following contribution

$$\begin{aligned} V_1(\varphi) &= \mathrm{i}\sum_{n=1}^{\infty}\frac{1}{2n}\int\frac{\mathrm{d}^4k}{(2\pi)^4}\left[\lambda\frac{1}{k^2 - m^2 + \mathrm{i}\varepsilon}\frac{\varphi^2}{2}\right]^n \\ &= -\tfrac{1}{2}\mathrm{i}\int\frac{\mathrm{d}^4k}{(2\pi)^4}\ln\left(1 - \frac{\tfrac{1}{2}\lambda\varphi^2}{k^2 - m^2 + \mathrm{i}\varepsilon}\right) \end{aligned} \tag{2.166}$$

The factor $(1/2n)(\frac{1}{2})^n$ can be easily understood by recalling the discussion at the end of Section 2.4: each diagram represents $[(2n-1)(2n-3)\ldots1](n-1)!$ Feynman diagrams each with the symmetry factor $\frac{1}{2}$; thus we get $(1/2n!)[(2n-1)(2n-3)\ldots1]$ $(n-1)!\frac{1}{2} = (1/2n)(\frac{1}{2})^n$. The integral in (2.166) exhibits the UV divergence typical for perturbative calculations in quantum field theories. Its evaluation requires setting up the renormalization programme and for further discussion of V_1 given by (2.166) we refer the reader to Section 12.2.

Our final, but very important, remark in this Section deals with the physical meaning of the effective potential. It can be shown (see e.g. Coleman 1974) that $V(\varphi)$ is the expectation value of the energy density in a certain state for which the expectation value of the field is φ. This interpretation of $V(\varphi)$ is not surprising: it is already suggested by the classical limit of $V(\varphi)$ and by e.g. (2.163) for the field vacuum expectation value $\langle 0|\Phi(x)|0\rangle = v$. If $V(\varphi)$ has several local minima, it is only the absolute minimum that corresponds to the true ground state of the theory.

2.7 Green's functions and the scattering operator

The Green's functions of quantum field theory can be used to determine the matrix elements of the operator S which are directly related to scattering amplitudes and therefore experimental results. Formally, the operator S can be obtained as an infinite time limit of the evolution operator in the interaction picture. In the interaction picture the time evolution of operators is governed by the free, H_0, part of the complete hamiltonian operator $H = H_0 + H_1$ of the system (from now on we put $\hbar = 1$)

$$\Phi(\mathbf{x}, t) = \exp(\mathrm{i}H_0 t)\Phi(\mathbf{x}, 0)\exp(-\mathrm{i}H_0 t) \tag{2.167}$$

while the state vectors obey the equation

$$\mathrm{i}\frac{\partial}{\partial t}\bigg|t\bigg\rangle = H_\mathrm{I}(t)|t\rangle; \quad H_\mathrm{I}(t) = \exp(\mathrm{i}H_0 t)H_\mathrm{I}\exp(-\mathrm{i}H_0 t) \tag{2.168}$$

with the formal solution

$$|t\rangle = U(t, t')|t'\rangle \tag{2.169}$$

$$U(t, t') = T\exp\left[-\mathrm{i}\int_{t'}^{t} \mathrm{d}t H_\mathrm{I}(t)\right] \tag{2.170}$$

We then define the scattering operator to be

$$S = U(\infty, -\infty) = T\exp\left[-\mathrm{i}\int_{-\infty}^{\infty} \mathrm{d}t H_\mathrm{I}(t)\right] \tag{2.171}$$

It transforms the 'incoming' ($t \to -\infty$) states into the 'outgoing' ($t \to +\infty$) ones.

The interaction hamiltonian $H_\mathrm{I}(t)$ in (2.170) can be expressed in terms of the interaction picture field operators. For simplicity, let us consider the case of a single scalar field, described by the lagrangian density

$$\mathscr{L} = \tfrac{1}{2}(\partial_\mu\Phi\partial^\mu\Phi - m^2\Phi^2) - V(\Phi) \tag{2.172}$$

or, equivalently, by the hamiltonian

$$H = \frac{1}{2}\int \mathrm{d}^3x\{\Pi^2(\mathbf{x}, t) + [\boldsymbol{\nabla}\Phi(x)]^2 + m^2\Phi^2(x)\} + \int \mathrm{d}^3x V(\Phi(x)) = H_0 + H_\mathrm{I} \tag{2.173}$$

where

$$\Pi(\mathbf{x}, t) = \partial\mathscr{L}/\partial\dot{\Phi}(\mathbf{x}, t).$$

Equation (2.167) implies that the interaction picture field operator obeys the Klein–Gordon equation of a free field

$$(\Box + m^2)\Phi(x) = 0 \tag{2.174}$$

so that at any moment of time we can write

$$\Phi(\mathbf{x}, t) = \int \frac{\mathrm{d}^3k}{(2\pi)^3 2k_0}[a(\mathbf{k})f_k(\mathbf{x}, t) + a^\dagger(\mathbf{k})f_k^*(\mathbf{x}, t)] \tag{2.175}$$

where f_k are the plane wave solutions of the Klein–Gordon equation

$$f_k(\mathbf{x}, t) = \exp[-\mathrm{i}(k_0 t_0 - \mathbf{k}\cdot\mathbf{x})], \quad k_0 = +(\mathbf{k}^2 + m^2)^{1/2} \tag{2.176}$$

and $a(\mathbf{k})$, $a^\dagger(\mathbf{k})$ are the annihilation and creation operators of scalar particles with three-momentum $\mathbf{k}$ satisfying $[a(\mathbf{k}), a^\dagger(\mathbf{k}')] = (2\pi)^3 2k_0\delta(\mathbf{k} - \mathbf{k}')$. We can now use (2.171), (2.173) and (2.175) to express S in terms of the annihilation and creation operators and calculate the matrix elements between states containing scalar particles.

When calculating the matrix elements it is convenient to deal with normal ordered products (in the interaction picture), with all creation operators standing to the left of the annihilation operators.

This is done by using Wick's theorem. When applied to the T-ordered product of m field operators Wick's theorem has the well-known form (Bjorken & Drell 1965)

$$T\Phi(x_1)\ldots\Phi(x_m) = :\Phi(x_1)\ldots\Phi(x_m): + \sum_{\substack{i,j\\ i\neq j}} (-\mathrm{i})G(x_i - x_j) :\prod_{k\neq i,j} \Phi(x_k): + \cdots$$

$$+ \sum_{i_1,j_1} \cdots \sum_{i_n,j_n} (-\mathrm{i})G(x_{i_1} - x_{j_1})\ldots(-\mathrm{i})G(x_{i_n} - x_{j_n})$$

$$* \begin{cases} 1 & \text{when} \quad m = 2n \\ :\Phi(x_l): & \text{when} \quad m = 2n+1,\ l \neq i_1 \neq \cdots \neq j_n \end{cases} \tag{2.177}$$

A more general and compact formulation is

$$TF[\Phi] = \exp\left[-\tfrac{1}{2}\mathrm{i}\int \frac{\delta}{\delta\Phi(x)} G(x-y) \frac{\delta}{\delta\Phi(y)} \mathrm{d}^4x\mathrm{d}^4y\right] :F[\Phi]: \tag{2.178}$$

where $F[\Phi]$ is some functional of the field operator Φ and $G(x-y)$ is the Green's function given by (2.91). Functional differentiation with respect to the operator Φ has been introduced here. It is a straightforward generalization of the usual functional differentiation: the only difference is that one must keep in mind that the field operators for different values of time do not commute. If we assume $F[\Phi]$ to be a product of four field operators: $\Phi(x_1)\Phi(x_2)\Phi(x_3)\Phi(x_4)$ and take the matrix element of (2.178) between the vacuum states, the result is exactly (2.94), the corollary of Wick's theorem used to derive the perturbation theory Feynman rules in the previous Section.

Now let us take the interaction hamiltonian to be

$$H_{\mathrm{I}}(t) = \int \mathrm{d}^3x\, V(\Phi) = -\int \mathrm{d}^3x J(x)\Phi(x) \tag{2.179}$$

where $J(x)$ is a c-number function. This corresponds to the case of scattering from an external source. We obtain, after performing the functional differentiation in (2.178):

$$T\exp\left(\mathrm{i}\int \mathrm{d}^4xJ\Phi\right) = \exp\left[\tfrac{1}{2}\mathrm{i}\int \mathrm{d}^4y\mathrm{d}^4xJ(x)G(x-y)J(y)\right] :\exp\left(\mathrm{i}\int \mathrm{d}^4xJ\Phi\right):$$

$$= W_0[J] :\exp\left(\mathrm{i}\int \mathrm{d}^4xJ\Phi\right): \tag{2.180}$$

where $W_0[J]$ is the previously introduced generating functional of the free field

Green's functions. We observe that

$$J(x)W_0[J] = -\mathrm{i}\int \mathrm{d}^4 y G^{-1}(x-y)\frac{\delta}{\delta J(y)} W_0[J] \tag{2.181}$$

so that (2.180) can be rewritten as follows

$$T\exp\left(\mathrm{i}\int \mathrm{d}^4 x J\Phi\right) = :\exp\left[\int \mathrm{d}^4 x \mathrm{d}^4 y \Phi(x) G^{-1}(x-y)\frac{\delta}{\delta J(y)}\right]: W_0[J] \tag{2.182}$$

Here $G^{-1}(x-y)$ is the inverse operator to $G(x-y)$

$$\int \mathrm{d}^4 y G^{-1}(x-y) G(y-z) = \delta(x-z) \tag{2.183}$$

i.e. the Klein–Gordon operator.

The generalization to arbitrary $V(\Phi)$ follows from the formula

$$T\exp\left[-\mathrm{i}\int \mathrm{d}^4 x V(\Phi)\right] = \exp\left[-\mathrm{i}\int \mathrm{d}^4 x V\left(\frac{1}{\mathrm{i}}\frac{\delta}{\delta J}\right)\right] T\exp\left(\mathrm{i}\int \mathrm{d}^4 x J\Phi\right)\bigg|_{J\equiv 0} \tag{2.184}$$

Consequently, the S operator of the interacting scalar field can be written in the form

$$\begin{aligned} S = T\exp\left[-\mathrm{i}\int \mathrm{d}^4 x V(\Phi)\right] &= :\exp\left[\int \mathrm{d}^4 x\, \mathrm{d}^4 y \Phi(x) G^{-1}(x-y)\frac{\delta}{\delta J(y)}\right]: \\ &\quad * \exp\left[-\mathrm{i}\int \mathrm{d}^4 x V\left(\frac{1}{\mathrm{i}}\frac{\partial}{\partial J}\right)\right] W_0[J]\bigg|_{J\equiv 0} \\ &= N^{-1}:\exp\left[\int \mathrm{d}^4 x\, \mathrm{d}^4 y \Phi(x) G^{-1}(x-y)\frac{\delta}{\delta J(y)}\right]: W[J]\bigg|_{J\equiv 0} \end{aligned} \tag{2.185}$$

We have been dealing with fields in the interaction picture but in $W[J]$ we can recognize, by (2.88), the generating functional of the previously introduced Green's functions of the interacting theory defined as the vacuum matrix elements of the T-ordered products of the field operator in the *Heisenberg* picture; see (2.54). Expanding the exponential we obtain

$$\begin{aligned} S = N^{-1}\sum_{n=0}^{\infty}\frac{1}{n!}\int & \mathrm{d}^4 x_1\, \mathrm{d}^4 y_1 \ldots \mathrm{d}^4 x_n\, \mathrm{d}^4 y_n G^{(n)}(y_1, \ldots, y_n) \\ & * \mathrm{i}G^{-1}(x_1-y_1)\ldots \mathrm{i}G^{-1}(x_n-y_n):\Phi(x_1)\ldots\Phi(x_n): \end{aligned} \tag{2.186}$$

The factor N^{-1} contains all the vacuum-to-vacuum transitions and will be omitted in the following.

The calculation of matrix elements is now straightforward. Let the initial state involve n particles and the final state m particles. We have (c, c' are the normalization constants)

$$\left.\begin{aligned} |\mathbf{k}_1\ldots\mathbf{k}_n\rangle &= ca^+(\mathbf{k}_1)\ldots a^+(\mathbf{k}_n)|0\rangle \\ \langle\mathbf{k}'_1\ldots\mathbf{k}'_m| &= c'\langle 0|a(\mathbf{k}'_1)\ldots a(\mathbf{k}'_m) \end{aligned}\right\} \tag{2.187}$$

It is convenient to assume that $\mathbf{k}'_i \neq \mathbf{k}_j$ for all i, j pairs. Only the term in the series with $:\Phi(x_1)\ldots\Phi(x_{n+m}):$ will then contribute. Of the $(n+m)$ field operators n will act as the annihilation and m as the creation operators: this gives a combinatorial factor $\binom{n+m}{m}$. Finally, there are $n!m!$ ways of matching the annihilation and creation operators with the external particles. The result is

$$\begin{aligned} &\langle\mathbf{k}'_1\ldots\mathbf{k}'_m|S|\mathbf{k}_1\ldots\mathbf{k}_n\rangle \\ &\quad = cc'\int \mathrm{d}^4x_1\ldots\mathrm{d}^4x_{n+m} f^{*\prime}_{k_1}(x_1)\ldots f_{k_n}(x_{n+m})\int \mathrm{d}^4y_1\ldots\mathrm{d}^4y_{n+m}\mathrm{i}G^{-1}(x_1-y_1)\cdots \\ &\quad * \mathrm{i}G^{-1}(x_{n+m}-y_{n+m})G^{(n+m)}(y_1,\ldots,y_{n+m}) \end{aligned} \tag{2.188}$$

With the usual invariant normalization

$$|\mathbf{k}_1\ldots\mathbf{k}_n\rangle = \frac{1}{(n!)^{1/2}}a^\dagger(\mathbf{k}_1)\ldots a^\dagger(\mathbf{k}_n)|0\rangle \tag{2.189}$$

this is just the on-shell limit of the Fourier transformed function

$$\begin{aligned} \langle\mathbf{k}'_1\ldots\mathbf{k}'_m|S|\mathbf{k}_1\ldots\mathbf{k}_n\rangle &= \frac{1}{(n!m!)^{1/2}}\lim_{k_i^2\to m^2} \\ &\quad * \prod_i\left[\frac{1}{\mathrm{i}}(k_i^2-m^2)\right]\tilde{G}^{(n+m)}(k'_1,\ldots,k'_m,-k_1,\ldots,-k_n) \end{aligned} \tag{2.190}$$

where

$$\tilde{G}^{(n)}(k_1,\ldots,k_n) = \int \mathrm{d}^4y_1\exp(\mathrm{i}k_1y_1)\ldots\int \mathrm{d}^4y_n\exp(\mathrm{i}k_ny_n)G^{(n)}(y_1\ldots y_n) \tag{2.191}$$

We see that the S matrix element is equal to the multiple on-shell residue of the Fourier transformed Green's function. Its connected part is proportional to the on-shell limit of the so-called connected truncated Green's function, obtained from $\tilde{G}^{(n)}_{\text{conn}}$ by factorizing out the exact two-point functions $\tilde{G}^{(2)}(k_i)$ for each of the external lines of $\tilde{G}^{(n)}_{\text{conn}}$.

At this point we must observe that in the derivation of (2.190) we have made the assumption (following from (2.175)) that the one-particle state created by $\Phi(x)$ from the vacuum has the standard normalization

$$\langle\mathbf{k}|\Phi(x)|0\rangle = \exp[\mathrm{i}(k_0x_0-\mathbf{k}\cdot\mathbf{x})] \tag{2.192}$$

with $\langle\mathbf{k}|$ normalized as in (2.189), that is

$$\langle\mathbf{k}| = \langle 0|a(\mathbf{k}) \tag{2.193}$$

or

$$\langle\mathbf{k}|\mathbf{k}'\rangle = (2\pi)^3 2k_0\delta(\mathbf{k}-\mathbf{k}')$$

Normalization (2.192) holds for the field operators both in the interaction and in the Heisenberg picture. Fields satisfying (2.192) we shall call the 'physical' fields. More general cases will be discussed in Chapter 4.

The above formal considerations leading to the Lehman–Symanzik–Zimmermann reduction theorem (2.190) can be supplemented by the following arguments, which by themselves can, actually, be taken as a proof of (2.190). Let us consider the Fourier transform (2.191) of the Green's function $G^{(n)}(y_1,\ldots y_n)=\langle 0|T\Phi(y_1)\ldots\Phi(y_n)|0\rangle$, where we have taken $\langle 0|0\rangle=1$, and assume that the field $\Phi(x)$ is normalized according to (2.192). Among the various terms which contribute to $\tilde{G}^{(n)}(k_1,\ldots,k_n)$ there is a time-ordering in which $\Phi(y_1)$ stands furthest to the left. We can insert a complete set of intermediate states between $\Phi(y_1)$ and the remaining time-ordered product getting

$$\tilde{G}^{(n)}(k_1,\ldots,k_n)=\sum\!\!\!\!\!\!\int_m \mathrm{d}^4y_1\ldots\mathrm{d}^4y_n \exp(\mathrm{i}k_1y_1)\ldots\exp(\mathrm{i}k_ny_n)$$
$$*\langle 0|\Phi(y_1)|m\rangle\langle m|T\Phi(y_2)\ldots\Phi(y_n)|0\rangle\Theta(y_1^0-\max\{y_2^0\ldots y_n^0\})+\cdots \tag{2.194}$$

The y_1 integration can now be carried out using

$$\langle 0|\Phi(y_1)|n\rangle=\langle 0|\Phi(0)|n\rangle\exp(-\mathrm{i}p_ny_1)$$

and the integral representation for the Θ-function

$$\Theta(y_1^0-t)=\frac{\mathrm{i}}{2\pi}\int_{-\infty}^{\infty}\frac{\mathrm{d}\omega\exp[-\mathrm{i}\omega(y_1^0-t)]}{\omega+\mathrm{i}\varepsilon}$$

One gets then the following

$$\int \mathrm{d}^4y_1\exp[\mathrm{i}(k_1-p_n)y_1]\Theta(y_1^0-t)$$
$$=\mathrm{i}\delta(\mathbf{k}_1-\mathbf{p}_n)(2\pi)^3\frac{1}{k_1^0-p_n^0+\mathrm{i}\varepsilon}\exp[\mathrm{i}(k_1^0-p_n^0)t]$$
$$=\mathrm{i}(2\pi)^3\delta(\mathbf{k}_1-\mathbf{p}_n)\frac{k_1^0+p_n^0}{k_1^2-p_n^2+\mathrm{i}\varepsilon}\exp[\mathrm{i}(k_1^0-p_n^0)t] \tag{2.195}$$

Thus, a single particle state of mass m contributes a pole at $k_1^2=m^2$ while no other state $|n\rangle$ and none of the remaining terms arising from other time-orderings can have such a pole. It is important to observe that the pole arises due the limit $y_1^0\to+\infty$ in the integral (2.195). We conclude therefore that

$$\lim_{k_1^2\to m^2}\tilde{G}^{(n)}(k_1,\ldots,k_n)=\frac{\mathrm{i}}{k_1^2-m^2}\langle\mathbf{k}_1,t=+\infty|T\Phi(y_2)\ldots\Phi(y_n)|0\rangle \tag{2.196}$$

$$\left(\int|1\rangle\langle 1|\equiv\int\frac{\mathrm{d}^3p}{(2\pi)^3 2p_0}|\mathbf{p}\rangle\langle\mathbf{p}|\right)$$

Similarly,

$$\lim_{k_1^2 \to m^2} \tilde{G}^{(n)}(-k_1, \ldots, k_n) = \frac{\mathrm{i}}{k_1^2 - m^2} \langle 0 | T\Phi(y_2) \ldots \Phi(y_n) | \mathbf{k}_1, t = -\infty \rangle \quad (2.197)$$

By repeating these steps we prove (2.190), remembering that

$$\langle \mathbf{k}'_1 \ldots \mathbf{k}'_m | S | \mathbf{k}_1 \ldots \mathbf{k}_n \rangle = \langle \mathbf{k}'_1 \ldots \mathbf{k}'_m, t = +\infty | \mathbf{k}_1 \ldots \mathbf{k}_n, t = -\infty \rangle$$

Problems

2.1 Derive (2.12) from (2.7) integrating over p_is first.

2.2 For a scalar field theory defined by a lagrangian

$$\mathscr{L} = \tfrac{1}{2}\partial^\mu \Phi(x) \partial_\mu \Phi(x) - V(\Phi(x))$$

and the corresponding equation of motion

$$\Box \Phi + V'(\Phi) = 0$$

derive the equation of motion for the Green's function $\langle 0 | T\Phi(x)\Phi(y) | 0 \rangle$:

$$\Box_x \langle 0 | T\Phi(x)\Phi(y) | 0 \rangle = -\langle 0 | TV'(\Phi(x))\Phi(y) | 0 \rangle - \mathrm{i}\delta^{(4)}(x-y)$$

using definitions (2.54) and (2.56) and the invariance of the generating functional (2.57) under an infinitesimal change of variables

$$\Phi(x) \to \Phi(x) + \varepsilon f(x)$$

where $f(x)$ is a function of x

$$\int \mathscr{D}\Phi \exp\left\{ \mathrm{i}S[\Phi + \varepsilon f] + \mathrm{i}\int \mathrm{d}^4x(\Phi + \varepsilon f)J \right\} = \int \mathscr{D}\Phi \exp\left\{ \mathrm{i}S[\Phi] + \mathrm{i}\int \mathrm{d}^4x \Phi J \right\}$$

Repeat the same in Euclidean space.

2.3 Using the rules of Section 2.4 calculate the following symmetry factors S in the $\lambda\Phi^4$ theory

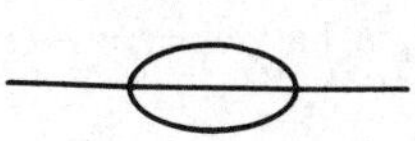

$S = \frac{1}{6}$

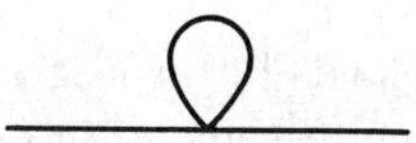

$S = \frac{1}{2}$

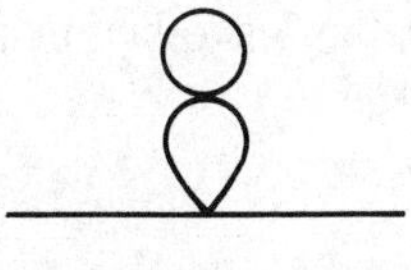

$S = \frac{1}{4}$

2.4 For real Grassmann variables η_i prove the formula

$$\int \mathrm{d}\eta_1 \ldots \mathrm{d}\eta_n \exp\left[\tfrac{1}{2}(\eta, A\eta)\right] = (\det A)^{1/2}$$

for any antisymmetric matrix A, where $(\eta, A\eta) = \sum_{i,j} \eta_i A_{ij} \eta_j$

2.5 Prove (2.154), (2.156) and (2.158).

2.6 Prove (2.178).

2.7 Using representation (2.10) calculate the Green's function for the harmonic oscillator

$$\langle x'|\exp[-\mathrm{i}(t'-t)H]|x\rangle = \left(\frac{\omega}{2\mathrm{i}\pi \sin \omega(t'-t)}\right)^{1/2} \exp[\mathrm{i}S(t'-t)]$$

where

$$H = -\tfrac{1}{2}(\partial^2/\partial x^2) + \tfrac{1}{2}\omega^2 x^2$$

$$S(\tau) = \frac{\omega}{2\sin\omega\tau}[(x'^2 + x^2)\cos\omega\tau - 2x'x]$$

Note that $S(\tau)$ is the classical action (along the classical path).

2.8 For a scalar field theory as in Problem 2.2 with $V(\Phi) = \lambda\Phi^4$ calculate the effective potential $\Gamma[\Phi]$ by evaluating the generating functional $W_{\mathrm{E}}[J]$ in Euclidean space

$$W_{\mathrm{E}}[J] = N_{\mathrm{E}} \int \mathscr{D}\Phi \exp\left\{-S_{\mathrm{E}}[\Phi] + \int J\Phi \mathrm{d}^4\bar{x}\right\}$$

by the method of steepest descent. The steps are

(*a*) expand the exponent around Φ_0 (often called the background field) which is the solution of the classical equation of motion in the presence of the external source $J(x)$

$$(-\bar{\partial}_\mu \bar{\partial}^\mu + m^2)\Phi_0 + V'(\Phi_0) - J = 0 \qquad \text{(I)}$$

to write

$$W_{\mathrm{E}}[J] = N_{\mathrm{E}} \exp\left\{-S_{\mathrm{E}}[\Phi_0] + \int J\Phi_0 \mathrm{d}^4x\right\} \int \mathscr{D}\Phi \exp\left\{-\frac{1}{2}\mathrm{d}^4\bar{x}\,\mathrm{d}^4\bar{y}\frac{\delta^2 S_{\mathrm{E}}[\Phi_0]}{\delta\Phi(\bar{x})\delta\Phi(\bar{y})}\right.$$

$$\left. *[\Phi(\bar{x}) - \Phi_0(\bar{x})][\Phi(\bar{y}) - \Phi_0(\bar{y})] + \cdots\right\} \qquad \text{(II)}$$

(*b*) in the classical approximation

$$W_{\mathrm{E}}[J] = N_{\mathrm{E}} \exp\left\{-S_{\mathrm{E}}[\Phi_0] + \int \mathrm{d}^4\bar{x} J\Phi_0\right\}$$

Evaluate $Z[J]$and $\Gamma[\Phi]$ explicitly by solving (I) perturbatively in powers of λ. Check that in this approximation $\Gamma[\Phi]$ is the classical action.

(*c*) calculate $\Gamma[\Phi]$ in the saddle point approximation. Evaluate the Gaussian integral in (II) by using $\det A = \exp \operatorname{Tr} \ln A$ and expanding the logarithm in powers of λ. Calculate the effective potential and prove (2.166).

2.9 For a matrix M acting in superspace

$$\begin{pmatrix} z \\ \eta \end{pmatrix}' = M\begin{pmatrix} z \\ \eta \end{pmatrix} = \begin{pmatrix} A & D \\ C & B \end{pmatrix}\begin{pmatrix} z \\ \eta \end{pmatrix}$$

where z and η are commuting and anticommuting coordinates, respectively

(*a*) write $\det M$ instead of $(\det M)^{-1}$ as a Gaussian integral

(*b*) show that the supertrace defined as

$$\operatorname{Tr} M = \operatorname{Tr} A - \operatorname{Tr} B$$

satisfies the cyclicity property

$$\operatorname{Tr}[M_1 M_2] = \operatorname{Tr}[M_2 M_1]$$

(*c*) show that

$$\operatorname{Tr}\ln M_1 M_2 = \operatorname{Tr}\ln M_1 + \operatorname{Tr}\ln M_2$$

Hint: use the Campbell–Backer–Hausdorff formula for ordinary matrices

$$\exp(A)\exp(B) = \exp\left\{A + B + \frac{1}{2!}[A,B] + \frac{1}{3!}\left(\tfrac{1}{2}[[A,B],B] + \tfrac{1}{2}[A,[A,B]]\right)\right.$$
$$\left. + \frac{1}{4!}[[A,[A,B]],B] + \cdots\right\}$$

the exponent and the logarithm of a matrix in superspace are defined by a series expansion, as for ordinary matrices.
(*d*) writing

$$M = \begin{bmatrix} A & 0 \\ C & 1 \end{bmatrix}\begin{bmatrix} 1 & A^{-1}D \\ 0 & B - CA^{-1}D \end{bmatrix}$$

and using property (*c*) prove that

$$\ln\det M = \operatorname{Tr}\ln M$$

and

$$\det(M_1 M_2) = \det M_1 \det M_2$$

(*e*) write the Jacobian of a transformation in superspace.

3

Feynman rules for Yang–Mills theories

3.1 The Faddeev–Popov determinant

Gauge invariance and the path integral

As discussed in Chapter 1 gauge quantum field theories are of special physical interest in our attempts to describe fundamental interactions. In this Chapter we shall derive Feynman rules for gauge theories. Consider a path integral over gauge field A_μ, corresponding to a physical, that is, gauge-invariant quantity

$$\int \mathscr{D}A_\mu f(A_\mu) \exp\left(\mathrm{i} \int \mathrm{d}^4x \mathscr{L} \right) \tag{3.1}$$

For brevity we write A_μ instead of A_μ^α, where α is the gauge group index. $\mathscr{L}$ is the lagrangian density and $f(A_\mu)$ denotes a gauge-invariant functional depending on the physical quantity under consideration. The integration measure $\mathscr{D}A_\mu$ we assume to be invariant under gauge transformations. That is, it must have the following property

$$\mathscr{D}A_\mu = \mathscr{D}A_\mu^g$$

where g is an arbitrary transformation from the gauge group. A_μ^g denotes the result of this transformation when applied to A_μ. The simplest formal *Ansatz* for the integration measure which has this property is $\mathscr{D}A_\mu = \prod_{\mu,\alpha,x} \mathrm{d}A_\mu^\alpha(x)$, the straightforward generalization of the integration measure introduced previously for scalar fields.

As explained before, the path integral in (3.1) runs over all possible configurations $A_\mu(x)$, which implies multiple counting of the physically equivalent configurations (equivalent up to a gauge transformation). Let us divide the configuration space $\{A_\mu(x)\}$ into the equivalence classes $\{A_\mu^g(x)\}$ called the orbits of the gauge group. An orbit of the group includes all the field configurations which result when all

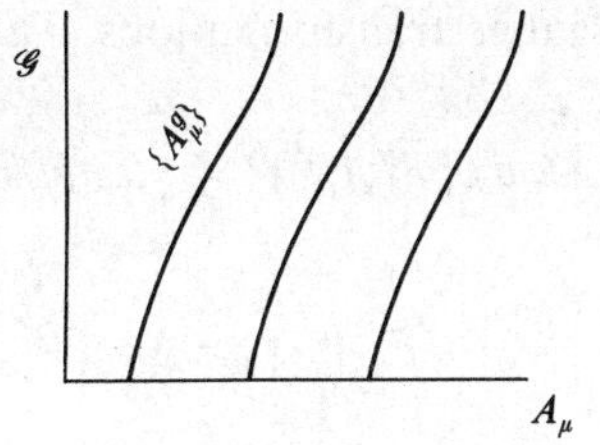

Fig. 3.1

possible transformations g from the gauge group $\mathscr{G}$ are applied to a given initial field configuration $A_\mu(x)$. This construction can be represented graphically as shown in Fig. 3.1. The integrand of (3.1) is constant along any orbit of the gauge group. Consequently, the integral as it stands is proportional to an infinite constant (the volume of the total gauge group $\mathscr{G}$). This is not an important difficulty in itself, because the infinity can always be cancelled by a normalization constant. The problem appears when we want to calculate (3.1) perturbatively because local gauge symmetry implies that the quadratic part of the gauge field action density has zero eigenvalues and therefore cannot be inverted in the configuration space $\{A_\mu(x)\}$ so that the propagator of the gauge field cannot be defined. One way to resolve this problem is to apply perturbation theory to the functional integral over the coset space of the orbits of the group (in other words, over the physically distinct field configurations) which follows from (3.1) after the infinite constant has been factorized out. We shall now describe a procedure by means of which this factorization can be achieved.

Let $\mathscr{D}g$ denote an invariant measure on the gauge group $\mathscr{G}$

$$\mathscr{D}g = \mathscr{D}(gg'); \qquad \mathscr{D}g = \prod_x \mathrm{d}g(x) \tag{3.2}$$

and let us introduce a functional $\Delta[A_\mu]$ defined by the following equation

$$1 = \Delta[A_\mu] \int \mathscr{D}g\delta[F[A_\mu^g]] \tag{3.3}$$

Here $\delta[f(x)]$ represents the product of the usual Dirac δ-functions: $\prod_x \delta(f(x))$, one at each space-time point. As for the functional $F[A_\mu]$, we assume that the equation

$$F[A_\mu^g] = 0 \tag{3.4}$$

has exactly one solution, g_0, for any initial field A_μ. In the configuration space $\{A_\mu(x)\}$ the equation $F[A_\mu] = 0$ must then define a surface that crosses any of the orbits of the group exactly once i.e., (3.4) defines a 'gauge'. Let us observe that the functional $F[A_\mu]$ should be of the general form $F^\alpha(x, [A_\mu])$, where α is the group index of the adjoint representation and x a space-time point. In fact to fix a gauge we need (at each space-time point) one equation for each parameter of the group.

$\Delta[A_\mu]$ is invariant under gauge transformations. This can be shown as follows

$$\Delta^{-1}[A_\mu^g] = \int \mathscr{D}g'\delta[F[A_\mu^{gg'}]] = \int \mathscr{D}(gg')\delta[F[A_\mu^{gg'}]]$$
$$= \int \mathscr{D}g''\delta[F[A_\mu^{g''}]] = \Delta^{-1}[A_\mu] \tag{3.5}$$

where we have used the invariance property of the group measure $\mathscr{D}g$. In effect, the functional $\Delta[A_\mu]$ depends only on the orbit to which A_μ belongs.

Our aim is to replace integration over all field configurations by integration restricted to the hypersurface $F[A_\mu]=0$. In this case each orbit will contribute only one field configuration and we shall have an integration over physically distinct fields. We start by inserting (3.3) (which is equal to one) under the path integral in (3.1). Next we change the order of integrations. The result is

$$\int \mathscr{D}g \int \mathscr{D}A_\mu \Delta[A_\mu] f(A_\mu)\delta[F[A_\mu^g]] \exp\{iS[A_\mu]\} \tag{3.6}$$

An important observation is that the complete expression under the $\int\mathscr{D}g$ integral is in fact independent of g. To show it we can use the gauge invariance of $\int\mathscr{D}A_\mu$, $\Delta[A_\mu]$, $f(A_\mu)$ and $S[A_\mu]$ to replace them by $\int\mathscr{D}A_\mu^g$, $\Delta[A_\mu^g]$, $f(A_\mu^g)$ and $S[A_\mu^g]$: the result

$$\int \mathscr{D}A_\mu^g \Delta[A_\mu^g] f(A_\mu^g)\delta[F[A_\mu^g]] \exp\{iS[A_\mu^g]\}$$

can be made manifestly g independent by a change in notation: $A_\mu^g \to A_\mu$. Consequently, the group integration $\int\mathscr{D}g$ factorizes out to produce an infinite constant: the volume of the full gauge group. We obtain finally

$$\left(\int \mathscr{D}g\right) \int \mathscr{D}A_\mu \Delta[A_\mu]\delta[F[A_\mu]] f(A_\mu) \exp\{iS[A_\mu]\} \tag{3.7}$$

which defines the theory as formulated in the $F[A_\mu]=0$ gauge. Unlike (3.1), (3.7) can be used as a starting point for perturbative calculations.

Faddeev–Popov determinant

We have still to calculate $\Delta[A_\mu]$. From (3.3) we obtain, after formal manipulations

$$\Delta^{-1}[A_\mu] = \int \mathscr{D}F \left(\det \frac{\delta F[A_\mu^g]}{\delta g}\right)^{-1} \delta[F]$$

that is

$$\Delta[A_\mu] = \det \left.\frac{\delta F[A_\mu^g]}{\delta g}\right|_{F[A_\mu^g]=0} \tag{3.8}$$

$\Delta[A_\mu]$ is usually called the Faddeev–Popov determinant. It is convenient to use the gauge invariance property of $\Delta[A_\mu]$ to choose A_μ which already satisfies the gauge condition $F[A_\mu]=0$. Then in (3.8) we can replace the constraint $F[A_\mu^g]=0$ by $g=1$, which simplifies the practical calculations

$$\Delta[A_\mu]=\det\left.\frac{\delta F[A_\mu^g]}{\delta g}\right|_{g=1}; \qquad F[A_\mu]=0 \tag{3.9}$$

Near $g=1$ we only have to deal with the infinitesimal transformations: $U(\Theta)=1-\mathrm{i}T^\alpha\Theta^\alpha(x)$ (where $\Theta^\alpha(x)\ll 1$) and the invariant group measure $\mathscr{D}g$ takes the simple form $\mathscr{D}\Theta\equiv\prod_{\alpha,x}\mathrm{d}\Theta^\alpha(x)$. It is easy to check the invariance property of $\mathscr{D}\Theta$

$$D\Theta=D\Theta'' \tag{3.10}$$

where $U(\Theta'')=U(\Theta)U(\Theta')$ and Θ' is arbitrary. For infinitesimal transformations $\Theta''\approx\Theta'+\Theta$ and therefore $\det(\mathrm{d}\Theta''/\mathrm{d}\Theta)=1$ so that (3.10) is proved.

We can now rewrite (3.9) in a more explicit form, with all relevant indices

$$\Delta[A_\mu]=\det\left.\frac{\delta F^\alpha(x,[A_\mu^\Theta])}{\delta\Theta^\beta(y)}\right|_{\Theta=0}; \qquad F^\alpha(x,[A_\mu])=0 \tag{3.11}$$

Here A_μ^Θ stands for A_μ^g and α,β are the gauge group indices. We have to calculate a determinant of a matrix in both space-time and the group indices $M^{\alpha\beta}(x,y)$. This matrix appears in the expansion of $F^\alpha(x,[A_\mu^\Theta])$ in powers of the infinitesimal parameters $\Theta^\beta(y)$

$$F^\alpha(x,[A_\mu^\Theta])=F^\alpha(x,[A_\mu])+\int\mathrm{d}^4yM^{\alpha\beta}(x,y)\Theta^\beta(y)+\cdots \tag{3.12}$$

so that

$$M^{\alpha\beta}(x,y)=\left.\frac{\delta F^\alpha(x,[A_\mu^\Theta])}{\delta\Theta^\beta(y)}\right|_{\Theta=0} \tag{3.13}$$

and

$$\Delta[A_\mu]=\det M; \qquad F[A_\mu]=0 \tag{3.14}$$

Recalling (1.41) for the infinitesimal transformation of the gauge field, we can also write

$$\begin{aligned} M^{\alpha\beta}(x,y)&=\int\mathrm{d}^4z\frac{\delta F^\alpha(x,[A_\mu])}{\delta A_\nu^\rho(z)}\frac{\delta A_\nu^\rho(z)}{\delta\Theta^\beta(y)}\\ &=\int\mathrm{d}^4z\frac{\delta F^\alpha(x,[A_\mu])}{\delta A_\nu^\rho(z)}\left(c_{\rho\beta\gamma}A_\nu^\gamma(z)-\frac{1}{g}\delta_{\rho\beta}\partial_\nu^{(z)}\right)\delta(z-y) \end{aligned} \tag{3.15}$$

For the subsequent applications the following observations will be helpful. First let us note that $\Delta[A_\mu]$ under a path integral like (3.7) can be replaced by $\det M$, because the condition $F[A_\mu]=0$ is already enforced by the δ-functional which fixes the gauge. Consider a class of gauge conditions of the form $F[A_\mu]-C(x)=0$ where $C(x)$ is an arbitrary function of a space-time point. For all gauge conditions belonging to this class $\det M$ is the same because $C(x)$ is unaffected by the gauge transformation from A_μ to A_μ^g. We can make use of this feature in order to replace the δ-functional in (3.7) by some other functional of the gauge condition, which may be more convenient for practical calculations. In the $F[A_\mu] - C(x) = 0$ gauge (3.7) becomes

$$\left(\int \mathscr{D}g\right) \int \mathscr{D}A_\mu \det M\delta[F[A_\mu] - C(x)] f(A_\mu) \exp\{iS[A_\mu]\} \tag{3.16}$$

Being gauge invariant, this is obviously independent of $C(x)$. We can integrate (3.16) functionally over $\int \mathscr{D}C$ with an arbitrary weight functional $G[C]$; the result will differ from (3.16) only by an overall normalization constant. Observing that the δ-functional in (3.16) is the only C-dependent term, we obtain

$$\int \mathscr{D}A_\mu \det M f(A_\mu) \exp\{iS[A_\mu]\} \int \mathscr{D}C\delta[F[A_\mu] - C]G[C]$$

$$= \int \mathscr{D}A_\mu \det M f(A_\mu) \exp\{iS[A_\mu]\} G[F[A_\mu]] \tag{3.17}$$

A popular choice for $G[C]$ is

$$G[C] = \exp\left\{-\frac{i}{2\alpha}\int d^4x[C(x)]^2\right\} \tag{3.18}$$

where α is a real constant. This is equivalent to replacing the original lagrangian density by

$$\mathscr{L}_{\text{eff}} = \mathscr{L} - \frac{1}{2\alpha}(F[A_\mu])^2 \tag{3.19}$$

which is no longer gauge invariant; consequently, perturbation theory may apply in this case. Notice also that in the limit $\alpha \to 0$ (3.18) approaches the δ-functional (up to the α-dependent normalization factor which for our purposes is irrelevant). The derivation of Feynman rules in gauges fixed by δ-functionals can often be simplified by taking this limit.

We have shown how a path integral representation, (3.1), of a gauge-invariant quantity can be rewritten in a fixed gauge, so that the perturbation theory may be set up. The rules of this perturbation theory will depend on the gauge, but the final results will not, because the quantity we calculate is gauge invariant by assumption. In the following we shall also deal with gauge-dependent quantities. In particular the generating functional $W[J]$ in the $F[A_\mu] = 0$ gauge is given by the following expression

$$W[J] = N\int \mathscr{D}A_\mu \Delta[A_\mu]\delta(F[A_\mu]) \exp\left[i\int d^4x(\mathscr{L} + J_\mu^\alpha A_\alpha^\mu)\right] \tag{3.20}$$

Due to the presence of fixed external sources J_μ^α this is not a gauge-invariant object and the same applies to its functional derivatives with respect to J, the

Green's functions of the gauge field. Although in a sense unphysical, the Green's functions are often useful at intermediate stages of calculations and physical quantities may be conveniently defined in terms of them. In particular, as we know from Section 2.7, an element of the S-matrix is obtained from the corresponding Green's functions by removing single-particle propagators corresponding to externals lines, taking the Fourier transform of the resulting 'amputated' Green's function and placing external momenta on the mass-shell. Although we have not discussed the renormalization programme yet, it is in order to mention here that the renormalized S-matrix elements are gauge independent. Equivalence of different gauges can be explicitly checked for various cases (see e.g. Abers & Lee 1973, Slavnov & Faddeev 1980, 't Hooft & Veltman 1973). One can show that only the renormalization constants attached to each external line depend on the gauge but that is of no consequence for the renormalized S-matrix. The other gauge-dependent terms do not contribute to the poles of the Green's functions at $p_i^2 = m^2$. These formal arguments are rigorous except for the fact that in a gauge theory, if not with a spontaneously broken gauge symmetry, the S-matrix suffers from IR divergences and may not even be defined. In such cases one can discuss IR regularized theory but the regulator must not break gauge invariance. IR dimensional regularization can be used as an intermediate step to prove gauge independence of the physical, finite, quantities.

Examples

As a simple example let us consider the case of QED. We shall calculate the free photon propagators $G_0^{\mu\nu}(x_1, x_2)$ corresponding to different gauge conditions. First we assume the so-called Lorentz gauge

$$\partial_\mu A^\mu = 0 \tag{3.21}$$

The $M(x, y)$ matrix defined by (3.13) is then of the form

$$M(x, y) = -\frac{1}{e}\partial^2\delta(x - y)$$

and the Faddeev–Popov determinant is independent of the field A_μ; consequently it can be included into the overall normalization factor, which is irrelevant. To deal with the δ-functional $\delta[\partial_\mu A^\mu]$ which fixes the gauge we shall make use of a decomposition of the field A_μ into the transverse A_μ^{T} and longitudinal A_μ^{L} parts

$$A_\mu^{\mathrm{T}} = P_{\mu\nu}A^\nu, \quad A_\mu^{\mathrm{L}} = (g_{\mu\nu} - P_{\mu\nu})A^\nu$$

where $P_{\mu\nu} = g_{\mu\nu} - \partial_\mu\partial_\nu/\partial^2$ is the projection operator. Clearly we have

$$\partial_\mu A_{\mathrm{T}}^\mu = 0$$

so that the gauge-fixing δ-functional does not affect the integration over the transverse component of the gauge field. The longitudinal fields are of the form $A_\mu^{\mathrm{L}} = \partial_\mu \Theta$ ('pure gauge'). The Lorentz gauge condition does not exactly set them to zero; instead, it requires that $\partial^2\Theta = 0$. This is the residual gauge freedom allowed by (3.21).

The existence of this residual gauge freedom contradicts our previous assumption that the gauge condition $F[A_\mu^g] = 0$ has exactly one solution g_0 for any A_μ. However, let us observe that the residual gauge freedom can be removed by an appropriate choice of the boundary conditions. In Euclidean space we can do it by assuming that the gauge field A_μ vanishes at Euclidean infinity. Such boundary conditions are in fact implicit in the usual derivation of perturbation theory rules from path integrals. There is also another sort of gauge ambiguity (the Gribov ambiguity) with more physical implications. In the case of the Lorentz gauge it appears only in non-abelian gauge theories. This is the case when the points of intersection of a group orbit with the $F = 0$ surface are at a finite distance from each other i.e. are separated by finite gauge transformations. Perturbation theory is, however, unaffected by this ambiguity.

Taking this all into account we obtain the following path integral formula for $W_0[J]$ in the Lorentz gauge (in the following the fermion degrees of freedom are always suppressed):

$$W_0[J] \sim \int \mathscr{D}A_\mu^{\mathrm{T}} \exp\left[\mathrm{i}\int \mathrm{d}^4x\left(\tfrac{1}{2}A_\mu^{\mathrm{T}}\partial^2 g^{\mu\nu}A_\nu^{\mathrm{T}} + J_\mu^{\mathrm{T}}A_{\mathrm{T}}^\mu\right)\right] \tag{3.22}$$

(the extra term $\frac{1}{2}\mathrm{i}\varepsilon A_\mu^{\mathrm{T}}g^{\mu\nu}A_\nu^{\mathrm{T}}$ in the action should always by kept in mind). Here we have made use of the fact that the lagrangian density of a free electromagnetic field can be written as follows

$$-\tfrac{1}{4}F_{\mu\nu}F^{\mu\nu} = \tfrac{1}{2}A_\mu\partial^2(g^{\mu\nu} - \partial^\mu\partial^\nu/\partial^2)A_\nu = \tfrac{1}{2}A_\mu^{\mathrm{T}}\partial^2 g^{\mu\nu}A_\nu^{\mathrm{T}} \tag{3.23}$$

Thus the longitudinal A_{L}^μ does not appear in the action and in the Lorentz gauge $\delta(\partial_\mu A_{\mathrm{L}}^\mu)$ makes the integration over $\mathscr{D}A_{\mathrm{L}}$ trivial. We can simply set $A_{\mathrm{L}}^\mu = 0$. We know that the J-dependence of a path integral like (3.22) is given by

$$W_0[J] = \exp\left[\tfrac{1}{2}\mathrm{i}\int \mathrm{d}^4x\,\mathrm{d}^4y\,J_\mu^{\mathrm{T}}(x)D^{\mu\nu}(x-y)J_\nu^{\mathrm{T}}(y)\right] \tag{3.24}$$

where $-D_{\mu\nu}(x-y)$ is the inverse of the quadratic action. In transverse space the unit operator is equivalent to $P_{\mu\nu}\delta(x-y)$, so that we have the following equation for $D_{\mu\nu}$

$$(\partial^2 - \mathrm{i}\varepsilon)D_{\mu\nu}(x-y) = -(g_{\mu\nu} - \partial_\mu\partial_\nu/\partial^2)\delta(x-y) \tag{3.25}$$

After a Fourier transformation this becomes

$$(-k^2 - \mathrm{i}\varepsilon)\tilde{D}_{\mu\nu}(k) = -(g_{\mu\nu} - k_\mu k_\nu/k^2)$$

so that

$$\tilde{D}_{\mu\nu}(k) = \frac{1}{k^2 + \mathrm{i}\varepsilon}\left(g_{\mu\nu} - \frac{k_\mu k_\nu}{k^2}\right) = \frac{\tilde{P}_{\mu\nu}}{k^2 + \mathrm{i}\varepsilon} \tag{3.26}$$

and

$$D_{\mu\nu}(x-y)=\int\frac{\mathrm{d}^4k}{(2\pi)^4}\frac{g_{\mu\nu}-k_\mu k_\nu/k^2}{k^2+\mathrm{i}\varepsilon}\exp[-\mathrm{i}k(x-y)] \tag{3.27}$$

The prescription of the k^2 pole in the $k_\mu k_\nu/k^2$ term can be arbitrary. Note also that in transverse space, as long as we deal only with transverse sources, we can always replace $P_{\mu\nu}$ by $g_{\mu\nu}$.

We can now write the formula for the free electromagnetic field propagator in the Lorentz gauge

$$G_0^{\mu\nu}(x_1,x_2)=\left(\frac{1}{\mathrm{i}}\right)^2\frac{\delta^2W_0[J]}{\delta J_\mu(x_1)\delta J_\nu(x_2)}\bigg|_{J\equiv0}=-\mathrm{i}D^{\mu\nu}(x_1-x_2) \tag{3.28}$$

where $D_{\mu\nu}(x_1-x_2)$ is given by (3.27).

A generalization of this result follows if we make use of another procedure for fixing the gauge, summarized in (3.17)–(3.19). With $F[A_\mu]=\partial_\mu A^\mu$ we obtain the effective lagrangian

$$\mathscr{L}_{\text{eff}}=-\tfrac{1}{4}F_{\mu\nu}F^{\mu\nu}-\frac{1}{2\alpha}(\partial_\mu A^\mu)^2=\tfrac{1}{2}A^\mu\left[g_{\mu\nu}\partial^2-\left(1-\frac{1}{\alpha}\right)\partial_\mu\partial_\nu\right]A^\nu \tag{3.29}$$

The quadratic action is now no longer proportional to a projection operator and therefore can be inverted over all configuration space. Again we can write

$$W_0[J]=\exp\left[\tfrac{1}{2}\mathrm{i}\int\mathrm{d}^4x\mathrm{d}^4yJ^\mu(x)D_{\mu\nu}(x-y)J^\nu(y)\right] \tag{3.30}$$

where $D_{\mu\nu}(x-y)$ satisfies the equation

$$[\partial^2g_{\mu\nu}-\partial_\mu\partial_\nu(1-1/\alpha)-\mathrm{i}\varepsilon]D^\nu_\lambda(x-y)=-g_{\mu\lambda}\delta(x-y) \tag{3.31}$$

with the solution

$$D_{\mu\nu}(x-y)=\int\frac{\mathrm{d}^4k}{(2\pi)^4}\left[\left(g_{\mu\nu}-\frac{k_\mu k_\nu}{k^2+\mathrm{i}\varepsilon}\right)+\alpha\frac{k_\mu k_\nu}{k^2+\mathrm{i}\varepsilon}\right]\frac{1}{k^2+\mathrm{i}\varepsilon}\exp[-\mathrm{i}k(x-y)] \tag{3.32}$$

In the limit $\alpha\to0$ this coincides with (3.27) (the Lorentz or Landau gauge). The case $\alpha=1$ corresponds to the Feynman gauge.

The free propagators for non-abelian gauge fields in covariant gauges are also given by (3.28) and (3.32). The only modification is that they are diagonal matrices in the non-abelian quantum number space.

Non-covariant gauges

Non-covariant gauges, depending on an arbitrary four-vector n_μ, are also frequently used. For instance the standard canonical quantization of QED uses the Coulomb gauge, also called the radiation gauge (Bjorken & Drell 1965), defined by the condition

$$\partial_\mu A^\mu - (n_\mu \partial^\mu)(n_\mu A^\mu) = 0, \quad n_\mu = (1,0,0,0) \tag{3.33}$$

In QCD one often uses axial gauges

$$n_\mu A^\mu_\alpha = 0, \quad n^2 = 0 \quad \text{or} \quad n^2 < 0, \quad \alpha = 1,\ldots,8 \tag{3.34}$$

which allow the quanta of the vector fields to be interpreted as partons. It is easy to check that in axial gauges the Faddeev–Popov determinant is A^α_μ independent even in a non-abelian theory.

Free propagators of the gauge field in the Coulomb and axial gauges can be formally obtained following the procedure summarized in (3.17)–(3.19), inverting the term of the action quadratic in the gauge fields and taking the limit $\alpha \to 0$. Only in this limit do the propagators behave as k^{-2}. For the Coulomb gauge one gets

$$\tilde{D}^{\alpha\beta}_{\mu\nu} = \frac{\delta^{\alpha\beta}}{k^2 + i\varepsilon}\left[g_{\mu\nu} - \frac{k\cdot n(k_\mu n_\nu + k_\nu n_\mu) - k_\mu k_\nu}{(k\cdot n)^2 - k^2} \right] \tag{3.35}$$

and for axial gauges

$$\tilde{D}^{\alpha\beta}_{\mu\nu} = \frac{\delta^{\alpha\beta}}{k^2 + i\varepsilon}\left[g_{\mu\nu} - \frac{k_\mu n_\nu + k_\nu n_\mu}{k\cdot n} + \frac{n^2}{(n\cdot k)^2} k_\mu k_\nu \right] \tag{3.36}$$

Propagator (3.35) can be split into the transverse propagator and the static Coulomb interaction term (Bjorken & Drell 1965). Our formal derivation of (3.36) neglects subtle problems raised by the presence of the $(n\cdot k)$ singularity. A more complete treatment of $n^2 < 0$ and $n^2 = 0$ gauges has been recently given in the framework of canonical quantization (Bassetto, Lazzizzera & Soldati 1985; Bassetto, Dalbosco, Lazzizzera & Soldati 1985). The $(n\cdot k)$ singularities are related to the residual gauge freedom still present after the axial gauge condition is imposed. In the space-like case $n^2 < 0$ one can show (Bassetto, Lazzizzera & Soldati 1985) that this residual gauge symmetry is directly connected to the spatial asymptotic behaviour of the potentials and can be eliminated by imposing a specific boundary condition. Choice of this condition gives rise to a specific prescription for the spurious singularity in the free boson propagator.

The treatment of the light-cone gauge $n^2 = 0$ is quite different (Bassetto *et al.* 1985). This time the residual gauge freedom involving functions which do not depend on the variable $n_\mu x^\mu$, cannot be eliminated: we are not allowed to impose boundary conditions which would interfere with the time evolution of the system. Nevertheless a hamiltonian canonical quantization leads to the well-defined prescription for the singularity $(n\cdot k)$:

$$\frac{1}{n\cdot k} = \frac{k_0 + k_3}{k^2 + k_\perp^2 + i\varepsilon}, \quad n = (1,0,0,1) \tag{3.37}$$

This prescription has been proposed earlier without any canonical justification as one which allows for Wick rotation (Mandelstam 1983; Leibbrandt 1984). In QCD calculations the principal value prescription has been also extensively used. With both kinds of prescription dimensional regularization is usually adopted (Pritchard & Stirling 1979 and Appendix A). However the general proof of the renormalizability of the theory in the light-cone gauge is still lacking.

3.2 Feynman rules for QCD

Calculation of the Faddeev–Popov determinant

In order to derive Feynman rules for a non-abelian gauge theory we have to take proper account of the Faddeev–Popov determinant whose presence under the functional integral (3.20) is a consequence of quantization with constraints: gauge-fixing conditions. For the sake of definiteness we choose a class of covariant gauges $\partial_\mu A^\mu = c(x)$ or, equivalently, assume (3.19). It then follows from (3.15) that the Faddeev–Popov matrix reads

$$M^{\alpha\beta}(x, y) = (-1/g)(\delta^{\alpha\beta}\partial^2 - gc^{\alpha\beta\gamma}A^\mu_\gamma(x)\partial_\mu)\delta(x - y) = (-1/g)(D_\mu\partial^\mu)^{\alpha\beta}\delta(x - y) \quad (3.38)$$

Relation (3.38) follows from (3.15) not only on the hypersurface $\partial_\mu A^\mu = 0$ but also for $\partial_\mu A^\mu = c(x)$ because $\det(\partial_\mu D^\mu) = \det(D_\mu\partial^\mu)$. The matrix $M^{\alpha\beta}$ depends on gauge fields A^γ_μ.

The standard method of dealing with the Faddeev–Popov determinant $\det M^{\alpha\beta}(x, y)$ is to replace it by an additional functional integration over some auxiliary complex fields $\eta(x)$ (ghost fields) which are Grassmann variables. Using the results of Section 2.5 we can write

$$\det M^{\alpha\beta}(x, y) = C\int \mathscr{D}\eta\mathscr{D}\eta^* \exp\left[\mathrm{i}\int \mathrm{d}^4x\mathrm{d}^4y\,\eta^{*\alpha}(x)M^{\alpha\beta}(x, y)\eta^\beta(y)\right] \quad (3.39)$$

where C is some unimportant constant (the factor $(-1/g)$ in relation (3.38) is also included in it) and the integration over η and η^* is equivalent to integration over the real and the imaginary part of the field η. Having (3.39) and the explicit form (3.38) of the matrix $M^{\alpha\beta}(x, y)$ we can take account of $\det M^{\alpha\beta}(x, y)$ in the framework of our general perturbative strategy for calculating various Green's functions. The Feynman rules for ghost fields follow in a straightforward way from (3.39) and (3.38). The ghost field propagator can be calculated from the generating functional $W^\eta_0[\beta]$

$$W^\eta_0[\beta] \sim \int \mathscr{D}_\eta\mathscr{D}^*_\eta \exp\left\{\mathrm{i}\int \mathrm{d}^4x[\eta^{*\alpha}(x)\delta^{\alpha\beta}\partial^2\eta^\beta(x) + \beta^{*\alpha}(x)\eta^\alpha(x) + \eta^{*\alpha}(x)\beta^\alpha(x)]\right\} \quad (3.40)$$

Using standard methods, the β-dependence of $W^\eta_0[\beta]$ reads explicitly

$$W^\eta_0[\beta] = \exp\left[\mathrm{i}\int \mathrm{d}^4x\mathrm{d}^4y\,\beta^{*\alpha}(x)\Delta^{\alpha\beta}(x - y)\beta^\beta(y)\right] \quad (3.41)$$

where

$$\Delta^{\alpha\beta}(x-y)=\delta^{\alpha\beta}\int\frac{d^4k}{(2\pi)^4}\frac{1}{k^2+i\varepsilon}\exp[-ik(x-y)] \tag{3.42}$$

One can define the 'propagator' of the ghost field η

$$G_0^{\eta\alpha\beta}(x,y)=\left(\frac{1}{i}\right)^2\frac{\delta}{\delta\beta^{*\alpha}(x)}W_0^{\eta}[\beta]\frac{\overleftarrow{\delta}}{\delta\beta^{\beta}(y)}\bigg|_{\beta=\beta^*=0} \tag{3.43}$$

and expanding (3.41) we get

$$G_0^{\eta\alpha\beta}(x,y)=-i\Delta^{\alpha\beta}(x-y) \tag{3.44}$$

Graphically, propagator $G_0^{\eta\alpha\beta}(x,y)$ will be denoted by

with an arrow pointing towards the source of η

Feynman rules

The full generating functional $W[J,\alpha,\beta]$ for QCD reads as follows:

$$\begin{aligned}W[J,\alpha,\beta]=N\int\mathscr{D}A_\mu^\alpha\mathscr{D}\Psi\mathscr{D}\bar{\Psi}\mathscr{D}\eta\mathscr{D}\eta^*\exp\Big\{&i[S[A,\Psi]+S^\eta[A,\eta]+S_G]\\&+i\int d^4x[J_\mu^\alpha(x)A_\alpha^\mu(x)+\beta^{*\alpha}(x)\eta^\alpha(x)+\eta^{*\alpha}(x)\beta^\alpha(x)\\&+\bar{\alpha}(x)\Psi(x)+\bar{\Psi}(x)\alpha(x)]\Big\}\end{aligned} \tag{3.45}$$

where

$$\left.\begin{aligned}S[A,\Psi]&=\int d^4x(-\tfrac{1}{4}G_{\mu\nu}^\alpha G_\alpha^{\mu\nu}+\bar{\Psi}i\gamma_\mu D^\mu\Psi-m\bar{\Psi}\Psi)\\S^\eta[A,\eta]&=\int d^4x\,\eta^{*\alpha}(x)[\delta^{\alpha\beta}\partial^2-gc^{\alpha\beta\gamma}A_\gamma^\mu(x)\partial_\mu]\eta^\beta(x)\\S_G&=-(1/2\alpha)\int d^4x[\partial^\mu A_\mu^\alpha]^2\end{aligned}\right\} \tag{3.46}$$

Spinors $\Psi,\bar{\Psi}$ describe quarks and are vectors in the three-dimensional colour space. We split the action into the term S_0 describing non-interacting theory (bilinear in fields and including S_G)

$$\begin{aligned}S_0[A,\Psi,\eta]=\int d^4x[&\tfrac{1}{2}A_\mu^\alpha(x)\delta^{\alpha\beta}\partial^2(g^{\mu\nu}-\partial^\mu\partial^\nu/\partial^2)A_\gamma^\beta(x)+\bar{\Psi}(x)(i\gamma_\mu\partial^\mu-m)\Psi(x)\\&+\eta^{*\alpha}(x)\delta^{\alpha\beta}\partial^2\eta^\beta(x)]+S_G\end{aligned} \tag{3.47}$$

and into the interaction term S_I

$$\begin{aligned}S_I[A,\Psi,\eta]=\int d^4x[&-gc_{\alpha\beta\gamma}(\partial_\mu A_\nu^\alpha)A^{\mu\beta}A^{\nu\gamma}-\tfrac{1}{4}g^2c_{\alpha\beta\gamma}c_{\alpha\rho\sigma}A_\gamma^\beta A_\nu^\gamma A^{\mu\rho}A^{\nu\sigma}\\&+g\bar{\Psi}\gamma_\mu A^{\mu\alpha}T^\alpha\Psi-gc_{\alpha\beta\gamma}A_\nu^\gamma\eta^{*\alpha}(\partial^\mu\eta^\beta)]\end{aligned} \tag{3.48}$$

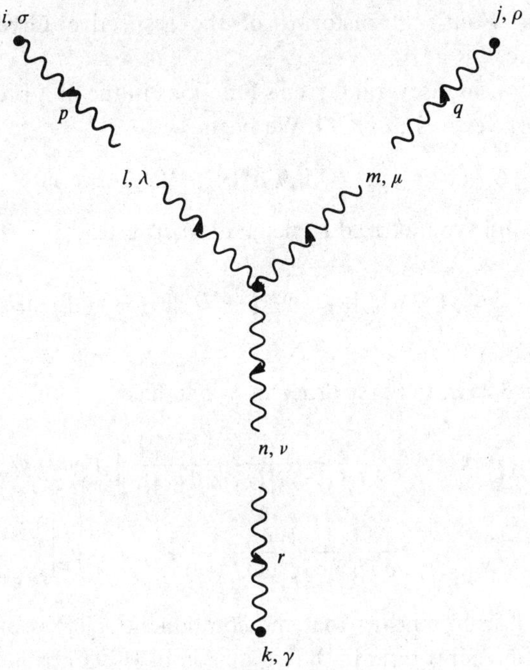

Fig. 3.2

Matrices T^α are $SU(3)$ colour generators in the fundamental representation. The 'free' generating functional reads

$$W_0 = \exp\left\{ \mathrm{i} \int \mathrm{d}^4x\,\mathrm{d}^4y \left[\tfrac{1}{2} J_\mu^\alpha(x) D_{\alpha\beta}^{\mu\nu}(x-y) J_\nu^\beta(y) + \bar{\alpha}(x)_\alpha S_{\alpha\beta}(x-y)\alpha_\beta(y) \right.\right.$$
$$\left.\left. + \beta^{*\alpha}(x)\Delta^{\alpha\beta}(x-y)\beta^\beta(y) \right] \right\} \tag{3.49}$$

where the Green's functions $D_{\alpha\beta}^{\mu\nu}$, $S_{\alpha\beta}$, and $\Delta^{\alpha\beta}$ are given by (3.32), (2.122) and (3.42), respectively. The propagators $\langle 0|TA_\mu^\alpha(x)A_\nu^\beta(y)|0\rangle$, $\langle 0|T\Psi^\alpha(x)\bar{\Psi}^\beta(y)|0\rangle$ and $\langle 0|T\eta^\alpha(x)\eta^{*\beta}(y)|0\rangle$ are obtained by multiplying the respective D, S and Δ-functions by $(-\mathrm{i})$.

Any Green's function in QCD can be calculated perturbatively from the expansion

$$\langle 0|TA_\mu^\alpha \ldots A_\nu^\beta \Psi \ldots \bar{\Psi}|0\rangle = \frac{\displaystyle\int \mathscr{D}[A\Psi\bar{\Psi}\eta\eta^*](A_\mu^\alpha \ldots A_\nu^\beta \Psi \ldots \bar{\Psi}) \sum_N \frac{(\mathrm{i}S_1)^N}{N!} \exp(\mathrm{i}S_0)}{\displaystyle\int \mathscr{D}[A\Psi\bar{\Psi}\eta\eta^*] \sum_N \frac{(\mathrm{i}S_1)^N}{N!} \exp(\mathrm{i}S_0)} \tag{3.50}$$

We recall that the presence of the denominator in (3.50) eliminates all vacuum-to-vacuum diagrams. One is usually interested in Green's functions in momentum space defined as Fourier transforms of the respective Green's functions in configuration space.

We shall now consider several Green's functions in the first order in S_I to derive Feynman rules for vertices in QCD. We begin with

$$G_{3g}(x_1, x_2, x_3) = \langle 0|TA^i_\sigma(x_1)A^j_\rho(x_2)A^k_\gamma(x_3)|0\rangle \tag{3.51}$$

which defines the full symmetrized triple gluon ($3g$) vertex by means of the relation

$$G_{3g}(x_1, x_2, x_3) = \int d^4y[-iD^{il}_{\sigma\lambda}(x_1 - y)][-iD^{jm}_{\rho\mu}(x_2 - y)][-iD^{kn}_{\gamma\nu}(x_3 - y)]\, V^{\lambda\mu\nu}_{lmn} \tag{3.52}$$

illustrated in Fig. 3.2. In the first order in S_I one has

$$G_{3g}(x_1, x_2, x_3) = \frac{1}{i^6}\frac{\delta^3}{\delta J^i_\sigma(x_1)\delta J^j_\rho(x_2)\delta J^k_\gamma(x_3)}\, i\int d^4y(-gc_{abc}\partial^{(a)}_\mu)$$
$$* \frac{\delta^3}{\delta J^a_\nu(y)\delta J^b_\mu(y)\delta J^{\nu c}(y)} W_0[J, \beta, \alpha]\Big|_{J=\beta=\alpha=0} \tag{3.53}$$

(the superscript 'a' in $\partial^{(a)}_\mu$ means that the component $A^a_\mu(y)$ is differentiated with respect to y) and the cubic term in the expansion of W_0 contributes to final result. Following the standard method we assign points x_1, x_2, x_3 and $y_1, y_2, y_3 \to y$ to the prototype diagram

For a connected diagram there are six different orderings in pairs joined by propagators. For each assignment there is a $2^3 3!$ symmetry factor which cancels with factors coming from the expansion of W_0. The factor $(1/i^6)i^3 = (-i)^3$ is included in the three propagators. The final result reads:

$$G_{3g}(x_1, x_2, x_3) = -igc_{abc}\int d^4y\partial^{(a)}_\mu \sum[-iD^{\nu ia}_\sigma(x_1 - y)]$$
$$*[-iD^{\mu jb}_\rho(x_2 - y)][-iD^{kc}_{\gamma\nu}(x_3 - y)]$$

where the sum denotes all permutations of sets of indices (a, ν), (b, μ) and (c, ν). Writing result (3.54) in the form of (3.52) we get

$$V^{lmn}_{\lambda\mu\nu} = -igc_{abc}(\delta^{la}\delta^{mb}\delta^{nc}g_{\lambda\nu}\partial^{(a)}_\mu + \delta^{la}\delta^{mc}\delta^{nb}g_{\lambda\mu}\partial^{(a)}_\nu + \delta^{lb}\delta^{ma}\delta^{nc}g_{\mu\nu}\partial^{(a)}_\lambda$$
$$+ \delta^{lb}\delta^{mc}\delta^{na}g_{\mu\nu}\partial^{(a)}_\lambda + \delta^{lc}\delta^{ma}\delta^{nb}g_{\lambda\mu}\partial^{(a)}_\nu + \delta^{lc}\delta^{mb}\delta^{na}g_{\lambda\nu}\partial^{(a)}_\mu) \tag{3.55}$$

In momentum space, we define $\tilde{G}(p, q, r)$ as

$$G_{3g}(x_1, x_2, x_3) = \int \frac{d^4p}{(2\pi)^4}\frac{d^4q}{(2\pi)^4}\frac{d^4r}{(2\pi)^4}\exp(-ipx_1)\exp(-iqx_2)\exp(-irx_3)\tilde{G}_{3g}(p, q, r) \tag{3.56}$$

and the $3g$ vertex $\tilde{V}^{lmn}_{\lambda\mu\nu}(p,q,r)$ by the relation

$$\tilde{G}_{3g}(p,q,r)=\tilde{G}_{2g}[p,p]^{il}_{\sigma\lambda}\ \tilde{G}_{2g}[q,q]^{jm}_{\rho\mu}\ \tilde{G}_{2g}[r,r]^{kn}_{\gamma\nu}\ \tilde{V}^{\lambda\mu\nu}_{lmn}(p,q,r) \tag{3.57}$$

where

$$\tilde{G}_{2g}[p,p]^{il}_{\sigma\lambda}=(-\mathrm{i})\delta_{il}\left[g_{\sigma\lambda}-\frac{p_\sigma p_\lambda}{p^2}(1-\alpha)\right]\frac{1}{p^2}$$

etc. Writing the propagators in relation (3.52) in terms of their Fourier transforms (3.57), differentiating according to (3.55), integrating over $\mathrm{d}y$ and finally comparing with (3.56) and (3.57) one gets the final result

$$\tilde{V}^{lmn}_{\lambda\mu\nu}(p,q,r)=(2\pi)^4\delta(p+q+r)gc_{lmn}[(p-r)_\mu g_{\lambda\nu}+(r-q)_\lambda g_{\mu\nu}+(q-p)_\nu g_{\lambda\mu}] \tag{3.58}$$

Similarly, one can derive the expression for the $4g$ vertex. The starting point is the Green's function

$$G_{4g}(x_1,x_2,x_3,x_4)=\langle 0|TA^i_\alpha A^j_\beta A^k_\gamma A^l_\delta|0\rangle$$

which in the first order in S_I defines the $4g$ vertex according to the following relation:

$$G_{4g}(x_1,\ldots,x_4)=\int \mathrm{d}^4y(-\mathrm{i})^4 D^{ia}_{\alpha\mu}(x_1-y)D^{jb}_{\beta\nu}(x_2-y)D^{kc}_{\gamma\sigma}(x_3-y)D^{ld}_{\delta\rho}(x_4-y)V^{\mu\nu\sigma\rho}_{abcd} \tag{3.59}$$

Writing

$$G_{4g}(x_1,\ldots,x_4)=\frac{1}{\mathrm{i}^8}\frac{\delta^4}{\delta J^i_\alpha\delta J^j_\beta\delta J^k_\gamma\delta J^l_\delta}\mathrm{i}\int \mathrm{d}^4y(-\tfrac{1}{4}g^2c_{stu}c_{swz})$$
$$*\frac{\delta^4}{\delta J^t_{\mu'}\delta J^u_{\nu'}\delta J^w_{\mu'}\delta J^z_{\nu'}(y)}W_0|_{J=\beta=\alpha=0} \tag{3.60}$$

one can calculate $V^{abcd}_{\mu\nu\sigma\rho}$. There are 24 different assignments of pairs of points $x_1,\ldots,x_4$, $y_1,\ldots,y_4$ to the prototype diagram. The additional $4!\ 2^4$ symmetry factor is cancelled by the same factor coming from the expansion of W_0. Remembering that the Lorentz contraction in the $4g$ vertex always occurs between colour indices t and w and u and z one gets the final result

$$V^{abcd}_{\mu\nu\sigma\rho}=-\mathrm{i}g^2[c_{abe}c_{cde}(g_{\mu\sigma}g_{\nu\rho}-g_{\mu\rho}g_{\nu\sigma})+c_{ace}c_{bde}(g_{\mu\nu}g_{\rho\sigma}-g_{\mu\rho}g_{\nu\sigma})$$
$$+c_{ade}c_{cbe}(g_{\mu\sigma}g_{\nu\rho}-g_{\mu\nu}g_{\rho\sigma})] \tag{3.61}$$

which is obviously also true in momentum space.

Following the functional approach one can also easily find that loop diagrams in Fig. 3.3 contribute with additional factors $\frac{1}{2}$, $\frac{1}{2}$ and $\frac{1}{6}$, respectively. In particular diagrams (*a*) and (*c*) occur in the second order in S_I; e.g. the expression corresponding to the first diagram is of the form

$$\frac{(\mathrm{i}S_\mathrm{I})^2}{2!}\frac{1}{4!}\frac{1}{2^4}\quad JDJ\quad JDJ\quad JDJ\quad JDJ \tag{3.62}$$

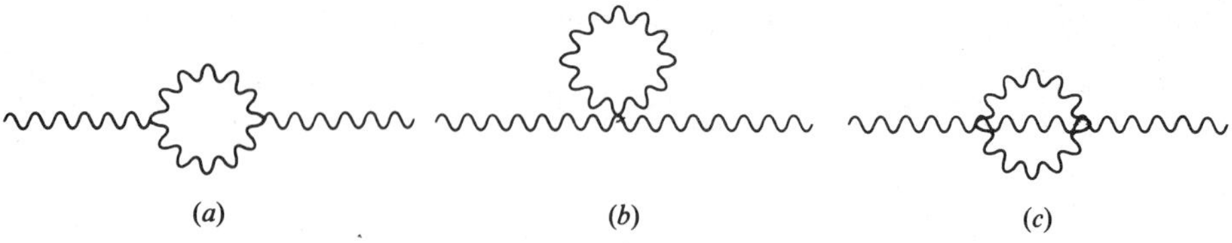
Fig. 3.3

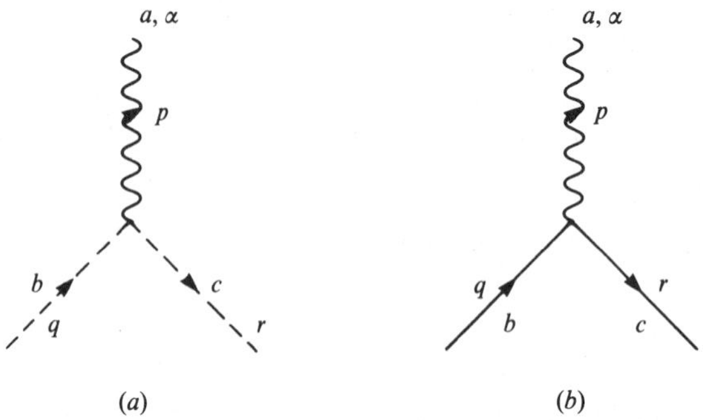

Fig. 3.4

The number of different assignments to the prototype diagram is $3 \times 2 \times 3 \times 2 \times 4!2^4$. Noticing that the product of two vertices gives us according to (3.58) 36 combinations we are left with an additional factor of $\frac{1}{2}$. In a similar way one can obtain the other factors.

The final task is to find expressions for the ghost and fermion vertices in Fig. 3.4. They can be obtained from the Green's functions $\langle 0|TA_\mu^i(x_1)\eta^j(x_2)\eta^{*k}(x_3)|0\rangle$ and $\langle 0|TA_\mu^i(x_1)\Psi^j(x_2)\bar{\Psi}^k(x_3)|0\rangle$ respectively, written as

$$\int d^4y[-iD_{\mu\alpha}^{ia}(x_1 - y)][-i\Delta^{jb}(y - x_2)][-i\Delta^{ck}(x_3 - y)]V_{abc}^{\alpha}(y) \tag{3.63}$$

and similarly for the $\bar{\Psi}\Psi A$ vertex. Following closely previous calculations we obtain in momentum space

Diagram (*a*): $$\tilde{V}_\alpha^{abc} = -gc_{abc}r_\alpha(2\pi)^4\delta(q - p - r) \tag{3.64}$$

Diagram (*b*): $$\tilde{V}_\alpha^{abc} = +ig\gamma_\alpha(T^a)_{bc}(2\pi)^4\delta(q - p - r) \tag{3.65}$$

Of course, Feynman rules for QED in the covariant gauge can be derived analogously. All the Feynman rules are collected in Appendix A.

3.3 Unitarity, ghosts, Becchi–Rouet–Stora transformation

Unitarity and ghosts

A spin one massless particle has only two physical polarization states $\varepsilon_\mu(\lambda, k)$, $\lambda = 1, 2$. Since the three four-vectors k_μ, $\varepsilon_\mu(\lambda, k)$ do not span four-dimensional space, the usual on-shell Lorentz condition $k\cdot\varepsilon=0$ for spin one massive particles is not enough to determine the photon polarization vectors uniquely. To make them unique we must sacrifice manifest Lorentz covariance and, for instance, choose another vector n_μ such that

$$k\cdot n\neq 0, \quad n\cdot\varepsilon=0 \tag{3.66}$$

Together with equations

$$k\cdot\varepsilon=0, \quad \varepsilon^*(\lambda_1,k)\varepsilon(\lambda_2,k)=-\delta_{\lambda_1\lambda_2} \tag{3.67}$$

conditions (3.66) specify the vectors $\varepsilon(\lambda,k)$ uniquely. The polarization sum reads

$$\sum_{\lambda=1,2}\varepsilon^*_\mu(\lambda,k)\varepsilon_\nu(\lambda,k)=-g_{\mu\nu}+\frac{k_\mu n_\nu+k_\nu n_\mu}{n\cdot k}-\frac{n^2k_\mu k_\nu}{(k\cdot n)^2}\equiv P_{\mu\nu} \tag{3.68}$$

In field theory every particle is associated with a field with definite transformation properties under the homogeneous Lorentz group. Thus the four-vector gauge field $A_\mu(x)$ actually represents only two physical degrees of freedom. An important point is that fixing the gauge is usually not sufficient to totally remove the unphysical degrees of freedom from the theory (an exception is the axial gauge with $n^2<0$). A well-known example of their presence in the theory is the Gupta–Bleuler formulation of QED in the covariant gauge $\partial_\mu A^\mu=0$. In general, such degrees of freedom can be seen explicitly if we construct the Fock space of the theory: a longitudinal photon state and a negative norm scalar photon state are present. A physically sensible theory can be defined in Hilbert space restricted to the transverse photons (see e.g. de Rafael 1979). However, this restriction does not prevent one from having unphysical states propagating as virtual states in the intermediate steps of perturbative calculations. Indeed, the gauge boson propagators are gauge dependent. The question then arises about the unitarity of the theory.

The requirement that the S-matrix is unitary implies that the scattering amplitude T defined as

$$S_{\rm if}=\delta_{\rm if}+{\rm i}(2\pi)^4\delta(p_{\rm i}-p_{\rm f})T_{\rm if} \tag{3.69}$$

satisfies the relation

$$T_{\rm if}-T^*_{\rm fi}={\rm i}\sum_n T_{{\rm i}n}T^*_{{\rm f}n}(2\pi)^4\delta(p_{\rm i}-p_n) \tag{3.70}$$

where the sum is taken over all physical states for which transitions to the initial and final states are allowed by quantum number conservation laws. For the forward elastic scattering the l.h.s. of (3.70) is just the imaginary part of the amplitude $T_{\rm ii}$ and in the general case it is the absorptive part of $T_{\rm if}$ (see e.g. Bjorken & Drell 1965, Section 18.12). The absorptive part of the scattering amplitude can be calculated perturbatively by means of the so-called Landau–Cutkosky rule: the contribution due to a Feynman diagram with a given intermediate state, Fig. 3.5,

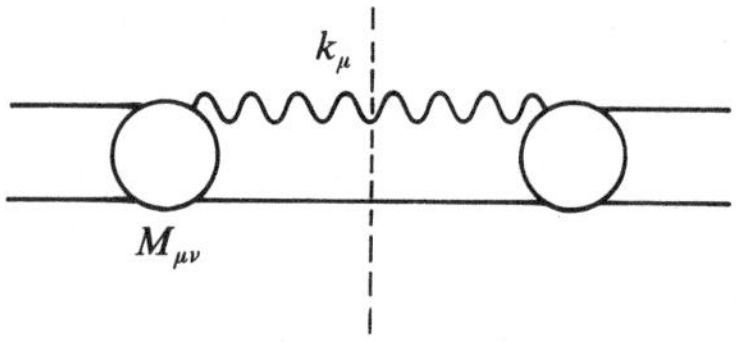

Fig. 3.5

is obtained from the Feynman amplitude by replacing the propagators in this intermediate state by their imaginary parts. For instance, for the gauge boson propagators in the Feynman gauge (see Section 3.1)

$$D_{\mu\nu}^{\alpha\beta} = \delta^{\alpha\beta} \frac{g_{\mu\nu}}{k^2 + i\varepsilon} \to -2\pi i \delta^{\alpha\beta} g_{\mu\nu} \delta(k^2) \Theta(k_0) \tag{3.71}$$

It is clear that in gauges which do not remove the unphysical degrees of freedom, like in the Feynman gauge, the unphysical intermediate states are included in our calculation. On the other hand we can calculate the absorptive part of the scattering amplitude from the r.h.s. of the unitarity relation (3.70). Here we sum over physical states only. Thus the unitarity condition demands that the unphysical intermediate states do not contribute to the amplitude when also calculated by means of the Landau–Cutkosky rule. In fact one can expect that gauge invariance, which reflects the presence of the redundant degrees of freedom, assures at the same time their decoupling. An explicit verification of this point may, however, be quite subtle as for solving the equations of motion we need a gauge-fixing term which breaks the manifest invariance. Nevertheless in QED there is no problem due to the fact that photons always couple to conserved currents

$$k_\mu M^{\mu\nu} = 0 \tag{3.72}$$

Therefore $P_{\mu\nu}$ given by (3.68) can be replaced by

$$P_{\mu\nu} \to -g_{\mu\nu} + \alpha k_\mu k_\nu \tag{3.73}$$

where α is an arbitrary quantity.

In non-abelian gauge theories, the situation is more complicated as can be illustrated by the standard example of the fermion–antifermion scattering in the lowest non-trivial order in the gauge-coupling constant (Feynman 1977, Aitchison & Hey 1982). The sum of the diagrams with only gauge particles in the intermediate state (Fig. 3.6) calculated in a covariant gauge (3.19) does not satisfy the unitarity condition (3.70) where the r.h.s. amplitudes T_{in} and T_{fn}^* are given by the diagrams in Fig. 3.7 with only physical gluons as the state n (Problem 3.1). However we have learned in the previous Sections that a consistent quantization of a non-abelian gauge theory requires an addition to the lagrangian of an extra term, the so-called ghost term (3.39). It turns out that these additional unphysical degrees of freedom cancel the gauge field unphysical polarization states exactly, leading to a perfectly

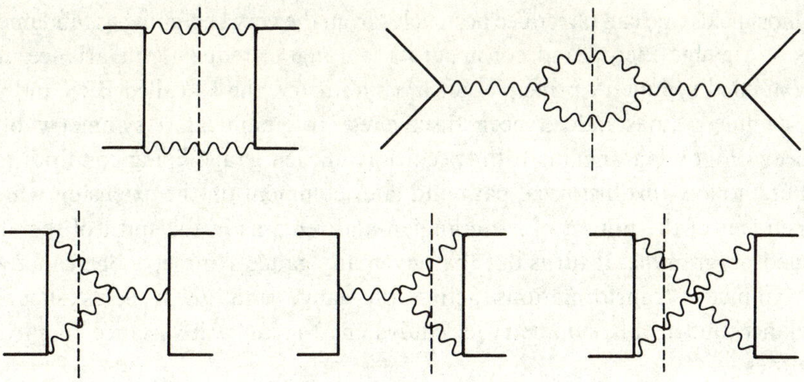

Fig. 3.6

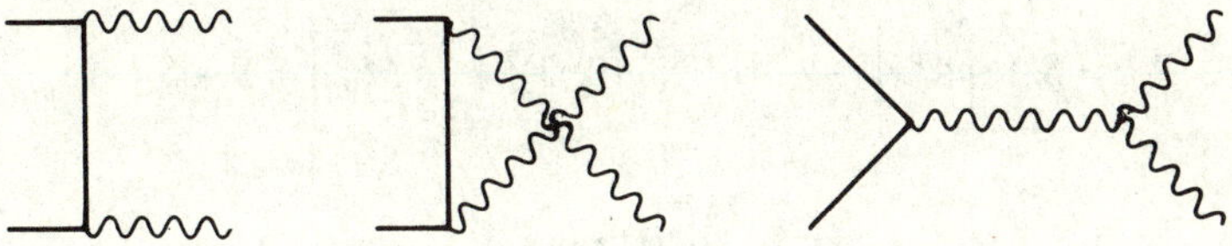

Fig. 3.7

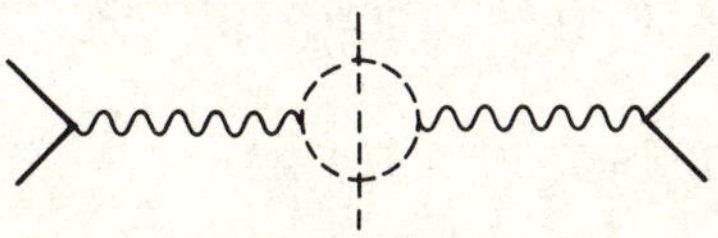

Fig. 3.8

unitary theory. In our specific example the cancellation occurs due to the contribution of the diagram with the ghost loop, shown in Fig. 3.8. An explicit perturbative calculation (Problem 3.1) shows that certain relations between the $f\bar{f}\to A^\alpha_\mu A^\beta_\nu$ and $f\bar{f}\to\eta^\alpha\eta^\beta$ amplitudes where A_μ and η are the gauge and ghost fields, respectively, are responsible for the cancellation. These are examples of Ward identities for a non-abelian gauge theory.

BRS and anti-BRS symmetry

We have introduced ghosts as a technical device to express the modification of the functional integral measure required to compensate for gauge dependence introduced by the gauge-fixing term. The discussion in the previous Subsection suggests, however, that it is very natural to adopt the point of view that ghosts, i.e. fields with unphysical statistics, are necessary to compensate for effects due to the quantum propagation of the unphysical states of the gauge fields. Provided one is able to express the gauge symmetry in a form involving both the classical and ghost fields one can introduce both fields from the very beginning as fundamental fields of a gauge theory and construct its lagrangian requiring invariance under this extended symmetry principle. Such a symmetry, the so-called BRS and anti-BRS symmetry, has indeed been discovered, originally as a symmetry of the Feddeev–Popov lagrangians with specifically chosen gauge-fixing conditions. We shall first follow this historical path and later comment on the extension which is independent of the notion of a lagrangian and remains in the spirit of the above outlined programme. It turns out that given any gauge symmetry one can always find symmetry transformations acting on gauge and ghost fields such that invariance under this symmetry is equivalent to gauge invariance of physical quantities.

To introduce BRS symmetry (Becchi, Rouet & Stora 1974, 1976) let us consider the action including the gauge-fixing term and the ghost term

$$S_{\text{eff}} = S + \int \mathrm{d}^4x \left[-\frac{1}{2\alpha}(F^\alpha[x, A(x)])^2 + \eta^* M\eta \right] = S + S_G + S_\eta \tag{3.74}$$

where

$$(M\eta)_\alpha = \int \mathrm{d}^4y M_{\alpha\beta}(x, y)\eta_\beta(y)$$

and

$$M^{\alpha\beta}(x, y) = \int \mathrm{d}^4z \frac{\delta F^\alpha[x, A(x)]}{\delta A^\rho_\mu(z)} D^{\rho\beta}_\mu(z)\delta(z - y) \tag{3.75}$$

and

$$D^{\rho\beta}_\mu = \partial_\mu \delta^{\rho\beta} - gc^{\rho\beta\gamma}A^\gamma_\mu$$

Thus

$$(M\eta)_\alpha = \delta F_\alpha[x, A(x)]$$

where δF_α is the variation of the functional F_α under a gauge transformation with $\Theta^\beta(x)$ replaced by $(-g\eta^\beta(x))$. The full action (3.74) is invariant under the global transformation

$$\left.\begin{aligned} \delta_{\text{BRS}}A^\alpha_\mu &= D^{\alpha\beta}_\mu \Theta\eta^\beta \equiv \Theta(\mathscr{s}A^\alpha_\mu) \\ \delta_{\text{BRS}}\Psi &= \mathrm{i}gT^\alpha\Theta\eta^\alpha\Psi \equiv \Theta(\mathscr{s}\Psi) \\ \delta_{\text{BRS}}\eta^\alpha &= -\tfrac{1}{2}gc^{\alpha\beta\gamma}\eta^\beta\eta^\gamma\Theta = \Theta(\mathscr{s}\eta^\alpha) \\ \delta_{\text{BRS}}\eta^{*\alpha} &= -(1/\alpha)\Theta F^\alpha \equiv \Theta(\mathscr{s}\eta^{*\alpha}) \end{aligned}\right\} \tag{3.76}$$

where Θ is a space-time-independent infinitesimal Grassmann parameter (note that it is not necessary to think of η and η^* in (3.39) as hermitian conjugates: see Problem 3.4). We observe that for the classical fields A_μ and Ψ the BRS transformation has the form of a gauge transformation with

$$\Theta^\alpha(x) = -g\Theta\eta^\alpha(x) \tag{3.77}$$

Therefore the BRS transformation leaves the original action invariant without the gauge-fixing and the ghost terms. Thus it remains to show that

$$\delta_{\mathrm{BRS}}(S_G + S_\eta) = 0 \tag{3.78}$$

This can be checked by a straightforward calculation. As intermediate steps we record the relation

$$\delta_{\mathrm{BRS}}(M\eta) = 0 \tag{3.79}$$

and the fact that the transformations on A_μ^α, Ψ and η are nilpotent (see Problem 3.2)

$$s^2 A_\mu^\alpha = s^2\Psi = s^2\eta^\alpha = 0 \tag{3.80}$$

(but $s^2\eta^{*\alpha} = (M\eta)^\alpha$). We note also that the measure $\mathscr{D}(A\Psi\eta\eta^*)$ is invariant under the BRS transformations.

The BRS symmetric lagrangian can also be written down in QED. In this case the ghost field is a free real scalar field and the covariant derivative in the first equation in (3.76) is replaced by the normal derivative (Problem 3.3).

The BRS transformation can be written down in a form more symmetric with respect to the fields η and η^* if we introduce an auxiliary field b. This also allows formulation of the notion of BRS symmetry in a lagrangian-independent way (transformations (3.76) depend on the gauge-fixing term of the lagrangian). This step is very important for promoting BRS symmetry to the fundamental symmetry of gauge theories. To be specific let us discuss the case of a covariant gauge. Using the freedom we have to choose any gauge condition without effecting the S-matrix we can take

$$\mathscr{L}_G = b^\alpha \partial_\mu A_\alpha^\mu + \tfrac{1}{2}\alpha b^\alpha b_\alpha \tag{3.81}$$

as our gauge-fixing condition. After taking into account the equation of motion for the field b we see that (3.81) is equivalent to the standard gauge fixing term $-(1/2\alpha)(\partial^\mu A_\mu^\alpha)^2$. Using the auxiliary field b the BRS transformation for η^* and b reads

$$s\eta^{*\alpha} = b^\alpha, \qquad sb^\alpha = 0 \tag{3.82}$$

The action of s on an arbitrary function of the fields follows from their action on field polynomials

$$s(AB) = (sA)B \pm AsB \tag{3.83}$$

where the minus sign occurs if there is an odd number of ghosts and antighosts in A. The nilpotency of BRS symmetry can now be expressed as

$$s^2 = 0 \tag{3.84}$$

It has also been discovered (Curci & Ferrari 1976, Ojima 1980) that in addition to BRS symmetry there is another symmetry which leaves the quantum Yang–Mills action (3.74) invariant provided the gauge-fixing functional F^α is linear in fields.

These anti-BRS transformations read

$$\left.\begin{aligned} \bar{s}A^\alpha_\mu &= D^{\alpha\beta}_\mu \eta^{*\beta} \\ \bar{s}\eta^{*\alpha} &= -\tfrac{1}{2}g c^{\alpha\beta\gamma}\eta^{*\beta}\eta^{*\gamma} \\ \bar{s}\eta^\alpha &= -b^\alpha - g c^{\alpha\beta\gamma}\eta^{*\beta}\eta^\gamma \\ \bar{s}b^\alpha &= -g c^{\alpha\beta\gamma}\eta^{*\beta}b^\gamma \\ \bar{s}\Psi &= \mathrm{i}g T^\alpha \eta^{*\alpha}\Psi \end{aligned}\right\} \tag{3.85}$$

and

$$s\bar{s} + \bar{s}s = \bar{s}^2 = 0 \tag{3.86}$$

We have introduced the BRS and anti-BRS transformations as symmetry transformations of a Yang–Mills lagrangian with a gauge-fixing term linear in fields. With the auxiliary field b interpreted as a Lagrange multiplier for the gauge-fixing condition, the symmetry generators s and $\bar{s}$ become independent of the notion of the lagrangian and satisfy the nilpotency relations (3.84) and (3.86). In terms of the fields, which are matrix-valued in the Lie algebra of the gauge group, we have

$$\left.\begin{aligned} sA_\mu &= D_\mu\eta & \bar{s}A_\mu &= D_\mu\eta^* \\ s\Psi &= -\eta\Psi & \bar{s}\Psi &= -\eta^*\Psi \\ s\eta &= -\tfrac{1}{2}[\eta,\eta] & \bar{s}\eta^* &= -\tfrac{1}{2}[\eta^*,\eta^*] \\ s\eta^* &= b & \bar{s}\eta &= -[\eta^*,\eta] - b \\ sb &= 0 & \bar{s}b &= -[\eta^*, b] \end{aligned}\right\} \tag{3.87}$$

An important next step is to promote full BRS (BRS or/and anti-BRS) symmetry to the fundamental symmetry of a gauge theory (Alvarez-Gaumé & Baulieu 1983). In fact, one can show that starting from a set of infinitesimal gauge transformations building up a closed algebra (possibly with field-dependent structure functions) with a Jacobi identity one can always build an associated nilpotent full BRS algebra. One can also prove that full BRS invariance of the lagrangian leads to the gauge independence of physics. Thus we are led to construct the quantum lagrangian of a gauge theory as the most general s- or/and $\bar{s}$-invariant and Lorentz-invariant function (with ghost number zero) of physical and ghost fields. This construction simultaneously generates gauge-fixing and ghost terms in the lagrangian. It is not just a formal improvement of the Faddeev–Popov approach since it also gives consistent quantum theories in cases for which the Faddeev–Popov method is inapplicable. Actually, the latter is a consistent approach only for four-dimensional Yang–Mills theories when one chooses a linear gauge condition. Specifying the general BRS-invariant lagrangian to this special case we indeed recover the lagrangian of Faddeev and Popov.†

In the general case, in a BRS-invariant lagrangian four-ghost or higher ghost interactions are present which cannot be obtained from the ordinary Faddeev–Popov procedure. Such couplings are in fact necessary for a consistent renor-

† In Problem 3.5 this is discussed for a less general lagrangian, which is simultaneously BRS *and* anti-BRS invariant.

malizable Yang–Mills theory if the gauge-fixing functional F^α is chosen to be non-linear in fields. Indeed, in such cases, one-loop calculation, based on the Faddeev–Popov lagrangian generating only quadratic ghost interactions, shows that the four-point ghost function is divergent. Thus, four-ghost interaction terms should have been introduced at the tree level. Also, in supergravity, quartic ghost interactions, absent in the Faddeev–Popov approach, have been shown to be necessary (Kallosch 1978, Nielsen 1978, Sterman, Townsend & van Nieuwenhuizen 1978). They are naturally generated by determining the quantum lagrangian from the requirement of its BRS invariance. Thus, full BRS invariance indeed emerges as the fundamental symmetry of the quantum theory associated with any given underlying gauge invariance.

Problems

3.1 Calculate in a covariant gauge the absorptive part of the amplitude corresponding to the sum of diagrams in Fig. 3.6 and Fig. 3.8 and check the unitarity condition (3.70).

3.2 Prove the invariance of (3.74) under transformation (3.76). Check (3.79) and (3.80). (Use the relation for the structure constants which follows from the commutation relations $[T^a, T^b] = \mathrm{i}c^{abc}T^c$ with $(T^a)_{bc} = -\mathrm{i}c^{abc}$.)

3.3 Check the invariance of the generalized QED lagrangian

$$\mathscr{L} = -\tfrac{1}{4}F_{\mu\nu}F^{\mu\nu} + \bar{\Psi}(\mathrm{i}\not{D} - m)\Psi - (1/2\alpha)(\partial_\mu A^\mu)^2 - \tfrac{1}{2}(\partial_\mu\omega)(\partial^\mu\omega)$$

where $\omega(x)$ is a free scalar field, under the transformation

$$\delta\Psi = -\mathrm{i}e\Theta\omega(x)\Psi(x)$$
$$\delta A_\mu = \Theta\partial_\mu\omega(x)$$
$$\partial\omega = (1/\alpha)\Theta\partial_\mu A^\mu$$

3.4 Note that the fields η and η^* in (3.39) can be any anticommuting fields. Check that, in particular, they can be taken as real and imaginary parts of η i.e. as hermitean fields. Check that due to the phase arbitrariness of the exponent in (3.39) the Faddeev–Popov term $\mathscr{L}_\eta$ can be made hermitean.

3.5 Check that the most general Yang–Mills and Lorentz scalar lagrangian which is BRS and anti-BRS invariant and has ghost number zero can be written as follows (Baulieu & Thierry Mieg 1982)

$$\mathscr{L} = -\tfrac{1}{4}(F_{\mu\nu})^2 - \tfrac{1}{2}(D_\mu\Phi)^2 + \bar{\Psi}\mathrm{i}\not{D}\Psi + \text{Yukawa terms} + \varepsilon^{\mu\nu\rho\sigma}F_{\mu\nu}F_{\rho\sigma}$$
$$+ V_{\text{inv}}(\Phi) + \mathfrak{s}\bar{\mathfrak{s}}(\tfrac{1}{2}A_\mu^2 + \beta\eta^*\eta + \gamma\langle\Phi\rangle\Phi) + \tfrac{1}{2}\alpha b^2$$

where α, β, γ are arbitrary gauge parameters and $\langle\Phi\rangle$ is the vacuum expectation value of the boson field Φ. When $\beta = 0$ one recovers the usual Faddeev–Popov lagrangian for linear gauges.

3.6 Use the superspace formulation of the Yang–Mills theory (Bonora & Tonin 1981, Hirshfeld & Leschke 1981) to construct the BRS- and anti-BRS-invariant lagrangians. Follow the same method in supersymmetric Yang–Mills theory (Falck, Hirshfeld & Kubo 1983) and in supergravity.

4

Introduction to the theory of renormalization

4.1 Physical sense of renormalization and its arbitrariness

Bare and 'physical' quantities

Let us consider a theory of interacting fields defined by a given lagrangian density; for definiteness, let us take the case of a scalar field with quartic coupling

$$\mathscr{L} = \tfrac{1}{2}(\partial_\mu \Phi_B)^2 - \tfrac{1}{2}m_B^2\Phi_B^2 - \frac{\lambda_B}{4!}\Phi_B^4 \tag{4.1}$$

We know already how to derive the Feynman rules which allow us to calculate the Green's functions of this theory in the form of the perturbative expansion in powers of the coupling constant λ_B. However, some of the loop momentum integrations in the resulting Feynman diagrams are divergent: consequently, the coefficients of the perturbative series are in general infinite. Does this mean that the lagrangian (4.1) cannot be used to define a theory?

In the following we shall show that this conclusion is not necessarily true. It is not the lagrangian (4.1) which is incorrect; rather, it is the 'bare' coupling constant λ_B which is not a correct expansion parameter. This conclusion is not so unnatural as it seems: the 'bare' parameters of the theory, like λ_B and the 'bare' mass m_B (as well as the Green's functions of the 'bare' field Φ_B) are not directly related to observable quantities. We shall now discuss this problem in some detail.

First let us recall the definition of the 'physical' field Φ_F given in Section 2.7:

$$\langle \mathbf{k}|\Phi_F(x)|0\rangle = \exp[\mathrm{i}(k_0x_0 - \mathbf{k}\cdot\mathbf{x})]; \qquad k_0 = (\mathbf{k}^2 + m_F^2)^{1/2} \tag{2.192}$$

The 'physical' field Φ_F is thus defined by a normalization condition on the one-particle state created by Φ_F. The on-shell truncated Green's functions of the physical field have a straightforward physical interpretation: they are equal to the appropriate S-matrix elements. Off-shell the equality also holds for analytic continuations.

The normalization of a field operator $\Phi(x)$ is directly related to the on-shell residue of its two-point Green's function in momentum space. This is convenient because we prefer to deal with Green's functions and not field operators throughout. Consider thus the two-point Green's function of the field Φ

$$G^{(2)}(x, y) = \langle 0|T\Phi(x)\Phi(y)|0\rangle \tag{4.2}$$

and insert the complete set of states (2.189) between the Φs. The contribution of the one-particle intermediate state is then of the form

$$\begin{aligned}
&\int \frac{d^3k}{(2\pi)^3 2k_0}[\langle 0|\Phi(x)|\mathbf{k}\rangle\langle\mathbf{k}|\Phi(y)|0\rangle\Theta(x_0 - y_0) \\
&\quad + \langle 0|\Phi(y)|\mathbf{k}\rangle\langle\mathbf{k}|\Phi(x)|0\rangle\Theta(y_0 - x_0)] \\
&= \int \frac{d^4k}{(2\pi)^3}\Theta(k_0)\delta(k^2 - m_F^2)\{\Theta(x_0 - y_0)\exp[-ik(x-y)] \\
&\quad + \Theta(y_0 - x_0)\exp[ik(x-y)]\}|\langle 0|\Phi(0)|\mathbf{k}\rangle|^2 \\
&= \int \frac{d^4k}{(2\pi)^4}\frac{i}{k^2 - m_F^2 + i\varepsilon}\exp[-ik(x-y)]|\langle 0|\Phi(0)|\mathbf{k}\rangle|^2
\end{aligned} \tag{4.3}$$

(the last step can be easily checked by calculating the r.h.s. first). Therefore, the Fourier transformed two-point function near the mass-shell behaves as

$$\tilde{G}^{(2)}(k) \underset{k^2 \to m_F^2}{\approx} \frac{i}{k^2 - m_F^2 + i\varepsilon}|\langle 0|\Phi(0)|\mathbf{k}\rangle|^2 \tag{4.4}$$

note that $\langle 0|\Phi(0)|\mathbf{k}\rangle$ is a constant from relativistic invariance. If $\Phi(x) = \Phi_F(x)$ it now follows from (2.192) that

$$\tilde{G}_F^{(2)}(k) \underset{k^2 \to m_F^2}{\approx} \frac{i}{k^2 - m_F^2 + i\varepsilon} \tag{4.5}$$

The result (4.5) for the 'physical' field Φ_F does not in general agree with the result of the perturbation theory calculation for 'bare' Green's function: $G_B^{(2)}(x, y) = \langle 0|T\Phi_B(x)\Phi_B(y)|0\rangle$ (with some regularization introduced to keep the loop integrals finite). We shall find instead that

$$\tilde{G}_B^{(2)}(k) \underset{k^2 \to m_F^2}{\approx} \frac{Z_3}{k^2 - m_F^2 + i\varepsilon} \tag{4.6}$$

where

$$Z_3 \neq 1$$

and is actually divergent. The 'bare' field Φ_B is thus not 'physical'.

We can, however, express the original field Φ_B in terms of the physical field Φ_F

$$\Phi_B = Z_3^{1/2}\Phi_F \tag{4.7}$$

which implies a simple relation between the Green's functions (connected or not):

$$G_F^{(n)}(x_1, \ldots, x_n) = Z_3^{-n/2}G_B^{(n)}(x_1, \ldots, x_n) \tag{4.8}$$

and its counterpart for the connected proper vertex functions

$$\Gamma_{\mathrm{F}}^{(n)}(x_1,\ldots,x_n)=Z_3^{n/2}\Gamma_{\mathrm{B}}^{(n)}(x_1,\ldots,x_n) \tag{4.9}$$

which also holds for the truncated Green's functions that appear in the expression (2.190) for the S-matrix elements. Consequently, the relation between the S-matrix and 'bare' Green's functions involves the renormalization constant Z_3. The 'bare' Green's functions by themselves are not observable. On the other hand, the 'physical' Green's functions are observable and must come out finite in perturbation theory. This is not yet the case in the 'bare' perturbation expansion in λ_{B}. Suppose that we define the coupling constant λ_{F} as the value of the connected proper vertex function $\tilde{\Gamma}_{\mathrm{F}}^{(4)}(p_i)$ at some specified point, the 'renormalization point', in momentum space. We can then attempt to express other physical Green's functions as perturbation series in powers of λ_{F}. In this way we would obtain the expression of one observable quantity, the physical Green's function, in terms of other observable quantities: λ_{F} and the physical mass. We expect that a relation between finite observables should be free from infinities. Consequently, the coefficients of the resulting perturbation expansion should be finite for observable quantities if the perturbation theory applies.

Let us denote the renormalization point by μ. By definition we have

$$\tilde{\Gamma}_{n}^{(4)}|_{\mu}=\lambda_{\mathrm{F}} \tag{4.10}$$

We can use the regularized bare perturbation expansion to calculate $\tilde{\Gamma}_{\mathrm{B}}^{(4)}$ at the same point. The result we shall write as

$$\tilde{\Gamma}_{\mathrm{B}}^{(4)}|_{\mu}=\frac{\lambda_{\mathrm{B}}}{Z_1} \tag{4.11}$$

which defines the new renormalization constant Z_1. Using (4.9) we now obtain

$$\lambda_{\mathrm{B}}=(Z_1/Z_{\mathrm{B}}^2)\lambda_{\mathrm{F}} \tag{4.12}$$

The physical mass m_{F}^2 is defined as the square of the four-momentum of a freely propagating particle. From the unitarity condition for the S-matrix, expressed in terms of the scattering amplitude T_{if}, (3.69) and (3.70), it then follows that the amplitude T_{if} for the process $i\to f$ which can proceed via a one-particle intermediate state has a pole at $p^2=m_{\mathrm{F}}^2$. In perturbation theory the general structure of the amplitude with a one-particle intermediate state corresponds to the diagram in Fig. 4.1 where the central bubble denotes the full propagator. Therefore the full propagator must have a pole at $p^2=m_{\mathrm{F}}^2$

$$\tilde{G}^{(2)}(p)=\frac{R}{p^2-m_{\mathrm{F}}^2}+\cdots$$

and the physical mass can also be defined as the value of p^2 for which the full propagator has a pole. Defined as a pole of the scattering amplitude, m_{F}^2 is an experimentally measurable parameter.

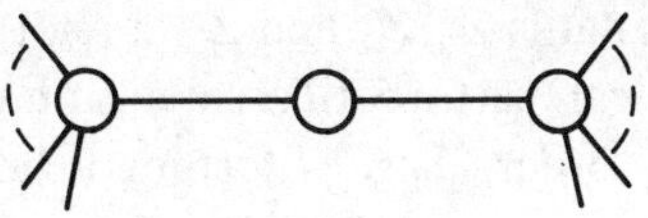

Fig. 4.1

Equivalently, we shall define m_F^2 as the position of the zero of the inverse propagator (see (2.152))

$$\tilde{\Gamma}^{(2)}(p^2 = m_F^2) = 0 \tag{4.13}$$

which can be either 'bare' or 'physical'. This definition is convenient because it avoids multiple poles present in $\tilde{G}^{(2)}(p)$ calculated to any finite order. Again, we can calculate m_F^2 in the regularized bare perturbation expansion and define the mass renormalization constant Z_0 by the relation

$$m_F^2 = (Z_3/Z_0)m_B^2 \tag{4.14}$$

Altogether, we have three independent renormalization constants: Z_0, Z_1 and Z_3 corresponding to three relations: (4.14), (4.12) and (4.7) between 'bare' and 'physical' quantities. Using these relations we can now rewrite the original lagrangian (4.1) in terms of new fields Φ_F and parameters m_F, λ_F

$$\begin{aligned}\mathscr{L} &= \tfrac{1}{2}(\partial_\mu\Phi_B)^2 - \tfrac{1}{2}m_B^2\Phi_B^2 - (\lambda_B/4!)\Phi_B^4 = \tfrac{1}{2}Z_3(\partial_\mu\Phi_F)^2 - \tfrac{1}{2}Z_0 m_F^2\Phi_F^2 \\ &- Z_1(\lambda_F/4!)\Phi_F^4 = \tfrac{1}{2}(\partial_\mu\Phi_F)^2 - \tfrac{1}{2}m_F^2\Phi_F^2 - (\lambda_F/4!)\Phi_F^4 \\ &+ \tfrac{1}{2}(Z_3 - 1)(\partial_\mu\Phi_F)^2 - \tfrac{1}{2}(Z_0 - 1)m_F^2\Phi_F^2 - (Z_1 - 1)(\lambda_F/4!)\Phi_F^4\end{aligned} \tag{4.15}$$

where

$$\Phi_B = Z_3^{1/2}\Phi_F, \quad m_B^2 = Z_0 Z_3^{-1} m_F^2, \quad \lambda_B = Z_1 Z_3^{-2}\lambda_F \tag{4.16}$$

The resulting 'renormalized' lagrangian we shall use as the basis for the formulation of the new perturbation expansion parametrized by the 'physical' quantities: m_F and λ_F. It is this 'physical' perturbation theory and its generalizations described below that will be used in calculations: we shall not refer to the 'bare' lagrangian explicitly.

Counterterms and the renormalization conditions

In (4.15) we have divided the lagrangian into two parts. The first part is identical to (4.1) with the original parameters replaced by physical ones. The perturbation theory based on this part of the lagrangian alone would give rise to the divergences analogous to those of the 'bare' perturbation expansion. We expect that these divergences will be cancelled order-by-order if we take into account the second, Z_i-dependent part of the lagrangian: the 'counterterms'. If this happens we say that the theory is renormalizable. The theory defined by (4.1) is renormalizable if considered in the space-time of dimension $d \leqslant 4$.

The renormalization constants Z_0, Z_1 and Z_3 are not observable quantities and expressed in the new 'physical' perturbation expansion in terms of λ_F and m_F can still be divergent order-by-order. This is fortunate because we need divergent counterterms to cancel the divergences. On the other hand this shows that we still need to use some regularization at the intermediate stage of calculations. After including the counterterms the regularization can be removed. The result is finite as long as we calculate 'physical' quantities. We shall return to the problem of divergences later on.

Now let us make the following important observation concerning the practical application of our results. We shall always use the 'physical' perturbation expansion and never refer to the 'bare' parameters at any stage of calculations. From this point of view to define Z_3, Z_1 and Z_0 by relations like (4.7), (4.12) and (4.14) is inconvenient because these relations involve 'bare' quantities. Instead, we shall fix the counterterms by imposing the 'renormalization conditions' on the physical Green's functions. We have, in fact, already used one such condition to define λ_F: it is (4.10). To fix all counterterms we need two more conditions, which we choose as follows

$$\frac{\partial}{\partial p^2}\tilde{\Gamma}_F^{(2)}(p)|_{p^2=m_F^2} = 1 \tag{4.17}$$

which implies

$$\tilde{\Gamma}_F^{(2)}(p) \underset{p^2 \to m_F^2}{\longrightarrow} p^2 - m_F^2$$

in accord with the fact that field Φ_F in (4.15) is assumed to be the 'physical' field; and

$$\Sigma_F(p)|_{p^2=m_F^2} = 0 \tag{4.18}$$

where

$$\tilde{\Gamma}_F^{(2)}(p) = p^2 - m_F^2 - \Sigma_F(p) \tag{4.19}$$

This choice is exactly equivalent to our previous definitions. The counterterms, present in the Feynman rules of the 'physical' perturbation expansion, can be determined from the renormalization conditions order-by-order. Observe that we now use m_F and λ_F as free parameters rather than m_B and λ_B: this is also more natural physically.

Arbitrariness of renormalization

We have outlined the procedure which replaces the original 'bare' perturbation expansion with divergent coefficients by a new perturbation expansion in which the coefficients are finite. How unique is this procedure? We have used the physical observability of new parameters and Green's functions as a guide in order to ensure that they are finite and in finite relation to each other. However, if the perturbation theory parametrized by m_F and λ_F is finite we would not expect to lose this property if we change the parametrization and use parameters which differ from m_F and λ_F by a finite amount. In other words once we have specified the counterterms which cancel the infinities we can make a finite change in them. We thus have an infinite, n-parameter class of different 'renormalization prescriptions' (n is the number of counterterms). This renormalization symmetry of the theory is very important, and not only because it allows one to choose the most convenient renormalization prescription; it is the basis of the renormalization group and the renormalization group equations.

Consider some examples. First of all we can change the renormalization point $\mu{:}\mu \to \mu'$. As a result, the value of λ_F changes: $\lambda_F \to \lambda'_F$. Also the renormalization constant Z_1, defined by (4.11), must change: $Z_1 \to Z'_1$. It is possible to do this transformation without affecting Z_3 or Z_0: the field Φ_F is still given by (4.7) and therefore does not change; however, it will now have to be expressed in terms of the new parameter λ'_F. The Green's functions are still 'physical' but have a new perturbation expansion, in powers of λ'_F.

So far we have kept to the original, 'physical' scheme. Let us now change the value of the mass renormalization constant $Z_0{:}Z_0 \to Z'_0$ keeping Z_3 unchanged. As m_B^2 is fixed and the physical content of the theory should not change we must conclude that the new mass parameter, the 'renormalized mass', defined as

$$m_R^2 = (Z_3/Z'_0)m_B^2 \tag{4.20}$$

is no longer the physical mass. In terms of the 'renormalization conditions' discussed before this may correspond to replacing (4.18) by

$$\Sigma_R(p)|_\mu = 0 \tag{4.21}$$

where now

$$\tilde{\Gamma}_R^{(2)} = p^2 - m_R^2 - \Sigma_R(p) \tag{4.22}$$

and μ is some renormalization point.

Similarly, we do not have to require that the coupling constant parameter is equal to the value of $\tilde{\Gamma}_F^{(4)}$ at any renormalization point. We can use any definition as long as the transition from λ_F to the new parameter λ_R is perturbatively finite: i.e. λ_R can be expressed in terms of λ_F in the λ_F-perturbation expansion, or vice versa. One important example of a renormalization procedure which makes use of this generalization is the minimal subtraction scheme, which we shall discuss in detail in the following Sections.

Finally we can relax the requirement of the 'physical' renormalization of the two-point renormalized Green's function. Instead of Φ_F we can use the 'renormalized' field Φ_R provided that their relative normalization is perturbatively finite. That is, we have

$$\Phi_R = \mathscr{Z}_R^{1/2}\Phi_F \tag{4.23}$$

where $\mathscr{Z}_R = Z_3^F/Z_3^R$ is finite in the perturbation expansion. The value of $\mathscr{Z}_R$ can be determined if we know the on-shell residue of the two-point renormalized

Green's function

$$\tilde{G}_{\mathrm{R}}^{(2)}(k) \underset{k^2 \to m_{\mathrm{F}}^2}{\approx} \frac{\mathrm{i}\mathscr{L}_{\mathrm{R}}}{k^2 - m_{\mathrm{F}}^2 + \mathrm{i}\varepsilon} \tag{4.24}$$

or, equivalently

$$\frac{\partial}{\partial k^2} \tilde{\Gamma}_{\mathrm{R}}^{(2)}(k)|_{k^2 = m_{\mathrm{F}}^2} = \frac{1}{\mathscr{L}_{\mathrm{R}}} \tag{4.25}$$

Due to the relation

$$G_{\mathrm{F}}^{(n)}(x_1, \ldots, x_n) = \mathscr{L}_{\mathrm{R}}^{-n/2} G_{\mathrm{R}}^{(n)}(x_1, \ldots, x_n) \tag{4.26}$$

the factors of $\mathscr{L}_{\mathrm{R}}$ will also appear in the relation between the renormalized Green's functions and the S-matrix elements. This does not make the renormalized Green's functions very 'unphysical' because the $\mathscr{L}_{\mathrm{R}}$ factors are finite order-by-order.

The relation between the S-matrix and the renormalized Green's functions has the following form

$$\langle \mathbf{k}_1' \ldots \mathbf{k}_m' | S | \mathbf{k}_1 \ldots \mathbf{k}_n \rangle = \frac{1}{(n!m!)^{1/2}} \lim_{k_i^2 \to m_{\mathrm{F}}^2} \prod_i \left[\frac{1}{\mathrm{i}} (k_i^2 - m_{\mathrm{F}}^2) \right]$$
$$* \mathscr{L}_{\mathrm{R}}^{-(n+m)/2} \tilde{G}_{\mathrm{R}}^{(n+m)}(k_1', \ldots, k_m', -k_1, \ldots, -k_n) \tag{4.27}$$

which generalizes (2.190).

Of course, the lagrangian (4.15) can be re-written in terms of the new renormalized fields and parameters

$$\mathscr{L} = \mathscr{L}(Z_3 \Phi_{\mathrm{R}}, (Z_0/Z_3) m_{\mathrm{R}}^2, (Z_1/Z_3^2)\lambda_{\mathrm{R}}, \ldots) = \mathscr{L}(\Phi_{\mathrm{R}}, m_{\mathrm{R}}^2, \lambda_{\mathrm{R}}, \ldots) + \Delta\mathscr{L} \tag{4.28}$$

where in the $\lambda\Phi^4$ theory

$$\Delta\mathscr{L} = \tfrac{1}{2}(Z_3 - 1)(\partial_\mu \Phi_{\mathrm{R}})^2 - \tfrac{1}{2}(Z_0 - 1) m_{\mathrm{R}}^2 \Phi_{\mathrm{R}}^2 - 1/4!(Z_1 - 1)\lambda_{\mathrm{R}} \Phi_{\mathrm{R}}^4 \tag{4.29}$$

The Feynman rules of the renormalized perturbation theory follow from the renormalized lagrangian (4.28). The counter-terms provide vertices, with Z-dependent coupling constants: $(Z_3 - 1)p^2$, $(Z_0 - 1)m_{\mathrm{R}}^2$, $(Z_1 - 1)\lambda_{\mathrm{R}}$. Note that the counter-terms are treated strictly perturbatively: when calculating the nth order correction it is enough to know the counterterms up to this order only.

The counterterms are constructed according to a renormalization prescription. One possibility is the 'physical' renormalization prescription described before. A more general case corresponds to specifying finite values for some renormalized Green's functions at a renormalization point. For instance (Σ_{R} is defined by (4.22))

$$\frac{\partial}{\partial k^2} \tilde{\Gamma}_{\mathrm{R}}^{(2)}(k)|_\mu = 1, \quad \Sigma_{\mathrm{R}}(k)|_\mu = 0, \quad \tilde{\Gamma}_{\mathrm{R}}^{(4)}(k)|_\mu = \lambda_{\mathrm{R}} \tag{4.30}$$

Clearly the number of the equations must be equal to the number of the renormalization constants.

Another possibility is to assume some trial form for the counterterms and adjust the parameters to make the renormalized Green's functions finite order-by-order.

This is the case in the minimal subtraction scheme. This procedure is particularly convenient because the counterterms can be determined from the divergent parts of the diagrams alone. We shall discuss this scheme in detail in Section 4.3, after the introduction of dimensional regularization.

Final remarks

Let us finish this Section with some remarks concerning the calculation of the scattering amplitudes in the renormalized perturbation expansion. As explained before, we determine the physical mass m_{F}^2 from the requirement that the inverse propagator $\tilde{\Gamma}_{\mathrm{R}}^{(2)}(k) = k^2 - m_{\mathrm{R}}^2 - \Sigma_{\mathrm{R}}(k)$ vanishes at $k^2 = m_{\mathrm{F}}^2$. We use this condition in preference to the one involving the propagator $\tilde{G}_{\mathrm{R}}^{(2)}(k)$. The reason is that the perturbation expansion for $\tilde{G}_{\mathrm{R}}^{(2)}(k)$ has the form

$$\tilde{G}_{\mathrm{R}}^{(2)}(k) = \frac{\mathrm{i}}{k^2 - m_{\mathrm{R}}^2} + \frac{\mathrm{i}}{k^2 - m_{\mathrm{R}}^2}(-\mathrm{i}\Sigma_{\mathrm{R}}(k))\frac{\mathrm{i}}{k^2 - m_{\mathrm{R}}^2} + \frac{\mathrm{i}}{k^2 - m_{\mathrm{R}}^2}(-\mathrm{i}\Sigma_{\mathrm{R}}(k))\frac{\mathrm{i}}{k^2 - m_{\mathrm{R}}^2}$$

$$*(-\mathrm{i}\Sigma_{\mathrm{R}}(k))\frac{\mathrm{i}}{k^2 - m_{\mathrm{R}}^2} + \cdots = \frac{\mathrm{i}}{k^2 - m_{\mathrm{R}}^2 - \Sigma_{\mathrm{R}}(k)} \tag{4.31}$$

corresponding to the diagrams

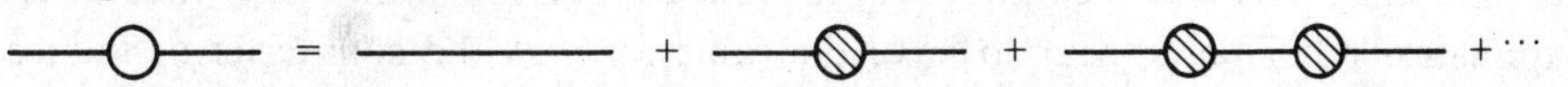

with the blobs on the r.h.s. denoting $-\mathrm{i}\Sigma_{\mathrm{R}}(k)$. We see that, to any finite order, $\tilde{G}^{(2)}(k)$ has multiple poles at $k^2 = m_{\mathrm{R}}^2$ instead of a single pole at $k^2 = m_{\mathrm{F}}^2$ required by unitarity. To obtain a single pole one must sum the whole geometric series (4.31).

Assume that $\tilde{\Gamma}_{\mathrm{R}}^{(2)}(k)$ calculated up to order n in some renormalization scheme is regular at $k^2 = m_{\mathrm{F}}^2$ and expand

$$\Sigma_{\mathrm{R}}(k^2) = \Sigma_0 + (k^2 - m_{\mathrm{F}}^2)\Sigma_1 + (k^2 - m_{\mathrm{F}}^2)^2\Sigma_2 + \cdots \tag{4.32}$$

where m_{F}^2 is understood as the value of k^2 at which the nth order approximation to $\tilde{\Gamma}_{\mathrm{R}}^{(2)}(k)$ vanishes; hence $\Sigma_0 = m_{\mathrm{F}}^2 - m_{\mathrm{R}}^2$. We obtain

$$\tilde{\Gamma}_{\mathrm{R}}^{(2)}(k) = (k^2 - m_{\mathrm{F}}^2)[1 - \Sigma_1 - (k^2 - m_{\mathrm{F}}^2)\Sigma_2 - \cdots] \tag{4.33}$$

and, using (4.25)

$$\mathscr{Z}_{\mathrm{R}} = \frac{1}{1 - \Sigma_1} \tag{4.34}$$

Consequently, (4.27) becomes

$$\langle \mathbf{k}'_1 \ldots \mathbf{k}'_m | S | \mathbf{k}_1 \ldots \mathbf{k}_n \rangle$$

$$= \frac{1}{(n!m!)^{1/2}} \lim_{k_i^2 \to m_{\mathrm{F}}^2} \prod_i \left(\frac{1}{\mathrm{i}}(k_i^2 - m_{\mathrm{F}}^2)\right)(1 - \Sigma_1)^{(n+m)/2}$$

$$* \tilde{G}_{\mathrm{R}}^{(n+m)}(k'_1, \ldots, k'_m, -k_1, \ldots, -k_n)$$

$$= \frac{1}{(n!m!)^{1/2}}(1 - \Sigma_1)^{-(n+m)/2}\tilde{G}_{\mathrm{R(trunc)}}^{(n+m)} \tag{4.35}$$

In the lowest non-trivial order (one-loop approximation) we can equivalently replace $(1-\Sigma_1)^{-(n+m)/2}$ by $(1+\frac{1}{2}(n+m)\Sigma_1)$ so that we then have the correction $(+\frac{1}{2}\Sigma_1)$ for each external line. We shall often refer to this rule in one-loop calculations.

4.2 Classification of the divergent diagrams

Structure of the UV divergences by momentum power counting

In the previous Section we have described the procedure of multiplicative renormalization. However, we have not given the proof that this procedure actually leads to finite results: we still have to show that a theory like $\lambda\Phi^4$ is renormalizable. In other words, we should still prove that the counterterms induced by the multiplicative renormalization, which are of the same general structure as vertices of the original lagrangian, are enough to cancel UV divergences due to loop momentum integrations. Here we shall give only a general idea of the proof. We shall concentrate on the structure of UV divergences which appear in Feynman diagrams of $\lambda\Phi^4$ perturbation theory (with counterterms not yet included) and use the results to determine the form of counterterms from the requirement that these divergences are cancelled. They appear to have the same form as the counterterms induced by the multiplicative renormalization described in Section 4.1. Some examples from QED and $\lambda\Phi^3$ theories will be also considered.

The basic blocks of a renormalization programme are the connected proper vertex functions $\Gamma^{(n)}$, also called the 1PI Green's functions. Let us recall that these are truncated Green's functions which remain connected when an arbitrary internal line is cut. In a renormalization programme these are important for two reasons. First, any other Green's function is made up of 1PI Green's functions joined together to form a tree, each loop included entirely in one of the 1PI parts, so that the momentum integrations in the two distinct parts are independent of each other. Consequently, all loops and therefore all divergences are present already in the 1PI Green's functions. The second reason is that the indentification of the necessary counterterms which are to be interpreted as additional vertices in the renormalized lagrangian is particularly simple if one looks at the divergences of the 1PI functions. This is because the connected proper vertex functions can themselves be understood as generalized vertices of an effective lagrangian which includes all loop corrections: the generating functional $\Gamma[\Phi]$.

The UV divergences we are concerned with come from the region of integration where some of the integration momenta are large. Some information about the possible structure of these divergences can be obtained by the simple procedure of counting the powers of momentum in the Feynman integrand. Suppose that all integration momenta become large simultaneously: $k_i \to \Lambda k_i$, $\Lambda \to \infty$. The integrand will then behave as a power of Λ, say Λ^{-N}. One defines the so-called

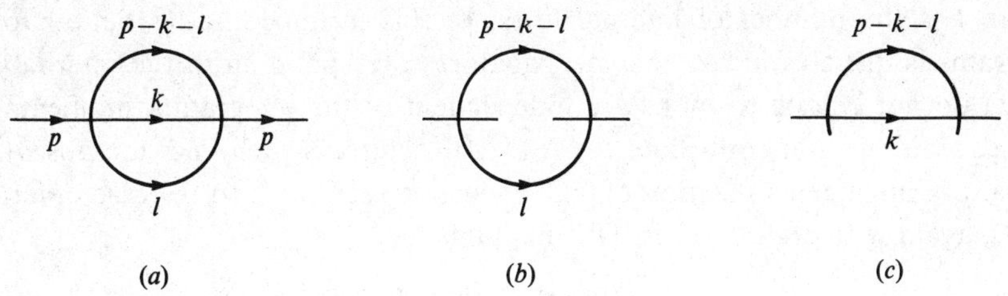

Fig. 4.2

superficial degree of divergence of a Feynman diagram

$$D = 4L - N \tag{4.36}$$

where L is the number of independent loops, each corresponding to a $\int d^4k_i$ momentum integration. For a diagram with V vertices and I internal lines the number of independent loops is $L = I - V + 1$.

Let us next assume that only some subset S of the integration momenta is large and of the order Λ. The integrand will now be proportional to another power of $\Lambda: \Lambda^{-N_S}$. We can introduce the superficial degree of divergence D_S of this subintegration: $D_S = 4L_S - N_S$, where L_S is the number of integration momenta belonging to S. It is easy to see that D_S can also be regarded as a superficial degree of divergence of a subdiagram built up of all the internal lines of the complete Feynman diagram with momenta depending on those in S: L_S is then the number of loops in this subdiagram. In this way we can assign a subdiagram and a degree of divergence to each loop subintegration. For example, in the case of the two-loop diagram (*a*) in Fig. 4.2 the subdiagrams (*b*) and (*c*) correspond to the subintegrations over l and k, respectively.

A theorem due to Weinberg states that a Feynman integral converges if the degree of divergence of the diagram as well as the degree of divergence associated with each possible subintegration over loop momenta is negative. This fundamental result will be the basis of our discussion. The first thing to note is that in $\lambda\Phi^4$ theory (and any other interacting quantum field theory in four dimensions) diagrams with $D \geqslant 0$ are present. In the above example diagram (*a*) has $D = 2$ while (*b*) and (*c*) both have $D = 0$.

Classification of divergent diagrams

To find out what counterterms are required we must examine all 1PI Green's functions with $D \geqslant 0$ which appear in the theory. Consider the case of a scalar field theory without derivative terms in the interaction ($\lambda\Phi^4$ is one example). Then, for any 1PI diagram (4.36) reads

$$D = 4L - N = 4L - 2I \tag{4.37}$$

where I is the number of internal lines. In this formula the restriction to 1PI diagrams is due to the fact that propagators corresponding to the internal lines which do not belong to loops are independent of the integration momenta and consequently do not contribute to N: in 1PI diagrams such lines are absent. We can also write a generalization of (4.37) which is applicable to the case of fermion QED, again with restriction to 1PI diagrams

$$D = 4L - 2I_B - I_F \tag{4.38}$$

Here I_B, I_F are the numbers of photon and fermion internal lines, respectively. Note that by replacing $4L$ with nL we obtain a generalization of (4.36), (4.37) and (4.38) to the case of an n-dimensional space.

We shall see that D is determined by the number of external lines of the diagram. The argument is based on dimensional analysis so let us first recall the dimensions of some quantities in quantum field theory. We set $\hbar = c = 1$ so that the action integral must be dimensionless while length and mass are measured in inverse units: $[x]^{-1} = [m]$. The canonical dimensions of Bose and Fermi fields are determined from the requirement that the kinetic terms in the action, which are respectively of the form $\int d^n x \Phi \partial^2 \Phi$ and $\int d^n x \bar{\Psi} i \not\partial \Psi$, are dimensionless. It follows that

$$[\Phi] = [m]^{(n-2)/2}, \quad [\Psi] = [m]^{(n-1)/2} \tag{4.39}$$

In the same manner the dimensions of a coupling constant follow from the requirement that the corresponding term in the action is dimensionless. Some examples are

$$\left.\begin{array}{lll} \lambda\Phi^3: & [m]^{-n}[\lambda][m]^{3(n-2)/2} = 1; & [\lambda] = [m]^{3-n/2} \\ \lambda\Phi^4: & [m]^{-n}[\lambda][m]^{4(n-2)/2} = 1; & [\lambda] = [m]^{4-n} \\ g\bar{\Psi}\not{A}\Psi: & [m]^{-n}[g][m]^{2(n-1)/2+(n-2)/2} = 1; & [g] = [m]^{2-n/2} \end{array}\right\} \tag{4.40}$$

where $[m]^{-n}$ is the dimension of the $\int d^n x$ integration measure. In four dimensions the quartic ($\lambda\Phi^4$) and QED ($g\bar{\Psi}\not{A}\Psi$) coupling constants are dimensionless while the cubic ($\lambda\Phi^3$) coupling constant has the dimension of mass. Assume that at large momenta the Bose and Fermi propagators behave respectively as $1/k^2$ and $1/k$. Then, for any 1PI Feynman diagram built up of vertices with dimensionless coupling constants only, the superficial degree of divergence coincides with the dimension of the diagram in the unit of mass. The exception is the possible lowest order no-loop contribution i.e. the single elementary vertex for which the superficial degree of divergence is undefined. As all contributions to a given Green's function must be of the same dimension it then follows that with dimensionless couplings, the superficial degree of divergence of a diagram is determined solely by the external lines. The canonical dimension of a 1PI diagram with B external boson lines and F external fermion lines is equal to the dimension of a coupling constant corresponding to the vertex of this structure, which is $[m]^{4-B-3F/2}$ in four dimensions. Consequently

$$D = 4 - B - \tfrac{3}{2}F \tag{4.41}$$

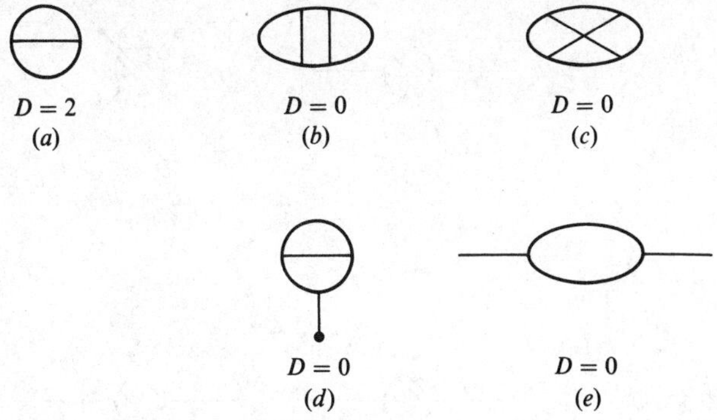

Fig. 4.3

As explained, this applies to 1PI diagrams with dimensionless couplings only. If the diagram also contains vertices with dimensional coupling constants of total dimension Δ in the units of mass (for example, $\Delta = K$ for K fermion mass insertions or cubic scalar couplings)

$$D = 4 - B - \tfrac{3}{2}F - \Delta \tag{4.42}$$

This formula has some important consequences. One is that in any theory with coupling constants all of positive dimension in the units of mass ($\lambda\Phi^3$ for example) the number of 1PI diagrams with $D \geqslant 0$ is finite. A theory like this is called superrenormalizable. In the case of the $\lambda\Phi^3$ theory the diagrams with $D \geqslant 0$ are as shown in Fig 4.3. Here (*a*), (*b*) and (*c*) are irrelevant vacuum diagrams: they cancel with the normalization factor of our Green's functions, (*d*) is a constant 'tadpole' which contributes to the vacuum expectation value $\langle\Phi\rangle$ of the Φ field and can be disregarded in the $\langle\Phi\rangle = 0$ case (diagrams of this kind will be important in theories with spontaneous symmetry breaking) and (*e*) produces an infinite mass renormalization counterterm which is the only divergent counterterm in $\lambda\Phi^3$ theory with $\langle\Phi\rangle = 0$. In particular there are no $D \geqslant 0$ 1PI diagrams with three external lines and consequently no divergent coupling constant renormalization counterterm.

On the other hand in a theory which involves, among others, any coupling constant of negative dimension, like $\lambda\Phi^4$ in more than four space-time dimensions or $G(\bar{\Psi}\Psi)^2$ in four-dimensional space, the perturbation expansion of any 1PI Green's function must include $D \geqslant 0$ diagrams when the order is high enough. Each 1PI Green's function requires new counterterms and new renormalization conditions: consequently, the theory is non-renormalizable. An important example is Fermi's theory of weak interactions.

Let us now return to theories with dimensionless coupling constants, so that (4.41) applies: all diagrams contributing to a specific 1PI Green's function have the same degree of divergence. For the $\lambda\Phi^4$ theory and for QED we have the $D \geqslant 0$ connected proper vertex functions shown in Fig. 4.4 (the triple photon

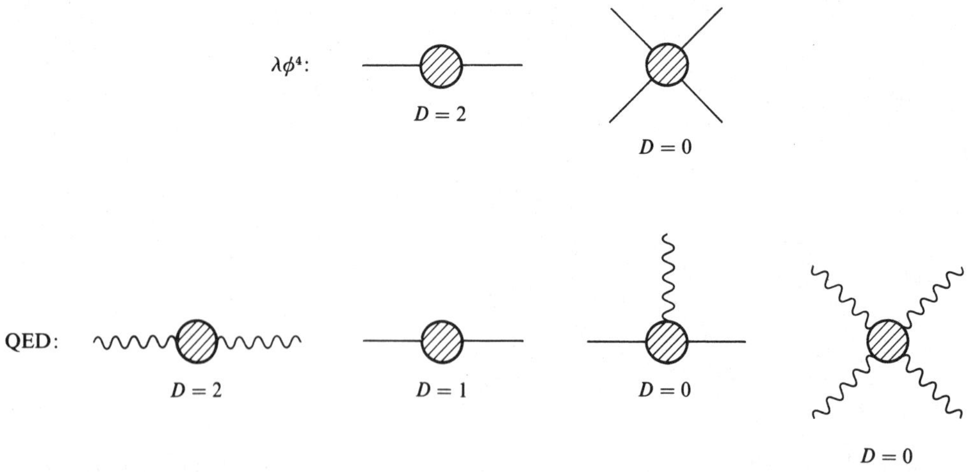

Fig. 4.4

proper vertex also has $D = 1$ but it vanishes because the photon has odd charge conjugation).

Necessary counterterms

To make the divergent diagrams finite we must find a way to lower the associated degrees of divergence. This can be achieved by making a subtraction in the Feynman integrand. Consider a 1PI Feynman diagram with $D \geqslant 0$. Let $\{f\}_D$ denote the sum of first $(D+1)$ terms of a Taylor expansion of the Feynman integrand f in powers of the external momenta with D equal to the superficial degree of divergence of the diagram considered. If we replace the integrand of this diagram by a subtracted expression

$$f - \{f\}_D \tag{4.43}$$

the superficial degree of divergence will become negative. A similar procedure should be applied to divergent subdiagrams.

In a renormalizable theory we should be able to understand these subtractions as due to the inclusion of counterterms. With some procedure for the regularization of divergent integrals we can calculate the Feynman integrals of the two terms in (4.43) separately. As the second term is a polynomial in the external momenta, after the integration it will produce a polynomial 'counterterm', or a combination of counterterms. Obviously the counterterms will diverge when the intermediate regularization is removed; however, from Weinberg's theorem, they must combine with the other part of (4.43) to give ultimately a finite result provided that we have dealt with all divergent subdiagrams. It still is a question whether the counterterms obtained by means of this procedure will have the structure and combinatorics implied by multiplicative renormalization.

Consider the case of the $\lambda\Phi^4$ scalar field theory. We then have two $D \geqslant 0$ Green's

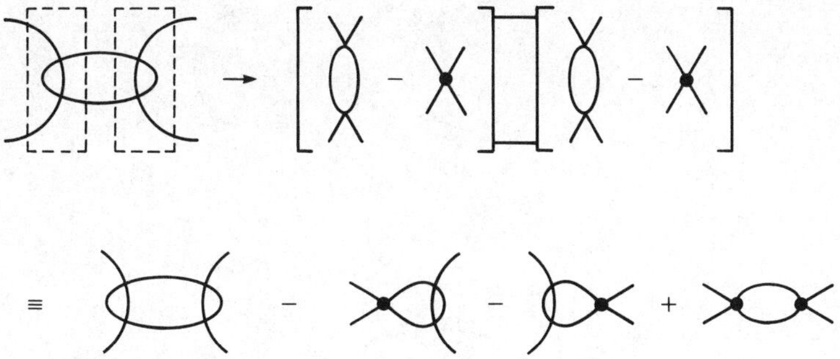

Fig. 4.5

functions: $\tilde{\Gamma}^{(2)}(p)$ with $D=2$ and $\tilde{\Gamma}^{(4)}(p_1,\ldots,p_4)$ with $D=0$. The two-point function has $D=2$ and consequently requires subtraction of the first three terms of the Taylor expansion of the integrand. This produces two counterterms

$$Ap^2+B$$

The third one, proportional to p_μ, vanishes because of relativistic invariance: all counterterms must be scalars. We recognize the counterparts of the wave-function renormalization constant (p^2 is equivalent to $-\partial^2$) and the mass counterterm. The four-point function has $D=0$ and only one (constant) counterterm which corresponds to the coupling constant counterterm. We conclude that the subtractions necessary to lower the degree of divergence of the $D \geqslant 0$ Green's functions are equivalent to including the counterterms of structure as implied by multiplicative renormalization of the original lagrangian.

Can these counterterms, if included in the interaction lagrangian, also deal with all divergences due to the $D \geqslant 0$ subdiagrams? Again (in the $\lambda\Phi^4$ case) these are the subdiagrams with two ($D=2$) or four ($D=0$) external lines, so that no new types of counterterm are apparently required. We can envisage the following procedure. Consider a diagram with no counterterm vertices which includes some $D \geqslant 0$ subdiagrams. We first identify all lowest order (one-loop) divergent subdiagrams. In each of these we make the appropriate subtractions. The effect of these subtractions should be equivalent to adding to the original diagram all diagrams obtained by replacing some of the considered lowest order subdiagrams by the corresponding lowest order counterterms. One example is shown in Fig 4.5. The boxes indicate the divergent subdiagrams we are considering. Note that in this case the dot 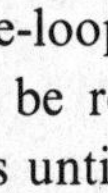 is not the complete lowest order coupling constant counterterm. This is because we make the subtraction in the indicated subdiagrams, and not in complete one-loop vertex functions which are shown in Fig. 4.6. This procedure should then be repeated with respect to higher order $D \geqslant 0$ subdiagrams and counterterms until all divergences are removed.

Unfortunately, this argument runs into difficulties: one problem is that we can not always make subtractions independently in different subdiagrams. For instance,

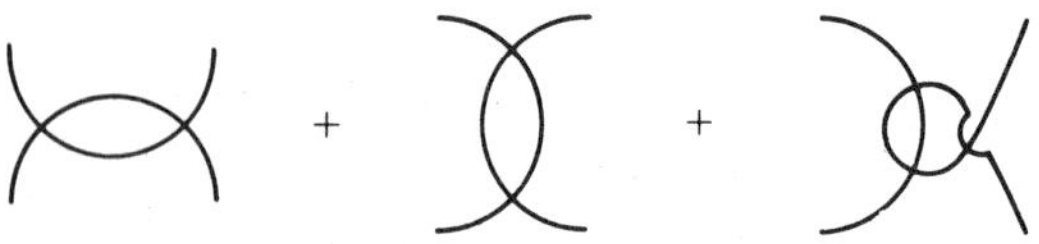

Fig. 4.6

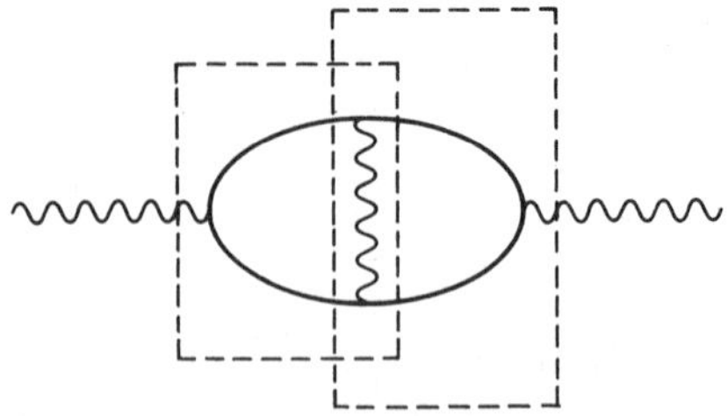

Fig. 4.7

subdiagrams (*b*) and (*c*) in Fig. 4.2 overlap each other. Another example of such 'overlapping divergences' is the QED diagram shown in Fig. 4.7 where the $D=0$ divergent subdiagram can be identified in two distinct ways. The problem of overlapping divergences requires more detailed treatment and we shall not discuss it here. However, the ultimate answer is positive: the counterterms in the renormalized lagrangian do indeed cancel all UV divergences.

We must also note another approach, the BPHZ construction (Bogoliubov–Parasiuk–Hepp and Zimmermann). In this case the subtractions are performed systematically by means of an 'R-operation' which is based on a somewhat different identification of subdiagrams.

4.3 $\lambda\Phi^4$: low order renormalization

Feynman rules including counterterms

In this Section we introduce in some detail the dimensional regularization method ('t Hooft & Veltman 1972, Bollini & Giambiagi 1972) and the minimal subtraction renormalization scheme using the $\lambda\Phi^4$ theory as an example. We choose the minimal subtraction scheme to illustrate our general consideration of Section 4.1 since this method of renormalization is particularly convenient in applications of the renormalization group equation: this is the so-called mass-independent renormalization, namely the renormalization constants Z_i do not depend on mass parameters.

The Feynman rules for the $\lambda\Phi^4$ theory are collected in Table 4.1. It also shows diagrams corresponding to the counterterms present in lagrangian (4.28). All counterterms are treated as interaction terms since we want to use counterterms order-by-order in the perturbation theory.

Table 4.1

propagator	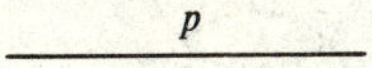	$\dfrac{\mathrm{i}}{p^2 - m^2 + \mathrm{i}\varepsilon}$
loop integration		$\displaystyle\int \frac{\mathrm{d}^4 k}{(2\pi)^4}$
vertex	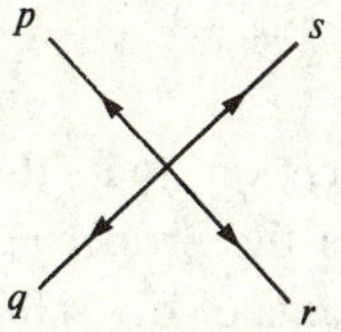	$-\mathrm{i}\lambda; p + q + r + s = 0$
symmetry factors S	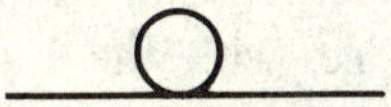	$S = \frac{1}{2}$
		$S = \frac{1}{6}$
	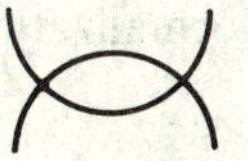	$S = \frac{1}{2}$
	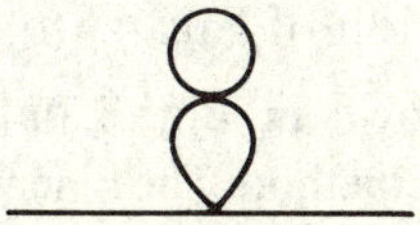	$S = \frac{1}{4}$
vertex counterterm	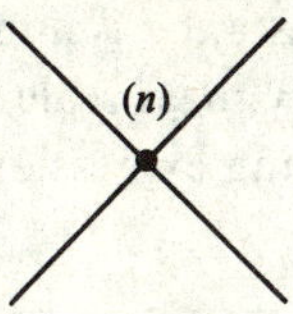	$-\mathrm{i}\lambda(Z_1 - 1)^{(n)}$
mass counterterm	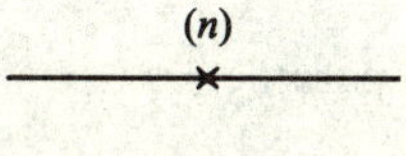	$-\mathrm{i}(Z_0 - 1)^{(n)} m^2$
wave-function renormalization counterterm	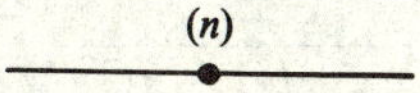	$+\mathrm{i}(Z_3 - 1)^{(n)} p^2$

Renormalization constants Z_i will be determined by the renormalization conditions in each order of the perturbation theory

$$Z_i = 1 + \sum_n (Z_i - 1)^{(n)}$$

where

$$(Z_i - 1)^{(n)} = f_i^{(n)} \lambda^n$$

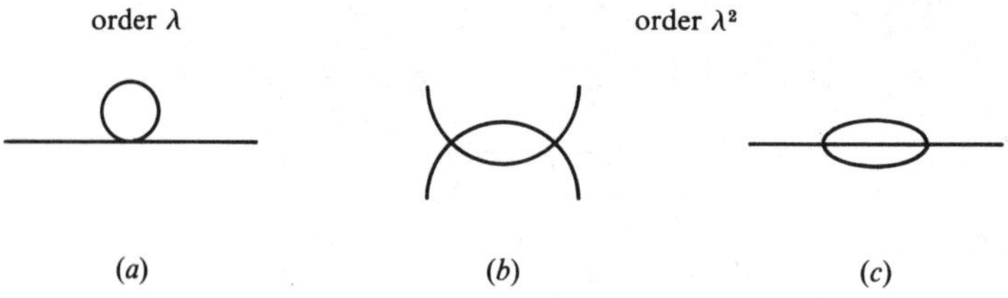

Fig. 4.8

It is clear from the previous Section that the main task for the lowest order renormalization is to calculate the proper (1PI) Green's functions corresponding to the three divergent diagrams of Fig. 4.8. The calculation will be based on the dimensional regularization method and we introduce it now briefly. We begin with vertex renormalization i.e. with the Fig. 4.8(*b*). In four space-time dimensions this is logarithmically divergent $\sim \int d^4k/k^4$. However if we do the integration in $4-\varepsilon$ dimensions the result is finite. This suggests the possibility of considering the analytic continuation of the amplitude to $4-\varepsilon$ dimensions (regularization), cancelling all terms divergent when $\varepsilon \to 0$ by properly adjusting the renormalization constants Z_i (renormalization) and continuing the result analytically to four dimensions.

Calculation of Fig. 4.8(*b*)

We first calculate the diagram shown in Fig. 4.8(*b*) in $4-\varepsilon$ dimensions and then discuss some subtle points of the method. The kinematics are defined in Fig. 4.9. In $n = 4-\varepsilon$ dimensions the coupling constant λ is dimensionful so it is convenient to write it as $\mu^{\varepsilon}\lambda$, where μ is an arbitrary mass scale and λ is dimensionless. The first diagram of Fig. 4.9 gives the following expression for the amplitude:

$$A_1 = \frac{1}{2}\int \frac{d^nk}{(2\pi)^n} \frac{i}{k^2 - m^2 + i\varepsilon} \frac{i}{(k-P)^2 - m^2 + i\varepsilon}(-i\lambda\mu^{\varepsilon})^2 \tag{4.44}$$

We introduce the standard Feynman parameters

$$\frac{1}{ab} = \int_0^1 \frac{dx}{[ax + b(1-x)]^2} \tag{4.45}$$

and change the variables as follows

$$k - Px \to k \tag{4.46}$$

which is legitimate for a convergent integral. We then get

$$A_1 = \tfrac{1}{2}\lambda^2\mu^{2\varepsilon}\int_0^1 dx \int \frac{d^nk}{(2\pi)^n} \frac{1}{[k^2 + sx(1-x) - m^2 + i\varepsilon]^2} \tag{4.47}$$

The integral over $dk_0 d^{n-1}k$ can be easily done after the Wick rotation. We note that the poles of the integrand in the k_0-plane are located as shown in Fig. 4.10,

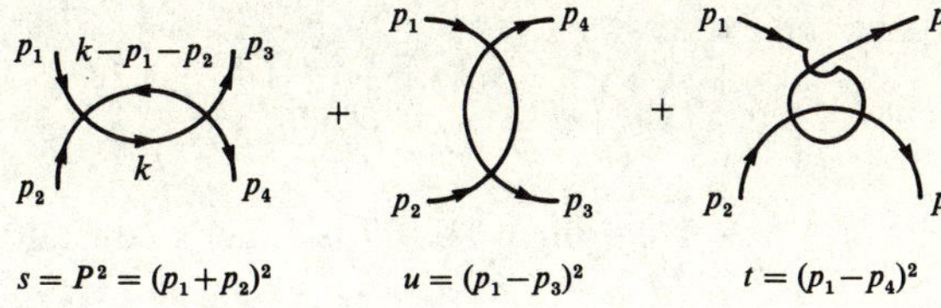

Fig. 4.9

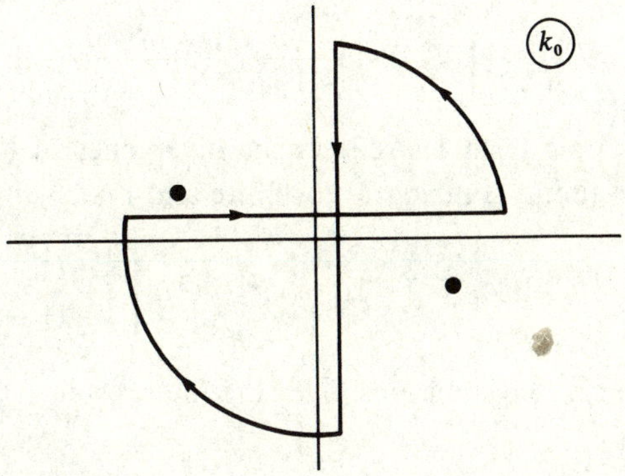

Fig. 4.10

since $k_0 = \pm[\mathbf{k}^2 - sx(1-x) + m^2 - \mathrm{i}\varepsilon]^{1/2}$. We can change the contour to integrate over the imaginary axis. Changing variables

$$\begin{aligned} &\mathrm{i}k_0 \to k_0 \\ &\mathrm{d}k_0 \to -\mathrm{i}\,\mathrm{d}k_0 \\ &k_0^2 - \mathbf{k}^2 \to -k_0^2 - \mathbf{k}^2 \\ &\int_{+\mathrm{i}\infty}^{-\mathrm{i}\infty} \mathrm{d}k_0 \to +\mathrm{i}\int_{-\infty}^{\infty} \mathrm{d}k_0 \end{aligned}$$

one gets the following expression

$$A_1 = \tfrac{1}{2}\lambda^2\mu^{2\varepsilon}\int_0^1 \mathrm{d}x \int \frac{\mathrm{d}^n k}{(2\pi)^n} \frac{(+\mathrm{i})}{[k^2 - sx(1-x) + m^2]^2} \tag{4.48}$$

where k is a vector in n-dimensional Euclidean space. The next step is to use the formula (in Euclidean space)

$$\int \frac{\mathrm{d}^n k}{(2\pi)^n} \frac{1}{(k^2 + b)^\alpha} = \frac{1}{(4\pi)^{n/2}} \frac{b^{n/2-\alpha}\Gamma(\alpha - \frac{1}{2}n)}{\Gamma(\alpha)} \tag{4.49}$$

which leads us to an almost final result

$$A_1 = \tfrac{1}{2}(\lambda\mu^\varepsilon)\mathrm{i}\frac{\lambda}{(4\pi)^{n/2}} \frac{\Gamma(\frac{1}{2}\varepsilon)}{\Gamma(2)} \int_0^1 \mathrm{d}x \left[\frac{m^2 - sx(1-x)}{\mu^2}\right]^{-\varepsilon/2} \tag{4.50}$$

To proceed we require the expression

$$\Gamma(x) \underset{x\to 0}{\longrightarrow} 1/x - \gamma + O(x)$$

where $\gamma = \lim_{n\to\infty}(1 + \frac{1}{2} + \cdots + 1/n - \ln n) = 0.577$ is Euler's constant, and we also use the expansion

$$a^{\varepsilon} = \exp(\varepsilon \ln a) = 1 + \varepsilon \ln a + O(\varepsilon^2)$$

In the limit $\varepsilon \to 0$ we get

$$A_1 = \mathrm{i}\lambda \frac{\lambda}{16\pi^2}\frac{1}{\varepsilon} - \mathrm{i}\lambda \frac{\lambda}{2(4\pi)^2}\left[\left[\int_0^1 \mathrm{d}x \ln \frac{m^2 - sx(1-x)}{\mu^2} + \gamma - \ln 4\pi\right]\right] + O(\varepsilon) \quad (4.51)$$

The presence of the pole term $1/\varepsilon$ reflects the divergence of the integral in four dimensions. The remaining contribution is finite and reads, after integration

$$A_1(s) = -\mathrm{i}\frac{\lambda^2}{2(4\pi)^2}\left\{\gamma - \ln 4\pi - 2 + 2(\tfrac{1}{4} - m^2/s)^{1/2} \ln\left[\frac{\frac{1}{2} + (\frac{1}{4} - m^2/s)^{1/2}}{\frac{1}{2} - (\frac{1}{4} - m^2/s)^{1/2}}\right]\right\} \quad (4.52)$$

The inclusion of the crossed diagrams gives the final result

$$A_1 = \mathrm{i}\lambda^2 \frac{3}{16\pi^2}\frac{1}{\varepsilon} + A_1(s) + A_1(t) + A_1(n) \quad (4.53)$$

Comments on analytic continuation to $n \neq 4$ dimensions

Before proceeding further a comment on the explored analytic continuation to $n \neq 4$ dimensions should be made. First, consider the space with one time-like dimension and $n - 1 \neq 3$ spacelike dimensions. The basic integral

$$I_n = \int \mathrm{d}^n k \frac{1}{(k^2 + b)^2}$$

makes sense for $n = 1, 2, 3$ and can be written as

$$I_n = \frac{1}{2}\int_{-\infty}^{\infty} \mathrm{d}k_0 \int_0^{\infty} \mathrm{d}r^2 (r^2)^{(n-3)/2} \frac{1}{(k^2 + b)^2} \int \mathrm{d}^{n-1}\Omega, \quad k^2 = k_0^2 - r^2 \quad (4.54)$$

where $\mathrm{d}^{n-1}\Omega$ is an element of the solid angle in $(n - 1)$ dimensions

$$\Omega^{n-1} = \int_0^{2\pi} \mathrm{d}\theta_1 \int_0^{\pi} \sin\theta_2\, \mathrm{d}\theta_2 \int_0^{\pi} \sin^2\theta_3\, \mathrm{d}\theta_3 \ldots \int_0^{\pi} \sin^{n-3}\theta_{n-2}\, \mathrm{d}\theta_{n-2} = \frac{2\pi^{(n-1)/2}}{\Gamma(\frac{1}{2}(n-1))}$$

Integral (4.54) can be used to define an analytic function I_n in the range $1 < n < 4$ which coincides with the original I_n for $n = 2, 3$ (the solid angle for a non-integer n is the analytic continuation of Ω^{n-1}, n integer). Furthermore, by partially integrating N times over $\mathrm{d}r^2$ one can find explicit expressions for the analytic continuation of I_n to $(1 - 2N) < n < 4$ dimensions. In addition we would like to be able to continue I_n to values $n > 4$. This can be achieved using the identity

$$1 = \frac{1}{2}\left(\frac{d}{dk_0}k_0 + \frac{d}{dr}r\right)$$

and by integrating each term by parts over k_0 and r, respectively. For $1 < n < 4$ the surface terms are zero and one gets

$$I_n \sim \int_{-\infty}^{\infty} dk_0 \int_0^{\infty} dr\, r^{n-2} \frac{(-2k_0^2 + 2r^2)}{(k^2+b)^3}$$
$$+ \frac{1}{2}\int_{-\infty}^{\infty} dk_0 \int_0^{\infty} dr\, r^{n-2} \frac{(n-2)}{(k^2+b)^2} = I'_n + \tfrac{1}{2}(n-2)I_n \tag{4.55}$$

so that

$$I_n = -\frac{1}{n-4} I'_n$$

where I'_n exists for $1 < n < 5$. The above procedure may be repeated to show that I_n is of the form

$$I_n = \Gamma(\tfrac{1}{2}(4-n))\tilde{I}_n \tag{4.56}$$

where $\tilde{I}$ is well behaved for arbitrarily large n. So one has constructed an analytic function I_n with simple poles at $n = 4, 6, 8, \ldots$. Dimensional regularization assumes the physical theory to be given by I_n for $n \to 4$.

Lowest order renormalization

Coming back to the result (4.53) one can now renormalize the amplitude by including counterterms. It is clear from the structure of the counterterm that the divergencies of (4.53) can be cancelled by the vertex counterterm in Fig. 4.11 i.e. by properly adjusting Z_1. The minimal subtraction prescription assumes that Z_i have only (multiple) pole terms in ε and the residua are fixed to cancel all the divergencies for $\varepsilon \to 0$. With such a prescription one then gets to order λ

$$(Z_1 - 1)^{(1)} = \frac{1}{\varepsilon}\frac{3\lambda}{16\pi^2} \tag{4.57}$$

and the renormalized (finite) amplitude of Fig. 4.8(b) is given by $A_1(s) + A_1(t) + A_1(u)$ of (4.53).

The other two diagrams of Fig. 4.8 require mass and wave-function renormalization. The amplitude of Fig. 4.8(a) reads

$$\tfrac{1}{2}(-i\lambda)\mu^{\varepsilon}\int \frac{d^n k}{(2\pi)^n}\frac{i}{k^2 - m^2 + i\varepsilon} = \tfrac{1}{2}(-i\lambda)\mu^{\varepsilon}\frac{1}{(4\pi)^2}(4\pi)^{\varepsilon/2}(m^2)^{1-\varepsilon/2}\Gamma(\tfrac{1}{2}\varepsilon - 1)$$
$$= \tfrac{1}{2}i\frac{2\lambda m^2}{(4\pi)^2}\frac{1}{\varepsilon} - \frac{1}{2}i\frac{\lambda m^2}{(4\pi)^2}\left(-1 + \gamma - \ln 4\pi + \ln\frac{m^2}{\mu^2}\right) \tag{4.58}$$

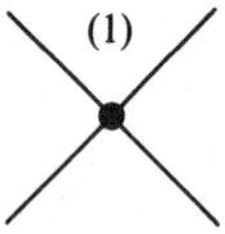

Fig. 4.11

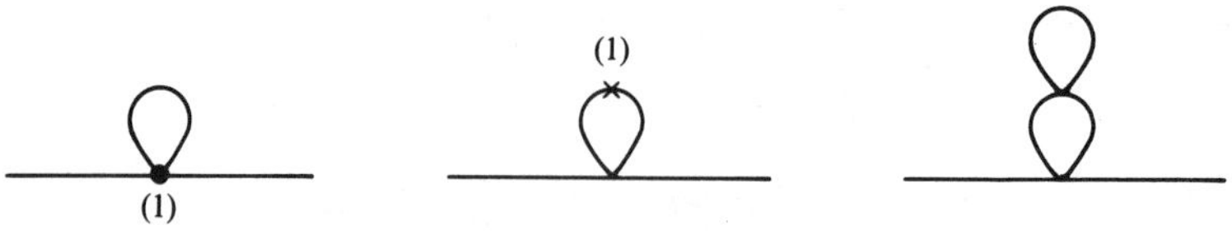

(a) (b) (c)

Fig. 4.12

The divergence is cancelled by the mass counterterm

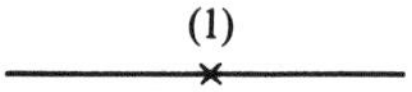

$\equiv (-\mathrm{i})(Z_0 - 1)^{(1)}m^2$

with

$$(Z_0 - 1)^{(1)} = \frac{\lambda}{16\pi^2}\frac{1}{\varepsilon} \tag{4.59}$$

in the minimal subtraction renormalization scheme.

The amplitude of Fig. 4.8(*c*) requires both mass and wave-function renormalization. It is of order λ^2 and therefore in order-by-order renormalization one must include simultaneously the other diagrams of order λ^2 given in Fig. 4.12. The divergencies which remain after adding the amplitude of Fig. 4.8(*c*) and those of Fig. 4.12(*a*)–(*c*) should be then cancelled by new λ^2 terms of the mass and wave-function renormalization counterterms

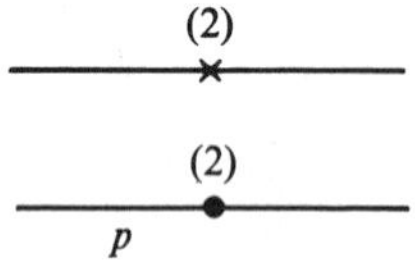

$\equiv -\mathrm{i}(Z_0 - 1)^{(2)}m^2$

$\equiv +\mathrm{i}(Z_3 - 1)^{(2)}p^2$

(notice that in $\lambda\Phi^4$ theory $(Z_3 - 1)^{(1)} = 0$).

We tabulate first the results for Fig. 4.12(*a*)–(*c*). Using (4.57) and (4.59) it is easy to get the following (we write down explicitly only pole terms)

Diagram 4.12(*a*):

$$\mathrm{i}\lambda^2\frac{m^2}{(16\pi^2)^2}\frac{3}{2}\left(\frac{2}{\varepsilon^2} - \frac{1}{\varepsilon}\ln\frac{m^2}{4\pi\mu^2} + \frac{1}{\varepsilon} - \frac{\gamma}{\varepsilon}\right) + O(\text{const.}) \tag{4.60a}$$

Diagram 4.12(*b*):

$$\mathrm{i}\lambda^2\frac{m^2}{(16\pi^2)^2}\frac{1}{2}\left(\frac{2}{\varepsilon^2} - \frac{1}{\varepsilon}\ln\frac{m^2}{4\pi\mu^2} - \frac{\gamma}{\varepsilon}\right) + O(\text{const.}) \tag{4.60b}$$

Diagram 4.12(*c*):

$$(-\mathrm{i})\lambda^2 \frac{m^2}{(16\pi^2)^2} \frac{1}{2}\left(\frac{2}{\varepsilon^2} - \frac{2}{\varepsilon}\ln\frac{m^2}{4\pi\mu^2} + \frac{1}{\varepsilon} - \frac{2\gamma}{\varepsilon}\right) + O(\text{const.}) \tag{4.60c}$$

The only non-trivial contribution is due to Fig. 4.8(*c*). The amplitude reads

$$I = \tfrac{1}{6}\lambda^2\mu^{2\varepsilon}\mathrm{i}\int \frac{\mathrm{d}^n k}{(2\pi)^n}\frac{\mathrm{d}^n l}{(2\pi)^n}\frac{1}{k^2 - m^2 + \mathrm{i}\varepsilon}\frac{1}{l^2 - m^2 + \mathrm{i}\varepsilon}\frac{1}{(p + l - k)^2 - m^2 + \mathrm{i}\varepsilon} \tag{4.61}$$

which is divergent for (i) l fixed, k large, (ii) k fixed, l large, (iii) both k and l large. This is what is called an overlapping divergence because several overlapping loop subintegrations lead to a divergent result. The calculation of amplitude (4.61) is lengthy (Collins 1974). The structure of pole terms turns out to be the following

$$I = \mathrm{i}\frac{\lambda^2}{(16\pi)^2}\left\{-\frac{m^2}{\varepsilon^2} + \frac{1}{\varepsilon}\left[m^2\ln\frac{m^2}{4\pi\mu^2} + \frac{1}{12}p^2 + (\gamma - \tfrac{3}{2})m^2\right]\right\}$$

Adding (4.60) one gets

$$i\frac{\lambda^2}{(16\pi^2)^2}\left(\frac{1}{\varepsilon^2}2m^2 + \frac{1}{\varepsilon}\frac{1}{12}p^2 - \frac{1}{\varepsilon}\frac{1}{2}m^2\right)$$

so that

$$\left.\begin{aligned} (Z_0 - 1)^{(2)} &= \frac{\lambda^2}{(16\pi^2)^2}\left(\frac{2}{\varepsilon^2} - \frac{1}{2}\frac{1}{\varepsilon}\right)\\ (Z_3 - 1)^{(2)} &= -\frac{\lambda^2}{12(16\pi^2)^2}\frac{1}{\varepsilon} \end{aligned}\right\} \tag{4.62}$$

Combining (4.59) with (4.62) one gets the following results for Z_0 and Z_3 (to order λ^2)

$$Z_0 = 1 + \frac{\lambda}{16\pi^2}\frac{1}{\varepsilon} + \frac{\lambda^2}{(16\pi^2)^2}\left(\frac{2}{\varepsilon^2} - \frac{1}{2}\frac{1}{\varepsilon}\right) \tag{4.65}$$

$$Z_3 = 1 - \frac{\lambda^2}{12(16\pi^2)^2}\frac{1}{\varepsilon} \tag{4.64}$$

In the literature one can also often find the constants Z_m and Z_λ defined by the relations

$$\left.\begin{aligned} m_\mathrm{B}^2 &= Z_m m_\mathrm{R}^2 \quad \text{i.e.} \quad Z_m = Z_0/Z_3\\ \lambda_\mathrm{B} &= Z_\lambda\lambda_\mathrm{R} \quad\ \ \text{i.e.} \quad Z_\lambda = Z_1/Z_3^2 \end{aligned}\right\} \tag{4.65}$$

We see in particular that to order λ^2

$$Z_m = 1 + \frac{1}{\varepsilon}\left[\frac{\lambda}{16\pi^2} - \frac{5}{12}\frac{\lambda^2}{(16\pi^2)^2}\right] + \frac{1}{\varepsilon^2}\frac{2\lambda^2}{(16\pi^2)^2} \tag{4.66}$$

Notice that there are logarithms of mass in the residua of the poles in each diagram. However they cancel in the final result for constants Z_i and we see explicitly that

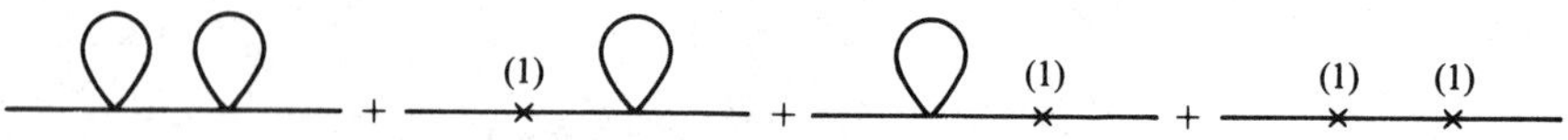

Fig. 4.13

at least in that order of perturbation theory the minimal subtraction scheme is indeed a mass-independent scheme.

We end this Section with two comments. Firstly, it should be remembered that one-particle reducible diagrams do not require new counterterms. Let us illustrate that fact with one example. Consider the diagrams of Fig. 4.13. We observe that the following relation holds.

Diagrams of Fig. 4.13 = 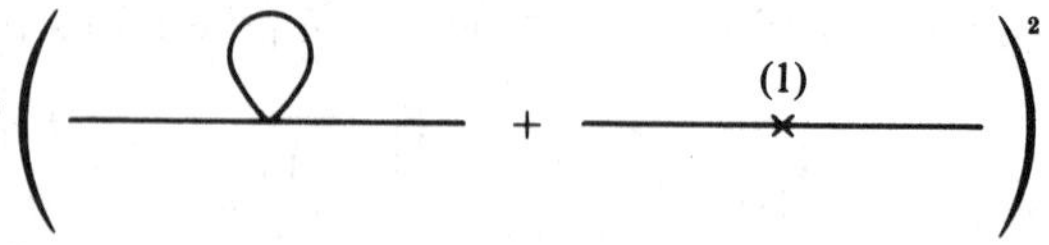

which illustrates our statement.

The second remark is devoted to the fact that the finite parts of Feynman integrals are not in general obtained by simply discarding poles and multiple poles at $n=4$ but have to be worked out by order-by-order renormalization procedure. Consider e.g. mass loop diagrams

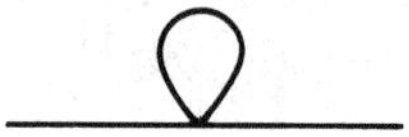 $\equiv (1/\varepsilon)f_1 + f_2 + \varepsilon f_3$

$\equiv -(1/\varepsilon)f_1$

so that in the second order

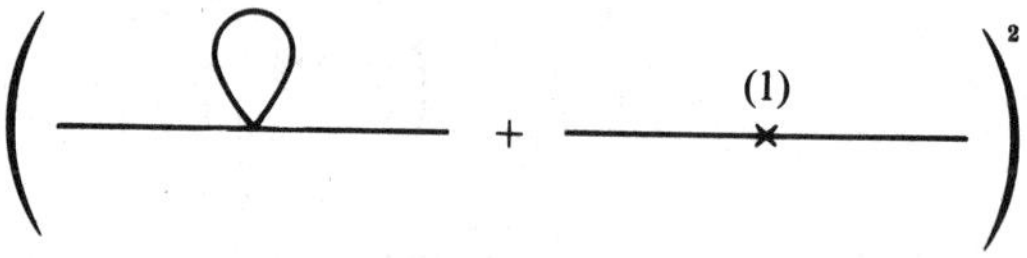

$\equiv f_2^2$

However, for the two-loop diagram alone we have

 $\equiv (1/\varepsilon^2)f_1^2 + f_2^2 + \varepsilon^2 f_3^2 + 2(1/\varepsilon)f_1 f_2 + 2f_1 f_3 + 2f_2 f_3 \varepsilon$

Discarding poles in the last result one obtains

$$f_2^2 + 2f_1 f_3 \neq f_2^2$$

which is wrong because we insist on unitarity being satisfied order-by-order in perturbation theory and therefore the finite part of the two-loop diagram should read f_2^2 as can be seen by considering the imaginary part of the amplitude, see (3.70).

5

Quantum electrodynamics

We begin our discussion of gauge theories with the simplest and the best known one: QED. So far most of the physical applications of gauge theories rely on perturbation theory, crucial for which is the renormalization programme described in Chapter 4 which is usually formulated in terms of the Green's functions. As we know from Chapter 3, to derive the Feynman rules for a gauge theory it is necessary to fix the gauge by adding some gauge-fixing conditions. All physical quantities, and the S-matrix elements in particular, are gauge independent. This fact, necessary for a consistent theory, has to be checked in each case. The Green's functions are not, however, physical objects and they are gauge dependent. Nevertheless, the gauge symmetry of the theory leads to certain constraints on the Green's functions: there exist relations between them known as the Ward–Takahashi identities. In terms of the Green's functions these are exactly the conditions which protect the physical equivalence of different gauges.

The QED lagrangian is based on the minimal coupling *Ansatz* and in terms of the bare fields and parameters it reads

$$\mathscr{L} = -\tfrac{1}{4}F^{\mathrm{B}}_{\mu\nu}F^{\mu\nu}_{\mathrm{B}} + \bar{\Psi}_{\mathrm{B}}(\mathrm{i}\partial\!\!\!/ - e_{\mathrm{B}}A\!\!\!/_{\mathrm{B}})\Psi - m_{\mathrm{B}}\bar{\Psi}_{\mathrm{B}}\Psi_{\mathrm{B}} + \mathscr{L}_{\mathrm{G}} \tag{5.1}$$

where $e < 0$ and $\mathscr{L}_{\mathrm{G}}$ is a gauge-fixing term. We will work in the class of covariant gauges, so that

$$\mathscr{L}_{\mathrm{G}} = -\frac{1}{2a_{\mathrm{B}}}(\partial_\mu A^\mu_{\mathrm{B}})^2$$

To begin, we consider the bare, regularized (that is finite) Green's functions. It is important that there exist regularization procedures which preserve the gauge symmetry of the theory. The 't Hooft–Veltman dimensional regularization defined for perturbative expansion (we can think, for instance, about perturbative expansion for the bare Green's function) belongs to this category. Thus, the results of our formal manipulations should always be understood as valid order-by-order in perturbation theory.

From the lagrangian (5.1) we can derive the Ward–Takahashi identities for the bare, regularized Green's functions of our theory. Using these identities one can then prove, allowing all parameters and fields to become renormalized, that the theory is renormalizable without destroying its gauge invariance. Leaving aside these formal considerations (consult e.g. Bogoliubov & Shirkov 1959, Slavnov & Faddeev 1980, Collins 1984) we simply rewrite the lagrangian (5.1) in terms of the renormalized quantities assuming the most general form consistent with gauge invariance after renormalization and we derive Ward identities for the renormalized Green's functions (actually, using these Ward identities one can also prove *a posteriori* that the renormalized Green's functions are finite).

The gauge invariant structure of the QED lagrangian is not destroyed when each gauge-invariant piece of it is multiplied by a constant. In addition the parameter e can be arbitrarily changed. Thus the most general gauge-invariant form for the lagrangian (5.1) written in terms of the renormalized quantities reads†

$$\mathscr{L} = -\tfrac{1}{4}Z_3 F_{\mu\nu}F^{\mu\nu} + Z_2\bar{\Psi}[i\partial\!\!\!/ - e(Z_1/Z_2)A\!\!\!/]\Psi - Z_0 m\bar{\Psi}\Psi - (1/2a)(\partial_\mu A^\mu)^2 \quad (5.2)$$

(as we shall see shortly, there is no need for a gauge-fixing counterterm). The relations between bare and renormalized quantities are as follows

$$A_\mu^{\mathrm{B}} = Z_3^{1/2} A_\mu, \quad \Psi_{\mathrm{B}} = Z_2^{1/2}\Psi, \quad e_{\mathrm{B}} = (Z_1/Z_2 Z_3^{1/2})e$$
$$m_{\mathrm{B}} = (Z_0/Z_2)m = (m - \delta m)/Z_2, \quad a_{\mathrm{B}} = Z_3 a$$

Apart from the gauge-fixing term, lagrangian (5.2) is invariant under the gauge transformations (1.27) and (1.30)

$$\left.\begin{aligned} \Psi'(x) &= \exp[-ie(Z_1/Z_2)\Theta(x)]\Psi(x) \\ A'(x) &= A_\mu(x) + \partial_\mu\Theta(x) \end{aligned}\right\} \quad (5.3)$$

If a free scalar ghost field is included the full lagrangian is invariant under BRS symmetry transformation (see Problem 3.3) on renormalized fields; the ghost field, being free, does not undergo renormalization. The renormalization constants Z_i must be fixed by imposing certain renormalization conditions. Our discussion in Chapter 4 of the renormalization programme applies to the present case. In addition it is clear that Z_1/Z_2 must be finite as all other quantities in (5.3) are finite (make this argument precise by considering BRS invariance and taking into account that the ghost field is not renormalized; compare with QCD, Section 8.1). Therefore $Z_1 = Z_2$ up to finite terms. This result will be derived formally using the Ward–Takahashi identities. Actually in most popular renormalization schemes $Z_1 = Z_2$ including finite terms.

Mainly for historical reasons, the so-called on-shell renormalization scheme is the one most commonly used in QED calculations. In this scheme we choose to interpret the coupling constant e, the mass parameter m and the renormalized fields which appear in the lagrangian (5.2) as the electric charge, the physical mass

† Strictly speaking one should include the scalar ghost field and refer to the BRS invariance (see Problem 3.3).

of the fermion and the 'physical' fields in the sense of Chapter 4, respectively. The renormalization conditions are then imposed so as to assure this interpretation of the renormalized quantities in any order of perturbation theory. Generally speaking, this can be achieved by demanding that in the on-shell limit, when all the four-momenta approach their on-shell values, in any order of perturbation expansion, the fermion and the photon propagator as well as the fermion–fermion–photon vertex function approach their zeroth order form. However, an infinite number of other renormalization prescriptions are possible. The renormalized quantities lose their previous interpretation but the theory can be formulated in terms of them in a totally equivalent way. In addition, in each order of perturbation theory there exist definite relations between sets of parameters corresponding to different renormalization schemes.

5.1 Ward–Takahashi identities

General derivation by the functional technique

We derive the Ward–Takahashi identities in QED using the functional technique. The generating functional for the full Green's functions is the following

$$W[J,\alpha,\bar{\alpha}] = \int \mathscr{D}(A_\mu \bar{\Psi}\Psi) \exp(\mathrm{i}S_{\mathrm{eff}}) \tag{5.4}$$

where

$$S_{\mathrm{eff}} = \int \mathrm{d}^4x [\mathscr{L}(A_\mu, \Psi) + J_\mu A^\mu + \bar{\alpha}\Psi + \bar{\Psi}\alpha] \tag{5.5}$$

and the lagrangian $\mathscr{L}(A_\mu, \Psi)$ is given by (5.2). Under the functional integral one can perform an infinitesimal gauge transformation

$$\left.\begin{aligned} \Psi' &= [1 - \mathrm{i}\bar{e}\Theta]\Psi \\ A'_\mu &= A_\mu + \partial_\mu \Theta \end{aligned}\right\} \tag{5.6}$$

where $\bar{e} = eZ_1/Z_2$. The integration measure is invariant under infinitesimal gauge transformation and the only gauge-dependent terms in the integrand are source terms and the gauge-fixing term. On the other hand a change of the integration variables in (5.4) cannot change the value of the integral. Therefore we conclude that

$$\delta W/\delta\Theta(y)|_{\theta=0} = 0 \tag{5.7}$$

where

$$\begin{aligned} \delta W[J,\alpha,\bar{\alpha},\Theta] = \mathrm{i} \int \mathrm{d}^4x \int & \mathscr{D}(A_\mu \bar{\Psi}\Psi) \exp(\mathrm{i}S_{\mathrm{eff}})[J_\mu(x)\partial^\mu\Theta(x) - \mathrm{i}\bar{e}\Theta(x)\bar{\alpha}(x)\Psi(x) \\ & + \mathrm{i}\bar{e}\Theta(x)\bar{\Psi}(x)\alpha(x) - (1/a)\partial_\mu^{(x)} A^\mu(x)\partial^2\Theta(x)] \end{aligned} \tag{5.8}$$

Integrating expression (5.8) by parts, neglecting the surface terms at infinity and remembering that functional integrals of derivatives of fields are derivatives of the Green's functions (see Section 2.3) we get

$$\frac{1}{a}\partial^2\partial_\mu \frac{1}{\mathrm{i}}\frac{\delta W}{\delta J_\mu(y)} + \partial_\mu J^\mu(y) W + e\bar{\alpha}(y)\frac{\delta W}{\delta \bar{\alpha}(y)} + \bar{e}\frac{\delta W}{\delta \alpha(y)}\alpha(y) = 0 \tag{5.9}$$

(remember that $(\vec{\delta}/\delta\alpha)W = -W(\overleftarrow{\delta}/\delta\alpha))$ or in terms of the generating functional $Z[J, \alpha, \bar{\alpha}]$ for the connected Green's functions ($W = \exp(\mathrm{i}Z)$)

$$\frac{1}{a}\partial^2\partial_\mu \frac{\delta Z}{\delta J_\mu(y)} + \partial_\mu J^\mu(y) + \mathrm{i}e\bar{\alpha}(y)\frac{\delta Z}{\delta \bar{\alpha}(y)} + \mathrm{i}\bar{e}\frac{\delta Z}{\delta \alpha(y)}\alpha(y) = 0 \tag{5.10}$$

Equation (5.10) is the general form of the Ward–Takahashi identities in QED, in a class of covariant gauges, for the connected Green's functions. It is also very useful to have the analogous relation for the 1PI Green's functions. The generating functional for the latter has been defined in Section 2.6. In the present case its arguments are the 'classical' boson field $A_\mu(x)$ and the 'classical' fermion fields $\Psi(x)$ and $\bar{\Psi}(x)$ (it is convenient to use the same notation for the arguments of Γ as for the fields in the lagrangian; see Section 2.6) so that

$$\Gamma[A_\mu(x), \Psi(x), \bar{\Psi}(x)] = Z[J, \alpha\bar{\alpha}] - \int \mathrm{d}^4x[J_\mu(x)A^\mu(x) + \bar{\alpha}(x)\Psi(x) + \bar{\Psi}(x)\alpha(x)] \tag{5.11}$$

where (left derivatives only)

$$\left.\begin{aligned} \frac{\delta Z}{\delta J_\mu} &= A_\mu, & \Psi &= \frac{\delta Z}{\delta \bar{\alpha}}, & \bar{\Psi} &= -\frac{\delta Z}{\delta \alpha} \\ J_\mu &= -\frac{\delta \Gamma}{\delta A^\mu}, & \bar{\alpha} &= \frac{\delta \Gamma}{\delta \Psi}, & \alpha &= -\frac{\delta \Gamma}{\delta \bar{\Psi}} \end{aligned}\right\} \tag{5.12}$$

In terms of the functional $\Gamma[A_\mu, \Psi, \bar{\Psi}]$ the identity (5.10) can be rewritten as follows

$$\frac{1}{a}\partial^2_{(y)}\partial^{(y)}_\mu A^\mu(y) - \partial^{(y)}_\mu \frac{\delta\Gamma}{\delta A_\mu(y)} + \mathrm{i}\bar{e}\frac{\delta\Gamma}{\delta\Psi(y)}\Psi(y) + \mathrm{i}e\bar{\Psi}(y)\frac{\delta\Gamma}{\delta\bar{\Psi}(y)} = 0 \tag{5.13}$$

In the following we shall illustrate the general relations (5.10) and (5.13) with several examples.

Examples

As the first example we consider the Ward identity for the longitudinal part of the connected photon propagator. Differentiating relation (5.10) with respect to $J_\nu(z)$ and putting $J = \alpha = \bar{\alpha} = 0$ one gets

$$-\frac{1}{a}\partial^2_{(y)}\partial^\mu_{(y)}\frac{\delta^2 Z}{\delta J^\mu(y)\delta J_\nu(z)} = \partial^\nu_{(y)}\delta(y-z) \tag{5.14}$$

Using (2.134) and Fourier transforming into momentum space we obtain the following relation

$$\mathrm{i}(1/a)k^2 k^\mu \tilde{G}_{\mu\nu}(k) = k_\nu \tag{5.15}$$

Here $\tilde{G}_{\mu\nu}(k)$ is the Fourier transform of the connected part of the photon propagator $G_{\mu\nu}(x_1 - x_2)$. The solution to (5.15) reads

$$\tilde{G}_{\mu\nu}(k) = -\mathrm{i}a(k_\mu k_\nu/k^4) + \tilde{G}^{\mathrm{T}}_{\mu\nu}(k) \tag{5.16}$$

where the transverse projection

$$\tilde{G}^{\mathrm{T}}_{\mu\nu}(k) = (g_{\mu\nu} - k_\mu k_\nu/k^2) f(k^2) \tag{5.17}$$

gives zero contribution to the relation (5.15). The Ward identity (5.15) constrains the full renormalized propagator to the form (5.16). We observe that the longitudinal (gauge-dependent) part is not altered by interactions (compare with (3.32)). It means, among other things, that the renormalization programme does not require any counterterms for the gauge fixing term.

Our next example is the well-known relation between the vertex and the fermion propagators. Differentiating the general identity (5.10) with respect to

$$\frac{\delta}{\delta\bar{\alpha}(x)}\frac{\delta}{\delta\alpha(z)} = -\frac{\delta}{\delta\alpha(z)}\frac{\delta}{\delta\bar{\alpha}(x)}$$

and setting $J = \bar{\alpha} = \alpha = 0$ one gets

$$\frac{1}{a}\partial^2_{(y)}\partial^{(y)}_\mu \frac{\delta^2}{\delta\bar{\alpha}(x)\delta\alpha(z)}\frac{\delta}{\delta J_\mu(y)} Z = \mathrm{i}\bar{e}\frac{\delta^2 Z}{\delta\alpha(z)\delta\bar{\alpha}(y)}\delta(y-x) + \mathrm{i}\bar{e}\frac{\delta^2 Z}{\delta\bar{\alpha}(x)\delta\alpha(y)}\delta(y-z) \tag{5.18}$$

Since

$$\langle 0|T\Psi(x)\bar{\Psi}(y)|0\rangle = \mathrm{i}\left.\frac{\delta^2 Z}{\delta\bar{\alpha}(x)\delta\alpha(y)}\right|_{J=\alpha=\bar{\alpha}=0}$$

and

$$\langle 0|TA_\mu(y)\Psi(x)\bar{\Psi}(z)|0\rangle = \left.\frac{\delta^3 Z}{\delta J_\mu(y)\delta\bar{\alpha}(x)\delta\alpha(z)}\right|_{J=\alpha=\bar{\alpha}=0}$$

relation (5.18) can be rewritten as follows

$$\begin{aligned}&-(1/a)\partial^2_{(y)}\partial^\mu_{(y)}\langle 0|TA_\mu(y)\Psi(x)\bar{\Psi}(z)|0\rangle\\ &= \bar{e}\langle 0|T\Psi(y)\bar{\Psi}(z)|0\rangle\delta(y-x) - \bar{e}\langle 0|T\Psi(x)\bar{\Psi}(y)|0\rangle\delta(y-z)\end{aligned} \tag{5.19}$$

This is the Ward identity in configuration space. Using translational invariance and Fourier transforming into momentum space

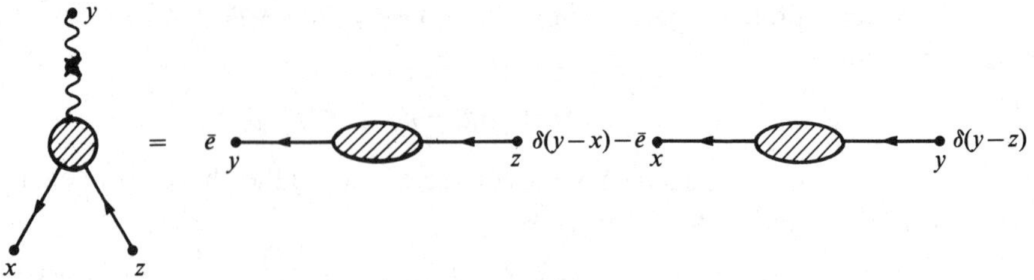

Fig. 5.1

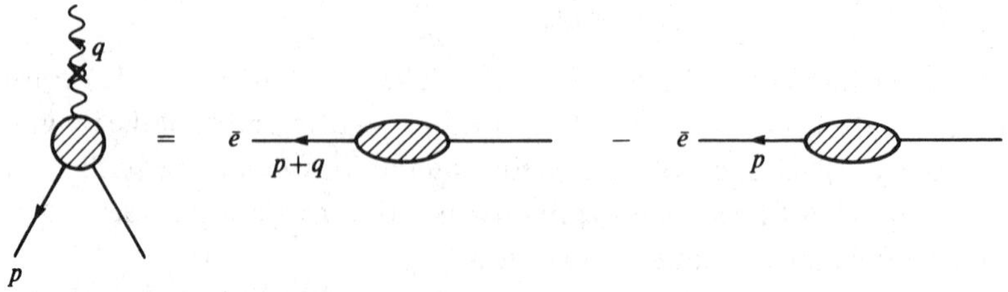

Fig. 5.2

$$\left.\begin{aligned}\langle 0|TA_\mu(y)\Psi(x)\bar{\Psi}(z)|0\rangle &= \langle 0|TA_\mu(y-z)\Psi(x-z)\bar{\Psi}(0)|0\rangle\\ &= \int\frac{\mathrm{d}^4p}{(2\pi)^4}\frac{\mathrm{d}^4q}{(2\pi)^4}\exp[-\mathrm{i}p(x-z)]\\ &\quad\times\exp[-\mathrm{i}q(y-z)]V_\mu(p,q)\\ \langle 0|T\Psi(y-z)\bar{\Psi}(0)|0\rangle &= \int\frac{\mathrm{d}^4k}{(2\pi)^4}\exp[-\mathrm{i}k(y-z)]\mathrm{i}S(k)\\ \langle 0|T\Psi(x-y)\bar{\Psi}(0)|0\rangle &= \int\frac{\mathrm{d}^4k}{(2\pi)^4}\exp[-\mathrm{i}k(x-y)]\mathrm{i}S(k)\end{aligned}\right\}\tag{5.20}$$

then inserting (5.20) into (5.19) and integrating over $\int\mathrm{d}x\,\mathrm{d}y\,\mathrm{d}z\exp(\mathrm{i}p'x)\exp(\mathrm{i}q'y)$ $\exp(\mathrm{i}k'z)$ to get rid of the δ-functions, we obtain the Ward identity in momentum space

$$-(1/a)q^2q^\mu V_\mu(p,q) = \bar{e}S(p+q) - \bar{e}S(p) \tag{5.21}$$

Graphically, relation (5.19) can be represented by Fig. 5.1 and relation (5.21) in momentum space by Fig. 5.2 where the cross denotes $-(1/a)\partial^2\partial^\mu$ or $-(1/a)q^2q^\mu$, respectively. Notice that the form (5.21) in momentum space can be immediately deduced from the graphical representation of relation (5.19) in configuration space.

The Ward identity (5.21) can also be written in terms of the 1PI Green's functions. We take the derivative of (5.13) with respect to

$$\frac{\delta}{\delta\bar{\Psi}(x)}\frac{\delta}{\delta\Psi(z)}$$

set $A_\mu = \bar{\Psi} = \Psi = 0$ and use the definitions (2.135) generalized to include fermions to get the following

$$\partial^\mu_{(x)} \Gamma^{(3)}_\mu(y, x, z) + i\bar{e}\Gamma^{(2)}(x, y)\delta(y - z) - i\bar{e}\Gamma^{(2)}(y, z)\delta(y - x) = 0 \tag{5.22}$$

By analogy with (5.21) and (5.19), in momentum space relation (5.22) reads

$$q_\mu \tilde{\Gamma}^{(3)}(p, q) = i\bar{e}\tilde{\Gamma}^{(2)}(p + q) - i\bar{e}\tilde{\Gamma}^{(2)}(p) \tag{5.23}$$

where the $\tilde{\Gamma}^{(n)}$s are Fourier transforms of the $\Gamma^{(n)}$s. We recall (2.152): $\tilde{\Gamma}^{(2)} = i[\tilde{G}^{(2)}(p)]^{-1} = +S^{-1}(k)$ where, e.g. for the free propagator, $S(k) = 1/(\not{k} - m)$.

As an interesting consequence of the Ward identities (5.21) or (5.23) we notice that the gauge parameter $\bar{e} = eZ_1/Z_2$ and consequently the ratio Z_1/Z_2 must be finite if the theory is renormalizable. Indeed all other quantities, including e, in these equations are the renormalized quantities. In perturbative calculations, using a regularization procedure which preserves gauge invariance, the Z_is are power series in the coupling constant $\alpha = e^2/4\pi$: $(Z_i - 1) = \sum_{n=1}^{\infty} f_i^{(n)} \alpha^n$. The coefficients $f_i^{(n)}$ contain pieces which are divergent in the limit where regularization is absent (e.g. when $\varepsilon \to 0$ in dimensional regularization) and possibly also finite pieces. Thus the finiteness of the ratio Z_1/Z_2 implies that order-by-order in perturbation theory $Z_1 = Z_2$ up to the finite terms. In practice, in all the commonly used renormalization schemes $Z_1 = Z_2$ including the finite terms. This is convenient (but not necessary) because the parameter e is then universal for all fermions in the theory. It is also worth remembering that only in the on-shell renormalization scheme is e the electric charge, that is $e^2/4\pi = 1/137$. This point will be discussed in more detail in Section 6.4. In addition, when $Z_1 = Z_2$ the counterterms form a gauge-invariant set and the Ward identities for the renormalized Green's functions are identical to those for the bare regularized Green's functions. The latter can of course be derived in a similar way to the renormalized Ward identities. The only difference is that one must start with the lagrangian expressed in terms of the bare quantities.

5.2 Lowest order QED radiative corrections by the dimensional regularization technique

General introduction

We recall the QED lagrangian ($e < 0$) in a covariant gauge

$$\begin{aligned}\mathscr{L} = &-\tfrac{1}{4}F_{\mu\nu}F^{\mu\nu} + \bar{\Psi}(i\not{\partial} - e\not{A})\Psi - m\bar{\Psi}\Psi - (1/2a)(\partial_\mu A^\mu)^2 - (Z_3 - 1)\tfrac{1}{4}F_{\mu\nu}F^{\mu\nu} \\ &+ (Z_2 - 1)\bar{\Psi}i\not{\partial}\Psi - (Z_0 - 1)\bar{\Psi}m\Psi - (Z_1 - 1)e\bar{\Psi}\not{A}\Psi\end{aligned} \tag{5.24}$$

(there is no counterterm to the gauge-fixing term: see Section 5.1). The Feynman rules are collected in Appendix A.

Dimensional regularization in QED in the presence of γ-matrices requires an extension of previous rules. We notice that e.g. in the relation

$$\gamma_\mu\gamma_\nu + \gamma_\nu\gamma_\mu = 2g_{\mu\nu}\mathbb{1}$$

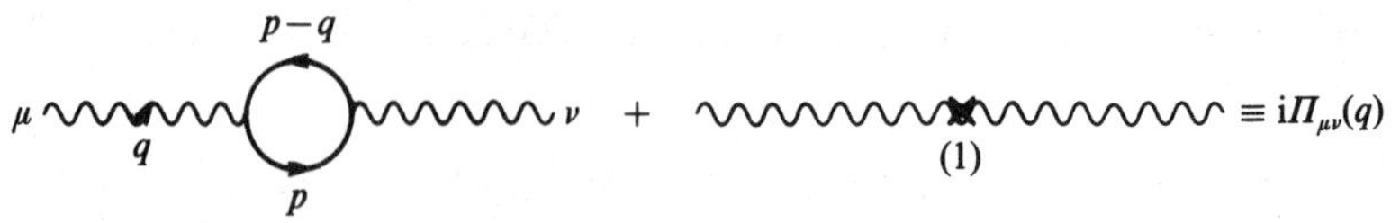

Fig. 5.3

the only space-time-dependent object is the tensor $g_{\mu\nu}$. In n dimensions we can assume

$$\mathrm{Tr}[\gamma_\mu\gamma_\nu] = \tfrac{1}{2}\mathrm{Tr}[\gamma_\mu\gamma_\nu + \gamma_\nu\gamma_\mu] = g_{\mu\nu}\,\mathrm{Tr}\,\mathbb{1} = 4g_{\mu\nu} \tag{5.25}$$

i.e. we still consider γ_μs as 4×4 matrices in the fermion and antifermion spin space, and define

$$g^\mu{}_\mu = g^{\mu\nu}g_{\mu\nu} = n$$

This convention is the one most frequently used in calculations but other conventions are also possible (Collins 1984). In particular, in supersymmetric theories a dimensional regularization method called dimensional reduction (Siegel 1979) is used which preserves supersymmetry. This is not the case for the standard dimensional regularization convention used here. Several other relations and integrals useful in the subsequent calculations are collected in Appendix A.

Vacuum polarization

We calculate the amplitude corresponding to the diagrams of Fig. 5.3. From consideration of Section 5.1 we know that only the transverse part of the photon propagator has higher order corrections, so on general grounds one can write it as follows:

$$\Pi_{\mu\nu}(q) = (q_\mu q_\nu - g_{\mu\nu}q^2)\Pi(q^2) \tag{5.26}$$

The scalar function $\Pi(q^2)$ is given by

$$g^{\mu\nu}\Pi_{\mu\nu}(q) = (1-n)q^2\Pi(q^2) \tag{5.27}$$

where from the Feynman rules we obtain

$$\Pi_{\mu\nu}(q) = (-\mathrm{i})e^2\int\frac{\mathrm{d}^n p}{(2\pi)^n}\frac{\mathrm{i}}{p^2-m^2}\frac{\mathrm{i}}{(p-q)^2-m^2}\mathrm{Tr}[\gamma_\mu(\not{p}+m)\gamma_\nu(\not{p}-\not{q}+m)]$$
$$+(Z_3-1)^{(1)}(q_\mu q_\nu - q^2 g_{\mu\nu}) \tag{5.28}$$

Therefore, after taking the trace

$$\mathrm{Tr}[\gamma_\mu\gamma_\alpha\gamma_\nu\gamma_\beta] = 4(g_{\mu\alpha}g_{\nu\beta} + g_{\mu\beta}g_{\nu\alpha} - g_{\mu\nu}g_{\alpha\beta})$$

and using the Feynman parametrization $1/ab = \int_0^1 \mathrm{d}x(1/[xa+(1-x)b]^2)$ one gets

$$(1-n)q^2\Pi(q^2) = \mathrm{i}e^2\int\frac{\mathrm{d}^n p}{(2\pi)^n}\int_0^1 \mathrm{d}x\frac{4(2-n)(p^2-p\cdot q)+4nm^2}{\{x(p^2-m^2)+(1-x)[(p-q)^2-m^2]\}^2}$$
$$+\,\text{counterterm} \tag{5.29}$$

Using the integrals (A.18)–(A.20) we get finally

$$\Pi(q^2) = \frac{\alpha}{\pi}(4\pi)^{\varepsilon/2}\Gamma(\tfrac{1}{2}\varepsilon)\int_0^1 \mathrm{d}x 2x(1-x)\left[\frac{m^2 - x(1-x)q^2 - \mathrm{i}\varepsilon}{\mu^2}\right]^{-\varepsilon/2}$$

$$= \frac{\alpha}{\pi}\left[\frac{1}{\varepsilon}\frac{2}{3} + \frac{1}{3}(\ln 4\pi - \gamma) - \int_0^1 \mathrm{d}x 2x(1-x)\ln\frac{m^2 - x(1-x)q^2 - \mathrm{i}\varepsilon}{\mu^2}\right]$$
$$+ O(\varepsilon) + (Z_3 - 1)^{(1)} \tag{5.30}$$

(we have taken $n = 4 - \varepsilon$, $\alpha = e^2/4\pi$, and α is dimensionless in $4 - \varepsilon$ dimensions: $\alpha \to \alpha\mu^\varepsilon$ where μ is an arbitrary mass parameter). The pole in ε reflects the UV divergence in four dimensions. It is cancelled by adding the Z_3 counterterm. In the minimal subtraction scheme

$$(Z_3 - 1)^{(1)} = -\frac{\alpha}{\pi}\frac{2}{3}\frac{1}{\varepsilon} \tag{5.31}$$

and the finite part of the vacuum polarization, given by (5.30), can easily be evaluated in the small q^2 and large q^2 limits and reads (up to constant terms)

$$\left.\begin{array}{ll} -\dfrac{\alpha}{3\pi}\ln\dfrac{m^2}{\mu^2} + \dfrac{\alpha}{\pi}\dfrac{q^2}{m^2}\dfrac{1}{15} + O\left(\dfrac{q^4}{m^4}\right) & q^2 \to 0 \\ -\dfrac{\alpha}{3\pi}\ln\dfrac{|q^2|}{\mu^2} \qquad |q^2| \gg m^2 & \end{array}\right\} \tag{5.32}$$

The renormalization scheme most frequently used in QED is the on-shell renormalization. In this case one demands:

$$\Pi^{\mathrm{ON}}(q^2) = 0 \quad \text{for} \quad q^2 = 0$$

Therefore, from (5.30)

$$(Z_3 - 1)^{(1)} = -\frac{\alpha}{\pi}\left[\frac{2}{3}\frac{1}{\varepsilon} + \tfrac{1}{3}(\ln 4\pi - \gamma) - \tfrac{1}{3}\ln\frac{m^2}{\mu^2}\right] \tag{5.33}$$

and

$$\Pi^{\mathrm{ON}}(q^2) = -\frac{\alpha}{\pi}2\int_0^1 \mathrm{d}x\, x(1-x)\ln\frac{m^2 - x(1-x)q^2 - \mathrm{i}\varepsilon}{m^2} \tag{5.34}$$

Asymptotically one obtains

$$\left.\begin{array}{l} \Pi^{\mathrm{ON}}(q^2) \underset{q^2 \to 0}{\longrightarrow} \dfrac{\alpha}{\pi}\dfrac{q^2}{m^2}\dfrac{1}{15} + O\left(\dfrac{q^4}{m^4}\right) \\ \Pi^{\mathrm{ON}}(q^2) \underset{-q^2/m^2 \to \infty}{\longrightarrow} \dfrac{\alpha}{\pi}\left[-\tfrac{1}{3}\ln\dfrac{|q^2|}{m^2} + \dfrac{5}{9} + O\left(\dfrac{m^2}{q^2}\right)\right] \end{array}\right\} \tag{5.35}$$

As a final comment are rewrite $\Pi^{\mathrm{ON}}(q^2)$ in the form of a dispersion integral (the dispersion technique will be discussed in some detail later). Firstly, we change

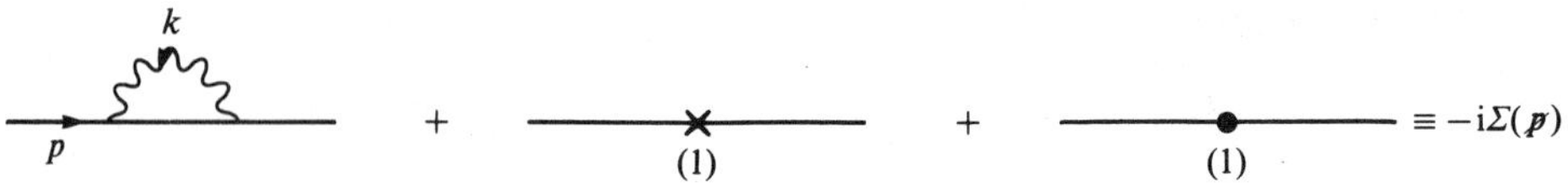

Fig. 5.4

variables in the integral (5.34) into $y = 1 - 2x$. Taking into account the symmetry $y \to -y$ of the integrand we can write

$$\Pi^{\mathrm{ON}}(q^2) = -\frac{\alpha}{\pi}\frac{1}{2}\int_0^1 \mathrm{d}y\left[\frac{\partial}{\partial y}(y - \tfrac{1}{3}y^3)\right]\ln\left[1 - \frac{q^2(1-y^2)}{4m^2 - \mathrm{i}\varepsilon}\right] \tag{5.36}$$

and integrate it by parts to get

$$\Pi^{\mathrm{ON}}(q^2) = \frac{\alpha}{\pi}\frac{1}{2}\int_0^1 \mathrm{d}y 2y(y - \tfrac{1}{3}y^3)\frac{q^2}{4m^2 - q^2(1-y^2) - \mathrm{i}\varepsilon} \tag{5.37}$$

Another change of variable $t = 4m^2/(1-y^2)$ gives the result

$$\Pi^{\mathrm{ON}}(q^2) = \frac{\alpha}{\pi}q^2\int_{4m^2}^{\infty}\frac{\mathrm{d}t}{t}\frac{1}{t - q^2 - \mathrm{i}\varepsilon}\frac{1}{3}\left(1 + \frac{2m^2}{t}\right)\left(1 - \frac{4m^2}{t}\right)^{1/2} \tag{5.38}$$

which is a so-called once subtracted dispersion relation.

Electron self-energy correction

We consider the electron self-energy truncated diagram and the necessary counterterms which are shown in Fig. 5.4. The proper two-point function $\Gamma^{(2)}(\not{p})$ reads

$$\Gamma^{(2)}(\not{p}) = -\mathrm{i}(\not{p} - m) - \mathrm{i}\Sigma(\not{p}, m) \tag{5.39}$$

The propagator is the inverse of the proper two-point function so we get

$$\hat{G}^{(2)}(\not{p}) = \frac{\mathrm{i}}{\not{p} - m - \Sigma} \tag{5.40}$$

The general structure of the electron self-energy correction is obviously the following

$$\Sigma(\not{p}, m) = A(p^2)I + B(p^2)\not{p} \tag{5.41}$$

It is often convenient to write it as an expansion

$$\Sigma(\not{p}, m) = \Sigma_0(\not{p} = m_F, m) + (\not{p} - m_F)\Sigma_1(\not{p} = m_F, m) + (\not{p} - m_F)^2\Sigma_2(\not{p}, m) \tag{5.42}$$

where m_F^2 is the zero of the inverse propagator to the order considered. We recall that Σ_1 (actually $\frac{1}{2}\Sigma_1$: see Section 4.1) appears as a correction to an external line of the on-shell scattering amplitude. It can be calculated as ($p^2 = \not{p}^2$)

$$\Sigma_1 = \left.\frac{\partial\Sigma}{\partial\not{p}}\right|_{\not{p}=m_F} = \left.\frac{\partial A}{\partial p^2}\right|_{p^2=m_F} 2m_F + B(m_F^2) + 2m_F^2\left.\frac{\partial B(p^2)}{\partial p^2}\right|_{p^2=m_F^2} \tag{5.43}$$

In this Section we shall calculate $\Sigma(\not{p})$ in the Feynman gauge (contrary to $\Pi_{\mu\nu}(q)$, the electron self-energy is gauge dependent). The calculation proceeds as follows. The Feynman integral reads.

$$-i\Sigma(\not{p}) = (-ie)^2 \int \frac{d^n k}{(2\pi)^n} \gamma^\mu \frac{-ig_{\mu\nu}}{k^2 + i\varepsilon} \frac{i}{\not{p} + \not{k} - m + i\varepsilon} \gamma^\nu + im(Z_0 - 1)^{(1)} + i(Z_2 - 1)^{(1)} \not{p} \tag{5.44}$$

As before we use dimensional regularization: $n = 4 - \varepsilon$. The functions $A(p^2)$ and $B(p^2)$ are given by

$$\left.\begin{aligned} \mathrm{Tr}\,\Sigma(\not{p}) &= 4A(p^2) \\ \mathrm{Tr}\,[\not{p}\Sigma(\not{p})] &= 4p^2 B(p^2) \end{aligned}\right\} \tag{5.45}$$

Introducing

$$N = \gamma^\mu(\not{p} + \not{k} + m)\gamma^\nu g_{\mu\nu}$$

we find (after some algebra)

$$\left.\begin{aligned} \mathrm{Tr}\,N &= 4nm \\ \mathrm{Tr}\,[\not{p}N] &= 4[p^2(2-n) + (p\cdot k)(2-n)] \end{aligned}\right\} \tag{5.46}$$

Next, we use Feynman parametrization for the denominators

$$\frac{1}{k^2} \frac{1}{(p+k)^2 - m^2} = \int_0^1 dx \frac{1}{[k^2 + xp\cdot 2k + x(p^2 - m^2)]^2} \tag{5.47}$$

and (A.18)–(A.20) to perform the integration over momentum k to get the following results

$$\left.\begin{aligned} A &= nmC \int_0^1 dx\, X - (Z_0 - 1)^{(1)} m \\ B &= (2-n)C \int_0^1 dx(1-x)X - (Z_2 - 1)^{(1)} \end{aligned}\right\} \tag{5.48}$$

where

$$C \equiv \frac{\alpha}{4\pi}\left(\frac{-1}{4\pi}\right)^{-\varepsilon/2} \Gamma(\tfrac{1}{2}\varepsilon)$$

$$X = \left[\frac{x(p^2 - m^2) - x^2 p^2}{\mu^2}\right]^{-\varepsilon/2}$$

Using standard tricks to single out the pole terms in ε we immediately find the renormalization constants in the minimal subtraction scheme

$$\left.\begin{aligned} (Z_0 - 1)^{(1)} &= \frac{2\alpha}{\pi}\frac{1}{\varepsilon} \\ (Z_2 - 1)^{(1)} &= -\frac{\alpha}{2\pi}\frac{1}{\varepsilon} \end{aligned}\right\} \tag{5.49}$$

The final (finite) expressions for A and B can then be easily written down, e.g.

$$A(p^2) = -\frac{\alpha}{\pi} m \left[\int \mathrm{d}x \ln \frac{xm^2 - x(1-x)p^2}{4\pi\mu^2} + \gamma + \tfrac{1}{2} \right] \tag{5.50}$$

and similarly for $B(p^2)$. We observe that A and B are IR finite (when $p^2 \to m^2$). However, their derivatives contributing to Σ_1 are not. Indeed, differentiating the regularized expression (5.48) with respect to p^2 one obtains the following result

$$\frac{\partial A}{\partial p^2} = nmC(-\tfrac{1}{2}\varepsilon) \int_0^1 \mathrm{d}x \frac{x(1-x)}{\mu^2} X^{1+2/\varepsilon} \tag{5.51}$$

which is of course UV finite (Z_i are constants) and in the limit $p^2 \to m^2$ gives (we can take $m_F^2 = m^2$ because we work in the first order in α)

$$\left.\frac{\partial A}{\partial p^2}\right|_{p^2=m^2} = n\frac{1}{m}\frac{\alpha}{4\pi}\left(1-\gamma\frac{\varepsilon}{2}\right)\left(\frac{m^2}{4\pi\mu^2}\right)^{-\varepsilon/2}\frac{\Gamma(-\varepsilon)\Gamma(2)}{\Gamma(2-\varepsilon)} \tag{5.52}$$

The integration over Feynman parameters has introduced a new (IR) singularity which has again been regularized by keeping $n = 4 - \varepsilon$, but this time with $\varepsilon < 0$ (changing ε from a negative to a positive value is legitimate since we already deal with a UV finite quantity). In order to keep trace of the origin of the singular terms here and in the following we shall often distinguish between $\varepsilon_{\mathrm{UV}}$ ($n = 4 - \varepsilon_{\mathrm{UV}}, \varepsilon_{\mathrm{UV}} > 0$) and $\varepsilon_{\mathrm{IR}}$ ($n = 4 + \varepsilon_{\mathrm{IR}}, \varepsilon_{\mathrm{IR}} > 0$).

The final result reads ($\varepsilon_{\mathrm{IR}} > 0$)

$$\left.\frac{\partial A}{\partial p^2}\right|_{p^2=m^2} = \frac{1}{m}\frac{\alpha}{4\pi}\left[4\left(\frac{1}{\varepsilon_{\mathrm{IR}}} + \tfrac{1}{2}\gamma - 1 + \tfrac{1}{2}\ln\frac{m^2}{4\pi\mu^2}\right) + 1\right] \tag{5.53}$$

and for $\partial B/\partial p^2$

$$\left.\frac{\partial B}{\partial p^2}\right|_{p^2=m^2} = -\frac{1}{2m^2}\frac{\alpha}{\pi}\left(\frac{1}{\varepsilon_{\mathrm{IR}}} - 1 + \tfrac{1}{2}\gamma + \tfrac{1}{2}\ln\frac{m^2}{4\pi\mu^2}\right) \tag{5.54}$$

In addition

$$B(p^2 = m^2) = \frac{\alpha}{4\pi}\left(\ln\frac{m^2}{4\pi\mu^2} + \gamma - 2\right) \tag{5.55}$$

so for Σ_1, we get

$$\Sigma_1 = \left.\frac{\partial \Sigma}{\partial \not{p}}\right|_{\not{p}=m} = \frac{\alpha}{\pi}\left(\frac{1}{\varepsilon_{\mathrm{IR}}} + \tfrac{3}{4}\ln\frac{m^2}{4\pi\mu^2} - 1 + \tfrac{3}{4}\gamma\right) \tag{5.56}$$

Our final remark is that although $B(p^2)$ does not have IR singularities (when $p^2 \to m^2$), it nevertheless has the so-called mass singularities (when $p^2 \to m^2 \to 0$).

Electron self-energy: IR singularities regularized by photon mass

The IR singularities encountered above are due to the zero mass of the photon. Instead of using the dimensional method one can regularize the IR singularities by giving the photon a small mass λ where eventually $\lambda \to 0$ (actually this is the most standard method).

For the electron self-energy correction we now have

$$-\mathrm{i}\Sigma(\not{p}) = -e^2 \int \frac{\mathrm{d}^n k}{(2\pi)^n} \gamma_\mu (\not{p} + \not{k} + m)\gamma^\mu \frac{1}{k^2 - \lambda^2} \frac{1}{(p+k)^2 - m^2}$$

Following the previous calculation, instead of (5.50) one gets:

$$\left.\begin{aligned} A(p^2) &= -m\frac{\alpha}{\pi}\int \mathrm{d}x \ln \frac{x^2p^2 - x(p^2 - m^2 + \lambda^2) + \lambda^2 - \mathrm{i}\varepsilon}{4\pi\mu^2} + \text{const.} \\ B(p^2) &= \frac{\alpha}{2\pi}\int \mathrm{d}x(1-x) \ln \frac{x^2p^2 - x(p^2 - m^2 + \lambda^2) + \lambda^2 - \mathrm{i}\varepsilon}{4\pi\mu^2} + \text{const.} \end{aligned}\right\} \quad (5.57)$$

These integrations can be done by writing the logarithm as $\ln p^2(x - x_1)(x - x_2)$ where x_1 and x_2 are the roots of the quadratic expressions. The derivatives of A and B can now be calculated in an elementary way. We get in particular

$$\left.\frac{\partial A}{\partial p^2}\right|_{p^2 = m^2} = m\frac{\alpha}{\pi}\frac{1}{2m^2}\ln\frac{m^2}{\lambda^2} + \text{const.} \quad (5.58)$$

$$\left.\frac{\partial B}{\partial p^2}\right|_{p^2 = m^2} = \frac{\alpha}{\pi}\left(-\frac{1}{4m^2}\right)\ln\frac{m^2}{\lambda^2} + \text{const.} \quad (5.59)$$

and finally

$$\Sigma_1 = \frac{\alpha}{\pi}\frac{1}{2}\ln\frac{m^2}{\lambda^2} + \frac{\alpha}{\pi}\frac{1}{4}\ln\frac{m^2}{4\pi\mu^2} + \text{const.} \quad (5.60)$$

Comparing with the dimensional regularization of IR singularities we observe the correspondence $2/\varepsilon_{\mathrm{IR}} \to \ln(m^2/\lambda^2)$.

On-shell vertex correction

We consider Fig. 5.5

$$\Lambda_\mu(p, p') = (-\mathrm{i}e)^2 \int \frac{\mathrm{d}^n k}{(2\pi)^n} \gamma_\alpha \frac{\mathrm{i}}{\not{p}' + \not{k} - m} \gamma_\mu \frac{\mathrm{i}}{\not{p} + \not{k} - m} \gamma_\beta (-\mathrm{i}g^{\alpha\beta}) \frac{1}{k^2} + (Z_1 - 1)^{(1)}\gamma_\mu$$
$$(5.61)$$

We work in the Feynman gauge and assume that the external electron lines are on-shell $p'^2 = p^2 = m^2$. Using the identity $\gamma^\mu \not{a} \gamma_\mu = (2-n)\not{a}$, we can write the integral as

$$\Lambda_\mu(p, p') - (Z_1 - 1)^{(1)}\gamma_\mu = -\mathrm{i}e^2 \int \frac{\mathrm{d}^n k}{(2\pi)^n} \frac{N}{[(p'+k)^2 - m^2][(p+k)^2 - m^2]k^2} \quad (5.62)$$

where

$$N = \gamma_\mu[4p\cdot p' + (n-2)k^2 + 4(p+p')\cdot k] - 4(p+p')_\mu \not{k} + k_\mu[4m - 2(n-2)\not{k}] \quad (5.63).$$

We now introduce Feynman parametrization for the propagators and then have

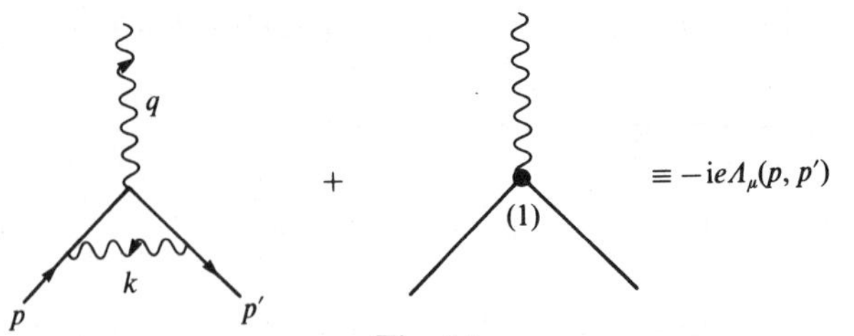

Fig. 5.5

to perform the following integrations

$$2\int_0^1 \mathrm{d}x\,\mathrm{d}y\,y\int\frac{\mathrm{d}^n k}{(2\pi)^n}\frac{[1, k_\alpha, k_\alpha k_\beta]}{[k^2+(p'xy+p(1-x)y)2k]^3} \tag{5.64}$$

Only the integral with two powers of k in the numerator is UV divergent (but IR finite). We regularize it by taking $n = 4 - \varepsilon_{\mathrm{UV}}$. The other integrals are at most only IR divergent, so we take $n = 4 + \varepsilon_{\mathrm{IR}}$ there. To simplify notation we write the final results in terms of the integrals

$$\left.\begin{aligned} I_1 &= \int_0^1 \frac{\mathrm{d}x}{-q^2x(1-x)+m^2} \qquad I_3 = \int_0^1 \frac{\mathrm{d}x\,x}{-q^2x(1-x)+m^2} \\ I_2 &= \int_0^1 \frac{\mathrm{d}x}{-q^2x(1-x)+m^2}\ln\frac{-q^2x(1-x)+m^2}{4\pi\mu^2} \end{aligned}\right\} \tag{5.65}$$

where $q = p - p'$ (we assume $q^2 \leqslant 0$) and μ^2 is an arbitrary mass scale. We get finally

$$\Lambda_\mu = \Lambda_\mu^{\mathrm{UV}} + \Lambda_\mu^{\mathrm{IR}}$$

where (in $n = 4 - \varepsilon_{\mathrm{UV}}$ dimensions)

$$\begin{aligned} \Lambda_\mu^{\mathrm{UV}} &= \gamma_\mu\frac{\alpha}{\pi}\left[\frac{1}{2\varepsilon_{\mathrm{UV}}} - \frac{1}{4}(2+\gamma) - \frac{1}{4}\int \mathrm{d}x\ln\frac{-q^2x(1-x)+m^2}{4\pi\mu^2}\right] \\ &\quad + \frac{\alpha}{\pi}\frac{m}{2}(p+p')_\mu I_3 + \gamma_\mu(Z_1-1)^{(1)} \end{aligned} \tag{5.66}$$

and (in $n = 4 + \varepsilon_{\mathrm{IR}}$ dimensions)

$$\begin{aligned} \Lambda_\mu^{\mathrm{IR}} &= \gamma_\mu\frac{\alpha}{4\pi}(2q^2-4m^2)\left[\left(\frac{1}{\varepsilon_{\mathrm{IR}}} + \tfrac{1}{2}\gamma\right)I_1 + \tfrac{1}{2}I_2\right] + \frac{\alpha}{\pi}\gamma_\mu(p+p')^2 I_3 \\ &\quad - \frac{\alpha}{\pi}(p+p')_\mu m(I_1 - I_3) \end{aligned} \tag{5.66a}$$

The above result is often rewritten in terms of the Dirac and Pauli formfactors

$$\Lambda_\mu(p,p') = \gamma_\mu F_1(q^2) + \tfrac{1}{2}(\mathrm{i}/m)\sigma_{\mu\nu}q^\nu F_2(q^2), \quad \sigma_{\mu\nu} = \tfrac{1}{2}\mathrm{i}[\gamma_\mu, \gamma_\nu] \tag{5.67}$$

Using the so-called Gordon decomposition

$$\bar{u}(p')\gamma_\mu u(p) - \tfrac{1}{2}(\mathrm{i}/m)\bar{u}(p')\sigma^{\mu\nu}q_\nu u(p) = \tfrac{1}{2}(1/m)\bar{u}(p')(p+p')^\mu u(p) \tag{5.68}$$

one gets the following

$$F_1(q^2) = \frac{\alpha}{\pi}\left[\frac{1}{2\varepsilon_{\rm UV}} - \frac{1}{4}(2+\gamma) - \frac{1}{4}\int {\rm d}x \ln\frac{-q^2x(1-x)+m^2}{4\pi\mu^2} + m^2 I_3\right] + (Z_1 - 1)^{(1)}$$
$$+ \frac{\alpha}{\pi}\left\{(\tfrac{1}{2}q^2 - m^2)\left[\left(\frac{1}{\varepsilon_{\rm IR}} + \tfrac{1}{2}\gamma\right)I_1 + \tfrac{1}{2}I_2\right] + (p+p')^2 I_3 - 2m^2(I_1 - I_3)\right\} \tag{5.69}$$

$$F_2(q^2) = \frac{\alpha}{\pi}m^2(2I_1 - 3I_3) \tag{5.70}$$

We can now make several observations. In the minimal subtraction scheme

$$(Z_1 - 1)^{(1)} = -\frac{\alpha}{2\pi}\frac{1}{\varepsilon_{\rm UV}} \tag{5.71}$$

and comparing with (5.49) we have $Z_1 = Z_2$ to this order. One can check that for the on-shell renormalization given by the conditions

$$\left.\frac{\partial\Sigma}{\partial \not{p}}\right|_{\not{p}=m} = 0; \qquad F_1(q^2 = 0) = 1$$

the relation $Z_1 = Z_2$ also holds. Indeed from (5.56) and (5.69) it follows that

$$Z_1 = Z_2 = 1 - \frac{\alpha}{\pi}\left(\frac{1}{2\varepsilon_{\rm UV}} + \tfrac{3}{4}\gamma + 1 - \tfrac{3}{4}\ln\frac{m^2}{4\pi\mu^2} - \frac{1}{\varepsilon_{\rm IR}}\right) \tag{5.72}$$

in this case (the renormalization constants now become IR divergent).

In our calculation we have reproduced the famous Schwinger result (Schwinger 1949) for the anomalous magnetic moment of the electron

$$F_2(q^2 = 0) = \alpha/2\pi \tag{5.73}$$

Notice that the radiative correction (5.73) for the anomalous magnetic moment is finite before renormalization. It depends on the renormalization scheme through the value of $\alpha(\mu^2)$ but that dependence is of the order $O(\alpha^2)$. This is an example of the general rule: all relations which hold in the tree approximation due to some restrictive assumptions on the couplings present in the lagrangian (in our case $F_2(0) = 0$ since $\sigma_{\mu\nu}F^{\mu\nu}$ coupling is assumed to be absent, although it satisfies the $U(1)$ gauge invariance of the theory) receive in higher orders only finite corrections if the theory with restrictions is multiplicatively renormalizable.

As the next point we discuss the structure of the leading logarithmic terms in our result for the vertex corrections in the limit $|-q^2| \gg m^2$. In the minimal subtraction scheme we get

$$-\frac{\alpha}{4\pi}\ln\frac{-q^2}{4\pi\mu^2}$$

as a finite contribution to F_1 from Λ_{UV} and†

$$\frac{\alpha}{\pi}\left[\frac{1}{\varepsilon_{IR}}\ln\frac{m^2}{-q^2}-\frac{1}{4}\ln^2\frac{m^2}{-q^2}+(\tfrac{1}{2}\gamma-1)\ln\frac{m^2}{-q^2}+\tfrac{1}{2}\ln\frac{-q^2}{4\pi\mu^2}\ln\frac{m^2}{-q^2}\right] \tag{5.74}$$

from Λ_{IR}. Here the first two terms are the dimensional regularization form of the famous double logarithms due to the simultaneous presence of IR and mass singularities.

The IR singularities can also be regularized by λ, the photon mass. A straightforward calculation of the leading double logarithms gives a result which corresponds to the replacement $2/\varepsilon_{IR}\to\ln(m^2/\lambda^2)$. As a final remark we notice that the double logarithm terms do not depend on the renormalization scheme (Z_i, in the on-shell renormalization, for instance, contains only single logarithms).

5.3 Massless QED

For some high energy reactions it is useful to have results in the $m=0$ approximation. We recalculate now in the Feynman gauge the mass and vertex corrections in this approximation. Consider first the self-energy correction. We have:

$$-\mathrm{i}\Sigma(\not{p})=-e^2\int\frac{\mathrm{d}^n k}{(2\pi)^n}N\frac{1}{k^2}\frac{1}{(p+k)^2} \tag{5.75}$$

where

$$N=(2-n)(\not{p}+\not{k})$$

(remember $\gamma^\mu\not{a}\gamma_\mu=(2-n)\not{a}$). Using Feynman parametrization and the integrals (A.18)–(A.20) one gets ($n=4-\varepsilon$ and $\alpha\to\alpha\mu^\varepsilon$):

$$\Sigma(\not{p})=\frac{\alpha}{\pi}\frac{1}{4}\not{p}(2-n)\left(\frac{1}{4\pi}\right)^{-\varepsilon/2}\left(\frac{-p^2}{\mu^2}\right)^{-\varepsilon/2}\Gamma(\tfrac{1}{2}\varepsilon)\int_0^1 x^{-\varepsilon/2}(1-x)^{1-\varepsilon/2}\,\mathrm{d}x$$
$$-(Z_2-1)^{(1)}\not{p}\equiv B(p^2)\not{p} \tag{5.76}$$

We can now proceed in either of two ways. One is to keep external lines off-shell $p^2<0$ to regularize the IR singularity. Then, in the minimal subtraction scheme

$$\Sigma(\not{p})=\not{p}\left(\frac{\alpha}{\pi}\frac{1}{4}\ln\frac{-p^2}{4\pi\mu^2}+\text{const}\right) \tag{5.77}$$

† It is easy to see that

$$I_1\approx\frac{-1}{-q^2}2\ln\frac{m^2}{-q^2},\qquad I_3=\tfrac{1}{2}I_1$$

$$I_2\approx\frac{-1}{-q^2}2\ln\frac{-q^2}{4\pi\mu^2}\ln\frac{m^2}{-q^2}+\frac{1}{-q^2}\left(\ln^2\frac{-q^2}{m^2}+\tfrac{1}{3}\pi^3\right)+O\left(\frac{1}{q^4}\right)$$

when $|q^2|\gg m^2$. In particular I_2 can be written in terms of the Spence functions. The logarithms can be also obtained simply by writing $\int_0^1=\int_0^\xi+\int_\xi^1$ and approximating the integrands in both integrals.

The other way is to take external lines on-shell $p^2=0$ and to use dimensional regularization of the IR singularity. Note that for massless QED the finite wave-function renormalization Σ_1 is given solely by $B(p^2=0)$ and not its derivative. We rewrite (5.76) in the following form

$$\Sigma(\not{p})=\frac{\alpha}{4\pi}(2-n)\not{p}\int_{-p^2}^{\infty}\frac{\mathrm{d}q^2}{q^2}\left(\frac{q^2}{4\pi\mu^2}\right)^{-\varepsilon/2}\Gamma(1+\tfrac{1}{2}\varepsilon)\int_0^1\mathrm{d}x\,x^{-\varepsilon/2}(1-x)^{1-\varepsilon/2}-(Z_2-1)^{(1)}\not{p} \tag{5.78}$$

For $p^2=0$ we split the integral into two: $\int_0^{\mu^2}+\int_{\mu^2}^{\infty}$ which are IR and UV divergent, respectively, when $\varepsilon\to 0$ (to regularize the first integral we work in $n=4+\varepsilon_{\mathrm{IR}}$ dimensions, so we take $\varepsilon\to-\varepsilon_{\mathrm{IR}}$). Finally, after cancelling the UV divergence by Z_2, we get† in the minimal subtraction scheme

$$B(p^2=0)=-\frac{\alpha}{4\pi}\frac{2}{\varepsilon_{\mathrm{IR}}} \tag{5.79}$$

where the pole in $\varepsilon_{\mathrm{IR}}$ reflects the IR singularity.

It is also straightforward to repeat the calculation of the vertex correction in the massless electron case. We have

$$\Lambda_\mu=-\mathrm{i}e^2\int\frac{\mathrm{d}^nk}{(2\pi)^n}\frac{N}{(p'+k)^2(p+k)^2k^2} \tag{5.80}$$

where

$$N=\gamma_\mu[4p\cdot p'+(n-2)k^2+4(p+p')\cdot k]-(p+p')_\mu 4\not{k}-2(n-2)k_\mu\not{k}$$

Repeating the steps which led us to the result (5.66), one now gets ($n=4-\varepsilon_{\mathrm{UV}}$)

$$\begin{aligned}\Lambda_\mu^{\mathrm{UV}}&=\gamma_\mu\frac{\alpha}{2\pi}\Gamma(1+\tfrac{1}{2}\varepsilon_{\mathrm{UV}})(2-\varepsilon_{\mathrm{UV}})\left(\frac{1}{\varepsilon_{\mathrm{UV}}}-1\right)\left(\frac{-q^2}{4\pi\mu^2}\right)^{-\varepsilon_{\mathrm{UV}}/2}\\&\quad*\int\mathrm{d}x\,\mathrm{d}y\,y^{1-\varepsilon_{\mathrm{UV}}}[x(1-x)]^{-\varepsilon_{\mathrm{UV}}/2}+\gamma_\mu(Z_1-1)^{(1)}\\&=\gamma_\mu\frac{\alpha}{2\pi}\left(\frac{-q^2}{4\pi\mu^2}\right)^{-\varepsilon_{\mathrm{UV}}/2}\frac{\Gamma(1-\frac{1}{2}\varepsilon_{\mathrm{UV}})}{\Gamma(1-\varepsilon_{\mathrm{UV}})}\frac{1}{\varepsilon_{\mathrm{UV}}}+\gamma_\mu(Z_1-1)^{(1)}\end{aligned} \tag{5.81}$$

and‡ ($n=4+\varepsilon_{\mathrm{IR}}$)

$$\begin{aligned}\Lambda_\mu^{\mathrm{IR}}&=\gamma_\mu\left(-\frac{\alpha}{2\pi}\right)\left(\frac{-q^2}{4\pi\mu^2}\right)^{\varepsilon_{\mathrm{IR}}/2}\Gamma(1-\tfrac{1}{2}\varepsilon_{\mathrm{IR}})\int\mathrm{d}y\,y^{\varepsilon_{\mathrm{IR}}-1}(1-y)\int\mathrm{d}x\,[x(1-x)]^{\varepsilon_{\mathrm{IR}}/2-1}.\\&=\gamma_\mu\left(-\frac{\alpha}{2\pi}\right)\left(\frac{-q^2}{4\pi\mu^2}\right)^{\varepsilon_{\mathrm{IR}}/2}\frac{\Gamma(1+\frac{1}{2}\varepsilon_{\mathrm{IR}})}{\Gamma(2+\varepsilon_{\mathrm{IR}})}\left(\frac{2}{\varepsilon_{\mathrm{IR}}}\right)^2\left[1+\frac{\pi^2}{6}(\tfrac{1}{2}\varepsilon_{\mathrm{IR}})^2\right]\end{aligned} \tag{5.82}$$

† This result is immediate if we note that $\int_{\mu^2}^{\infty}-(Z_2-1)$ is already UV finite and we can write it in $(4+\varepsilon_{\mathrm{IR}})$ dimensions by changing $\varepsilon\to-\varepsilon_{\mathrm{IR}}$. Then the $\int_{\mu^2}^{\infty}$ part cancels the $\int_0^{\mu^2}$ part and the final result is simply $-(Z_2-1)$ written in terms of $\varepsilon_{\mathrm{IR}}$.

‡ $\Gamma(1-\varepsilon)\cong\Gamma(1)-\Gamma'(1)\varepsilon+\frac{1}{2}\Gamma''(1)\varepsilon^2=1+\gamma\varepsilon+\frac{1}{2}(\gamma^2+\frac{1}{6}\pi^2)\varepsilon^2$

The double pole structure in $\Lambda_\mu^{\mathrm{IR}}$ is due to the simultaneous occurrence of IR (massless photons) and mass (massless electrons) singularities. Both have been regularized by working in $n=4+\varepsilon_{\mathrm{IR}}$ dimensions. We note also that, once $(Z_1-1)^{(1)}$ is explicitly determined e.g. in the minimal subtraction scheme, we can rewrite $\Lambda_\mu^{\mathrm{UV}}$ (which is of course finite) in $n=4+\varepsilon_{\mathrm{IR}}$ dimensions by changing $\varepsilon_{\mathrm{UV}}\to-\varepsilon_{\mathrm{IR}}$. Then the full vertex is written in $(4+\varepsilon_{\mathrm{IR}})$ dimensions.

5.4 Dispersion calculation of $O(\alpha)$ virtual corrections in massless QED, in $(4\mp\varepsilon)$ dimensions

It is instructive to repeat the calculation of the electron self-energy and vertex corrections using dispersion relations. The dispersive technique has been extensively used in particle physics. We do not attempt here its systematic discussion, referring the interested reader to other books (Bjorken & Drell 1965; Eden, Landshoff, Olive & Polkinghorne 1966). Our purpose is mainly a technical one: to show yet another method of calculating Feynman diagrams which in some cases happens to be instructive.

As we know, in $(4-\varepsilon)$ dimensions Feynman amplitudes are UV convergent. So we can always write for them the dispersion relation without subtraction. However, the calculated amplitude has, of course, poles in ε and we have to perform next the standard renormalization procedure, i.e. to add counterterms.

Self-energy calculation

The dispersion relation for the function $B(p^2)$ defined by (5.41) reads:

$$B(q^2=0)=\frac{1}{\pi}\int\frac{\mathrm{d}p^2}{p^2}\,\mathrm{Im}\,B(p^2) \tag{5.83}$$

The imaginary part of a Feynman diagram amplitude is obtained by the replacement

$$\frac{1}{p^2-m^2+\mathrm{i}\varepsilon}\to-2\pi\mathrm{i}\delta(p^2-m^2)\Theta(p_0)\equiv-2\pi\mathrm{i}\delta_+(p^2-m^2)$$

in the Feynman integral (Landau–Cutkosky rule). Therefore for the diagram shown in Fig. 5.6 we can write

$$\mathrm{Im}\,B(p^2)=-\tfrac{1}{2}e^2(2-n)\Theta(p^2)\int\mathrm{d}\Phi_2(1-p\cdot k/p^2) \tag{5.84}$$

where

$$\mathrm{d}\Phi_2=\frac{\mathrm{d}^nk}{(2\pi)^n}2\pi\delta_+(k^2)\frac{\mathrm{d}^nl}{(2\pi)^n}2\pi\delta_+(l^2)(2\pi)^n\delta^{(n)}(k+l-p)$$

$$=\frac{\mathrm{d}^nk}{(2\pi)^n}2\pi\delta_+(k^2)2\pi\delta_+((p-k)^2) \tag{5.85}$$

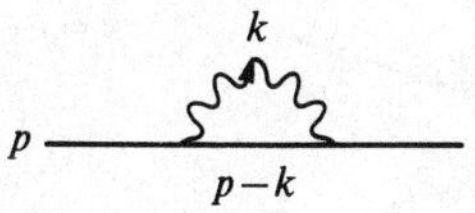

Fig. 5.6

is the two-body phase space element in n dimensions. In the (k, l) centre of mass system it can be written as follows in terms of momenta $p = k + l$ and $r = \frac{1}{2}(k - l)$

$$d\Phi_2 = \frac{1}{(2\pi)^{n-2}} d^n r \delta(\tfrac{1}{4}p^2 + pr + r^2)\delta(\tfrac{1}{4}p^2 - pr + r^2) = \frac{1}{(2\pi)^{n-2}} d^n r \tfrac{1}{2}\delta(\tfrac{1}{4}p^2 + r^2)\delta(pr)$$
$$= \frac{1}{(2\pi)^{n-2}} d^{n-1}\mathbf{r} \frac{1}{2p_0} \delta(\tfrac{1}{4}p^2 - \mathbf{r}^2) \tag{5.86}$$

We now introduce polar coordinates for the Euclidean vector $\mathbf{r}$ in $(n-1)$ dimensions

$$d^{n-1}\mathbf{r} = |\mathbf{r}|^{n-2} d|\mathbf{r}| d\Omega_{n-1} = |\mathbf{r}|^{n-2} d|\mathbf{r}| d\theta (\sin\theta)^{n-3} d\Omega_{n-2} \tag{5.87}$$

where $0 \leqslant \theta \leqslant \pi$ and due to an azimuthal symmetry of the problem we integrate over all but one angle θ to get

$$\int d^{n-1}\mathbf{r} = (\mathbf{r}^2)^{(n-3)/2} d\mathbf{r}^2 (\sin\theta)^{n-3} d\theta \frac{\pi^{(n-2)/2}}{\Gamma(\frac{1}{2}(n-2))} \tag{5.88}$$

For the phase space element integrated over angles one then gets

$$d\Phi_2 = \frac{\pi^{1-\varepsilon/2}}{(2\pi)^{2-\varepsilon}} \frac{1}{\Gamma(1-\frac{1}{2}\varepsilon)} \frac{1}{4}\left(\frac{p^2}{4}\right)^{-\varepsilon/2} (\sin\theta)^{n-3} d\theta \tag{5.89}$$

(we specify $n = 4 - \varepsilon$ and remember that $r_0^2 = 0$ and $p_0 = (p_2)^{1/2}$ in cms). Finally one can write

$$d\theta(\sin\theta)^{n-3} = dz(1-z^2)^{-\varepsilon/2} = dy 2[4y(1-y)]^{-\varepsilon/2}$$

where $z = \cos\theta$ and $y = \frac{1}{2}(1+z)$ to get

$$d\Phi_2 = \frac{1}{8\pi}\left(\frac{p^2}{4\pi}\right)^{-\varepsilon/2} \frac{1}{\Gamma(1-\frac{1}{2}\varepsilon)} dy[y(1-y)]^{-\varepsilon/2} \tag{5.90}$$

We are now ready to calculate $\operatorname{Im} B(p^2)$ and $B(q^2 = 0)$. Since in cms $p\cdot k = \frac{1}{2}p^2$ we obtain (as usual, in $n = 4 - \varepsilon$ dimensions, we define dimensionless α by $\alpha \to \alpha\mu^\varepsilon$).

$$B(q^2 = 0) = -\frac{\alpha}{4\pi} \frac{(2-n)}{\Gamma(1-\frac{1}{2}\varepsilon)} \frac{1}{2} \int_0^\infty \frac{dp^2}{p^2}\left(\frac{p^2}{4\pi\mu^2}\right)^{-\varepsilon/2} \int_0^1 dy[y(1-y)]^{-\varepsilon/2} \tag{5.91}$$

which is equivalent to (5.78).

Vertex calculation

Calculation of the vertex correction proceeds similarly. We shall limit ourselves to the UV finite part Λ_μ^{IR} to understand better the origin of the IR and mass

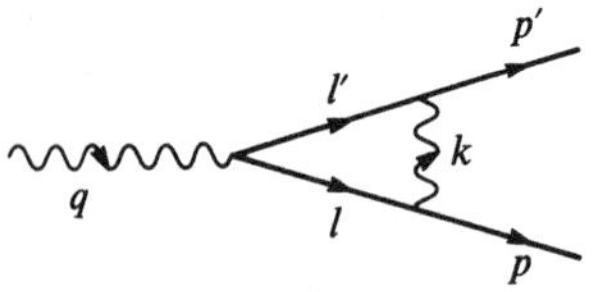

$$k = l - p = p' - l'$$

$$k^2 = -2l\cdot p = -2l'\cdot p'$$

$$p\cdot k = p\cdot l = -\tfrac{1}{2}k^2 = -p'\cdot k$$

Fig. 5.7

singularities. Writing

$$\Lambda_\mu^{\mathrm{IR}} = \gamma_\mu 2q^2 \Lambda(q^2) \tag{5.92}$$

and using the kinematical variables defined in Fig. 5.7 we have

$$\mathrm{Im}\,\Lambda(q^2) = -e^2 \int \mathrm{d}\Phi_2 \left[1 + \frac{4k\cdot(p-p')}{2q^2}\right]\frac{1}{k^2} \tag{5.93}$$

(the term proportional to $\not{k}$ when integrated must give $A\not{p} + B\not{p}'$, so it does not contribute to the on-shell limit) where

$$\begin{aligned}\mathrm{d}\Phi_2 &= \frac{\mathrm{d}^n k}{(2\pi)^n} 2\pi\delta_+((p+k)^2) 2\pi\delta_+((p'+k)^2)\\ &= \frac{\mathrm{d}^n l}{(2\pi)^n} 2\pi\delta_+(l^2) \frac{\mathrm{d}^n l'}{(2\pi)^n} 2\pi\delta_+(l'^2)(2\pi)^n \delta^{(n)}(q+l+l')\end{aligned} \tag{5.94}$$

Therefore

$$\mathrm{Im}\,\Lambda(q^2) = -e^2 \int \mathrm{d}\Phi_2 \left(\frac{1}{k^2} + \frac{2}{q^2}\right) \tag{5.95}$$

The next step is to use the expression (5.90) for $\mathrm{d}\Phi_2$. In the present case $r = \frac{1}{2}(l - l')$ and in the (l, l') cms system ($\mathbf{l} + \mathbf{l}' = \mathbf{p} + \mathbf{p}' = 0$) $\mathbf{r} = \mathbf{l}$. Taking the z axis in the direction $\mathbf{p}$ we have

$$k^2 = -2l\cdot p = -2l_0 p_0 + 2\mathbf{l}\cdot\mathbf{p} = -2(\tfrac{1}{2}q_0)^2 + 2(\tfrac{1}{2}q_0)z = -\tfrac{1}{2}q^2(1-z)$$

where $z = \cos\theta$ and finally ($n = 4 + \varepsilon$)

$$\begin{aligned}\Lambda(Q^2) = &\frac{\alpha}{4\pi}\frac{1}{4\pi\mu^2}\frac{1}{\Gamma(1+\frac{1}{2}\varepsilon)} \int_0^\infty \frac{\mathrm{d}q^2}{q^2 - Q^2}\left(\frac{q^2}{4\pi\mu^2}\right)^{\varepsilon/2 - 1}\\ &* \int_0^1 \mathrm{d}y \left(\frac{1}{1-y} - 2\right)[y(1-y)]^{\varepsilon/2}\end{aligned} \tag{5.96}$$

The integral over $\mathrm{d}y$ gives $[\Gamma(1+\frac{1}{2}\varepsilon)\Gamma(\frac{1}{2}\varepsilon)/\Gamma(1+\frac{1}{2}\varepsilon) - \Gamma^2(1+\frac{1}{2}\varepsilon)/\Gamma(2+\varepsilon)]$ and it is divergent for $\varepsilon \to 0$. The divergence comes from the region $y \approx 1$ and therefore $z \approx 1$. In our specific reference frame $z = 1$ corresponds to $\mathbf{l} \parallel \mathbf{p}$ i.e. the exchanged photon has to have zero momentum. To integrate over q^2 (for $Q^2 < 0$)

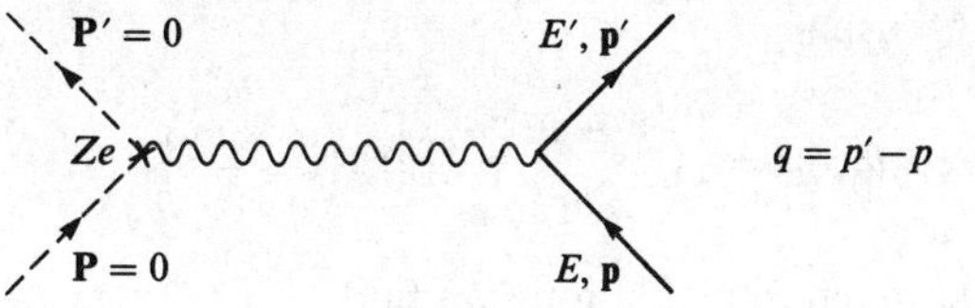

Fig. 5.8

we change variables twice: $q^2 = -Q^2 x$ and $u = 1/(1+x)$. We then get the final result

$$\int_0^\infty \frac{\mathrm{d}q^2}{q^2 - Q^2} (q^2)^{\varepsilon/2-1} = (-Q^2)^{\varepsilon/2-1} \Gamma(1-\tfrac{1}{2}\varepsilon)\Gamma(\tfrac{1}{2}\varepsilon) \tag{5.97}$$

The pole in ε is due to the lower limit of integration over q^2, i.e. due to a massless electron–positron intermediate state (mass singularity). Finally using (5.92) we get $(n = 4 + \varepsilon)$

$$\Lambda_\mu^{\mathrm{IR}} = \gamma_\mu \left(-\frac{\alpha}{2\pi}\right)\left(\frac{-Q^2}{4\pi\mu^2}\right)^{\varepsilon/2} \frac{2}{\varepsilon}\Gamma(1-\tfrac{1}{2}\varepsilon)\left[\frac{\Gamma(1+\frac{1}{2}\varepsilon)\Gamma(\frac{1}{2}\varepsilon)}{\Gamma(1+\varepsilon)} - \frac{\Gamma^2(1+\frac{1}{2}\varepsilon)}{\Gamma(2+\varepsilon)}\right] \tag{5.98}$$

which can be seen to coincide with (5.82).

5.5 Coulomb scattering and the IR problem

Corrections of order α

In the Born approximation the scattering of an electron from a static Coulomb potential

$$A_0(x) = -Ze/4\pi|\mathbf{x}|, \quad A_i(x) = 0$$

is described by the diagram in Fig. 5.8. According to the general rules† the cross section reads

$$\mathrm{d}\sigma_0^{(0)} = \frac{E'}{|\mathbf{p}'|} 2\pi\delta(E - E') \frac{\mathrm{d}^3\mathbf{p}'}{(2\pi)^3} |M_0^{(0)}|^2 \tag{5.99}$$

where

$$|M_0^{(0)}|^2 = \frac{1}{2}\sum_{\mathrm{pol}} \left| \left(\frac{m}{E'}\right)^{1/2} \bar{u}(p')\gamma_0 \left(\frac{m}{E}\right)^{1/2} u(p) \right|^2 \frac{Z^2 e^4}{|\mathbf{q}|^4} \tag{5.100}$$

and $-Ze/|\mathbf{q}|^2$ is the Fourier transform of the $A_0(x)$. (We use the static approximation to the general formulae

$$\text{flux} \frac{1}{|\mathbf{v}_1 - \mathbf{v}_2|} \rightarrow \frac{E}{|\mathbf{p}|} = \frac{E'}{|\mathbf{p}'|}$$

$$(2\pi)^4 \delta^{(4)}(P + p - P' - p') \frac{\mathrm{d}^3\mathbf{P}'}{(2\pi)^3} \rightarrow 2\pi\delta(E - E')$$

† We use the conventions as in Bjorken & Drell (1964); in particular our spinors are normalized so that there is a factor $(m/E)^{1/2}$ for each spin $\frac{1}{2}$ particle wave function. Projection operators are $\Lambda = (\not{p} + m)/2m$.

for $\mathbf{P} = \mathbf{P}' = 0$.) As usual

$$\frac{1}{2}\sum_{\text{pol}} |\bar{u}(p')\gamma_0 u(p)|^2 = \frac{1}{2}\left(\frac{\not{p}' + m}{2m}\right)_{\delta\alpha} (\gamma_0)_{\alpha\beta} \left(\frac{\not{p} + m}{2m}\right)_{\beta\lambda} (\gamma_0)_{\lambda\delta} = \tfrac{1}{2}\mathrm{Tr}[\cdots] \quad (5.101)$$

Performing the necessary calculations we find

$$\mathrm{d}\sigma_0^{(0)} = \tfrac{1}{4}Z^2\alpha^2 \frac{1}{|\mathbf{p}|^4 \sin^4(\frac{1}{2}\theta)/E^2}[1 - \beta^2 \sin^2(\tfrac{1}{2}\theta)]\mathrm{d}\Omega \mathrm{d}E' \delta(E - E') \quad (5.102)$$

where $\beta = |\mathbf{p}|/E$ and $\mathbf{p}\cdot\mathbf{p}' = 2|\mathbf{p}|^2 \cos\theta$. We shall now study real and virtual corrections of order α to the Born cross section.

Virtual corrections of order α are easily included using our results for the electron self-energy, vacuum polarization and the vertex. The amplitude now reads

$$\begin{aligned} M_0^{(1)} &= \bar{u}(p')(1 + \tfrac{1}{2}\Sigma_1)[\gamma_0 + \Lambda_0(p, p')](1 + \tfrac{1}{2}\Sigma_1)u(p)[1 - \Pi(q^2)] \\ &\approx \bar{u}(p')\gamma_0 u(p)[1 + \Sigma_1 + F_1(q^2)] \end{aligned} \quad (5.103)$$

In the final expression (5.103) we have neglected the terms $O(\alpha^2)$ and the corrections which are regular in the IR limit: $\Pi(q^2)$ and the Pauli formfactor $F_2(q^2)$. (In the amplitude for *bremsstrahlung* of a real photon we shall only be interested in the limit of a soft photon.)

Including virtual corrections to order α we obtain therefore

$$\mathrm{d}\sigma_0^{(1)} = \mathrm{d}\sigma_0^{(0)}[1 + 2\Sigma_1 + 2F_1(q^2)] \quad (5.104)$$

where using the results of Section 5.3

$$2\Sigma_1 = 2\frac{\alpha}{\pi}\left(\frac{1}{\varepsilon_{\text{IR}}} + \frac{3}{4}\ln\frac{m^2}{4\pi\mu^2} - 1 + \tfrac{3}{4}\gamma\right) \quad (5.105)$$

(one-quarter (one-half) of the $\ln(m^2/4\pi\mu^2)$ comes from the dimensional regularization of the UV (IR) divergence) and

$$\begin{aligned} 2F_1(q^2) = &-2\frac{\alpha}{\pi}\frac{1}{4}\ln\frac{-q^2}{4\pi\mu^2} - 2\frac{\alpha}{\pi}\left[\frac{1}{\varepsilon_{\text{IR}}}\ln\frac{-q^2}{m^2} + \frac{1}{4}\ln^2\frac{m^2}{-q^2}\right. \\ &\left. + (\tfrac{1}{2}\gamma - 1)\ln\frac{-q^2}{m^2} + \frac{1}{2}\ln\frac{-q^2}{4\pi\mu^2}\ln\frac{-q^2}{m^2}\right] \end{aligned} \quad (5.106)$$

The above expressions are valid for $|q^2| \gg m^2$ and in the minimal subtraction scheme. The first term in $F_1(q^2)$ comes from the dimensional regularization of the UV divergences. Together with the analogous term in Σ_1 it gives $(\alpha/4\pi)\ln(-q^2/m^2)$. The remaining μ^2 dependence is due only to our method of regularization of the IR singularity and should cancel out in every cross section free of such singularities.

The origin of the IR singularity encountered in our cross section $\mathrm{d}\sigma_0^{(1)}$ is in the masslessness of the photon, or in other words in the long range nature of the electromagnetic interactions. In consequence the system has a high degree of degeneracy: a final state consisting of a single electron state is not distinguishable by any measurement from an $e+n$ soft $(k \to 0)$ γs state. Therefore the previously defined cross section $d\sigma_0$ is not a physically meaningful (measurable) quantity. One feels that a summation over experimentally indistinguishable final states is necessary. Following Bloch and Nordsieck and working to order α we introduce a 'measureable' cross section σ as follows

$$\sigma = \int_0^{\Delta E} d\mathscr{E} \frac{d}{d\mathscr{E}}(\sigma_0^{(1)} + \sigma_{1\gamma}^{(1)}) \tag{5.107}$$

where ΔE is the energy resolution of the detection equipment, $\mathscr{E} = E' - E$ is the energy of the undetected real photon, E and E' are, as before, the energies of the initial and final electrons in the laboratory system, and $d\sigma_{1\gamma}$ is the differential cross section with a photon in the final state. We shall explicitly check that the cross section σ is free of the IR singularity. Later we shall discuss this problem in more general terms.

Working to order α we have to include $d\sigma_0^{(1)}$ given by (5.104) and $d\sigma_{1\gamma}^{(1)}$ which is the cross section for a single photon emission given by the Feynman diagrams shown in Fig. 5.9. Using the standard rules we get the following

$$d\sigma_{1\gamma}^{(1)} = |M_{1\gamma}^{(1)}|^2 \frac{1}{|\mathbf{v}|} \frac{d^3\mathbf{p}'}{(2\pi)^3} \frac{d^3\mathbf{k}}{2k_0(2\pi)^3} 2\pi\delta(E - E' - k_0) \tag{5.108}$$

where in the soft photon approximation (momentum k neglected in the numerators of the Feynman expressions)

$$|M_{1\gamma}^{(1)}|^2 = |M_0^{(0)}|^2 e^2 \left[\frac{2p\cdot p'}{k\cdot p\, k\cdot p'} - \frac{m^2}{(k\cdot p)^2} - \frac{m^2}{(k\cdot p')^2} \right] \tag{5.109}$$

In (5.109) the amplitude $|M_0^{(0)}|^2$ is defined by (5.100) and the sum over photon polarizations has been taken. We observe that the corrections due to the soft photon emission factorize. The first term comes from the interference of the two diagrams shown in Fig. 5.9, the next two are just the respective amplitudes squared. We now proceed as follows

$$\int_0^{\Delta E} d\mathscr{E} \frac{d\sigma_{1\gamma}^{(1)}}{d\mathscr{E}} = \int_0^{\Delta E} d\mathscr{E} \int \frac{d^3\mathbf{k}}{2k_0(2\pi)^3} 2\pi\delta(E - E' - k_0) \frac{E}{|\mathbf{p}|} \frac{|\mathbf{p}'|^2 d\Omega}{(2\pi)^3} |M_{1\gamma}^{(1)}|^2$$

$$\cong \int_0^{\Delta E} d\mathscr{E} \frac{d\sigma_0^{(0)}}{d\mathscr{E}} \int_0^{|k|<\Delta E} \frac{d^3\mathbf{k}}{2|k|(2\pi)^3} e^2 \left[\frac{2p\cdot p'}{k\cdot p\, k\cdot p'} - \frac{m^2}{(k\cdot p)^2} - \frac{m^2}{(k\cdot p')^2} \right] \tag{5.110}$$

where the last step is legitimate in the soft photon limit (we denote $|k| = |\mathbf{k}| = k_0$). At this point we get for the cross section, σ, the result

$$\frac{d\sigma}{d\Omega} = \frac{d\sigma_0^{(0)}}{d\Omega} [1 + 2\Sigma_1 + 2F_1(q^2) + I_1 + I_2 + I_3] \tag{5.111}$$

Fig. 5.9

where, e.g.

$$I_1 = e^2 \int_0^{|k|<\Delta E} \frac{d^3\mathbf{k}}{2|k|(2\pi)^3} \frac{2p\cdot p'}{k\cdot p\, k\cdot p'} \tag{5.112}$$

and I_2 and I_3 are self-evident. Our last task is to calculate the corrections I_i. They are IR divergent and for consistency with the calculation of virtual corrections we must regularize them by working in $(4 + \varepsilon_{IR})$ dimensions. We replace

$$\frac{d^3\mathbf{k}}{2|k|(2\pi)^3} \to \frac{d^{n-1}\mathbf{k}}{2|k|(2\pi)^{n-1}}$$

where $|k| = (k_1^2 + \cdots + k_{n-1}^2)^{1/2}$ and introduce spherical coordinates in n dimensions (see (5.88)):

$$\int d^{n-1}\mathbf{k} = \int d|k|\,|k|^{n-2} \sin^{n-3}\theta_1 \sin^{n-4}\theta_2 \ldots \sin\theta_{n-3} d\theta_1 \ldots d\theta_{n-2}$$

$$= 2\frac{\pi^{n/2-1}}{\Gamma(\frac{1}{2}n-1)} \int d|k|\,|k|^{n-2} \int_{-1}^{1} dz(1-z^2)^{n/2-2} \tag{5.113}$$

where $z = \cos\theta_1$ and the integration over the remaining angles has been done in the last step (assuming the integrand depends only on θ_1). Let us consider I_2 first. Writing $(k\cdot p)^2 = |k|^2(E - |\mathbf{p}|z)^2$ we can easily perform the necessary integrations. We see that the probability of a soft photon emission into an interval $d|k|$ is proportional to $d|k|/|k|$ which is easy to understand on dimensional grounds (soft emission does not depend on the electron momentum p). We see also that the dominant contribution to I_2 comes from that region of phase space where $\mathbf{k} \parallel \mathbf{p}$. So, that part of the radiation tends to be collimated around the direction of motion of the initial electron. The final result reads

$$I_2 = -\frac{\alpha}{\pi}\left[\frac{1}{\varepsilon_{IR}} + \frac{1}{2}\ln\frac{(\Delta E)^2}{4\pi\mu^2} - \frac{1}{2}\ln\frac{4E^2}{m^2} + \text{const}\right] \tag{5.114}$$

For I_3 we get the same result with E changed into E'. To calculate the term I_1 it is convenient to choose the reference frame in which $\mathbf{p} + \mathbf{p}' = 0$ (the Breit frame). In this frame

$$p\cdot k = |\tilde{k}|(\tilde{E} - |\tilde{\mathbf{p}}|\cos\theta)$$
$$p'\cdot k = |\tilde{k}|(\tilde{E} + |\tilde{\mathbf{p}}|\cos\theta)$$

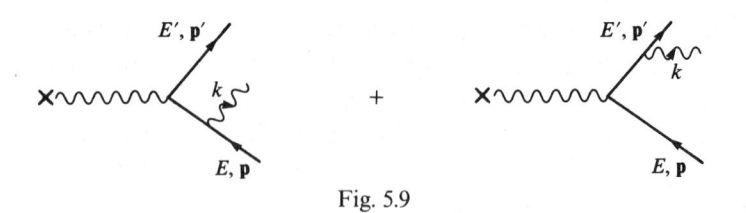

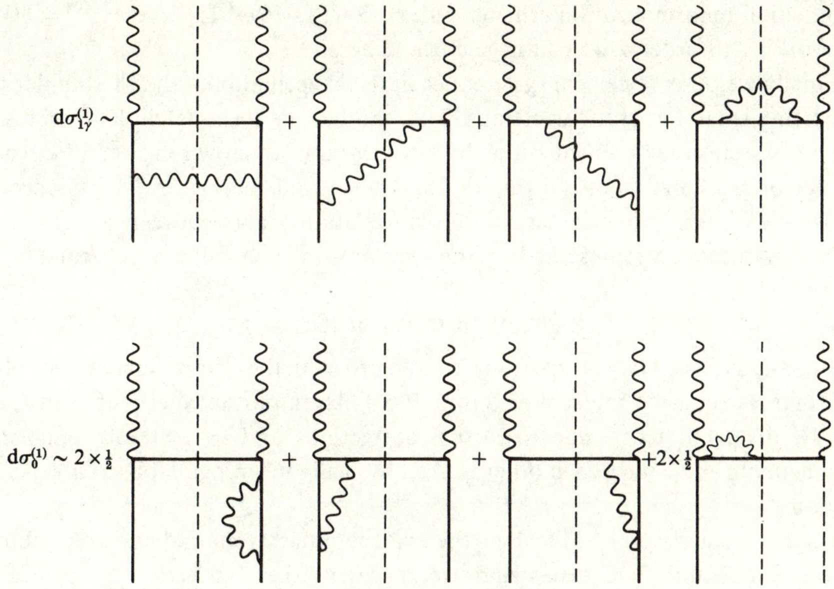

Fig. 5.10

where the sign $\tilde{}$ denotes quantities in the Breit frame and the condition $|k| < \Delta E$ in the laboratory system translates into $|\tilde{k}| < \Delta E \sin(\frac{1}{2}\theta)$ where θ is the scattering angle in the laboratory (indeed $p_{\mathrm{i}}\cdot p_{\mathrm{f}} = p\cdot p' = 2E^2 \sin^2(\frac{1}{2}\theta)$ in the laboratory and $p_{\mathrm{i}}\cdot p_{\mathrm{f}} = 2\tilde{E}^2$ in the Breit frame; so $\tilde{E} = E\sin(\frac{1}{2}\theta)$). In the Breit system the integration over the angle can be performed with help of Spence's functions and we get

$$I_1 \cong 2\frac{\alpha}{\pi}\left[\frac{1}{\varepsilon_{\mathrm{IR}}}\ln\frac{-q^2}{m^2} + \frac{1}{2}\ln\frac{(\Delta E)^2}{4\pi\mu^2}\ln\frac{-q^2}{m^2} + \frac{1}{4}\ln^2\frac{m^2}{-q^2} + \tfrac{1}{2}\gamma\ln\frac{-q^2}{m^2} + f(\theta)\right] \tag{5.115}$$

for $|q^2| \gg m^2$. Adding all the contributions we obtain the final result

$$\frac{d\sigma}{d\Omega} = \frac{d\sigma_0^{(0)}}{d\Omega}\left\{1 + \frac{\alpha}{\pi}\left[\ln\frac{-q^2}{(\Delta E)^2} + \ln\frac{(\Delta E)^2}{-q^2}\ln\frac{-q^2}{m^2} + \frac{3}{2}\ln\frac{-q^2}{m^2} + f(\theta) + \text{const}\right]\right\} \tag{5.116}$$

(remember that $-q^2 \cong 4E^2$). The cross section $d\sigma/d\Omega$ is indeed free of the IR singularity and depends on the experimental resolution ΔE. The transition probability decreases with decreasing ΔE. Our calculation is valid for $|(\alpha/\pi)\ln[-q^2/(\Delta E)^2]| \ll 1$. Otherwise the correction becomes large and we must include higher order contributions which, as we shall see later, exponentiate. The results presented here are for $|q^2| \gg m^2$. The limit $|q^2| \ll m^2$ can also be easily obtained. It should be noted that although the final result (5.116) does not, of course, depend on the method of regularization of the IR singularity, the splitting into real and virtual contributions does. The coupling constant α is defined by the minimal subtraction renormalization scheme and $\alpha = \alpha^{ON}[1 + O(\alpha^{ON})]$ where $\alpha^{ON} = 1/137$. So working to order α we can neglect the difference.

Finally we observe certain systematics in the cancellation of the IR singularities between virtual and real corrections. This can best be formulated if we represent the cross section for real emissions by cut diagrams as shown in Fig. 5.10, where the factor $\frac{1}{2}$ follows from the rule of Section 4.1 and the factor 2 takes account of two identical interference terms. The cancellation occurs between contributions to the cross section represented by the diagrams in each column, separately.

IR problem to all orders in α

The previous results can be generalized to all orders in the electromagnetic coupling constant α (Yennie, Frautschi & Suura 1961). An important fact is that in QED the IR divergent terms due to virtual corrections and to real soft emmisions exponentiate and cancel each other in the physically meaningful (observable) cross section.

Let us consider a process in which there are some electrons and non-soft photons in the initial and final states and we represent their momenta by p and p', respectively. Let the lowest order matrix element for our process be $M_0^{(0)}(p, p')$ and let $M_0^{(n)}(p, p')$ be the contribution corresponding to all diagrams in which there are in addition n virtual photons. The complete matrix element is then

$$M_0(p, p') = \sum_{n=0}^{\infty} M_0^{(n)}(p, p') \tag{5.117}$$

It has been shown (Yennie, Frautschi & Suura 1961) that the $M^{(n)}$s have the following structure

$$\left.\begin{aligned} M_0^{(0)} &= m_0 \\ M_0^{(1)} &= m_0 \alpha B + m_1 \\ M_0^{(2)} &= m_0 (\alpha B)^2/2 + m_1 \alpha B + m_2 \\ &\vdots \\ M_0^{(n)} &= \sum_{i=0}^{n} m_{n-i} (\alpha B)^i / i! \end{aligned}\right\} \tag{5.118}$$

where m_j is an IR finite function (independent of n) of order α^j relative to $M_0^{(0)}$ and each factor αB contains the IR contribution from one virtual photon. Thus we get

$$M_0 = \exp(\alpha B) \sum_{j=0}^{\infty} m_j = \exp(\alpha B) \hat{M}_0 \tag{5.119}$$

and the IR divergent exponent can be read off from our lowest order calculation, (5.104)–(5.106). An important intermediate step in proving (5.119) is to convince oneself that IR divergences originate from only those virtual photons which have both ends terminating on external lines. Actually, this is to be expected on physical grounds since long wave photons should not see charge distribution at short distance.

For the Coulomb elastic scattering we get from (5.119) and (5.106), in the so-called double logarithmic approximation,

$$M_{\mathrm{el}} = \exp\left(-\frac{\alpha}{4\pi}\ln\frac{-q^2}{m^2}\ln\frac{-q^2}{\lambda^2}\right)\hat{M}_{\mathrm{el}} \tag{5.120}$$

where $\hat{M}$ is free of IR singularities. To write (5.120) we have changed the dimensional IR regularization into a cut-off by a small photon mass $\lambda: 1/\varepsilon_{\mathrm{IR}} \to \frac{1}{2}\ln(m^2/\lambda^2)$ (see Section 5.2) in order to follow the traditional notation and in the exponent we have only retained terms quadratic in large logarithms $\ln(-q^2/m^2)$ and $\ln(-q^2/\lambda^2)$. The exponent in (5.120) is called the on-shell electron formfactor in the double logarithmic approximation. We observe that due to the IR divergences encountered in order-by-order calculation the complete amplitude for purely elastic scattering is zero in the limit $\lambda \to 0$. This is just a reflection of the fact that electron scattering is always accompanied by soft photon radiation and the cross section for elastic scattering is not a physical observable.

Let us now consider an inclusive cross section summed over any number of undetected real photons in the final state, with total energy $\mathscr{E}$ such that $\mathscr{E} \leqslant \Delta E$ where ΔE is the energy resolution of the detectors. The differential cross section for emission of m undetected real photons with total energy $\mathscr{E}$ reads

$$\frac{\mathrm{d}\sigma_m}{\mathrm{d}\mathscr{E}} = \exp(2\alpha B)\frac{1}{m!}\int\prod_{i=1}^{m}\frac{\mathrm{d}^3\mathbf{k}_i}{2k_{i0}}\delta(\mathscr{E} - \sum_i k_{i0})\rho_m(p, p', k_1 \ldots k_m) \tag{5.121}$$

where p and p' represent, as before, momenta of all particles in the initial state and of the detected particles in the final state. Virtual corrections to all orders in α are included in (5.121) and, as mentioned before, each σ_m is zero due to the IR divergences. However, an inclusive cross section defined as

$$\frac{\mathrm{d}\sigma}{\mathrm{d}\mathscr{E}} = \sum_{m=0}^{\infty}\frac{\mathrm{d}\sigma_m}{\mathrm{d}\mathscr{E}} \tag{5.122}$$

turns out to be free of IR singularities and thus it is a finite, physical quantity.

Soft real photons must terminate exclusively on external lines in order to contribute IR singularities. Writing (Yennie, Frautschi & Suura 1961)

$$\delta(\mathscr{E} - \sum_i k_{i0}) = \frac{1}{2\pi}\int_{-\infty}^{\infty}\mathrm{d}y\exp(\mathrm{i}y\mathscr{E} - \mathrm{i}y\sum_i k_{i0}) \tag{5.123}$$

using a generalization of (5.109)

$$\rho_m(p, p', k_1 \ldots k_m) = \hat{\sigma}_0\mathrm{p}(k_1)\ldots\mathrm{p}(k_m) \tag{5.124}$$

where $\mathrm{p}(k_i)$ is given by the expression in the square brackets in (5.109) and, as in our order α calculation, neglecting for soft photon radiation the energy–

momentum conservation constraints one gets the following result

$$\frac{d\sigma}{d\mathscr{E}} = \exp[2\alpha(B+\tilde{B})]\hat{\sigma}_0 \frac{1}{2\pi}\int_{-\infty}^{\infty} dy \exp[iy\mathscr{E} + D(y)] \tag{5.125}$$

where

$$2\alpha\tilde{B} = \int^{|k|<\mathscr{E}} \frac{d^3\mathbf{k}_i}{2k_{i0}} p(k_i) \tag{5.126}$$

and

$$D = \int^{|k|<\mathscr{E}} \frac{d^3\mathbf{k}_i}{2k_{i0}} p(k_i)[\exp(-iyk_{i0}) - 1] \tag{5.127}$$

Thus all IR singularities due to real emissions are collected in the exponential factor $\exp(2\alpha\tilde{B})$, where $\tilde{B}$ is given by our order α calculation, and they cancel with virtual IR singularities in B giving together (see (5.116)):

$$2\alpha(B+\tilde{B}) = \frac{\alpha}{\pi}\left(\ln\frac{-q^2}{\mathscr{E}^2} + \ln\frac{\mathscr{E}^2}{-q^2}\ln\frac{-q^2}{m^2}\right) \tag{5.128}$$

The cross section $d\sigma/d\mathscr{E}$ is finite, as expected. For a detector resolution ΔE the integration over $\mathscr{E}$ up to ΔE is necessary. One should also remember that the IR finite part $\hat{\sigma}_0$ contains, in general, terms like $\ln(q^2/\mu^2)$ where q^2 is some characteristic for the process four-momentum squared and μ^2 is the renormalization point. The coupling constant α in (5.121) is $\alpha = \alpha(\mu^2)$.

Problems

5.1 Derive the Ward identities (5.15) and (5.21) using the invariance of the QED Green's functions under the BRS transformations (Problem 3.3). Hint: apply the BRS transformation to the Green's functions

$$\langle 0|TA_\mu(x)\omega(y)|0\rangle = 0$$

and

$$\langle 0|T\Psi(x)\bar{\Psi}(y)\omega(z)|0\rangle = 0$$

where $\omega(x)$ is the free auxiliary scalar field

5.2 Derive the Ward identity for the truncated connected Green's function of the Compton amplitude. For the forward Compton amplitude and in the limit of the photon momentum $k \to 0$ (the Thomson limit) express the amplitude in terms of the electron propagator. Show that in the on-shell renormalization scheme the coupling constant e can be defined by means of the Compton amplitude in the Thomson limit.

5.3 Study the singularity structure of the photon propagator in the one-loop approximation in the complex p^2-plane and derive the Landau–Cutkosky rule. Hint: calculate the difference $P(p^2 + i\Delta) - P(p^2 - i\Delta)$ where $P(p^2) = p^2\Pi(p^2)$ and $\Pi(p^2)$ is defined by (5.26): choose the cms frame $p = (p_0, 0)$, study the pole structure of the integrand as a function of k_0 and integrate over dk_0 first.

5.4 Calculate the off-shell electron formfactor (Sudakov form-factor) in the double logarithmic approximation

$$\Gamma_\mu(p_1, p_2, q) = \gamma_\mu \exp\left(-\frac{\alpha}{2\pi}\ln\frac{|q^2|}{|p_1^2|}\ln\frac{|q^2|}{|p_2^2|}\right) \qquad |q^2| \gg |p_1^2|, |p_2^2| \gg m^2$$

5.5 Calculate the on-shell (see (5.120)) and the off-shell electron formfactors in the double logarithmic approximation using the light-like gauge $n \cdot A = 0$ where $n^2 = 0$. Compare with the calculation in the Feynman gauge.

6
Renormalization group

6.1 Renormalization group equation (RGE)

Derivation of the RGE

The spirit of the renormalization group approach lies in the observation discussed in Chapter 4 that in a specific theory the renormalized constants such as the couplings or the masses are mathematical parameters which can be varied by changing arbitrarily the renormalization prescription. Once the infinities of the theory have been subtracted out by a renormalization prescription R one is still free to perform further finite renormalizations resulting in each case in a different effective renormalization R'. Each renormalization prescription can be interpreted as a particular reordering of the perturbative expansion and expressing it in terms of the new renormalized constants. The latter are, of course, in each case related differently to physical constants (e.g. the mass defined as the pole of the propagator is a physical constant) which are directly measurable and therefore renormalization invariant. A change in renormalization prescription is compensated by simultaneous changes of the parameters of the theory so as to leave, by construction, all exact physical results renormalization invariant. (In practice there is a residual renormalization scheme dependence in each order of perturbation theory.)

Most often, and also in this Section, only subsets of arbitrary renormalization prescription transformations which can be parametrized by a single mass scale parameter μ are discussed. The parameter can be e.g. the value of the subtraction point in the μ-subtraction schemes or the dimensionful parameter μ in the minimal subtraction prescription. For a single-mass-scale-dependent subset of renormalization transformations we derive in the following a differential equation which controls the changes in the renormalized parameters induced by changing μ: the RGE. Actually, the set of renormalization prescription transformations $\{R\}$ does not always have a group structure so the name is partly due to historical reasons.

Let us assume a particular renormalization scheme R and call Γ_R the R-renormalized 1PI Green's functions. Their relation to the bare Green's functions

Γ_B will be

$$\Gamma_R = Z(R)\Gamma_B$$

where $Z(R)$ denotes the appropriate product of renormalization constants defined within the R scheme. If we choose another renormalization scheme R' then,

$$\Gamma_{R'} = Z(R')\Gamma_B$$

Obviously, there exists a relation between both renormalized Green's functions, namely

$$\Gamma_{R'} = Z(R', R)\Gamma_R$$

where

$$Z(R', R) = Z(R')/Z(R)$$

Let us now consider the set of all possible $Z(R', R)$ for arbitrary R and R'. Among the elements of this set there is the composition law

$$Z(R'', R) = Z(R'', R')Z(R', R) \tag{6.1}$$

and, in addition, to each element $Z(R', R)$ there is an inverse

$$Z^{-1}(R', R) = Z(R, R')$$

and we can define the unit element as

$$Z(R, R) = 1$$

The composition law (6.1) is not always defined, however, for arbitrary pairs of Zs: the product

$$Z(R_i, R_j)Z(R_k, R_l) \tag{6.2}$$

is not in general an element of the set $Z(R', R)$, unless $R_j = R_k$. A simple example of the above considerations is provided by the photon wave-function renormalization constant Z_3 (de Rafael 1979).

Suppose we work in the μ-subtraction scheme and we want to relate two renormalizations performed at subtraction points μ_1 and μ_2. The relevant quantity will be $Z(\mu_1, \mu_2)$ which in the one-electron loop approximation reads

$$Z(\mu_2, \mu_1) = 1 + \frac{\alpha}{2\pi}\int_0^1 \mathrm{d}x\, 2x(1-x)\ln\frac{\mu_2^2 x(1-x) + m^2}{\mu_1^2 x(1-x) + m^2} \tag{6.3}$$

This representation of Z obeys the composition law (6.1) but the group multiplication law (6.2) is not obeyed, unless $m = 0$.

In this Section we shall discuss the RGE in renormalization schemes such that renormalization constants Z_i are mass independent (mass-independent renormalization schemes). These include the minimal subtraction prescription and, of course, in a trivial way any prescription for a massless theory.

The RGE in e.g. $\lambda\Phi^4$ theory can be derived as follows. We begin with the relation between the renormalized and bare 1PI Green's functions in the theory regularized by the dimensional method i.e. in $(4-\varepsilon)$ dimensions

$$\Gamma_R^{(n)}(p_i, \lambda_R, m_R, \mu, \varepsilon) = Z_3^{n/2}\Gamma_B^{(n)}(p_i, \lambda_B, m_B, \varepsilon) \tag{6.4}$$

(for the definition of the renormalization constants Z_i see Section 4.1). The finite limit (for $\varepsilon \to 0$) of (6.4) exists since the theory is renormalizable. The bare Green's function $\Gamma_B^{(n)}(p_i, \lambda_B, m_B, \varepsilon)$ is in an obvious way independent of the renormalization scale μ. The RGE will be derived by differentiating relation (6.4) with respect to the renormalization scale μ for fixed λ_B, dimensionful in $(4-\varepsilon)$ dimensions, m_B and ε. In the mass-independent renormalization scheme we have

$$Z_i = Z_i(\lambda_R, \varepsilon)$$

where λ_R is defined as dimensionless i.e.

$$\lambda_B \mu^{-\varepsilon} = Z_3^{-2} Z_1 \lambda_R$$

Similarly, we have

$$m_B^2 = Z_3^{-1} Z_0 m_R^2$$

Differentiating (6.4) and using $\mathrm{d}\Gamma_B/\mathrm{d}\mu = 0$ one obtains

$$\begin{aligned}
&\mu\left(\frac{\partial}{\partial\mu} + \frac{\mathrm{d}\lambda_R}{\mathrm{d}\mu}\frac{\partial}{\partial\lambda_R} + \frac{\mathrm{d}m_R}{\mathrm{d}\mu}\frac{\partial}{\partial m_R}\right)\Gamma_R^{(n)}(p_i, \lambda_R, m_R, \mu, \varepsilon) \\
&= \left(\tfrac{1}{2} n Z_3^{n/2-1} \mu \frac{\mathrm{d}}{\mathrm{d}\mu} Z_3\right)\Gamma_B^{(n)}(p_i, \lambda_B, m_B, \varepsilon) \\
&= \left(\tfrac{1}{2} n \frac{1}{Z_3} \mu \frac{\mathrm{d}}{\mathrm{d}\mu} Z_3\right)\Gamma_R^{(n)}(p_i, \lambda_R, m_R, \mu, \varepsilon)
\end{aligned} \tag{6.5}$$

We can define now the following functions

$$\beta(\lambda_R, \varepsilon) = \mu\frac{\mathrm{d}}{\mathrm{d}\mu}\lambda_R = \lambda_B \mu^{-\varepsilon}\left(\mu \frac{\mathrm{d}}{\mathrm{d}\mu} Z_3^2 Z_1^{-1}\right) - \varepsilon\mu^{-\varepsilon}\lambda_B Z_3^2 Z_1^{-1} \underset{\varepsilon\to 0}{\longrightarrow} \beta(\lambda_R) \tag{6.6}$$

$$\gamma(\lambda_R, \varepsilon) = \tfrac{1}{2}\mu\frac{\mathrm{d}}{\mathrm{d}\mu}\ln Z_3 \underset{\varepsilon\to 0}{\longrightarrow} \gamma(\lambda_R) \tag{6.7}$$

$$\gamma_{\mathrm{m}}(\lambda_R, \varepsilon) = \frac{\mu}{m_R}\frac{\mathrm{d}m_R}{\mathrm{d}\mu} = \frac{1}{2}\mu\frac{\mathrm{d}}{\mathrm{d}\mu}\ln(Z_3 Z_0^{-1}) \underset{\varepsilon\to 0}{\longrightarrow} \gamma_{\mathrm{m}}(\lambda_R) \tag{6.8}$$

Taking the limit $\varepsilon \to 0$ in (6.5) (finite limits of relations (6.5)–(6.8) exist since the theory is renormalizable) we obtain the final result

$$\left[\mu\frac{\partial}{\partial\mu} + \beta(\lambda_R)\frac{\partial}{\partial\lambda_R} + \gamma_{\mathrm{m}}(\lambda_R) m_R \frac{\partial}{\partial m_R} - n\gamma(\lambda_R)\right]\Gamma_R^{(n)}(p_i, \lambda_R, m_R, \mu) = 0 \tag{6.9}$$

It is important to remember that only in mass-independent renormalization schemes or for $m=0$ the functions β, γ_{m} and γ depend on λ_R only. In general they also can depend on the ratio m_R/μ. This happens, for instance, in the μ-subtraction scheme.

Solving the RGE

Equation (6.9) describes the change of renormalized parameters and multiplicative scaling factors for the renormalized Green's functions which compensates a change of the renormalization scale μ. To solve (6.9) suppose for a moment that $\gamma(\lambda_R)\equiv 0$. We then have a homogeneous equation $\mu(\mathrm{d}/\mathrm{d}\mu)\Gamma_R^{(n)}(p_i, \lambda_R, m_R, \mu)=0$ which can be written in the form (subscript R in λ_R and m_R will be omitted from now on):

$$\left[-\frac{\partial}{\partial t}+\beta(\lambda)\frac{\partial}{\partial\lambda}+\gamma_{\mathrm{m}}(\lambda)m\frac{\partial}{\partial m}\right]\Gamma_R^{(n)}(p_i, \lambda, m, \mu_0\mathrm{e}^{-t})=0 \tag{6.10}$$

where $\mu=\mathrm{e}^{-t}\mu_0$, and μ_0 is some fixed scale. We recall also that the parameters λ and m are renormalized at μ. As we shall see the meaning of this equation is that the dependence of the function $\Gamma_R^{(n)}$ on t and hence on μ can be expressed solely through the functions $\bar{\lambda}(t,\lambda)$, $\bar{m}(t,\lambda,m)$ defined by the equations

$$\left.\begin{aligned} &\int_{\lambda}^{\bar{\lambda}(t,\lambda)} \mathrm{d}x/\beta(x)=t \\ \bar{m}(t,\lambda,m)=m\exp&\left[\int_{\lambda}^{\bar{\lambda}(t,\lambda)}\mathrm{d}x(\gamma_{\mathrm{m}}(x)/\beta(x))\right]=m\exp\left[\int_0^t \gamma_{\mathrm{m}}(\bar{\lambda}(t'))\,\mathrm{d}t'\right]\end{aligned}\right\} \tag{6.11}$$

Taking the partial derivative $\partial/\partial t$ one can rewrite equations (6.11) as differential equations

$$\left.\begin{aligned}\partial\bar{\lambda}(t,\lambda)/\partial t&=\beta(\bar{\lambda})\\ \partial\bar{m}(t,\lambda,m)/\partial t&=\gamma_{\mathrm{m}}(\bar{\lambda})\bar{m}\end{aligned}\right\} \tag{6.12}$$

with the boundary conditions

$$\bar{\lambda}(0,\lambda)=\lambda, \qquad \bar{m}(0,\lambda,m)=m$$

We also observe that equations (6.11) themselves can be written in the form of the RGEs; for instance differentiating $\bar{\lambda}(t,\lambda)$ given by the first of equations (6.11) with respect to λ one gets

$$\left[-\frac{\partial}{\partial t}+\beta(\lambda)\frac{\partial}{\partial\lambda}\right]\bar{\lambda}(t,\lambda)=0 \tag{6.13}$$

Using (6.13) we can easily check that any function $f(\lambda,t,\mu_0)=f(\bar{\lambda}(\lambda,t),\mu_0)$ is μ independent and satisfies the equation

$$\mu\frac{\mathrm{d}}{\mathrm{d}\mu}f=\left(-\frac{\partial}{\partial t}+\beta(\lambda)\frac{\partial}{\partial\lambda}\right)f(\bar{\lambda}(t,\lambda),\mu_0)=0 \tag{6.14}$$

Therefore the solution to (6.10) can be written as follows

$$\Gamma_R^{(n)}(p_i, \lambda, m, \underbrace{\mu_0 e^{-t}}_{\mu}) = \Gamma_R^{(n)}(p_i, \bar{\lambda}(t), \bar{m}(t), \underbrace{\mu_0}_{\mu e^t}) \tag{6.15}$$

where $\bar{\lambda}(t)$ stands for $\bar{\lambda}(t, \lambda)$. The solution to the inhomogeneous (6.9) reads

$$\Gamma_R^{(n)}(p_i, \lambda, m, \mu) = \Gamma_R^{(n)}(p_i, \bar{\lambda}(t), \bar{m}(t), \mu e^t) \exp\left[-n\int_0^t \gamma(\bar{\lambda}(t'))\, dt'\right] \tag{6.16}$$

as can be checked by applying to it the differential operator of (6.9). Equation (6.16) relates a Green's function $\Gamma_R^{(n)}(p_i, \lambda, m, \mu)$, with the renormalized parameters equal to λ and m at the renormalization scale μ, to the Green's function $\Gamma_R^{(n)}(p_i, \bar{\lambda}(t), \bar{m}(t), \mu e^t)$ where $\bar{\lambda}(t)$ and $\bar{m}(t)$ can be understood as the renormalized parameters corresponding to the scale μe^t.

The renormalized parameters are renormalization scale dependent. Any physical quantity P (renormalization scheme independent) can be calculated in terms of them: $P = P(m, \lambda, \mu)$. Being by definition renormalization scheme independent P must satisfy the homogeneous RGE:

$$\mathscr{R}P(\lambda, m, \mu) \equiv \left[\mu\frac{\partial}{\partial\mu} + \beta(\lambda)\frac{\partial}{\partial\lambda} + \gamma_{\rm m}(\lambda)m\frac{\partial}{\partial m}\right]P(\lambda, m, \mu) = 0 \tag{6.17}$$

We shall now check, using (6.9), that this is indeed the case for the physical mass defined as the position of a pole in a propagator and for renormalized S-matrix elements. Let us consider the scalar propagator $\Delta(p^2)$ which satisfies the equation

$$[\mathscr{R} + 2\gamma(\lambda)]\Delta(p^2, \lambda, m, \mu) = 0 \tag{6.18}$$

Assuming that Δ has a pole at $p^2 = m_{\rm F}^2$ we can write a Laurent expansion for Δ

$$\Delta(p^2, \lambda, m, \mu) = \frac{R}{p^2 - m_{\rm F}^2} + \tilde{\Delta}(p^2, \lambda, m, \mu) \tag{6.19}$$

The position of the pole, $m_{\rm F}(\lambda, m, \mu)$ and the residue of the pole $R(\lambda, m, \mu)$ themselves satisfy RGEs which can be derived by using (6.19) in (6.18) and equating the residua of the poles at $p^2 = m_{\rm F}^2$

$$\left.\begin{aligned} \mathscr{R}m_{\rm F}^2(m, \lambda, \mu) &= 0 \\ [\mathscr{R} + 2\gamma(\lambda)]R(\lambda, m, \mu) &= 0 \end{aligned}\right\} \tag{6.20}$$

We consider next an arbitrary n-scalar particle S-matrix element. This is obtained from the 1PI Green's function $\Gamma^{(n)}$ by going to the physical poles $p_i^2 = m_i^2$ and multiplying by $R^{n/2}$ (see (4.27)). Since $\Gamma^{(n)}$ satisfies (6.9) and m_i^2 and R satisfy (6.20), we have

$$\mathscr{R} \lim_{p_i^2 \to m_i^2} R^{n/2}\Gamma^{(n)} = \lim_{p_i^2 \to m_i^2} \mathscr{R}R^{n/2}\Gamma^{(n)} = \lim_{p_i^2 \to m_i^2} [n\gamma(\lambda) - n\gamma(\lambda)]R^{n/2}\Gamma^{(n)} = 0 \tag{6.21}$$

One must stress that it is an exact S-matrix element which is renormalization scheme independent. In perturbation theory truncated at order N there is always a residual renormalization scheme dependence $O(\lambda^{N+1})$.

Green's functions for rescaled momenta

The RGE is used most frequently in a somewhat different context, namely to discuss the change in the Green's functions when all momenta are rescaled $p_i \to \rho p_i$ (but for fixed μ). Let us take a Green's function of dimension D:$\Gamma_R^{(n)} \sim [M]^D$. On dimensional grounds it can be therefore written as follows

$$\Gamma_R^{(n)}(\rho p_i, m, \lambda, \mu) = \mu^D f(\rho^2 p_i \cdot p_j/\mu^2, m/\mu, \lambda) \tag{6.22}$$

We observe that $\Gamma_R^{(n)}$ is a homogeneous function of order D of variables m, ρ and μ. Using the basic property of such functions we can write for it the following equation

$$\left(\rho\frac{\partial}{\partial\rho} + m\frac{\partial}{\partial m} + \mu\frac{\partial}{\partial\mu} - D\right)\Gamma_R^{(n)}(\rho p_i, m, \lambda, \mu) = 0 \tag{6.23}$$

Combining (6.9) and (6.23) one gets ($t = \ln\rho$)

$$\left[-\frac{\partial}{\partial t} + \beta(\lambda)\frac{\partial}{\partial\lambda} + (\gamma_{\mathrm{m}} - 1)m\frac{\partial}{\partial m} - n\gamma(\lambda) + D\right]\Gamma_R^{(n)}(\rho p_i, m, \lambda, \mu) = 0 \tag{6.24}$$

which is the result we are looking for.

We notice first that for $m = 0$ and for a non-interacting theory ($\beta \equiv \gamma \equiv 0$) the solution to (6.24) reads

$$\Gamma_R^{(n)}(\rho p_i, \lambda, \mu) = \rho^D \Gamma_R^{(n)}(p_i, \lambda, \mu) \tag{6.25}$$

i.e. the Green's functions exhibit canonical scaling (see Chapter 7). The general solution to (6.24) can be written using arguments similar to those used to solve (6.9) and it reads

$$\Gamma_R^{(n)}(\mathrm{e}^t p_i, \lambda, m, \mu) = \Gamma_R^{(n)}(p_i, \bar{\lambda}(t), \bar{\bar{m}}(t), \mu)\exp\left[Dt - n\int_0^t \gamma(\bar{\lambda}(t'))\,\mathrm{d}t'\right] \tag{6.26}$$

where

$$\bar{\bar{m}}(t) = \bar{m}(t)/\alpha$$

and $\bar{\lambda}(t)$ and $\bar{m}(t)$ satisfy (6.11). This solution can also be obtained directly by the following steps

$$\begin{aligned}\Gamma_R^{(n)}(\mathrm{e}^t p_i, \lambda, m, \mu) &= \Gamma_R^{(n)}(\mathrm{e}^t p_i, \bar{\lambda}(t), \bar{m}(t), \mathrm{e}^t\mu)\exp\left[-n\int\gamma(t')\,\mathrm{d}t'\right] \\ &= \Gamma_R^{(n)}(\mathrm{e}^t p_i, \bar{\lambda}(t), \mathrm{e}^t\mathrm{e}^{-t}\bar{m}(t), \mathrm{e}^t\mu)\exp\left(-n\int\gamma\,\mathrm{d}t'\right) \\ &= \rho^D\Gamma_R^{(n)}(p_i, \bar{\lambda}(t), \bar{\bar{m}}(t), \mu)\exp\left(-n\int\gamma\,\mathrm{d}t'\right)\end{aligned}$$

The first step corresponds to a change of the renormalization point according to (6.16) and the last step is a change of the mass scale by a factor ρ. We notice that (6.26) contrary to (6.16), relates Green's functions which both have the same renormalization scale μ, but different values of the renormalized couplings: λ, m in the l.h.s. and $\bar{\lambda}(t)$, $\bar{m}(t)$ in the r.h.s. One can say, therefore, that they are Green's functions of different theories.

Result (6.26) gives us Green's functions for the rescaled momenta $e^t p_i$ (off-mass-shell) in terms of the 'effective' parameters $\bar{\lambda}(t)$ and $\bar{m}(t)$ which are solutions of (6.11). To use those results in physical applications one must know the functions $\beta(\lambda)$, $\gamma(\lambda)$ and $\gamma_{\mathrm{m}}(\lambda)$ which can only be calculated in perturbation theory.

Strictly speaking the RGE solution (6.26) applies in the deep Euclidean region, $p_i^2 < 0$, away from possible IR problems and new thresholds associated with time-like momenta.

RGE in QED

We choose to work in the set of mass-independent renormalization schemes and in a covariant gauge defined by the gauge-fixing term: $-(1/a)(\partial_\mu A^\mu)^2$. The RGE can then be derived following the steps outlined previously and we get

$$\left[\mu\frac{\partial}{\partial\mu} + \beta(\alpha, a)\frac{\partial}{\partial\alpha} + \gamma_{\mathrm{m}}(\alpha, a)m\frac{\partial}{\partial m} + \delta(\alpha, a)\frac{\partial}{\partial a} - n_\gamma\gamma_\gamma(\alpha, a) - n_{\mathrm{F}}\gamma_{\mathrm{F}}(\alpha, a)\right]\Gamma^{(n_\gamma, n_{\mathrm{F}})}(p_i, \alpha, m, \mu, a) = 0 \tag{6.27}$$

where in analogy to (6.6)–(6.8)

$$\left.\begin{aligned}
\beta(\alpha, a) &= \lim_{\varepsilon\to 0}\beta(\alpha, a, \varepsilon) = \lim_{\varepsilon\to 0}\mu\frac{\mathrm{d}}{\mathrm{d}\mu}\alpha \\
\gamma_{\mathrm{m}}(\alpha, a) &= \lim_{\varepsilon\to 0}\mu\frac{\mathrm{d}}{\mathrm{d}\mu}\ln Z_{\mathrm{m}} \\
\gamma_\gamma(\alpha, a) &= \lim_{\varepsilon\to 0}\tfrac{1}{2}\mu\frac{\mathrm{d}}{\mathrm{d}\mu}\ln Z_3 \\
\gamma_{\mathrm{F}}(\alpha, a) &= \lim_{\varepsilon\to 0}\tfrac{1}{2}\mu\frac{\mathrm{d}}{\mathrm{d}\mu}\ln Z_2
\end{aligned}\right\} \tag{6.28}$$

Derivatives are taken for fixed α_{B} and ε where $\alpha = e^2/4\pi$, $\alpha_{\mathrm{B}} = \mu^\varepsilon Z_3^{-1}\alpha$ in $n = 4 - \varepsilon$ dimensions, $m_{\mathrm{B}} = Z_{\mathrm{m}}^{-1}m$, $A_{\mu\mathrm{B}} = Z_3^{1/2}A_\mu$, $\Psi_{\mathrm{B}} = Z_2^{1/2}\Psi$ (for definitions of the renormalization constants see Chapter 5). The only new element is the μ dependence of the gauge-fixing parameter $a = Z_3^{-1}a_{\mathrm{B}}$ which is responsible for the term $\delta(\alpha, a)\partial/\partial a$ with

$$\delta(\alpha, a) = \lim_{\varepsilon\to 0}\mu\frac{\mathrm{d}a}{\mathrm{d}\mu} = \lim_{\varepsilon\to 0}\left(-a\mu\frac{\mathrm{d}}{\mathrm{d}\mu}\ln Z_3\right) \tag{6.29}$$

The solution to (6.27) is obtained by the obvious modification of (6.16) with the running parameters given by (6.11) and, in addition

$$\partial\bar{a}(t,\alpha,a)/\partial t=\delta(\bar{\alpha},\bar{a}),\quad \bar{a}(0,\alpha)=a \tag{6.30}$$

Using our previous experience we can also derive the RGE analogous to (6.26) for QED.

6.2 Calculation of the renormalization group functions β, γ, γ_m

We address ourselves to the problem of calculating functions $\beta(\lambda)$, $\gamma(\lambda)$ and $\gamma_m(\lambda)$ in a mass-independent renormalization scheme. We recall the definitions (6.6)–(6.8), e.g.

$$\beta(\lambda)=\lim_{\varepsilon\to 0}\beta(\lambda,\varepsilon)=\lim_{\varepsilon\to 0}\mu\frac{d}{d\mu}\lambda\Big|_{\text{fixed }\lambda_B,\varepsilon} \tag{6.31}$$

In perturbation theory we can calculate the renormalization constants Z_i which relate λ to λ_B

$$\lambda_B=\mu^{\varepsilon}Z_{\lambda}\lambda=\mu^{\varepsilon}\left[1+\sum_{\nu}\frac{a_{\nu}(\lambda)}{\varepsilon^{\nu}}\right]\lambda \tag{6.32}$$

where for instance $Z_\lambda=Z_3^{-2}Z_1$ for the $\lambda\Phi^4$ theory. The coefficients $a_\nu(\lambda)$ are the series in λ

$$\begin{aligned}a_1(\lambda)&=a_{11}\lambda+a_{12}\lambda^2+\cdots+a_{1n}\lambda^n\\ &\vdots\\ a_m(\lambda)&=a_{m1}\lambda+a_{m2}\lambda^2+\cdots+a_{mn}\lambda^n\end{aligned}$$

Using the definition (6.31) we get

$$\beta(\lambda,\varepsilon)=\mu\frac{d}{d\mu}\lambda_B\mu^{-\varepsilon}Z_{\lambda}^{-1}=-\varepsilon\lambda-\lambda Z_{\lambda}^{-1}\mu\frac{d}{d\mu}Z_{\lambda} \tag{6.33}$$

Since

$$\mu\frac{d}{d\mu}Z_\lambda=\mu\frac{d\lambda}{d\mu}\frac{dZ_\lambda}{d\lambda}=\beta(\lambda,\varepsilon)\frac{dZ_\lambda}{d\lambda}$$

we can rewrite (6.33) in the following form

$$\beta(\lambda,\varepsilon)Z_\lambda+\varepsilon\lambda Z_\lambda+\beta(\lambda,\varepsilon)\lambda dZ_\lambda/d\lambda=0 \tag{6.34}$$

Substituting expansion (6.32) in (6.34) we look for the solution for $\beta(\lambda,\varepsilon)$ in the form of a series $\beta(\lambda,\varepsilon)=\sum_\nu\beta_\nu\varepsilon^\nu$ which is regular at $\varepsilon=0$ (since our theory is renormalizable a finite limit of $\beta(\lambda,\varepsilon)$, for $\varepsilon\to 0$, exists). Equating the residua of the poles in ε we obtain the following final result

$$\left.\begin{aligned}\beta(\lambda,\varepsilon)&=-\varepsilon\lambda+\beta(\lambda)\\ \beta(\lambda)&=\lambda^2\frac{da_1}{d\lambda}\\ \lambda^2\frac{da_{\nu+1}}{d\lambda}&=\beta(\lambda)\frac{d}{d\lambda}(\lambda a_\nu)\end{aligned}\right\} \tag{6.35}$$

We arrive, therefore, at two important conclusions

(i) $\beta(\lambda)$ is totally determined by the residua of the simple poles in ε of Z_λ.
(ii) The residua of the higher order poles of Z_λ are totally determined by $a_1(\lambda)$. This reflects an important aspect of renormalization theory. In each successive order of perturbation theory only one new divergence arises i.e. the simple pole in ε. There will, of course, be multiple poles, up to $1/\varepsilon^N$ in the Nth order but their residua are determined by lower order perturbation expansion. One can easily check in particular that $a_{mn}=0$ for $m<n$.

In a quite analogous way we also find the results for $\gamma(\lambda)$ and $\gamma_m(\lambda)$

$$\gamma(\lambda) = -\tfrac{1}{2}\lambda \frac{d}{d\lambda} Z_3^{(1)}(\lambda) \tag{6.36}$$

where

$$Z_3 = 1 + \sum_\nu \frac{Z_3^{(\nu)}}{\varepsilon^\nu}$$

and

$$\gamma_m(\lambda) = \tfrac{1}{2}\lambda \frac{dZ_m^{(1)}}{d\lambda} \tag{6.37}$$

where

$$m_B^2 = Z_m m^2, \qquad Z_m = 1 + \sum_\nu \frac{Z_m^{(\nu)}}{\varepsilon^\nu}$$

($Z_m = Z_0 Z_3^{-1}$ in the $\lambda\Phi^4$ theory).

Using the results of Section 4.3 for the renormalization constants Z_i in the $\lambda\Phi^4$ theory

$$\left.\begin{aligned} Z_1 &= 1 + \frac{3\lambda}{16\pi^2\varepsilon} \\ Z_0 &= 1 + \frac{\lambda}{16\pi^2\varepsilon} - \frac{\lambda^2}{2(16\pi^2)\varepsilon} + \frac{2\lambda^2}{(16\pi^2)^2\varepsilon^2} \\ Z_3 &= 1 - \frac{\lambda^2}{12(16\pi^2)\varepsilon} \end{aligned}\right\} \tag{6.38}$$

and with help of (6.35), (6.36) and (6.37) one obtains the following renormalization group functions in the $\lambda\Phi^4$ theory

$$\left.\begin{aligned} \beta(\lambda) &= \frac{3\lambda^2}{16\pi^2} + O(\lambda^3) \\ \gamma_m(\lambda) &= \frac{1}{2}\frac{\lambda}{16\pi^2} - \frac{5}{12}\frac{\lambda^2}{6(16\pi^2)^2} + O(\lambda^3) \\ \gamma(\lambda) &= \frac{1}{12}\left(\frac{\lambda}{16\pi^2}\right)^2 + O(\lambda^3) \end{aligned}\right\} \tag{6.39}$$

For QED the relation (6.32) is replaced by $\alpha_B = \mu^\varepsilon Z_\alpha \alpha$ with $Z_\alpha = Z_3^{-1}$. Using (5.31) we get

$$Z_\alpha = 1 + \frac{\alpha}{\pi}\frac{2}{3}\frac{1}{\varepsilon} + \left(\frac{\alpha}{\pi}\right)^2 b_2 + \cdots$$

and therefore, from (6.35)

$$\beta(\alpha) = \alpha\left[\frac{\alpha}{\pi}\frac{2}{3} + \left(\frac{\alpha}{\pi}\right)^2 b_2 + \cdots\right] \tag{6.40}$$

Though not calculated explicitly, the second order contribution $(\alpha/\pi)^2 b_2$ has been singled out for the sake of the discussion later. Note also that $\beta(e)$ defined as $\beta(e) = \mu(\mathrm{d}/\mathrm{d}\mu)e$ is given by $\beta(e) = 2\pi\beta(\alpha)/e$. Therefore

$$\beta(e) = \frac{e^3}{16\pi^2}\frac{4}{3} + \frac{e^5}{(16\pi^2)^2} 8b_2 + \cdots \tag{6.41}$$

6.3 Fixed points; effective coupling constant

Fixed points

The main use of the renormalization group is in discussing the large, or small, momentum behaviour of a quantum field theory which is determined, as follows from (6.26), by the behaviour of the 'effective' coupling constant $\bar{\lambda}(t)$ as $t \to \pm\infty$. The latter is determined by

$$\int_{\lambda}^{\bar{\lambda}(t,\lambda)} \mathrm{d}x/\beta(x) = t \tag{6.11}$$

where $\beta(\lambda)$ can be calculated perturbatively from (6.35). Equation (6.11) provides a basis for a physically very important classification of different theories. We shall discuss here only theories with one coupling constant. First we assume (6.11) to be valid in the whole range $-\infty < t < \infty$. Otherwise the renormalization scale μ could not vary arbitrarily and the theory would need some natural cut-off Λ. A renormalizable theory with perturbative expansion valid only in a certain energy range cut-off by a scale parameter Λ (if e.g. new interactions switch on at a mass scale Λ) is not necessarily uninteresting physically and we shall come back to this point later.

The solution $\bar{\lambda}(t, \lambda)$ of (6.11) must for $t \to \infty$ and $t \to -\infty$ approach the zero of $\beta(x)$ nearest to λ or go to infinity if there is no zero to approach. A zero of $\beta(x)$ is called a fixed point. Let $\bar{\bar{\lambda}}$ be the fixed point nearest to λ. If

$$\lim_{t\to\infty} \bar{\lambda}(t, \lambda) = \bar{\bar{\lambda}} \equiv \lambda_+ \tag{6.42}$$

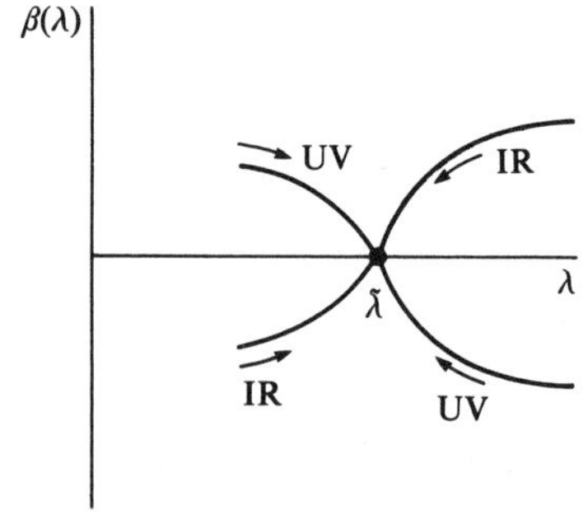

Fig. 6.1

it is called a UV stable fixed point; if

$$\lim_{t \to -\infty} \bar{\lambda}(t, \lambda) = \bar{\lambda} \equiv \lambda_- \tag{6.43}$$

it is called an IR stable fixed point.

In both cases one says that λ is in the domain of attraction of the fixed point $\bar{\lambda}$. A domain is a region which lies between two zeros of β. Of course, for two different values of the coupling λ lying in two different domains one in general has two different theories (or two different phases of one theory). The nature of the fixed point is determined, if it is a simple zero, by the sign of the derivative $\beta'(\lambda)$ at $\lambda = \bar{\lambda}$. From (6.11) and Fig. 6.1 it is clear that for

$\beta'(\lambda)|_{\bar{\lambda}} < 0$, $\bar{\lambda}$ is a UV stable fixed point;
$\beta'(\lambda)|_{\bar{\lambda}} > 0$, $\bar{\lambda}$ is an IR stable fixed point.

In other words, both for $\lambda < \bar{\lambda}$ and $\lambda > \bar{\lambda}$, to satisfy (6.11) the effective coupling $\bar{\lambda}(t, \lambda) \to \bar{\lambda}$ for $t \to \infty$ if $\beta'(\lambda)|_{\bar{\lambda}} < 0$, and $\bar{\lambda}(t, \lambda) \to \bar{\lambda}$ for $t \to -\infty$ if $\beta'(\lambda)|_{\bar{\lambda}} > 0$. (If $\bar{\lambda}$ is a multiple zero of $\beta(\lambda)$, of odd order M, then the $\bar{\lambda}$ is a UV (IR) stable fixed point for $\beta(\lambda)/(\lambda - \bar{\lambda})^M < 0$ (> 0).)

It is obvious but important to notice that for all single coupling theories the value $\lambda = 0$ is a fixed point ($\beta(0) = 0$ in perturbation theory). A theory is said to be asymptotically free if $\lambda = 0$ is a UV stable fixed point and to be IR stable if it is an IR stable fixed point. All single coupling theories are either asymptotically free or IR stable. In an asymptotically free theory the effective coupling vanishes for infinite momenta. For theories with several couplings the origin may be UV stable for some couplings and IR stable for others. The $\lambda\Phi^4$ theory (with $\lambda \geqslant 0$) and QED are IR stable ($\beta(\lambda)$ is positive for small λ) whereas QCD is asymptotically free ($\beta(\lambda)$ is negative for small λ). Of course, for a given theory, the only place that we can compute $\beta(\lambda)$ in perturbation theory is near $\lambda = 0$ and its behaviour for increasing λ is essentially a speculation. Consequently, the behaviour of the coupling constant $\bar{\lambda}(t, \lambda)$ for $t \to \infty$ in $\lambda\Phi^4$ and QED, and for $t \to -\infty$ in QCD also remains, strictly speaking, a speculation. In Fig. 6.2 we show some examples of what $\beta(\lambda)$ might look like.

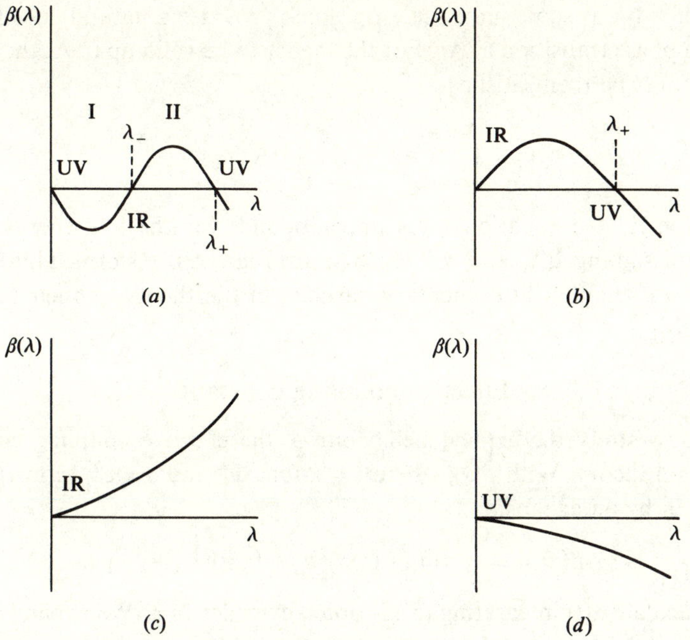

Fig. 6.2

In Fig. 6.2(*a*) we have distinguished two domains of attraction. If λ is in region I, the theory is asymptotically free and for $t \to -\infty$ (low momentum) λ approaches the fixed point λ_-. On the other hand, if λ is in the region II then the theory is not asymptotically free; instead $\bar{\lambda}(t, \lambda) \to \lambda_+$ when $t \to \infty$. Fig. 6.2(*b*) shows what might happen in QED: IR stability and fixed point λ_+ for $t \to \infty$. Fig. 6.2(*c*) shows another alternative for QED and Fig. 6.2(*d*) shows what we hope happens in QCD. In Fig. 6.2(*d*) there is no IR stable fixed point. To satisfy (6.11)

$$\int_{\lambda}^{\bar{\lambda}(-\infty,\lambda)} \mathrm{d}x/\beta(x) = -\infty$$

$\bar{\lambda}(t, \lambda)$ must diverge when $t \to -\infty$ and $\beta(x)$ must be such that $\int_{\lambda}^{\infty} \mathrm{d}x/\beta(x)$ is divergent. If the effective coupling becomes infinite in the IR region one often refers to this situation as IR slavery. It is generally believed (but not proved) that IR slavery in QCD is responsible for confinement.

Finally, suppose that the β-function is always positive (as in Fig. 6.2(*c*)) and that it diverges at infinity in such a way that

$$\int_{\lambda}^{\infty} \mathrm{d}x/\beta(x) = t_{\mathrm{c}} < \infty$$

It is likely that this situation occurs in the $\lambda\Phi^4$ theory (see (6.39)). For such a theory the renormalized perturbative expansion cannot be consistent for all energies (unless $\lambda(\mu_0) = 0$ for any μ_0). The theory must be modified by new interactions which switch on at some mass scale Λ_c which provides a natural cut-off or must undergo a phase transition at Λ_c. For the theory to be valid up to Λ_c the following equation must be then satisfied

$$\int_{\lambda(\mu_0)}^{\bar{\lambda}(\Lambda_c,\lambda)} dx/\beta(x) = \ln(\Lambda_c/\mu_0) \tag{6.44}$$

For a given Λ_c the equation gives upper bounds for the low energy coupling constants (assuming $\bar{\lambda}(\Lambda_c, \lambda) = \infty$) which in turn, can be transformed into an upper bound for the number of elementary particles in the theory or their masses (see Problem 6.4).

Effective coupling constant

We will now study the explicit behaviour of the effective coupling constant in perturbation theory. With $\beta(\alpha)$, where $\alpha = \lambda$ for $\lambda\Phi^4$ and $\alpha = e^2/4\pi$ for QED and QCD, given by an expansion

$$\beta(\alpha) = \alpha[(\alpha/\pi)b_1 + (\alpha/\pi)^2 b_2 + (\alpha/\pi)^3 b_3 + \cdots] \tag{6.45}$$

we can calculate $\alpha(t)$ integrating (6.12) order-by-order in α. We expand

$$\alpha(t) = \sum_{m=1} a_m(t)\alpha^m, \quad a_1 = 1 \tag{6.46}$$

and

$$\beta(\alpha(t)) = \sum_{n=1} \alpha^{n+1}(t) b_n/\pi^n \tag{6.47}$$

Therefore

$$\sum_{m=2} \frac{da_m(t)}{dt}\alpha^m = \sum_{n=1} \frac{b_n}{\pi^n}\left[\alpha + \sum_{m=2} a_m(t)\alpha^m\right]^{n+1} \tag{6.48}$$

Comparing the coefficients at the same powers of α and integrating over t one gets

$$\begin{aligned} a_2(t) &= b_1 t/\pi \\ a_3(t) &= (b_1 t/\pi)^2 + b_2 t/\pi \\ &\vdots \\ a_n(t) &= (b_1 t/\pi)^{n-1} + O(t^{n-2}) \end{aligned} \tag{6.49}$$

so that

$$\alpha(t) = \alpha\left[1 + \sum_{n=1} (b_1 t\alpha/\pi)^n + O(t^{n-1}\alpha^n)\right] \tag{6.50}$$

For large $t = \frac{1}{2}\ln(\mu^2/\mu_0^2)$ or $t = \frac{1}{2}\ln(p^2/p_0^2)$, depending on whether we use (6.16) or (6.26), we can consider the so-called leading logarithm approximation in which we neglect terms $O(t^{n-1}\alpha^n)$ as compared to the terms $O(t^n\alpha^n)$. In this approximation the effective coupling constant reads

$$\alpha(t) = \frac{\alpha}{1 - (\alpha/\pi)b_1 t}, \quad \alpha = \alpha(t = 0) \tag{6.51}$$

The same result can be obtained by expanding $1/\beta(x)$ in the integral (6.11) in powers of x and directly integrating over x. This way one gets the series

$$b_1 t = \left[\frac{\pi}{\alpha} - \frac{\pi}{\alpha(t)}\right] + B_1 \ln\frac{\alpha}{\alpha(t)} + \sum_{n=1} \frac{C_n}{n}\left\{\left[\frac{\alpha(t)}{\pi}\right]^n - \left(\frac{\alpha}{\pi}\right)^n\right\} \tag{6.52}$$

where

$$B_n = b_{n+1}/b_1$$

and

$$C_n = (-1)^{n+1} \begin{vmatrix} B_1 & B_2 & \cdots & & B_{n+1} \\ 1 & B_1 & \cdots & & B_n \\ 0 & 1 & B_1 & \cdots & B_{n-1} \\ \vdots & & \ddots & \ddots & \vdots \\ 0 & 0 & 0 & 1 & B_1 \end{vmatrix}$$

Truncating $1/\beta(x)$ at a given order in x, gives us the corresponding approximate solution for $\alpha(t)$. In the first approximation ($b_1 \neq 0$, $b_n = 0$ for $n > 1$) we recover the leading-logarithm-result (6.51). In the second approximation ($b_1 \neq 0$, $b_2 \neq 0$, $b_n = 0$ for $n > 2$) we get the following

$$\alpha(t) = \alpha_1(t)\left[1 - \frac{\alpha_1(t)}{\pi}\frac{b_2}{b_1}\ln\left(1 - \frac{\alpha}{\pi}b_1 t\right)\right] \tag{6.53}$$

where $\alpha_1(t)$ is called the leading-logarithm-result (6.51).

In QED $b_1 = +\frac{2}{3}$ (Section 5.2) and $b_2 = \frac{1}{2}$ (de Rafael & Rosner 1974). Taking the limit $t \to -\infty$ in (6.53),

$$\alpha(t) \underset{t\to-\infty}{=} -\frac{\pi}{b_1 t} + \frac{\pi b_2}{b_1^3 t^2}\ln\frac{1}{t} + O\left(\frac{1}{t^2}\right) \tag{6.54}$$

we see explicitly that the point $\alpha = 0$ is an IR stable fixed point. In QCD $b_1 < 0$ and $b_2 < 0$, therefore, $\alpha = 0$ is a UV stable fixed point.

Another renormalization group parameter is the running mass $\bar{m}(t, m, \lambda)$

$$\bar{m}(t, m, \lambda) = m \exp\left\{\int_\lambda^{\bar{\lambda}(t,\lambda)} [\gamma_{\rm m}(x)/\beta(x)]\,dx\right\} \tag{6.55}$$

The behaviour of the Green's functions for rescaled momenta $e^{+t}p_i$ is controlled by $e^{-t}\bar{m}(t)$. In particular, in a theory with the UV fixed point λ_+ one gets

$$e^{-t}\bar{m}(t) = m\exp\{-t[1 - \gamma_{\rm m}(\lambda_+)]\}\exp\left\{-\int_\lambda^{\bar{\lambda}(t)} [dx/\beta(x)][\gamma_{\rm m}(\lambda_+) - \gamma_{\rm m}(x)]\right\} \tag{6.56}$$

Whether in the high energy limit masses can be neglected or not depends on $\gamma_{\rm m}(\lambda_+)$ and on the integral in the exponent. For asymptotically free theories $\lambda_+ = 0$ (and $\gamma_{\rm m}(0) = 0$) so only the integral is responsible for the large t limit of $\bar{m}(t)$, and e.g. for QCD mass corrections are suppressed by $e^{-t}t^p$ as $t \to \infty$, with $p > 0$.

6.4 Renormalization scheme and gauge dependence of the RGE parameters

Renormalization scheme dependence

The last problem we would like to discuss in this Chapter is the renormalization scheme dependence within the class of mass-independent renormalization procedures and the gauge dependence of the perturbative results for the functions β, γ_m and γ. The parameters of one renormalization scheme can be determined in terms of those in another scheme. If λ is the coupling in one scheme, another one will give a coupling λ'

$$\left.\begin{aligned}\lambda' = F(\lambda) &= \lambda + O(\lambda^2) \quad \text{for } \lambda\Phi^4 \\ &= \lambda + O(\lambda^3) \quad \text{for trilinear coupling theories}\end{aligned}\right\} \tag{6.57}$$

In the tree approximation the coupling is scheme independent. The wave-function renormalization Z_3 and mass renormalization Z_m will be given in the new scheme by

$$Z'_m(\lambda') = Z_m(\lambda)F_m(\lambda), \quad Z'_3(\lambda') = Z_3(\lambda)F_3(\lambda) \tag{6.58}$$

where $F_i(\lambda) = 1 + O(\lambda^2)$ (for the sake of definiteness we restrict our discussion to the case of the trilinear coupling theories (QED, QCD); modifications for $\lambda\Phi^4$ are trivial). For small enough λ we can assume that $dF/d\lambda \neq 0$ and $F_i \neq 0$. We get therefore

$$\begin{aligned}\beta'(\lambda') &= \mu\frac{d}{d\mu}\bigg|_{\lambda_{B,\varepsilon}} \lambda' = (dF/d\lambda)\beta(\lambda) \\ \gamma'(\lambda') &= \tfrac{1}{2}\mu\frac{d}{d\mu}\bigg|_{\lambda_{B,\varepsilon}} \ln Z'_3(\lambda') = \gamma(\lambda) + \frac{1}{2}\left(\frac{d}{d\lambda}\ln F_3\right)\beta(\lambda) \\ \gamma'_m(\lambda') &= \gamma_m(\lambda) + \frac{1}{2}\left(\frac{d}{d\lambda}\ln F_m\right)\beta(\lambda)\end{aligned} \tag{6.59}$$

Using (6.59) the following results can easily be obtained (let $\bar{\lambda}$ and $\bar{\lambda}' = F(\bar{\lambda})$ be fixed points in both schemes)

$$\text{(i)} \quad \frac{d\beta'}{d\lambda'}\bigg|_{\bar{\lambda}'} = \frac{d\beta}{d\lambda}\bigg|_{\bar{\lambda}} \tag{6.60}$$

$$\begin{aligned}&\text{(ii) If } \beta(\lambda) = -2b_0\lambda^3 - 2b_1\lambda^5 + \cdots \\ &\text{then } \beta'(\lambda') = -2b_0\lambda'^3 - 2b_1\lambda'^5 + \cdots\end{aligned} \tag{6.61}$$

i.e. the first two coefficients of the expansion for β are for mass-independent schemes, renormalization scale independent. Properties (i) and (ii) are important to ensure that the nature of the fixed points (asymptotic freedom, IR stability) is scheme independent.

$$\text{(iii)} \quad \gamma'(\bar{\lambda}') = \gamma(\bar{\lambda}), \quad \gamma'_m(\bar{\lambda}') = \gamma_m(\bar{\lambda})$$

This again must be so since the values of γ and γ_{m} at a fixed point determine the scaling behaviour and masses, and have physical significance.

(iv) if $\quad \gamma(\lambda)=\gamma_0\lambda^2+O(\lambda^4)$

then $\quad \gamma'(\lambda')=\gamma_0\lambda'^2+O(\lambda'^4)$

and similarly for $\gamma_{\mathrm{m}}(\lambda)$.

This is again important since γ_0 and γ_{m}^0 determine the leading behaviour for $t\to\pm\infty$ of the exponents in (6.26) and (6.11), respectively (scale factors for the field and the mass).

We stress that these properties hold only in the class of mass-independent renormalization schemes. In particular within this class of renormalization schemes and in the leading logarithm approximation $\alpha(q^2)$ is scheme independent. In mass-dependent schemes the renormalization constants Z_i and consequently the coefficients b_i, γ_i depend on the ratio m/μ and obviously cannot be equal to those in the mass-independent schemes.

Effective α in QED

It is of some interest to find a link between the effective coupling constant in QED and the 'on-shell' $\alpha^{\mathrm{ON}}=1/137$. To this end we consider the photon propagator renormalized in some mass-independent renormalization scheme

$$\alpha D^{\mu\nu}(q)=-\mathrm{i}\left(g^{\mu\nu}-\frac{q^\mu q^\nu}{q^2}\right)\frac{\alpha d(q^2/\mu^2,\alpha)}{q^2}-\alpha(1-a)\mathrm{i}\frac{q^\mu q^\nu}{(q^2)^2} \tag{6.62}$$

where $\alpha=\alpha(\mu)$ and

$$d(q^2/\mu^2,\alpha)=\frac{1}{1+\Pi(q^2/\mu^2,\alpha)}$$

The function Π is the scalar factor of the vacuum polarization tensor $\Pi_{\mu\nu}(q)$ defined in (5.26). An important observation is that in QED the quantity

$$\alpha_{\mathrm{eff}}(q^2)=\alpha d(q^2/\mu^2,\alpha)=(\alpha/Z_3)Z_3 d(q^2/\mu^2,\alpha)=\alpha_{\mathrm{B}}d_{\mathrm{B}}(q^2) \tag{6.63}$$

is renormalization scheme independent because the bare quantities are. Incidentally, we see that in QED the product $\alpha_{\mathrm{B}}d_{\mathrm{B}}(q^2)$ is finite. Relation (6.63) does not hold in QCD because then the Ward identities do not imply $Z_1=Z_2$. Thus, in QCD the relation between $\alpha d(q^2,\mu^2,\alpha)$ and $\alpha_{\mathrm{B}}d_{\mathrm{B}}(q^2)$ involves uncancelled renormalization constants; $\alpha d(q^2,\mu^2,\alpha)$ is then renormalization scheme dependent and $\alpha_{\mathrm{B}}d_{\mathrm{B}}(q^2)$ is not finite.

Turning again to QED we observe that

$$\alpha^{\mathrm{ON}}=\lim_{q^2\to 0}\alpha_{\mathrm{eff}}(q^2) \tag{6.64}$$

Indeed, in the on-shell renormalization scheme $\Pi^{\mathrm{ON}}(q^2=0)=0$ and given the scheme independence of α_{eff} defined by (6.63) we arrive at the result (6.64). It remains to find the relation between the $\alpha_{\text{eff}}(q^2)$ and the running $\alpha(q^2)$ defined as the solution to (6.11). We recall first that

$$\alpha(q^2) \underset{q^2\to 0}{=} 0$$

(it is an IR fixed point) whereas according to (6.64)

$$\alpha_{\text{eff}}(q^2) \underset{q^2\to 0}{=} \alpha^{\text{ON}}$$

Since the product $\alpha(\mu)d(q^2/\mu^2, \alpha(\mu))$ is renormalization scheme independent we can change $\mu^2 \to q^2$ to get

$$\alpha_{\text{eff}}(q^2) = \alpha(q^2)d(1, \alpha(q^2)) \tag{6.65}$$

Using (5.32) for $\Pi(q^2/\mu^2, \alpha(\mu^2))$ in the first order in α and with $\mu^2 = q^2$ one gets

$$\alpha_{\text{eff}}(q^2) \underset{q^2\to\infty}{=} \alpha(q^2) \tag{6.66}$$

This result actually holds to any order in α in the leading approximation because (see Problem 6.5)

$$\Pi(q^2/\mu^2, \alpha(\mu^2)) = A_1\alpha(\mu^2)\ln(|q^2|/\mu^2) + \sum_{n=2} A_n\alpha^n(\mu^2)\ln^{n-1}(|q^2|/\mu^2)$$
$$+ \text{non-leading terms}$$

The value of $\alpha_{\text{eff}}(q^2)$ in the limit $q^2 = 0$ can be obtained in terms of $\alpha(q^2)$ from the knowledge of the first terms of the perturbation expansion again using (5.32) as the finite limit of the expression

$$\alpha_{\text{eff}}(q^2) \underset{q^2\to 0}{=} \frac{\alpha(q^2)}{1 + [\alpha(q^2)/3\pi]\ln(|q^2|/m^2)} = \frac{0}{0}\,. \tag{6.67}$$

The inverse relation to first order in $\alpha^{\text{ON}} = \alpha_{\text{eff}}(0)$ reads

$$\alpha(q^2) \underset{\substack{q^2\to 0 \\ \alpha^{\text{ON}}\ln(|q^2|/m^2)\ll 1}}{=} \alpha^{\text{ON}}\left(1 + \frac{\alpha^{\text{ON}}}{3\pi}\ln\frac{|q^2|}{m^2}\right) \tag{6.68}$$

Gauge dependence of the β-function

In QED the $\beta(\alpha)$-function is gauge independent. This follows immediately from the fact that the renormalization constant Z_3 is gauge independent. It renormalizes the gauge-invariant operator $F_{\mu\nu}F^{\mu\nu}$ or, in other words, the transverse part of the propagator, which is gauge invariant (see also Problem 6.6).

In non-abelian gauge theories the renormalization programme is more complicated (see Section 8.1) and $\beta(\alpha)$ is in general gauge dependent. Nevertheless one can prove that (i) in any renormalization scheme the coefficient b_1 in the expansion

$$\beta(\alpha) = \alpha\left[b_1\frac{\alpha}{\pi} + b_2\left(\frac{\alpha}{\pi}\right)^2 + \cdots\right] \tag{6.69}$$

is gauge independent, (ii) in the minimal subtraction scheme $\beta(\alpha)$ is gauge independent.

To prove the first theorem we note that if we change the gauge we might change $\alpha(\mu)$ to $\alpha'(\mu)$; both can be defined e.g. as the value of the three-point-function at some value μ^2 of the kinematical invariants or according to the minimal subtraction scheme. But we can expand

$$\alpha'(\mu) = \alpha(\mu) + O(\alpha^2) \tag{6.70}$$

$\alpha' = \alpha$ in the zeroth order but the coefficients of the higher order terms might depend in general on the gauge parameter a. Now we consider the change of both $\alpha'(\mu)$ and $\alpha(\mu)$ under a change of the renormalization point μ to calculate $\beta'(\alpha')$ and $\beta(\alpha)$

$$\beta'(\alpha') = \mu\frac{\mathrm{d}}{\mathrm{d}\mu}\alpha'\bigg|_{\alpha_{\mathrm{B}},a_{\mathrm{B}},\varepsilon\,\text{fixed}} = \mu\frac{\mathrm{d}}{\mathrm{d}\mu}[\alpha(\mu) + O(\alpha^2)] = \beta(\alpha) + O(\alpha^3) \tag{6.71}$$

because $\alpha(\mu + \mathrm{d}\mu) = \alpha(\mu) + O(\alpha^2)$. This proves theorem (i).

Theorem (ii) can be proved as follows: consider the bare coupling constant α_{B} of the theory. By its definition it is a fixed, gauge-independent parameter. Therefore

$$\frac{\mathrm{d}}{\mathrm{d}a}\alpha_{\mathrm{B}}\bigg|_{\varepsilon\,\text{fixed}} = \frac{\mathrm{d}a_{\mathrm{B}}}{\mathrm{d}a}\frac{\mathrm{d}}{\mathrm{d}a_{\mathrm{B}}}\alpha_{\mathrm{B}} = 0 \tag{6.72}$$

and using the relation $\alpha_{\mathrm{B}} = \mu^{\varepsilon} Z_{\alpha}\alpha$ we have

$$0 = \frac{\mathrm{d}}{\mathrm{d}a}(Z_{\alpha}\alpha)\bigg|_{\alpha_{\mathrm{B}},\varepsilon\,\text{fixed}} = Z_{\alpha}\frac{\mathrm{d}}{\mathrm{d}a}\alpha + \alpha\frac{\mathrm{d}}{\mathrm{d}a}Z_{\alpha} \tag{6.73}$$

In the minimal subtraction scheme

$$Z_{\alpha} = 1 + Z_{\alpha}^{(1)}(\alpha, a)/\varepsilon + Z_{\alpha}^{(2)}(\alpha, a)/\varepsilon^2 + \cdots \tag{6.74}$$

and in particular in this scheme Z_{α} does not contain any constant term $C(\alpha, a)$ which would destroy the following argument. Inserting the expansion (6.74) into (6.73) we get

$$\frac{\mathrm{d}\alpha}{\mathrm{d}a} + \frac{1}{\varepsilon}\left(\frac{\mathrm{d}\alpha}{\mathrm{d}a}Z_{\alpha}^{(1)} + \alpha\frac{\mathrm{d}Z_{\alpha}^{(1)}}{\mathrm{d}a}\right) + O\left(\frac{1}{\varepsilon^2}\right) = 0 \tag{6.75}$$

The only way this equation can be satisfied is $\mathrm{d}\alpha/\mathrm{d}a = 0$ and $\mathrm{d}Z_{\alpha}^{(1)}/\mathrm{d}a = 0$. Thus in the minimal subtraction scheme the renormalized coupling constant $\alpha(\mu)$ is gauge independent and consequently the same is true for the $\beta(\alpha) = \mu(\mathrm{d}/\mathrm{d}\mu)\alpha$.

Problems

6.1 Calculate the renormalization group functions β, γ, γ_m in QED in the momentum-subtraction renormalization scheme.

6.2 Consider a two-coupling theory: non-abelian gauge theory with a scalar multiplet transforming according to the representation R of the gauge group. Write down the system of coupled RGEs for the gauge coupling g and for the meson self-coupling λ, to the lowest order in perturbation theory, and study the effect of the meson self-interaction on the asymptotic freedom of the theory. (See also Gross 1976.)

6.3 Consider the QED of n_h 'heavy' fermions Ψ_h, each with non-vanishing mass M, and n_l 'light' fermions Ψ_l each with zero mass. Obtain the effective field theory for the light fields A_μ and Ψ_l, in the one-loop approximation by 'integrating out' the heavy fermions from the Green's functions with light field external legs: take the photon propagator as an example and check the validity of the Appelquist–Carazzone decoupling theorem in the on-shell renormalization scheme (Appelquist & Carazzone 1975) and in the minimal subtraction scheme (Ovrut & Schnitzer 1981). In the latter case understand the role of a finite renormalization of the coupling constant in absorbing the effect of the heavy mass. Study the RGE for the Green's functions of the effective theory.

6.4 Consider the standard electroweak theory based on the gauge group $SU(2) \times U(1)$ with the gauge symmetry spontaneously broken (see Section 11.1) by one complex scalar doublet. Adopting the point of view expressed by (6.44) and using the lowest order result for the β-function estimate the upper bound for the Higgs particle mass (Dashen & Neuberger 1983). Assume that for consistency the relation $m_H \leqslant \Lambda_c$ must be satisfied.

6.5 Using the RGE in a mass-independent renormalization scheme show that the vacuum polarization in QED, (5.26), reads

$$\Pi(q^2/\mu^2, \alpha(\mu^2)) = A_1\alpha(\mu^2)\ln(|q^2|/\mu^2) + \sum_{n=2} A_n\alpha^n(\mu^2)\ln^{n-1}(|q^2|/\mu^2) + \text{non-leading terms}$$

where the coefficients A_n are determined in terms of the b_1 and b_2 of the β-function.

6.6 Using the functional integral representation for the bare full two-photon Green's function in QED show by taking the derivative with respect to the gauge parameter a that the transverse part of the propagator is gauge independent.

7

Scale invariance and operator product expansion

7.1 Scale invariance

Scale transformations

An important concept in quantum field theory is that of scale invariance. Among others, it is closely related to the famous Bjorken scaling in the deep inelastic lepton–hadron scattering. Scale transformation for the coordinates is defined as follows

$$x \to x' = e^{-\varepsilon} x \tag{7.1}$$

where ε is a real number. Obviously, all such transformations form an abelian group. Scale invariance is the invariance under the group of scale transformations.

If the configuration variables are scaled down as in (7.1) local fields $\Phi(x)$ (in this Section we denote by $\Phi(x)$ any scalar field $\varphi(x)$ or spinor field $\Psi(x)$) will be in general subject to a unitary transformation $u(\varepsilon)$

$$\Phi(x) \to \Phi'(x) = U(\varepsilon)\Phi(x)U^{-1}(\varepsilon) = T(\varepsilon)\Phi(e^{\varepsilon}x) \tag{7.2}$$

where $T(\varepsilon)$ is a finite dimensional representation of the group. We assume it to be fully reducible and therefore we can write

$$T(\varepsilon) = \exp(d_{\Phi}\varepsilon) \tag{7.3}$$

where the constant d_{Φ} is called the scale dimension of the field $\Phi(x)$.

Let us consider first a free massless field theory. Its lagrangian does not contain dimensionful constants and should be invariant under the transformation (7.2). For free fields one has the canonical equal-time commutation relations

$$\left.\begin{aligned} [\varphi(\mathbf{x},t), \dot{\varphi}(\mathbf{y},t)] &= i\delta(\mathbf{x}-\mathbf{y}) \\ \{\Psi(\mathbf{x},t), \Psi^{\dagger}(\mathbf{y},t)\} &= \delta(\mathbf{x}-\mathbf{y}) \end{aligned}\right\} \tag{7.4}$$

and the scale dimension d_{Φ} is then defined so that the commutation relations (7.4) remain invariant under the scale transformation. Transforming the fields in relations (7.4) according to (7.2) and (7.3) i.e.

$$\Phi(x) = \exp(-d_\Phi \varepsilon)\Phi'(e^{-\varepsilon}x) \tag{7.5}$$

one easily sees that invariance of (7.4) implies that these so-called canonical scale dimensions are $d_\varphi = 1$ for scalar fields and $d_\Psi = \frac{3}{2}$ for spinor fields. We see that the canonical scale dimensions of fields coincide with their ordinary dimensions defined on purely dimensional grounds (see Section 4.2).

In an infinitesimal form the scale transformation (7.2) is as follows

$$\Phi'(x) = U(\varepsilon)\Phi(x)U^{-1}(\varepsilon) = \Phi(x) + \varepsilon(d_\Phi + x_\mu(\partial/\partial x_\mu))\Phi(x) + O(\varepsilon^2) \tag{7.6}$$

For interacting field theories one still has the canonical commutation relations (7.4) for bare fields. Defining scale dimensions for bare fields as canonical ones, one can easily check that a lagrangian which does not contain dimensionful constants is invariant under the scale transformation (7.2) (of course, strictly speaking, it is the action integral which is scale invariant). Take, for instance, the following lagrangian density

$$\mathscr{L} = i\bar{\Psi}\partial\!\!\!/\Psi + \tfrac{1}{2}(\partial_\mu\varphi)(\partial^\mu\varphi) + g\bar{\Psi}\gamma_5\Psi\varphi - (\lambda/4!)\varphi^4 \tag{7.7}$$

Using (7.6) and the values $d_\varphi = 1$, $d_\Psi = \frac{3}{2}$ it is easy to calculate

$$\begin{aligned}\delta\mathscr{L}(x) &= \mathscr{L}'(x) - \mathscr{L}(x) = \mathscr{L}(\Phi'(x)) - \mathscr{L}(\Phi(x)) \\ &= \sum_{\Phi=\varphi,\Psi} \frac{\partial\mathscr{L}}{\partial\Phi}\delta\Phi + \frac{\partial\mathscr{L}}{\partial(\partial\Phi/\partial x_\mu)}\delta\left(\frac{\partial\Phi}{\partial x_\mu}\right) \\ &= \varepsilon(4 + x_\mu(\partial/\partial x_\mu))\mathscr{L}(x)\end{aligned} \tag{7.8}$$

where we have used the relation

$$\delta\left(\frac{\partial}{\partial x_\nu}\Phi(x)\right) = \frac{\partial}{\partial x_\nu}(\Phi'(x) - \Phi(x)) = \varepsilon\left(d_\Phi + 1 + x\cdot\frac{\partial}{\partial x}\right)(\partial/\partial x_\nu)\Phi(x)$$

This result is expected on general grounds since in a lagrangian which does not contain dimensionful constants each term has a canonical dimension equal to its ordinary dimension which is four. Therefore the corresponding action integral is invariant under scale transformations

$$\delta I = \int d^4x\,\delta\mathscr{L}(x) = 0 \tag{7.9}$$

(one can also integrate explicitly using (7.8) and neglecting, as usual the surface terms at infinity). Terms like $m^2\Phi^2(x)$, if added to (7.7), would, of course, break the scale invariance of the lagrangian as their scale dimension is two. Notice the difference between scale dimension and ordinary dimension which is four for the term $m^2\Phi^2$; scale transformation is a transformation of dynamical variables (the fields) but not of the dimensionful parameters such as masses since we want to stay in the framework of the same physical theory. Terms like $m^2\Phi^2(x)$ break scale invariance explicitly. However, it is already worth recognizing at this point that in quantum field theory described by a scale-invariant lagrangian the scale invariance is in general, also broken. This is due to the necessity of renormalization which introduces a dimensionful parameter.

Turning back to an exactly scale-invariant world we can introduce the generator D of the scale transformation

$$U(\varepsilon) = \exp(\mathrm{i}\varepsilon D) \tag{7.10}$$

and study its commutation relations with generators of the Poincaré group (for conventions see Problem 7.1)

$$\left.\begin{aligned} &U(a) = \exp(\mathrm{i}a^\mu P_\mu) \\ \text{and}\quad & \\ &U(\omega) = \exp(-\tfrac{1}{2}\mathrm{i}\omega^{\mu\nu} M_{\mu\nu}) \end{aligned}\right\} \tag{7.11}$$

By comparing two sequences of dilatation and translation

$$\Phi(x) \underset{\varepsilon}{\rightarrow} \exp(d_\Phi \varepsilon)\Phi(\mathrm{e}^\varepsilon x) \underset{a_\mu}{\rightarrow} \exp(d_\Phi \varepsilon)\Phi(\mathrm{e}^\varepsilon x + a_\mu)$$

and

$$\Phi(x) \underset{a_\mu}{\rightarrow} \Phi(x + a_\mu) \underset{\varepsilon}{\rightarrow} \exp(d_\Phi \varepsilon)\Phi(\mathrm{e}^\varepsilon(x + a_\mu))$$

we conclude that

$$U(\varepsilon)U(a) = U(\mathrm{e}^\varepsilon a)U(\varepsilon)$$

or

$$\mathrm{i}[D, P_\mu] = P_\mu \tag{7.12}$$

Similarly, one gets

$$U(\omega)U(\varepsilon) = U(\varepsilon)U(\omega)$$

and

$$[D, M_{\mu\nu}] = 0 \tag{7.13}$$

From (7.6) we get the following differential representation of the generator D

$$\mathrm{i}[D, \Phi(x)] = (d_\Phi + x_\mu(\partial/\partial x_\mu))\Phi(x) \tag{7.14}$$

Dilatation current

The dilatation generator D can be written in terms of the Noether dilatation current $D_\mu(x)$. The conserved dilatation current can be deduced by comparing an infinitesimal change, obtained without the use of the Euler–Lagrange equations of motion, under the scale transformation of the lagrangian density of a scale invariant theory (7.8)

$$\delta\mathscr{L} = \varepsilon\left(4 + x_\mu \frac{\partial}{\partial x_\mu}\right)\mathscr{L} = \varepsilon\frac{\partial}{\partial x_\mu}(x_\mu \mathscr{L})$$

with an alternate expression for $\delta\mathscr{L}$ which does use Euler–Lagrange equations

$$\delta\mathscr{L} = (\partial/\partial x_\mu)(\Pi_\mu \delta\Phi) \quad \Phi = (\varphi, \Psi) \tag{7.15}$$

where Π_μ is the conjugate momentum. The Noether dilatation current is therefore the following

$$D_\mu(x) = \Pi_\mu(d_\Phi + x_\nu(\partial/\partial x_\nu))\Phi(x) - x_\mu\mathscr{L}$$

or

$$D_\mu(x) = \Pi_\mu d_\Phi \Phi(x) + x^\nu \Theta^{\text{can}}_{\mu\nu} \tag{7.16}$$

where

$$\Theta^{\text{can}}_{\mu\nu} = \Pi_\mu \frac{\partial}{\partial x^\nu}\Phi(x) - g_{\mu\nu}\mathscr{L}$$

is the canonical energy–momentum tensor. The current is conserved in the presence of the symmetry, but it may, of course, be defined even in the absence of scale invariance.

The dilatation current can be written in a more compact way if we introduce a modified energy–momentum tensor $\Theta_{\mu\nu}$. Let us consider first a scalar field theory and define $\Theta_{\mu\nu}$ as follows:

$$\Theta_{\mu\nu} = \Theta^{\text{can}}_{\mu\nu} - \tfrac{1}{6}(\partial_\mu\partial_\nu - g_{\mu\nu}\Box)\varphi^2 \tag{7.17}$$

The new tensor has the following properties

$$\Theta_{\mu\nu} = \Theta_{\nu\mu}, \quad \partial^\mu\Theta_{\mu\nu} = 0$$

and

$$P_\mu = \int d^3x \Theta^{\text{can}}_{0\mu} = \int d^3x \Theta_{0\mu}$$

$$M_{\mu\nu} = \int d^3x (x_\mu\Theta_{0\nu} - x_\nu\Theta_{0\mu})$$

i.e. $\Theta_{\mu\nu}$ is a legitimate energy–momentum tensor of our theory. Using the relation $\Pi_\mu = \partial_\mu\varphi$ it is easy to check that in terms of $\Theta_{\mu\nu}$ the D_μ of (7.16) ($d_\Phi = 1$) reads

$$D_\mu = x^\nu\Theta_{\mu\nu} - \tfrac{1}{6}\partial_\nu(x_\mu\partial^\nu - x^\nu\partial_\mu)\varphi^2$$

The last term is the divergence of an antisymmetric tensor, so it does not contribute to $\partial^\mu D_\mu$ and to D. Thus we define

$$D_\mu(x) = x^\nu\Theta_{\mu\nu}(x) \tag{7.18}$$

and

$$\partial^\mu D_\mu(x) = \Theta_\mu{}^\mu \tag{7.19}$$

An important property of the new tensor is that its trace vanishes

$$\Theta_\mu{}^\mu = 0 \tag{7.20}$$

for a scale-invariant scalar field theory (and

$$\Theta_\mu{}^\mu = m^2\varphi^2 \tag{7.20a}$$

if a mass term is added) while the trace of $\Theta_{\mu\nu}^{\text{can}}$ contains singular derivative terms. The new tensor has been invented by Callan, Coleman & Jackiw (1970) when looking for an object with the 'soft' divergences (7.20a) in the context of searching for an energy–momentum tensor whose matrix elements between physical states are finite. In a renormalizable field theory the finiteness of matrix elements of the T-ordered products of local currents is an additional requirement, which in general must be separately verified (see, for instance, Section 10.1). Actually, the proof of Callan, Coleman & Jackiw of the finiteness of matrix elements of the $\Theta_{\mu\nu}$ is incomplete because it ignores the non-soft contribution (7.46) to the $\Theta_\mu{}^\mu$ from the anomalous scale invariance breaking; see Adler, Collins & Duncan (1977) for the complete discussion.

Equation (7.17) can be generalized to include fields with non-zero spin (Callan, Coleman & Jackiw 1970):

$$\Theta_{\mu\nu} = T_{\mu\nu} - \tfrac{1}{6}(\partial_\mu\partial_\nu - g_{\mu\nu}\Box)\varphi^2 \tag{7.21}$$

where

$$T_{\mu\nu} = \Pi_\mu\partial_\nu\Phi - g_{\mu\nu}\mathscr{L} + \tfrac{1}{2}\partial^\lambda(\Pi_\lambda\Sigma_{\mu\nu}\Phi - \Pi_\mu\Sigma_{\lambda\nu}\Phi - \Pi_\nu\Sigma_{\lambda\mu}\Phi) \tag{7.22}$$

is the conventional symmetric energy–momentum tensor† and where all fields of the theory are assembled into a vector Φ whereas the scalar fields are denoted by φ. The spin matrix $\Sigma_{\mu\nu}$ is defined by the transformation properties of the fields under an infinitesimal Lorentz transformation

$$\delta\Phi = -\tfrac{1}{2}(x_\mu\partial_\nu\Phi - x_\nu\partial_\mu\Phi + \Sigma_{\mu\nu}\Phi)\omega^{\mu\nu} \tag{7.23}$$

and reads, e.g. $\Sigma_{\mu\nu} = 0$ for spin 0 fields, $\Sigma_{\mu\nu} = -\tfrac{1}{2}\mathrm{i}\sigma_{\mu\nu} = \tfrac{1}{4}[\gamma_\mu, \gamma_\nu]$ for spin one-half fields. In terms of $T_{\mu\nu}$ the generators of the Lorentz transformation are

$$M_{\mu\nu} = \int \mathrm{d}^3x(x_\mu T_{0\nu} - x_\nu T_{0\mu})$$

The tensor $\Theta_{\mu\nu}$ defined by (7.21) has the same properties as the tensor (7.17) defined for scalar fields only. In particular its trace is a measure of scale invariance breaking.

Conformal transformations

It turns out that many scale-invariant theories are invariant under a larger symmetry group of transformations, the so-called conformal group, which in

† It can be constructed by noting that from conservation of the angular momentum current $\partial_\lambda\mathscr{M}^{\lambda\mu\nu} = 0$ where

$$\mathscr{M}^{\lambda\mu\nu} = x^\mu T^{\lambda\nu}_{\text{can}} - x^\nu T^{\lambda\mu}_{\text{can}} + \Pi^\lambda\Sigma^{\mu\nu}\Phi$$

and

$$M^{\mu\nu} = \int \mathrm{d}^3x\mathscr{M}^{0\mu\nu}$$

we get

$$T^{\mu\nu}_{\text{can}} - T^{\nu\mu}_{\text{can}} + \partial_\lambda(\Pi^\lambda\Sigma^{\mu\nu}\Phi) = 0$$

Equation (7.22) follows if we require $\partial_\mu T^{\mu\nu} = \partial_\nu T^{\mu\nu} = 0$

addition to translations, Lorentz transformations and scale transformation also contains special conformal transformations. A conformal (i.e. angle-preserving) transformation must leave invariant the ratio

$$\frac{\mathrm{d}x^\alpha \,\mathrm{d}y_\alpha}{|\mathrm{d}x||\mathrm{d}y|}$$

So it must have the property that

$$\begin{aligned}\mathrm{d}s'^2 &= g_{\alpha\beta}\,\mathrm{d}x'^\alpha\,\mathrm{d}x'^\beta = g_{\alpha\beta}(\partial x'^\alpha/\partial x^\gamma)(\partial x'^\beta/\partial x^\delta)\mathrm{d}x^\gamma \mathrm{d}x^\delta \\ &= f(x) g_{\alpha\beta}\,\mathrm{d}x^\alpha \mathrm{d}x^\beta \equiv f(x)\,\mathrm{d}s^2\end{aligned} \tag{7.24}$$

where $f(x)$ is a certain scalar function.
For infinitesimal transformations

$$x'^\mu = x^\mu - \varepsilon^\mu(x)$$

one therefore obtains the following condition

$$\partial_\nu \varepsilon_\mu(x) + \partial_\mu \varepsilon_\nu(x) = -h(x) g_{\mu\nu} \tag{7.25}$$

where $h(x) = f(x) - 1$. This equation gives

$$\begin{aligned}(n-2)\partial_\alpha \partial_\beta h(x) &= 0 \\ \partial^\mu \partial_\mu \varepsilon_\alpha(x) &= \tfrac{1}{2}(n-2)\partial_\alpha h(x)\end{aligned}$$

where n is the dimension of the space-time. We see that for $n > 2$,† $h(x)$ is at most linear in x and, consequently, $\varepsilon_\mu(x)$ is at most quadratic in x. The general solution for $\varepsilon_\mu(x)$ satisfying our constraints reads

$$\varepsilon^\mu(x) = a^\mu + \varepsilon x^\mu + \omega^{\mu\nu} x_\nu + (-2b\cdot x x^\mu + b^\mu x^2); \; \omega^{\mu\nu} = -\omega^{\nu\mu} \tag{7.26}$$

It includes (listed are the finite transformations)

translations	$x'^\mu = x^\mu - a^\mu,$	$h(x) = 0$
scale transformation	$x'^\mu = \mathrm{e}^{-\varepsilon} x^\mu,$	$h(x) = 2\varepsilon$
Lorentz transformations	$x'^\mu = \Lambda^{\mu\nu}(\omega) x_\nu,$	$h(x) = 0$
special conformal transformations	$x'^\mu = \dfrac{x^\mu - b^\mu x^2}{1 - 2b\cdot x + b^2 x^2},$	$h(x) = 4b\cdot x$

which form together the 15-parameter conformal group (see Problem 7.1). Conformal invariance implies scale invariance since the commutator of an infinitesimal conformal transformation and an infinitesimal translation contains a scale transformation. However the opposite is not true and examples of field theories which are scale invariant but not conformally invariant can be constructed. Scale invariance implies conformal invariance if, and only if, there exists in the considered theory a symmetric traceless conserved energy–momentum tensor so that $\partial^\mu D_\mu = \Theta^\mu{}_\mu = 0$

†The case $n = 2$, interesting for statistical mechanics and string theories, requires a separate discussion; see Problem 7.2.

It can be checked that such a $\Theta_{\mu\nu}$ exists in all renormalizable field theories of fields of spin $\leqslant 1$. In this case the currents

$$(j^{\varepsilon})^{\mu}(x) = \Theta^{\mu\nu}(x)\varepsilon_{\nu}(x) \tag{7.27}$$

are conserved for any $\varepsilon_{\nu} = \varepsilon_{\nu}(x, a^{\alpha}, \varepsilon, \omega^{\alpha\beta}, b^{\alpha})$ satisfying (7.25) (check it) and indeed scale invariance goes together with the special conformal symmetry. From (7.26) and (7.27) we find explicitly the following four generators of the special conformal transformations

$$K^{\alpha} = \int d^3x (j^{\alpha})^0 = \int d^3x (x^2 \Theta^{\alpha 0} - 2x^{\alpha} x_{\nu} \Theta^{\nu 0}) \tag{7.28}$$

For the transformation of fields under special conformal transformations see Problem 7.1.

7.2 Broken scale invariance

General discussion

Scale invariance cannot be an exact symmetry of the real world. If it were all particles would have to be massless or their mass spectra continuous. Indeed, it follows from commutation relation (7.12) that

$$[D, P^2] = -2iP^2$$

or, exponentiating,

$$\exp(i\varepsilon D) P^2 \exp(-i\varepsilon D) = \exp(2\varepsilon) P^2 \tag{7.29}$$

Acting on a single particle state $|p\rangle$ with four-momentum $p\,(P^2|p\rangle = p^2|p\rangle)$ one gets

$$P^2 \exp(-i\varepsilon D)|p\rangle = \exp(2\varepsilon) p^2 \exp(-i\varepsilon D)|p\rangle \tag{7.30}$$

i.e. the state $\exp(-i\varepsilon D)|p\rangle$ is an eigenstate of P^2 with eigenvalue $\exp(2\varepsilon)p^2$. If we assume in addition that the vacuum is unique under scale transformations (that is if scale invariance is not spontaneously broken)

$$\exp(-i\varepsilon D)|0\rangle = |0\rangle$$

then we conclude that

$$\exp(-i\varepsilon D)|p\rangle = \exp(-i\varepsilon D) a^{\dagger}(p)|0\rangle = \exp(d_a) a^{\dagger}(e^{\varepsilon}p)|0\rangle \tag{7.31}$$

where $a^{\dagger}(p)$ is the creation operator for the considered particle with momentum p and d_a is its dimension. Result (7.31) means that the state $\exp(-i\varepsilon D)|p\rangle$ is a quantum of the same field as the state $|p\rangle$ but with a rescaled momentum (if the vacuum was not unique then the state $\exp(-i\varepsilon D)|p\rangle$ would belong to a different Hilbert space than the state $|p\rangle$ and our final conclusion would be avoided) and therefore, by (7.30), all particles must be massless or the mass spectrum must be continuous.

We will learn in the next Chapters that there are several ways to account for the symmetry breaking. The simplest idea for symmetry breaking is that of 'approximate' symmetries. One assumes that there are terms in the lagrangian (e.g. mass terms in the case of scale invariance) which violate the symmetry, but they are 'small'. Another concept is that of spontaneous symmetry breaking. Here the dynamical equations are completely symmetric, the Noether currents associated with the symmetry are conserved and the Ward identities are still valid, but the ground state is asymmetric. We believe this idea to be realized for chiral symmetry in strong interactions. Finally there is a third way: anomalous breaking of symmetries. Whenever a classical theory possesses a symmetry, but there is no way of quantizing the theory so as to preserve that symmetry, we say that there are anomalies in the conservation equations for the symmetry currents. More technically, quantum field theoretical calculations must be regularized to avoid infinities and anomalies occur when there does not exist a regularization procedure which respects the classical symmetry and the corresponding Ward identity. Of course, a given symmetry may be broken by several of these mechanisms, simultaneously.

In quantum field theory, even with a scale-invariant lagrangian, the anomalous scale invariance breaking is quite expected. Scale invariance requires that there be no dimensionful parameters whereas the regularization of the quantum theory is effected by introducing a dimensionful cut-off or dimensionful coupling constants in the dimensional regularization procedure. Equivalently, an unavoidable renormalization procedure necessarily introduces a scale at which the theory is renormalized and this breaks scale invariance.

In the rest of this Section we shall study the 'anomalous' response of the theory to scale transformations by comparing the naive Ward identities with the renormalization group approach. For the discussion of the spontaneous breaking of scale invariance we refer the reader to the literature. In QCD the effects of the anomalous explicit breaking of scale symmetry are dominant, and there is no remaining consequence of the original classical scale symmetry.

Anomalous breaking of scale invariance

Ward identities for the dilatation current (trace identities) can be obtained from the general formula (10.11)

$$\frac{\partial}{\partial y_\mu}\langle 0|TD_\mu(y)\Phi(x_1)\ldots\Phi(x_n)|0\rangle = \langle 0|T\Theta_\mu{}^\mu(y)\Phi(x_1)\ldots\Phi(x_n)|0\rangle$$
$$-\mathrm{i}\sum_i \delta(x_i-y)\langle 0|T\Phi(x_1)\ldots\delta\Phi(x_i)\ldots\Phi(x_n)|0\rangle \tag{7.32}$$

where $\delta\Phi$s are given by (7.14). By integrating with respect to variable y the l.h.s. vanishes: it gives surface terms which can be neglected if there are no zero mass particles coupled to $\Theta_\mu{}^\mu$; in reality the scale invariance may be spontaneously broken but simultaneously the corresponding Goldstone bosons (dilatons) get masses due to anomalies; integrating with respect to y is equivalent to taking the Fourier transform and letting $k \to 0$. We then obtain

$$\int d^4y \langle 0| T\Theta_\mu{}^\mu(y)\Phi(x_1)\ldots\Phi(x_n)|0\rangle = \mathrm{i}\sum_i \langle 0| T\Phi(x_1)\ldots\delta\Phi(x_i)\ldots\Phi(x_n)|0\rangle$$

$$= \mathrm{i}\left(nd_\Phi + x_1^\mu \frac{\partial}{\partial x_1^\mu} + \cdots + x_n^\mu \frac{\partial}{\partial x_n^\mu} \right)\langle 0| T\Phi(x_1)\ldots\Phi(x_n)|0\rangle \tag{7.33}$$

In the last step of (7.33) we have brought the derivatives outside the T-product which results in further equal-time commutator terms which, however, cancel in pairs. For example, for $n = 2$ we have

$$\langle 0| T\Phi(x_1)x_2^\mu \frac{\partial}{\partial x_2^\mu}\Phi(x_2)|0\rangle + \langle 0| Tx_1^\mu \frac{\partial}{\partial x_1^\mu}\Phi(x_1)\Phi(x_2)|0\rangle$$

$$= \left(x_1^\mu \frac{\partial}{\partial x_1^\mu} + x_2^\mu \frac{\partial}{\partial x_2^\mu} \right)\langle 0| T\Phi(x_1)\Phi(x_2)|0\rangle + x_1^0\delta(x_1^0 - x_2^0)\langle 0|[\Phi(x_1), \Phi(x_2)]|0\rangle$$

$$+ x_2^0\delta(x_1^0 - x_2^0)\langle 0|[\Phi(x_2), \Phi(x_1)]|0\rangle \tag{7.34}$$

Defining the momentum-dependent Green's functions

$$(2\pi)^4\delta\left(\sum_1^n p_i\right)G^{(n)}(p_1,\ldots,p_{n-1})$$

$$= \int dx_1\ldots dx_n \exp\left(\mathrm{i}\sum_i p_i x_i\right)\langle 0| T\Phi(x_1)\ldots\Phi(x_n)|0\rangle \tag{7.35}$$

and analogously for $G_\Theta^{(n)}(k, p_1,\ldots,p_{n-1})$ as the Fourier transform of the $\langle 0| T\Theta_\mu{}^\mu(y)\Phi(x_1)\ldots\Phi(x_n)|0\rangle$ we get from (7.33) the following Ward identity

$$\left[-\sum_{i=1}^{n-1} p_i \frac{\partial}{\partial p_i} + n(d_\Phi - 4) + 4 \right] G^{(n)}(p_1,\ldots,p_{n-1}) = -\mathrm{i}G_\Theta^{(n)}(0, p_1,\ldots,p_{n-1}) \tag{7.36}$$

Summation over $(n - 1)$ momenta and the additional factor 4 in the square bracket are due to the fact that, effectively, we have only $(n - 1)$ four-dimensional integrations in (7.35). We observe that for $d_\Phi = d^{\mathrm{can}}$ the $(nd_\Phi - 4n + 4)$ is just the dimension D of the Green's function $G^{(n)}(p_1,\ldots,p_{n-1})$ $(G^{(n)} \sim M^D)$. In addition if we parametrize all momenta by the scaling factor $p_i = \rho\hat{p}_i$ so that $\sum_i p_i\partial/\partial p_i = \rho\partial/\partial\rho$ we can rewrite (7.36) as follows ($t = \ln\rho$)

$$[-\partial/\partial t + D]G^{(n)}(\mathrm{e}^t p_1,\ldots,\mathrm{e}^t p_{n-1}) = -\mathrm{i}G_\Theta^{(n)}(0, \mathrm{e}^t p_1,\ldots,\mathrm{e}^t p_{n-1}) \tag{7.37}$$

From (7.37) one can derive analogous equations for the 1PI Green's functions $\Gamma^{(n)}(p_1,\ldots,p_{n-1})$ and $\Gamma_\Theta^{(n)}(0, p_1,\ldots,p_{n-1})$

$$[-\partial/\partial t + D]\Gamma^{(n)}(\mathrm{e}^t p_i) = -\mathrm{i}\Gamma_\Theta^{(n)}(0, \mathrm{e}^t p_i) \tag{7.38}$$

where $D = 4 - nd_\Phi$ is the dimension of the 1PI Green's function $\Gamma^{(n)}$.

Equations (7.37) and (7.38) are the Ward identities following from the Noether relation $\partial_\mu D^\mu = \Theta_\mu{}^\mu$ and from the canonical field transformation rule (7.14). In particular, for $\Theta_\mu{}^\mu = 0$, i.e. for a scale-invariant lagrangian, the solution to (7.38) reads

$$\Gamma^{(n)}(\rho p_i) = \rho^D \Gamma^{(n)}(p_i) \tag{7.39}$$

In this case the Green's functions exhibit canonical scaling. However we have already mentioned that there is no reason to expect these identities to be valid in quantum field theory since the scale invariance is always broken by a dimensionful cut-off. The response of a quantum field theory to the scale transformation is given by the RGE (6.24).

Comparing (6.24) with the Ward identity (7.38) we see that the previously mentioned anomalous breaking of canonical scale invariance occurs in quantum field theory. Either the dilatation current has an anomalous divergence, or the field transforms anomalously, or both. For easy reference we recall the solution to (6.24)

$$\Gamma^{(n)}(\mathrm{e}^t p_i, \lambda, m, \mu) = \exp(Dt)\Gamma^{(n)}(p_i, \bar{\lambda}(t), \bar{m}(t), \mu)\exp\left[-n\int_0^t \gamma(\bar{\lambda}(t'))\,\mathrm{d}t'\right] \tag{7.40}$$

where

$$\int_\lambda^{\bar{\lambda}(t)} \mathrm{d}x/\beta(x) = t \tag{7.41}$$

and

$$\bar{m}(t) = (1/\rho)m\exp\left[\int_0^t \gamma_\mathrm{m}(\bar{\lambda}(t'))\,\mathrm{d}t'\right]$$

$$= (1/\rho)m\exp\left\{\int_\lambda^{\bar{\lambda}(t)} \mathrm{d}x\,[\gamma_\mathrm{m}(x)/\beta(x)]\right\} \tag{7.41a}$$

Consider first a massless $m = 0$ theory with a conserved Noether dilatation current $\partial_\mu D^\mu = 0$. As mentioned in Section 6.1 only for $\beta(\lambda) = \gamma(\lambda) = 0$ i.e. for a non-interacting theory do we recover the canonical scaling, namely solution (7.39) to (7.38). Imagine now a theory with a UV fixed point λ_+. The exponential factor in (7.40) can then be written as

$$\exp\left[-n\int_0^t \gamma(\bar{\lambda}(t'))\,\mathrm{d}t'\right] = \rho^{-n\gamma(\lambda_+)+\varepsilon(t)}$$

where

$$\varepsilon(t) = -(1/t)\int_\lambda^{\bar{\lambda}(t)} \mathrm{d}x\,n[\gamma(x) - \gamma(\lambda_+)]/\beta(x) \tag{7.42}$$

If the integral defining $\varepsilon(t)$ is convergent then $\varepsilon(t) = O(1/t)$ and the theory is asymptotically, for $t \to \infty$, scale invariant. Apart from a trivial case when λ is exactly equal to λ_+ and the RGE can then be solved directly and gives (7.39) with $D \to D - n\gamma(\lambda_+)$, this happens when λ_+ is a simple zero of $\beta(x)$: $\beta(\lambda_+) = 0$, $\beta'(\lambda_+) < 0$ (we assume that $\gamma(x)$ is differentiable). The asymptotic behaviour of $\Gamma^{(n)}(\mathrm{e}^t p_i, \lambda, \mu)$

is obtained by expanding $\bar{\lambda}(t)$ about λ_+. The leading term is

$$\Gamma^{(n)}(e^t p_i, \lambda, \mu) \sim \rho^{D-n\gamma(\lambda_+)}\Gamma^{(n)}(p_i, \lambda_+, \mu) \tag{7.43}$$

The function $\gamma(\lambda_+)$ is called the anomalous dimension of the field. The leading corrections to (7.43) arise from $O(\bar{\lambda}(t) - \lambda_+)$ terms in the expansion of $\Gamma^{(n)}(p_i, \bar{\lambda}(t), \mu)$ and of the exponential factor (7.42). More specifically one gets

$$\Gamma^{(n)}(\rho p_i, \lambda, \mu) = \rho^{D-n\gamma(\lambda_+)} \exp\left\{ - \int_\lambda^{\lambda_+} dx[\gamma(x) - \gamma(\lambda_+)]n/\beta(x)\right\}$$
$$* \Gamma^{(n)}(p_i, \lambda_+, \mu)[1 + O(\rho^{-|\beta'(\lambda_+)|})] \tag{7.44}$$

Asymptotic scale invariance, in particular, is not a feature of asymptotically free gauge theories. In this case $\beta(x) = -bx^3 + O(x^5)$ and for most of the interesting operators $\gamma(x) - \gamma(0) = 2cx^2 + O(x^4)$. So the integral (7.42) is logarithmically divergent and it causes logarithmic deviations from the asymptotic scale invariance (7.43) (remember that $\bar{\lambda}^2(t) \sim 1/2bt$)

$$\exp\left(\int_\lambda^{\bar{\lambda}(t)} \frac{2c}{b}\frac{dx}{x}\right) \sim (2b\lambda^2 t)^{-c/b} \tag{7.45}$$

We also note an operator equation for the anomalous divergence of the dilatation current in gauge theories reflecting the anomalous scale invariance breaking

$$\Theta_\mu{}^\mu = (\beta(\lambda)/2\lambda^3)G^a_{\mu\nu}G^{\mu\nu}_a \tag{7.46}$$

Rigorous proof of relation (7.46) is rather lengthy (Adler, Collins & Duncan 1977; Collins, Duncan & Joglekar 1977; Nielsen 1977). We see again that only for $\lambda = \lambda_+$ do we recover exact scale invariance, possibly with anomalous dimensions.

We now extend our discussion to theories with scale invariance explicitly broken by mass terms. The identity (for bosons, to be specific)

$$m\frac{\partial}{\partial m}\frac{i}{p^2 - m^2} = \frac{i}{p^2 - m^2}(-2im^2)\frac{i}{p^2 - m^2} \tag{7.47}$$

implies that the operation $m\,\partial/\partial m$ with external momenta and the coupling constant λ fixed (we assume that there is only one mass and one dimensionless coupling constant) is equivalent to the insertion of a new vertex $-im^2\Phi^2$ at zero-momentum transfer. Thus we again see that for $\beta = \gamma_m = \gamma = 0$ (6.24) and (7.37) are identical, with $\Theta_\mu{}^\mu = m^2\Phi^2$.

The mass effects in general are summarized in the solutions (7.40) and (7.41) to the RGE. The $\rho \to \infty$ limit of (7.40) is controlled by the asymptotic behaviour of both $\bar{\lambda}(t)$ and $\bar{m}(t)$ (we recall that $t = \ln\rho$)

$$\bar{m}(\rho) = m\rho^{\gamma_m(\lambda_+)-1}\exp\left\{\int_\lambda^{\bar{\lambda}} dx[\gamma_m(x) - \gamma_m(\lambda_+)]/\beta(x)\right\} \xrightarrow[\rho\to\infty]{} m\rho^{\gamma_m(\lambda_+)-1} * \{\text{logs of } \rho\} \tag{7.48}$$

Now, for the mass effects to be asymptotically non-leading it is essential that

$$\bar{m}(\rho) \xrightarrow[\rho\to\infty]{} 0 \quad \text{or} \quad \gamma_{\mathrm{m}}(\lambda_+) < 1$$

Introducing the scale dimension $d_{\Delta m}$ of the mass operator Δm

$$d_{\Delta m} = \begin{cases} 3 + \gamma_{\mathrm{m}}(\lambda_+) & \Delta m = m\bar{\Psi}\Psi \\ 2 + 2\gamma_{\mathrm{m}}(\lambda_+) & \Delta m = \Phi^2 m^2 \end{cases}$$

we get

$$d_{\Delta m} < 4 \tag{7.49}$$

as the condition for 'softly' broken scale invariance (Wilson 1969*a*). Otherwise the mass insertion would effectively correspond to a hard operator and would influence the asymptotic behaviour of the Green's functions. Equation (7.49) is automatically satisfied if the theory is asymptotically free because the relevant anomalous dimension γ_{m} vanishes.

7.3 Dimensional transmutation

The phenomenon of dimensional transmutation is closely related to scale invariance broken by the renormalization of a quantum field theory. To introduce this concept let us consider any dimensionless observable quantity $A(Q)$ which can depend only on one dimensionful variable Q e.g. $A(Q) = Q^2 \cdot \sigma_{\mathrm{tot}}$. If we assume our theory to be described by a scale-invariant lagrangian (no dimensionful parameters) then from dimensional analysis we must conclude that

$$A(Q) = \text{const.} \tag{7.50}$$

Indeed, dimensional analysis tells us that e.g. a function $f(x, y)$ which depends on two massive variables x, y and which is dimensionless and whose definition does not involve any massive constants must be a function of the ratio x/y only. So if it does not depend on y it must be a constant. As we know from our experience with quantum field theories the conclusion (7.50) is in general wrong. The reason why pure dimensional analysis breaks down in quantum field theory is this: our quantity A actually depends also on the dimensionless free parameter(s) of the theory e.g. bare coupling constant(s): $A = A(Q, g_{\mathrm{B}})$. Moreover the predictions of the theory cannot be expressed directly in terms of g_{B} because the theory requires renormalization. The physics is independent of the details of the renormalization prescription but not of the necessity of renormalization which produces an effective scale. Dimensional transmutation is a process which exploits the above to introduce dimensionful parameters into the predictions of the theory.

An important illustration of these general considerations is provided by the behaviour of the running coupling constant $\alpha(Q^2)$ in QCD (we forget here about a subtlety: $\alpha(Q^2)$ is not really an observable; it depends on the definition of the renormalization scheme). One-parameter ambiguity for $\alpha(Q^2)$ reflects itself in the fact that the theory specifies the first derivative of $\alpha(Q^2)$ rather then $\alpha(Q^2)$ itself

$$\mathrm{d}\alpha/\mathrm{d}t = \beta(\alpha) \tag{7.51}$$

where $Q = \mathrm{e}^t Q_0$, $\alpha = g^2/4\pi$ and $\beta(\alpha) = (b_1/\pi)\alpha^2 + O(\alpha^3)$. Integrating (7.51) we get

$$\ln(Q/Q_0) = F(\alpha(Q)) - F(\alpha(Q_0))$$

where

$$F(x) = \int \mathrm{d}x/\beta(x)$$

Thus, an effective scale Q_0 appears as a constant of integration of the RGE (7.51). If we choose Q_0 so that $F(\alpha(Q_0)) = 0$ and call it Λ then

$$\alpha(Q) = F^{-1}[\ln(Q/\Lambda)] \tag{7.52}$$

and in the leading logarithm approximation

$$\alpha(Q) = -\pi/[b_1 \ln(Q/\Lambda)] \tag{7.53}$$

Λ, defined by the condition $F(\alpha(\Lambda)) = 0$, is the confinement scale: $\alpha(\Lambda) = \infty$. Equation (7.53) can be also rewritten in the following form

$$\Lambda/Q = \exp\{-\pi/[|b_1|\alpha(Q)]\} \tag{7.53a}$$

which tells us that given the strong coupling constant $\alpha(Q)$ at some Q the theory predicts the confinement scale Λ. Dimensionless $\alpha(Q)$ and dimensionful Λ can be traded for each other.

7.4 Operator product expansion (OPE)

Short distance expansion

The OPE technique originates in studies of the relevance of scale invariance in quantum field theory (Wilson 1969*a*). In the previous Section we have discussed how mass terms and renormalizable interactions break this invariance which may however be reestablished, at least up to logarithmic corrections, in the limit of momenta going to infinity. In position space this corresponds to short distances, that is to vanishing separation for the operators in the matrix elements. OPE and scale invariance arguments allow us to determine the short distance singularity structure of operator products. The physical importance of studying operator products at short distance will be discussed later. Here we shall introduce the technique of the OPE.

According to Wilson's hypothesis a product of two local operators $A(x)$ and $B(y)$, when x_μ is near to y_μ, may be written as

$$A(x)B(y) = \sum_n C^n_{AB}(x-y)O_n(\tfrac{1}{2}(x+y)) \tag{7.54†}$$

† The product also can be written in terms of, say, $O_n(y)$ by expanding $O_n(\frac{1}{2}(x+y))$ in a Taylor series about y.

where $\{O_n(x)\}$ is a complete set of hermitean local normal-ordered operators and $\{C^n_{AB}(x)\}$ is a set of c-number functions. The expansion is valid in the weak sense i.e. when sandwiched between physical states. Relation (7.54) is true in free-field theory (See Problem 7.4) and in any order of perturbation theory for interacting fields (Zimmermann 1970). Assuming scale invariance to be an approximate symmetry at short distance one gets the leading behaviour of the Cs for small values of their arguments, hopefully up to at most logarithmic corrections, in terms of the scale dimensions of the operators. Indeed, commuting the dilatation generator D with (7.54) we get

$$\left[D, A(x)B(y) - \sum_n C^n_{AB}(x-y)\, O_n((x+y)/2)\right] = 0 \tag{7.55}$$

and in an exactly scale-invariant theory

$$\sum_n \left[d_n - d_A - d_B - (x-y)\frac{\partial}{\partial(x-y)}\right] C^n_{AB}(x-y) O_n(\tfrac{1}{2}(x+y)) = 0 \tag{7.56}$$

where d_n, d_A and d_B are scale dimensions of the operators O_n, A and B, respectively. Since the operators O_n can be chosen to be independent we deduce that

$$(x-y)\frac{\partial}{\partial(x-y)} C^n_{AB}(x-y) = (d_n - d_A - d_B)\, C^n_{AB}(x-y) \tag{7.57}$$

and therefore C^n_{AB} is homogeneous of degree $d_n - d_A - d_B$ in its argument

$$C^n_{AB}(x) \sim x^{d_n - d_A - d_B} \tag{7.58}$$

Asymptotic scale invariance at short distance implies that the relation (7.58) holds at least in the limit $x \to 0$. In QCD asymptotic scale invariance is broken by logarithmic corrections (see Section 7.2) so the relation (7.58) is modified but only by logarithmic terms. We see that the most singular contribution to $A(x)B(y)$ as $x \to y$ is given by the operator O_n having the lowest dimension. We have learned in Section 7.2 that the scale dimensions d_A, d_B and d_n are not in general equal to the canonical dimensions of the corresponding operators. Anomalous dimensions (see (7.43)) can be calculated in perturbation theory. However, for several interesting operators like conserved currents or currents of softly broken symmetries, i.e broken by operators with scale dimensions $d < 4$, the anomalous dimensions vanish (see Section 10.1).

The Lorentz transformation properties of the Wilson coefficients $C^n_{AB}(x)$ require them to be polynomials in the components of x times scalar functions of x^2: $C^n_{AB} = C^n_{AB}(x, x^2)$. We have to determine prescriptions for handling light-cone singularities of the coefficients C^n_{AB} for different types of products: time-ordered products, simple products, commutators etc. To do this let us refer to free scalar field theory. For instance, for the T-product of two currents $J(x) = :\Phi(x)\Phi(x):$ in free scalar field theory we have (see Problem 7.4)

$$TJ(x)J(0) = -2\Delta_F^2(x,m^2)I + 4i\Delta_F(x,m^2){:}\Phi(x)\Phi(0){:}$$
$$+ {:}\Phi^2(x)\Phi^2(0){:} \underset{x\to 0}{=} -2\Delta_F^2(x,m^2)I + 4i\Delta_F(x,m^2)J(0)$$
$$+ 4i\Delta_F(x,m^2)x_\mu{:}\Phi(0)(\partial^\mu\Phi)(0){:} + \cdots \tag{7.59}$$

where

$$\Delta_F = -i\langle 0|T\Phi(x)\Phi(0)|0\rangle = \frac{1}{(2\pi)^4}\int d^4k\exp(-ikx)\frac{1}{k^2-m^2+i\varepsilon}$$
$$\underset{x^2\to 0}{=} -\frac{i}{4\pi^2}\frac{1}{x^2-i\varepsilon} \tag{7.60}$$

is the free particle propagator in position space. Notice that ${:}\Phi(x)\Phi(0){:}$ may be expanded in a Taylor series since it has no singularity. Thus for T-products we conclude that scalar factors in Cs depend on $(x^2 - i\varepsilon)$. Since

$$TA(x)B(0) = A(x)B(0)\Theta(x_0) + B(0)A(x)\Theta(-x_0)$$

and

$$\frac{1}{-x^2+i\varepsilon x_0}\Theta(x_0) + \frac{1}{-x^2-i\varepsilon x_0}\Theta(-x_0) = -\frac{1}{x^2-i\varepsilon} \tag{7.61}$$

for simple products we have the following prescription $C^n_{AB}(x,x^2) \equiv C^n_{AB}(x, x^2 - i\varepsilon x_0)$. For commutators $[A(x), B(0)]$ we then get $(1/(x \pm i\varepsilon) = P(1/x) \mp i\pi\delta(x))$

$$\frac{1}{x^2-i\varepsilon x_0} - \frac{1}{x^2+i\varepsilon x_0} = 2\pi i\delta(x^2)\varepsilon(x_0) = \varepsilon(x_0)2i\,\mathrm{Im}\frac{1}{x^2-i\varepsilon} \tag{7.62†}$$

and by differentiating both sides with respect to x^2

$$\left(\frac{1}{-x^2+i\varepsilon x_0}\right)^n - \left(\frac{1}{-x^2-i\varepsilon x_0}\right)^n = -\frac{2\pi i}{(n-1)!}\delta^{(n-1)}(x^2)\varepsilon(x_0) \tag{7.63}$$

where

$$\varepsilon(x) = \begin{cases} 1 & \text{for } x>0 \\ -1 & \text{for } x<0 \end{cases}$$

In a general case when the light-cone singularity (7.63) is $(x^2)^{-p}$ with some non-integer power p the difference

$$(-x^2+i\varepsilon x_0)^{-p} - (-x^2-i\varepsilon x_0)^{-p}$$

appearing in a commutator $[A(x), B(0)]$ vanishes for space-like separation $x^2 < 0$: the prescription $x^2 \to x^2 - i\varepsilon x_0$, by placing the cut along the positive real axis, is consistent with causality.

Using the expansion of the form

$$[A(x), B(0)] = \sum_n \tilde{C}^n_{AB}O_n(0) \tag{7.64}$$

† Note in particular that:

$\Delta = -i\langle 0|[\Phi(x), \Phi(0)]|0\rangle = \varepsilon(x_0)2\mathrm{Im}(i\Delta_F) = -(1/2\pi)\varepsilon(x_0)\delta(x^2)$

where

$$\tilde{C}^n_{AB} = C^n_{AB}(x, x^2 - \mathrm{i}\varepsilon x_0) - C^n_{AB}(x, x^2 + \mathrm{i}\varepsilon x_0)$$

and letting $x_0 \to 0$ one can study the equal-time commutators. This is interesting because the appearence in the commutator of the so-called Schwinger terms (proportional to derivatives of the δ-functions) is linked to the singularity structure of operator products at short distances. Following Wilson (1969*a*) we can see that the coefficients $\tilde{C}^n_{AB}$ are equivalent to a sum of δ-functions

$$\tilde{C}^n_{AB} = {}_0F^n_{AB}(x_0)\delta^{(3)}(\mathbf{x}) + {}_1\mathbf{F}^n_{AB}(x_0)\cdot\nabla\delta^{(3)}(\mathbf{x}) + \cdots \tag{7.65}$$

where

$${}_0F^n_{AB}(x_0) = \int \mathrm{d}^3x \tilde{C}^n_{AB}(x_0, \mathbf{x})$$

$${}_1\mathbf{F}^n_{AB}(x_0) = -\int \mathrm{d}^3x \tilde{C}^n_{AB}(x_0, \mathbf{x})\mathbf{x}$$

etc. To get (7.65) one considers an integral

$$\int \mathrm{d}^3x \tilde{C}^n_{AB}(x_0, \mathbf{x})\rho(\mathbf{x}) = \int \mathrm{d}^3x \tilde{C}^n_{AB}(x_0, \mathbf{x})(\rho(0) + \mathbf{x}\cdot\nabla\rho(0) + \cdots) \tag{7.66}$$

where $\rho(\mathbf{x})$ is a differentiable trial function and the Taylor expansion is justified for small x_0 because $\tilde{C}^n_{AB}$ vanishes for $|\mathbf{x}| \geqslant |x_0|$. Equation (7.65) follows immediately from (7.66) and from the definitions of the δ-function and its derivatives. The dependence of the functions ${}_iF^n_{AB}(x_0)$ on x_0 is determined by dimensional analysis to be

$${}_iF^n_{AB}(x_0) \sim x_0^{-2p_n + 3 + i} \tag{7.67}$$

when

$$\tilde{C}^n_{AB} \sim x^{-2p_n}$$

The equal-time commutator is obtained in the limit $x_0 \to 0$. Thus for $p > 1.5$ the coefficient of the δ-function is infinite and for $p > 2$ derivatives of the δ-functions appear in the commutator. Some terms may vanish due to rotational invariance. As an example let us take the product of two electromagnetic currents

$$J_\mu(x) = :\bar{\Psi}(x)\gamma_\mu\Psi(x): \tag{7.68}$$

In free fermion field theory one has (See Problem 7.4)

$$J_\mu(x)J_\nu(0) \sim \frac{x^2 g_{\mu\nu} - 2x_\mu x_\nu}{\pi^4(x^4 - \mathrm{i}\varepsilon x_0)^4}\mathbb{1} + O(x^{-3}) \tag{7.69}$$

The first term gives the familiar Schwinger term in the commutator $[J_0(x), J_i(0)]$ (but not for two J_0s):

$$[J_0(\mathbf{x}, 0), J_i(0)] \sim C_{ij}\partial_j\delta^{(3)}(\mathbf{x}) \tag{7.70}$$

where C_{ij} is quadratically divergent. We will see that the vacuum expectation value of this term is related to the e^+e^- annihilation cross-section calculated under the assumption that the leading short distance singularities are those of the free-field theory. Such terms are usually not revealed by direct calculation of the commutator of two operators based on equal-time canonical commutation relations between fields and their conjugate momenta.

Light-cone expansion

The OPE can be extended to the light-cone region as well. It is obvious, however, that terms in (7.54) which are finite in the limit $x_\mu \to 0$ may be singular in the limit $x^2 \to 0$ with x_μ finite. Therefore we must expect an infinity of terms to contribute to the leading singularity on the light-cone, in contrast to the situation at $x = 0$ where the leading behaviour is given by one term.

Let us look in more detail at the expansion of a product of two Lorentz scalar operators, for instance two currents in free scalar field theory, (7.59). As mentioned before the Lorentz tensor operators $(\partial_\mu \Phi)(0)$, $(\partial_\mu \partial_\nu \Phi)(0)$ etc. cannot be neglected on the light-cone and the general structure of the expansion is as follows

$$A(\tfrac{1}{2}x)B(-\tfrac{1}{2}x) = \sum_n C^n_{AB}(x^2) x^{\mu_1} \ldots x^{\mu_j} O^n_{\mu_1 \ldots \mu_j}(0) \tag{7.71}$$

We have taken the bases $O^n_{\mu_1 \ldots \mu_j}$ to be symmetric traceless tensors with j Lorentz indices; they are irreducible tensors of spin j. We see that the behaviour of the expansion for small x^2 is determined by scalar functions of x^2, the $C^n_{AB}(x^2)$ in the neighbourhood of $x = 0$. (Note that we use the same notation for complete Wilson coefficients in (7.54) and for their scalar part in (7.71). This should not lead to any confusion.) As long as there are no extra divergences introduced by the summation over n, short-distance behaviour and light-cone behaviour are connected

$$C^n_{AB}(x^2) = (x^2)^{(d_n - j_n - d_A - d_B)/2} + \text{higher orders in } x^2 \tag{7.72}$$

so that the degree of singularity on the light-cone is determined by

$$\tau_n = d_n - j_n \tag{7.73}$$

The difference (dimension of an operator – its spin) is called twist. The leading singularity on the light-cone comes from operators with the lowest twist. In free-field theory there exist infinite series of operators with fixed twist, since operating on a field with a derivative raises the spin and the dimension by one unit simultaneously. The scalar field Φ and the fermion field Ψ as well as their derivatives have twist one. Examples of twist two operators are

$$\Phi^* \overleftrightarrow{\partial}_{\mu_1} \overleftrightarrow{\partial}_{\mu_2} \ldots \overleftrightarrow{\partial}_{\mu_j} \Phi$$

and

$$\bar{\Psi} \gamma_\mu \overleftrightarrow{\partial}_{\mu_1} \ldots \overleftrightarrow{\partial}_{\mu_j} \Psi$$

When dimensions are modified by the interactions it is not obvious that they are correlated with the spin in this way. Nevertheless, one usually takes the free-field theory classification of operators as a guide in writing down the OPE.

As an exercise let us expand

$$\Phi^*\left(-\frac{x}{2}\right)\Phi\left(\frac{x}{2}\right) = \Phi^*(0)\left[1 - \overleftarrow{\partial}_{\mu_1}\frac{x^{\mu_1}}{2} + \frac{1}{2!}\overleftarrow{\partial}_{\mu_1}\overleftarrow{\partial}_{\mu_2}\frac{x^{\mu_1}}{2}\frac{x^{\mu_2}}{2} - \cdots\right]$$

$$* \left[1 + \frac{x^{\nu_1}}{2}\overrightarrow{\partial}_{\nu_1} + \frac{1}{2!}\frac{x^{\nu_1}}{2}\frac{x^{\nu_2}}{2}\overrightarrow{\partial}_{\nu_1}\overrightarrow{\partial}_{\nu_2} + \cdots\right]\Phi(0)$$

$$= \sum_n \frac{1}{n!}\frac{x^{\mu_1}}{2}\frac{x^{\mu_2}}{2}\cdots\frac{x^{\mu_n}}{2}\Phi^*(0)\overleftrightarrow{\partial}_{\mu_1}\overleftrightarrow{\partial}_{\mu_2}\cdots\overleftrightarrow{\partial}_{\mu_n}\Phi(0) \tag{7.74}$$

Inserting (7.74) into (7.59) we get the leading terms in the light-cone expansion of the product of two scalar currents in free-field theory. In a similar way we can obtain the light-cone expansion for two electromagnetic currents.

7.5 The relevance of the light-cone

Electron–positron annihilation

We briefly discuss now the relevance of the light-cone for high energy phenomena. Consider first the process of the electron–positron annihilation into hadrons. It is a process widely investigated experimentally and interesting in many respects as a laboratory for important discoveries in particle physics. We focus only on the total cross section which is of the order of a few tens of nanobarns and is comparable to the cross section for the $e^+e^- \to \mu^+\mu^-$. In the one-photon exchange approximation (See Fig. 7.1) the amplitude is

$$T_n = e^2\bar{v}(p_+)\gamma_\mu u(p_-)(1/q^2)\langle n|J_\mu(0)|0\rangle \tag{7.75}$$

where J_μ is the hadronic electromagnetic current. Thus the total cross section is determined by the tensor

$$W_{\mu\nu}(q) = \sum_n (2\pi)^4\delta(q - p_n)\langle 0|J_\mu(0)|n\rangle\langle n|J_\nu(0)|0\rangle$$

$$= \int d^4x \exp(iqx)\langle 0|J_\mu(x)J_\nu(0)|0\rangle \tag{7.76}$$

J_μ is conserved and therefore one may write

$$W_{\mu\nu}(q) = \rho(q^2)(q^2 g_{\mu\nu} - q_\mu q_\nu) \tag{7.77}$$

It can be checked that the total annihilation cross section for unpolarized initial leptons reads

$$\sigma(q^2) = 8\pi^2\alpha^2\rho(q^2)/q^2 \tag{7.78}$$

To see what happens when $q^2 \to \infty$ we observe first that $W_{\mu\nu}(q)$ can be written in

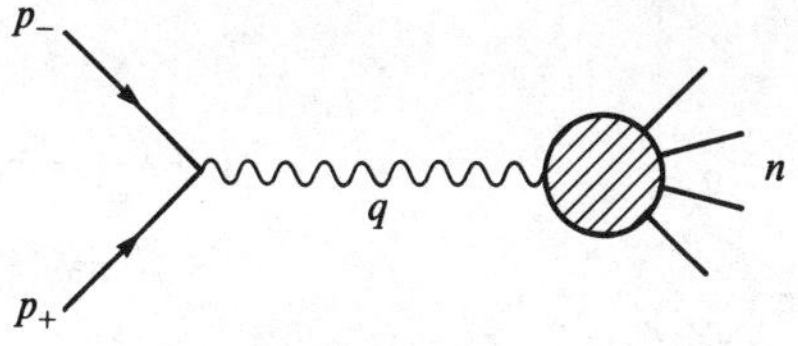

Fig. 7.1

terms of the current commutator

$$W_{\mu\nu}(q) = \int d^4x \exp(iqx)\langle 0|[J_\mu(x), J_\nu(0)]|0\rangle \qquad (7.79)$$

Equation (7.79) differs from (7.76) by the term

$$-(2\pi)^4\delta(q+p_n)\langle 0|J_\nu(0)|n\rangle\langle n|J_\mu(0)|0\rangle$$

which is zero because q_0 and p_n^0 are positive. Since $q^2 > 0$ we may choose a frame in which q has only a time component and use the fact that the commutator vanishes for $x^2 < 0$. Then in this frame

$$W_{\mu\nu}(q) = \int_{-\infty}^{\infty} dx_0 \exp[i(q^2)^{1/2}x_0] \int_{|\mathbf{x}|<x_0} d^3x\langle 0|[J_\mu(x), J_\nu(0)]|0\rangle \qquad (7.80)$$

Since the dominant contribution to the integral over x_0 comes from regions with the least rapid oscillations this Fourier transform is determined by the singularity of the commutator at $x = 0$. In a free-field theory the leading singularity of the commutator originates from the term (7.69) which is responsible for the Schwinger term (7.70). Using (7.63) and the relation

$$\int d^4x \exp(ikx)\delta^{(n)}(x^2)\varepsilon(x_0) = \frac{2^{2-2n}}{(n-1)!}\pi^2 i(k^2)^{n-1}\Theta(k^2)\varepsilon(k_0) \qquad (7.81)$$

we easily find (for a single fermion with charge ± 1)

$$W_\mu{}^\mu(q) \underset{q^2\to\infty}{=} (1/2\pi)q^2 \qquad (7.82)$$

and consequently

$$\sigma(q^2) \to 4\pi\alpha^2/3q^2 \qquad (7.83)$$

This result is not surprising: a free-field theory is asymptotically scale invariant and the cross section must behave like $1/q^2$ on dimensional grounds.

Deep inelastic hadron leptoproduction

Let us now consider another famous process: production of hadrons in lepton–hadron collisions with very large momentum transfer. It is shown in Fig. 7.2 which also contains some definitions of kinematical variables: k, k' are lepton four-momenta; P is the four-momentum of the hadronic target (usually a proton); E, E'

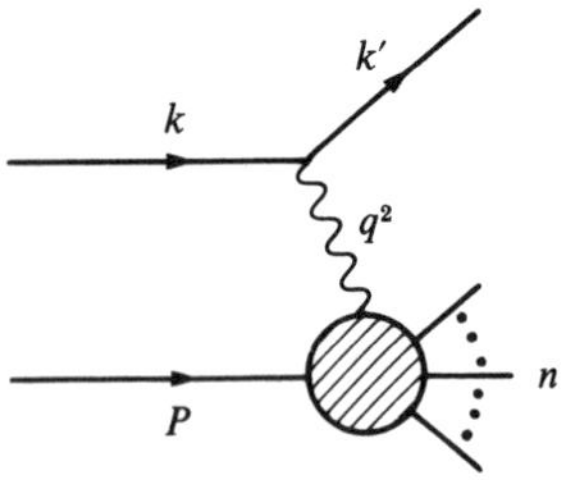

$$q^2 = (k' - k)^2 = -4EE' \sin^2(\tfrac{1}{2}\theta)$$

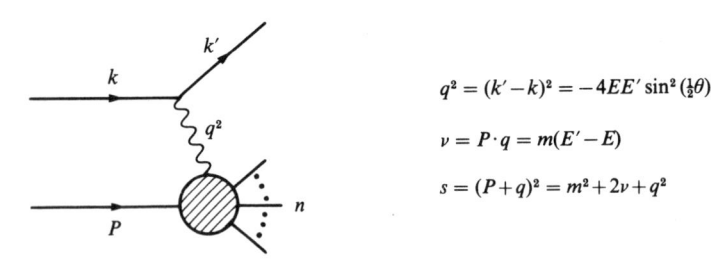

$$\nu = P \cdot q = m(E' - E)$$

$$s = (P+q)^2 = m^2 + 2\nu + q^2$$

Fig. 7.2

and θ are the lepton energies and lepton scattering angle in the laboratory frame, respectively. As we shall see it is also very useful to define the so-called Bjorken variable

$$x_{\rm BJ} = -q^2/2P\cdot q \tag{7.84}$$

For elastic scattering

$$s = m^2 \Rightarrow 2P\cdot q + q^2 = 0 \Rightarrow x_{\rm BJ} = 1$$

In terms of the Bjorken variable, for fixed q^2, the physical region for the process is therefore

$$0 < x_{\rm BJ} < 1 \tag{7.85}$$

To be specific we consider the process of electroproduction. As in case of the e^+e^- annihilation into hadrons in the one-photon exchange approximation the cross section for this process is determined by the tensor

$$\begin{aligned} W_{\mu\nu}(P,q) &= \sum_n (2\pi)^4 \delta(q + P - p_n) \langle P|J_\mu(0)|n\rangle\langle n|J_\nu(0)|P\rangle \\ &= \int \mathrm{d}^4x \exp(iqx) \langle P|J_\mu(x)J_\nu(0)|P\rangle \\ &= \int \mathrm{d}^4x \exp(iqx) \langle P|[J_\mu(x), J_\nu(0)]|P\rangle \end{aligned} \tag{7.86}$$

where J_μ is the hadron electromagnetic current. The passage from product to commutator can be justified in a way similar to that for the annihilation reaction. We collect together some properties of the tensor $W_{\mu\nu}(P,q)$, important for further discussion. If we introduce the forward Compton scattering amplitude, for photons with mass q^2,

$$T_{\mu\nu}(q^2, P\cdot q) = \mathrm{i}\int \mathrm{d}^4x \exp(iqx) \langle P|TJ_\mu(x)J_\nu(0)|P\rangle \tag{7.87}$$

then it follows from the spectral representations (see also (7.62)) that

$$W_{\mu\nu}(q^2, P\cdot q) = 2\,\mathrm{Disc}\, T_{\mu\nu}(q^2, P\cdot q) \tag{7.88}$$

where Disc $T_{\mu\nu}$ is the discontinuity of the $T_{\mu\nu}$ along its branch cuts. The amplitude

$T_{\mu\nu}$ has normal threshold cuts in $s=(P+q)^2$ and $u=(P-q)^2$ for $m^2<s<\infty$ and $m^2<u<\infty$. They correspond to a cut in x_{BJ} for $|x_{\mathrm{BJ}}|<1$. Note that only for $0<x_{\mathrm{BJ}}<1$ is the $W_{\mu\nu}(q^2,P\cdot q)$ given by the deep inelastic cross section.

Now we are ready to discuss the relevance of the light-cone in our process. Let us choose a frame in which

$$P=(m,0,0,0),\quad q=(q_0,0,0,q_3)$$

and introduce the so-called light-cone variables

$$x_\pm = x_0 \pm x_3,\quad \mathbf{x}_\perp=(x_1,x_2)$$

$$q_\pm=\tfrac{1}{2}(q_0\pm q_3)=(\nu/2m)[1\pm(1-m^2q^2/\nu^2)^{1/2}],\quad \nu=q_0m$$

Then

$$W_{\mu\nu}(P,q)=\int \mathrm{d}x_-\exp(\mathrm{i}q_+x_-)\int \mathrm{d}x_+\exp(\mathrm{i}q_-x_+)$$
$$*\int_{x_\perp^2\leqslant x_+x_-}\mathrm{d}^2x_\perp\langle P|[J_\mu(x),J_\nu(0)]|P\rangle \tag{7.89}$$

where the limit on the $\mathbf{x}_\perp$ integration comes from the vanishing of the commutator for $x^2=x_+x_--x_\perp^2<0$. We see that in the limit

$$\left.\begin{matrix} q_+\to\infty \\ q_-\ \text{fixed}\end{matrix}\right\} \tag{7.90}$$

the behaviour of $W_{\mu\nu}$ is determined by the behaviour of the integrand for $x_-\to 0$, x_+ finite i.e. in the region with the least oscillations. Since the integrand is non-vanishing only for $x^2<x_+x_-$ (causality), $x_-\to 0$ for x_+ finite corresponds to $x^2\to 0$ (but not to $x_\mu\to 0$) and $W_{\mu\nu}$ is determined by the singularity of the current commutator on the light-cone. In the limit (7.90) $q^2\to-\infty$, $\nu\to\infty$ and x_{BJ} is fixed ($x_{\mathrm{BJ}}=-q^2/2\nu$). The idea of studying the limit (7.90) is due to Bjorken (1969) and it has turned out to be one of the most creative ideas in modern elementary particle physics.

For simplicity let us study the scalar current analogue of the electroproduction tensor

$$W(q^2,P\cdot q)=\int \mathrm{d}^4x\exp(\mathrm{i}qx)\langle P|[J(x),J(0)]|P\rangle \tag{7.91}$$

in the Bjorken limit. Spin complications are not essential for general orientation. Apart from the c-number contribution which is irrelevant for the considered matrix element (first term in (7.59)) the leading singularities of the operator products on the light-cone are given by the twist two operators

$$\langle P|[J(x),J(0)]|P\rangle=\frac{1}{\pi^2}2\pi\mathrm{i}\delta(x^2)\varepsilon(x_0)\sum_n\frac{1}{n!}x_{\mu_1}\ldots x_{\mu_n}$$
$$*\langle P|\Phi(0)\partial_{\mu_1}\ldots\partial_{\mu_n}\Phi(0)|P\rangle+\text{less singular terms} \tag{7.92}$$

where we have used (7.63) and (7.74). For the realistic case of particles with spin, twist two operators are again the leading ones. The general form of the matrix element in (7.92) is

$$\langle P|O_{\mu_1\ldots\mu_n}|P\rangle = A_n P_{\mu_1}\ldots P_{\mu_n} + \delta_{\mu_1\mu_2}P_{\mu_3}\ldots P_{\mu_n} + \cdots \tag{7.93}$$

Terms having at least one Kronecker delta may be ignored because when contracted with the x_μs in (7.92) they give additional factors x^2 and therefore are less singular. So we have

$$\langle P|[J(x), J(0)]|P\rangle = \frac{2i}{\pi}\delta(x^2)\varepsilon(x_0)f(xP) + \text{less singular terms} \tag{7.94}$$

where

$$f(xP) = \sum_n \frac{1}{n!}(x\cdot P)^n A_n$$

Defining the Fourier transform

$$f(xP) = (1/2\pi)\int d\xi \exp(i\xi x\cdot P)\tilde{f}(\xi)$$

and using the identity†

$$i\int d^4x \exp(ikx)\delta(x^2)\varepsilon(x_0) = (2\pi)^2\varepsilon(k_0)\delta(k^2) \tag{7.95}$$

we get

$$W(q^2, P\cdot q) = 4\int d\xi\delta((\xi P + q)^2)\varepsilon(q_0 + \xi P_0)\tilde{f}(\xi)$$

The roots of the argument of the δ-function are

$$\xi_\pm = -\frac{\nu}{m^2} \pm \frac{\nu}{m^2}\left(1 - \frac{q^2m^2}{\nu^2}\right)^{1/2} \xrightarrow[\text{BJ}]{} \frac{-q^2}{2\nu}, -\frac{2\nu}{m^2} \tag{7.96}$$

where 'BJ' means Bjorken limit. As it has already been explained $W(q^2, P\cdot q)$, being the imaginary part of the amplitude for forward off-mass-shell Compton scattering, is non-vanishing for $|-q^2/2\nu| < 1$. Thus only the ξ_+ is relevant and in terms of the Bjorken variable our result reads

$$\nu W(q^2, P\cdot q) = 2\tilde{f}(x_{\text{BJ}}) \tag{7.97}$$

† Identity (7.95) follows from the spectral representation for the free-field commutator (Bjorken & Drell 1965) and the relation (7.62):

$$\langle 0|[\Phi(x), \Phi(0)]|0\rangle = -\frac{1}{(2\pi)^3}\int d^4k \exp(ikx)\delta(k^2 - m^2)\varepsilon(k_0)$$

$$= \left(\frac{i}{2\pi}\right)\delta(x^2)\varepsilon(x_0)$$

In the free-field theory with twist two operators giving the leading singularity on the light-cone we obtain the famous Bjorken scaling for the structure function $W(q^2, P\cdot q)$.

An important final observation is that only even spin operators are allowed in the expansion (7.92). Indeed W is real (it is a cross section) so $\tilde{f}(\xi)$ must be real and therefore $f(xP)$ must be even in x.

Wilson coefficients and moments of the structure function

Use of the OPE is by no means limited to the free-field theory case. Its validity for renormalized operators has been proved in perturbation theory and therefore it can be used to obtain further information about the scaling limit of the structure functions, beyond that contained in the hypothesis of canonical dimensions. An important step in this direction is to establish some relation between the Wilson coefficients and measurable quantities which is not limited to the canonical scaling case. The standard procedure is as follows. Consider first the forward scattering amplitude for scalar currents

$$T(q^2, P\cdot q) = \mathrm{i}\int \mathrm{d}^4x \exp(\mathrm{i}qx)\langle P|TJ(\tfrac{1}{2}x)J(-\tfrac{1}{2}x)|P\rangle \tag{7.98}$$

The structure function (7.91) is the discontinuity of T, see (7.88). Instead of (7.92) let us consider

$$TJ(\tfrac{1}{2}x)J(-\tfrac{1}{2}x) \underset{x^2\to 0}{\approx} \sum_{n,k} C_n^k(x^2)x_{\mu_1}\ldots x_{\mu_n}O_k^{\mu_1\ldots\mu_n} \tag{7.99}$$

where the sum is taken over all twist two operators O_k and all even spins n. Using (7.93) and defining

$$\frac{2^n q^{\mu_1}\ldots q^{\mu_n}}{(-q^2)^{n+1}}\tilde{C}_n^k(q^2) = \mathrm{i}\int \mathrm{d}^4x \exp(\mathrm{i}qx)x^{\mu_1}\ldots x^{\mu_n}C_n^k(x^2) \tag{7.100}$$

the leading contribution to $T(q^2, P\cdot q)$ in the Bjorken limit can be written as

$$T(q^2, P\cdot q) = \frac{1}{-q^2}\sum_{n,k} x_{\mathrm{BJ}}^{-n}\tilde{C}_n^k(q^2)A_n^k \tag{7.101}$$

Note that $\tilde{C}_n^k(q^2)$ are defined so that they are dimensionless for twist two operators in the expansion of the product of two currents with dimension $d_J = 3$. Therefore in the case of canonical scaling $\tilde{C}_n^k(q^2) = \tilde{C}_n^k$ are constants. The coefficients $\tilde{C}_n^k(q^2)$ can be related to the moments of the structure function $W(q^2, P\cdot q)$ if we integrate $T(q^2, x_{\mathrm{BJ}})$ over x_{BJ} for fixed q^2. Remembering that it is an analytic function of x_{BJ} in the complex x_{BJ}-plane with a branch cut along the real axis for $|x_{\mathrm{BJ}}| < 1$ we integrate over the contour shown in Fig. 7.3. Since the discontinuity of the T is just $\frac{1}{2}W(q^2, P\cdot q)$ we have ($x \equiv x_{\mathrm{BJ}}$)

$$\frac{1}{2\pi\mathrm{i}}\oint_C \mathrm{d}x x^{m-1}T(q^2, x) = \frac{1}{4\pi}\int_{-1}^{1} \mathrm{d}x x^{m-1}W(q^2, x) \tag{7.102}$$

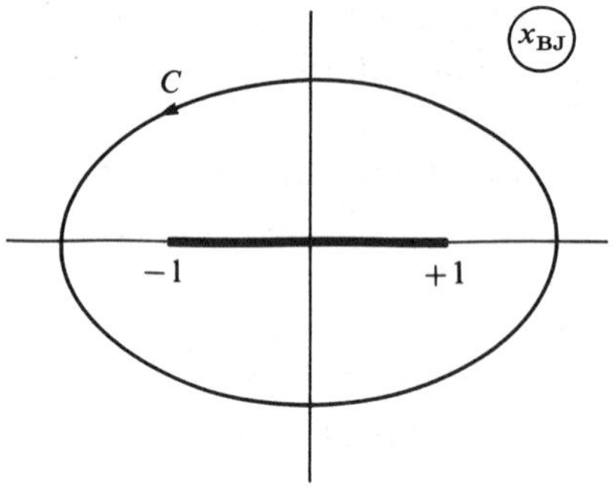

Fig. 7.3

and on the other hand

$$\frac{1}{2\pi i}\oint_C dx x^{m-1} T(q^2, x) = \frac{1}{-q^2}\sum_k \tilde{C}^k_m(q^2) A^k_m \tag{7.103}$$

Finally we get

$$M_m(q^2) = (1/2\pi)\int_{-1}^{1} dx x^m \nu W(q^2, x) = \sum_k \tilde{C}^k_m(q^2, \mu^2, g) A^k_m(\mu^2, g) \tag{7.104}$$

In (7.104) we have written down explicitly the dependence on the renormalization point and the coupling constant.

Thus we get a relationship between moments of the deep inelastic structure function and the coefficient functions. Using the crossing properties of the structure function under $q \to -q$ or $x \to -x$

$$W(q^2, x) = -W(q^2, -x) \tag{7.105}$$

we can rewrite the relation (7.104) for even m in terms of the integral over the physical region in the electroproduction process, $0 < x < 1$. We also note that (7.104) is meaningful for values of m such that the integral converges. This may not be the case for low moments: for fixed q^2 the limit $x \to 0$ corresponds to $\nu \to \infty$ where $\nu = P\cdot q \approx s$ and s is the cms energy squared in the forward Compton amplitude. At high energy the imaginary part of this amplitude is expected to behave as s^α, $\alpha \approx 1$, so that $W(q^2, x) \sim x^{-1}$.

An important virtue of the OPE is that it allows us to factorize out the q^2 dependence in quantities like M_m which depend on q^2 and $P^2 = m^2$. This dependence can then be studied by means of the RGE. Similar analysis is not possible directly for $M_n(q^2, P^2 = m^2)$ because the scale transformation $q \to \rho q$, $P \to \rho P$ relates M_ns for different q^2s and m^2s.

As we know, exact Bjorken scaling gives

$$\tilde{C}^k_n(q^2) = \text{const.} \tag{7.106}$$

Interactions can modify this behaviour. In the next Section we shall study the q^2 dependence of the Wilson coefficients using RGE.

7.6 Renormalization group and OPE

Renormalization of composite operators

We are used by now to working with Green's functions involving not only elementary fields but also composite operators, e.g.

$$G_{\mathbf{O}}^{(n)}(x_1,\ldots,x_n,x) = \langle 0|T\Phi(x_1)\ldots\Phi(x_n)\mathbf{O}(x)|0\rangle \tag{7.107}$$

or in momentum space

$$(2\pi)^4\delta(p_1+\cdots+p_n+p)\tilde{G}_{\mathbf{O}}^{(n)}(p_1,\ldots,p_n,p)$$
$$= \int \mathrm{d}^4x \exp(\mathrm{i}px)\prod_{i=1}^{n}\mathrm{d}^4x_i \exp(\mathrm{i}p_ix_i)G_{\mathbf{O}}^{(n)}(x_1,\ldots,x_n,x) \tag{7.108}$$

The generating functional formalism and perturbation theory rules can be extended to such cases if we introduce sources $\Delta(x)$ coupled to operators $\mathbf{O}(x)$. The functional

$$W[J,\Delta] = \int \mathscr{D}\Phi \exp\left[\mathrm{i}S + \mathrm{i}\int \mathrm{d}^4x(J\Phi + \Delta\mathbf{O})\right] \tag{7.109}$$

generates these new Green's functions. Connected Green's functions are obtained from the functional Z and the Legendre transformation performed on sources J (only) generates 1PI Green's functions with $\mathbf{O}$ insertions. As usual one has to take derivatives with respect to Δ and then set $\Delta = 0$. Feynman rules for vertices involving elementary fields and the composite operators $\mathbf{O}(x)$ can be derived in exactly the same way as in Chapter 2 or Chapter 3. The new Green's functions calculated perturbatively, in general, require renormalization. The counterterms present in the lagrangian for the renormalization of the Green's functions involving only elementary fields are not sufficient for eliminating divergences in the Green's functions with composite operator insertions. Additional operator renormalization constants are needed

$$\mathbf{O}^{\mathrm{B}}(x) = Z_{\mathrm{O}}\mathbf{O}^{\mathrm{R}}(x) \tag{7.110}$$

If we want to express $\mathbf{O}^{\mathrm{R}}$ in terms of the renormalized fields the wave-function renormalization constants will explicitly appear, e.g.

$$(\Phi_{\mathrm{B}}^2)_{\mathrm{B}} = Z_{\Phi^2}(\Phi_{\mathrm{B}}^2)_{\mathrm{R}} = Z_{\Phi^2}Z_3(\Phi_{\mathrm{R}}^2)_{\mathrm{R}} \tag{7.111}$$

Only in exceptional cases like conserved currents or currents of softly broken symmetries does $Z_{\mathrm{O}} = 1$ (see Section 10.1). Relation (7.110) corresponds to a replacement

$$\Delta(x)\mathbf{O}^{\mathrm{B}}(\Phi_{\mathrm{B}}^i(x)) = \Delta\mathbf{O}^{\mathrm{R}}(\Phi_{\mathrm{R}}^i) + \Delta(Z_{\mathrm{O}}Z_3^{i/2} - 1)\mathbf{O}^{\mathrm{R}}(\Phi_{\mathrm{R}}^i) \tag{7.112}$$

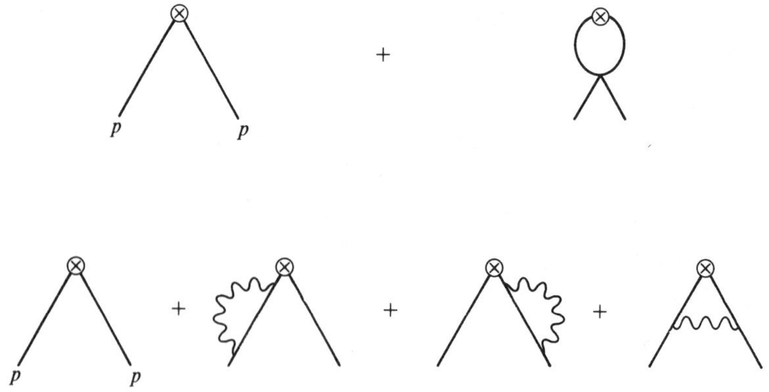

Fig. 7.4

in the generating functional, (7.109). We have taken $\mathbf{O}(x)$ to be of the ith order in the elementary fields $\Phi(x)$. The last term in (7.112) is the new counterterm. The Z_O can be calculated in perturbation theory by calculating the divergent Green's functions with operator insertions. The superficial degree of divergence (see Section 4.2) D_O of $\Gamma^{(n)}_O$ differs from D of $\Gamma^{(n)}$ and reads

$$D_O = D + (d_O^{\text{can}} - 4) \tag{7.113}$$

where d_O^{can} is the canonical dimension of $\mathbf{O}$. As an example let us consider $\lambda\Phi^4$ theory and discuss a single insertion of $\mathbf{O}(x) = \frac{1}{2}\Phi^2(x)$ (Itzykson & Zuber 1980) or an insertion of $\mathbf{O}(x) = \bar{\Psi}(x)\Psi(x)$ in QCD. The divergent Green's functions with this insertion are $\Gamma^{(2)}_{\Phi^2}$ and $\Gamma^{(2)}_{\bar{\Psi}\Psi}$ with $D_O = 0$. It is an easy exercise to calculate Z_O in these simple cases in the lowest order of perturbation theory. The relevant Feynman diagrams are shown in Fig. 7.4 where the vertex $\otimes$ is in both cases just one and for simplicity we can take the momenta of the external legs to be equal.

Of course, new renormalization constants require new renormalization conditions. For instance, we may impose

$$\Gamma^{(2)}_{\Phi^2}(0,0) = 1 \qquad \text{and} \qquad \Gamma^{(2)}_{\bar{\Psi}\Psi}(0,0) = 1 \tag{7.114}$$

in agreement with the lowest order.

If an operator $\mathbf{O}$ is multiplicatively renormalized as in (7.110) then the 1PI n-point Green's functions with a single $\mathbf{O}$ insertion satisfy

$$\Gamma^{(n)}_{O,R}(p_1, \ldots, p_n, p, \lambda, \mu, m) = Z_O^{-1} Z_3^{n/2} \Gamma^{(n)}_{O,B}(p_1, \ldots, p_n, p, \lambda_B, m_B) \tag{7.115}$$

Simple multiplicative renormalization, (7.110), is actually not always sufficient. For instance, it may not eliminate divergences from some Green's functions involving more then one insertion of $\mathbf{O}$. A well-known example is the vacuum expectation value of two electromagnetic currents $J_\mu(x)$ which requires a subtraction although $Z_J = 1$ (see Section 10.1).

Another complication arises for the following reason. Using the arguments of Section 4.2 concerning the necessary counterterms in the standard renormalization programme we can convince ourselves that an operator in the lagrangian with dimension d requires in general all possible counter-terms with dimensions $\leqslant d$ allowed by the symmetries of the bare lagrangian. If we limit our discussion to Green's functions involving only one insertion of an operator $\mathbf{O}$ the same is true for the renormalization of such composite operators. Therefore if there exist several operators with the same quantum numbers as the operator $\mathbf{O}$ and with their canonical dimensions lower or equal to the dimension of the operator $\mathbf{O}$, they will mix with $\mathbf{O}$ under the renormalization. Thus, (7.110) takes a matrix form

$$\mathbf{O}_n^{\mathrm{B}} = (Z_{\mathrm{O}})_{nk}\mathbf{O}_k \tag{7.116}$$

and correspondingly

$$\Gamma^{(n)}_{\mathrm{O}_n,\mathrm{R}}(p_1,\ldots,p_n,p,\lambda,\mu,m) = (Z_{\mathrm{O}}^{-1})_{nk} Z_3^{n/2} \Gamma^{(n)}_{\mathrm{O}_k,\mathrm{B}}(p_1,\ldots,p_n,p,\lambda_{\mathrm{B}},m_{\mathrm{B}}) \tag{7.117}$$

Note that the matrix Z_{ij} is not symmetric: $Z_{ij}=0$ if $\dim \mathbf{O}_i < \dim \mathbf{O}_j$.

RGE for Wilson coefficients

We now turn back to the OPE and study the q^2 dependence of the Wilson coefficient functions by means of the RGE. Let us recall that in general

$$A(\tfrac{1}{2}x)B(-\tfrac{1}{2}x) \underset{x^2\to 0}{\approx} \sum_i C_i(x, g(\mu), \mu)\mathbf{O}_i(0) \tag{7.118}$$

and therefore for any Green's function with insertion of the product AB we have

$$\Gamma^{(n)}_{AB}(q, p_1,\ldots,p_n; g(\mu),\mu) = \sum_i \tilde{C}_i(q,\mu)\Gamma^{(n)}_{\mathrm{O}_i}(0, p_1,\ldots,p_n, g(\mu),\mu) \tag{7.119}$$

From (7.117) we can derive for Green's functions with an operator insertion the RGE (6.24)

$$\left\{\left[-\frac{\partial}{\partial t} + \beta(g)\frac{\partial}{\partial g} - n\gamma + D\right]\delta_{ij} + \gamma_{\mathrm{O}}^{ij}\right\}\Gamma^{(n)}_{\mathrm{O}_j}(\mathrm{e}^t p_i, g(\mu),\mu) = 0 \tag{7.120}$$

where

$$\gamma_{\mathrm{O}}^{ij} = -\left(\mu\frac{\mathrm{d}}{\mathrm{d}\mu}(Z_{\mathrm{O}})^{ik}\right)(Z_{\mathrm{O}}^{-1})^{kj}$$

D is the dimension of the Green's function (equal to the canonical dimension of all operators involved) and the mass dependence has been neglected. By applying the differential operator

$$\mathscr{D} = -\frac{\partial}{\partial t} + \beta(g)\frac{\partial}{\partial g}$$

to (7.119) and expanding $\Gamma^{(n)}_{AB}$, again using (7.119) we get

$$\left[-\frac{\partial}{\partial t} + \beta(g)\frac{\partial}{\partial g} + D_A + D_B - D_{\mathrm{O}} + \gamma_A + \gamma_B - \gamma_{\mathrm{O}}^{ij}\right]\tilde{C}_i(\mathrm{e}^t q, g(\mu),\mu) = 0 \tag{7.121}$$

where D_A, D_B and $D_{\mathbf{O}}$ are canonical dimensions of A, B and $\mathbf{O}$ respectively and γ_A, γ_B and $\gamma_{\mathbf{O}}^{ij}$ are their anomalous dimensions. We have assumed here that operators A and B do not mix with other operators when undergoing renormalization. This is in particular trivially true for the electromagnetic currents. The solution to the RGE for the Wilson coefficient is analogous to (6.26). Let us write it down for some specific cases. Firstly take $\tilde{C}_i$ dimensionless i.e. $D_A + D_B - D_{\mathbf{O}} = 0$. In addition put $\gamma_A = \gamma_B = 0$ as for conserved currents. For the no-mixing case $\gamma_{\mathbf{O}}^{ij} = \delta^{ij}\gamma_i$ we then have

$$\tilde{C}_i(e^t q, g(\mu), \mu) = \exp[-\gamma_i(g_+)t] \exp\left\{-\int_g^{\bar{g}(t)} dx[\gamma_i(x) - \gamma_i(g_+)]/\beta(x)\right\} \tilde{C}_i(q, \bar{g}(t), \mu) \tag{7.122}$$

To proceed further one has to calculate $\tilde{C}_i(q, \bar{g}(t), \mu)$, $\gamma_i(g)$ and $\beta(g)$ in perturbation theory. In asymptotically free theories the UV fixed point $g_+ = 0$, $\beta(x) = (2/16\pi^2)b_1 x^3 + \cdots$ and $\gamma_i(x) = (4/16\pi^2)\gamma_i^0 x^2 + \cdots$. Putting $q^2 = \mu^2$, $t = \frac{1}{2}\ln(q^2/\mu^2)$ and expanding

$$\tilde{C}_i(\mu, \bar{g}(t), \mu) = \tilde{C}_i(\mu, 0, \mu)[1 + C_i^1 \bar{g}^2(t)/4\pi + \cdots]$$

in the leading order in g^2 we get

$$\tilde{C}_i(q, g(\mu), \mu) = \tilde{C}_i(\mu, 0, \mu)[\bar{g}^2(t)/g^2(\mu^2)]^{-\gamma_i^0/b_1} \tag{7.123}$$

Thus for the moments of the structure function we have (from (7.104) when $k = 1$ i.e. if there is only one twist-two operator in expansion (7.99))

$$\frac{M_i(q^2)}{M_i(q_0^2)} = \left[\frac{\bar{g}^2(q^2)}{\bar{g}^2(q_0^2)}\right]^{-\gamma_i^0/b_1} = \left[\frac{\ln(q^2/\Lambda^2)}{\ln(q_0^2/\Lambda^2)}\right]^{+\gamma_i^0/b_1} \tag{7.124}$$

In the last step we have used (7.53). Equation (7.124) is valid for both q_0^2 and q^2 large as compared to Λ^2. For a non-zero UV fixed point $g_+ \neq 0$ (7.122) gives a power-like violation of scaling

$$\tilde{C}_i(q, g(\mu), \mu) \approx (q^2/\mu^2)^{-\gamma_i(g_+)} \tilde{C}_i(\mu, g_+, \mu) \tag{7.125}$$

In the case of operator mixing the solution to the RGE can be written in the matrix form (take $g_+ = 0$)

$$\hat{C}(q, g(\mu), \mu) = \left\{T_g \exp\left[\int_{\bar{g}(q^2)}^{g(\mu^2)} dx\,\hat{\gamma}(x)/\beta(x)\right]\right\} \hat{C}(\mu, \bar{g}(q^2), \mu) \tag{7.126}$$

where

$$\hat{C} = \begin{bmatrix} C_1 \\ \vdots \\ C_n \end{bmatrix}, \quad \hat{\gamma} = [\gamma_{ij}]$$

and the T_g ordering, necessary since $[\hat{\gamma}(x_1), \hat{\gamma}(x_2)] \neq 0$, is defined as follows

$$T_g \exp\left[\int_{\bar{g}}^{g} dx \frac{\hat{\gamma}(x)}{\beta(x)}\right] = 1 + \int_{\bar{g}}^{g} dx \frac{\hat{\gamma}(x)}{\beta(x)} + \frac{1}{2!}\int_{\bar{g}}^{g} dx \int_{\bar{g}}^{x} dy \frac{\hat{\gamma}(x)}{\beta(x)}\frac{\hat{\gamma}(y)}{\beta(y)} + \cdots \tag{7.127}$$

OPE beyond perturbation theory

Strictly speaking, the use of the OPE to study the moments of the structure function in deep inelastic lepton–hadron scattering, which is discussed in this Section, goes beyond perturbation theory. This is because in the realistic case, when the strong interactions are described by QCD, the matrix elements of the operators taken between hadronic states contain non-perturbative effects responsible for the binding of quarks and gluons into hadrons. Nevertheless we assume the validity of the OPE and moreover that the coefficient functions C_n^{AB} are calculable in QCD perturbation theory. This approach is implicitly based on the parton model assumption discussed in Section 8.4. It is also in agreement with the possibility, very suggestive as the physical sense of the OPE formalism, that the contribution from short distances can be separated from that of large distances. One can then expect that to a good approximation the coefficient functions are calculable in perturbation theory while the non-perturbative effects are accounted for by the matrix elements of the operators O_n.

The same idea underlies the use of the OPE to account for the non-perturbative structure of the QCD vacuum, pioneered by Shifman, Vainshtein & Zakharov (1979) and extensively discussed in the recent years. Non-perturbative effects manifest themselves through non-trivial vacuum expectation values of various operators which vanish in perturbation theory

$$\langle 0|:O_n:|0\rangle = 0 \quad \text{if} \quad O_n \neq I$$

The best known examples are quark and gluon condensates

$$\langle 0|:\bar{\Psi}\Psi:|0\rangle \quad \text{and} \quad \langle 0|:G^a_{\mu\nu}G_a^{\mu\nu}:|0\rangle$$

the first one is responsible for the spontaneous breaking of chiral symmetry in QCD (Chapter 9). This use of the OPE has been successful phenomenologically but its theoretical status and possible limitations are still subject to some discussion (see, for instance, Novikov *et al.* 1984 and references therein).

The coefficient functions $C_n^{AB}(q)$ can be found by calculating the appropriate Feynman diagrams (Problem 7.5). Important for this programme is the theorem that the coefficient functions in the OPE manifestly reflect the symmetries of the lagrangian whether or not they are symmetries of the vacuum (Bernard *et al.* 1975).

Problems

7.1 (*a*) Check that the set of 15 infinitesimal transformations defined by (7.26) is closed under commutation. Check the differential representation of the generators

$$\mathrm{i}[P_\mu, f(x)] = \partial_\mu f$$
$$\mathrm{i}[M^{\mu\nu}, f(x)] = (x^\mu\partial^\nu - x^\nu\partial^\mu) f$$
$$\mathrm{i}[D, f(x)] = x^\mu\partial_\mu f$$
$$\mathrm{i}[K_\mu, f(x)] = (x^2 g_{\mu\nu} - 2x_\nu x_\mu)\partial^\nu f$$

where $f(x)$ is a dimensionless scalar function and

$$f'(x) = Uf(x)U^{-1}$$

where

$$U(a) = \exp(\mathrm{i}a_\mu P^\mu), \quad U(\varepsilon) = \exp(\mathrm{i}\varepsilon D),$$
$$U(\omega) = \exp(-\tfrac{1}{2}\mathrm{i}\omega^{\mu\nu}M_{\mu\nu}), \quad U(b) = \exp(\mathrm{i}b^\mu K_\mu)$$

for $x'^\mu = x^\mu - \varepsilon^\mu(x)$.

(*b*) Check the algebra

$$[M_{\mu\nu}, D] = [P_\mu, P_\nu] = [K_\mu, K_\nu] = [D, D] = 0$$
$$[P_\mu, D] = \mathrm{i}P_\mu$$
$$[K_\mu, D] = \mathrm{i}K_\mu$$
$$[M_{\mu\nu}, K_\lambda] = \mathrm{i}(g_{\mu\lambda}K_\nu - g_{\nu\lambda}K_\mu)$$
$$[K_\mu, P_\nu] = 2\mathrm{i}(-g_{\mu\nu}D + M_{\mu\nu})$$
$$[M^{\mu\nu}, P^\lambda] = -\mathrm{i}(g^{\mu\lambda}P^\nu - g^{\nu\lambda}P^\mu)$$
$$[M^{\alpha\beta}, M^{\mu\nu}] = -\mathrm{i}(g^{\alpha\mu}M^{\beta\nu} - g^{\beta\mu}M^{\alpha\nu} + g^{\alpha\nu}M^{\mu\beta} - g^{\beta\nu}M^{\mu\alpha})$$

(*c*) Check the infinitesimal transformation law for a field Φ under the special conformal transformation

$$\delta_\mu\Phi(x) = (x^2\partial_\mu - 2x_\mu x_\nu\partial^\nu - 2x_\mu d_\Phi - 2x^\nu\Sigma_{\mu\nu})\Phi(x)$$

7.2 Study conformal symmetry in two-dimensional quantum field theories (Belavin, Polyakov & Zamolodchikov 1984).

7.3 When the dilatation current is not conserved one may still define a time-dependent dilatation charge by

$$D(t) = \int \mathrm{d}^3x x_\mu \Theta^{0\mu}(x) = tP^0 + \int \mathrm{d}^3x x_i \Theta^{0i}(x)$$

Check that $D(t)$ generates the same dilatation transformation (7.14) on the fields as in the scale-invariant theory (use the canonical formula (7.16) for $D_\mu(x)$). Obtain the commutators of $D(t)$ with the Poincaré generators

$$\mathrm{i}[D(t), P^\alpha] = P^\alpha - g^{\alpha 0}\int \mathrm{d}^3x \partial_\mu D^\mu(x)$$

$$\mathrm{i}[D(t), M^{\alpha\beta}] = \int \mathrm{d}^3x(g^{\alpha 0}x^\beta - g^{\beta 0}x^\alpha)\partial_\mu D^\mu(x)$$

7.4 (*a*) Using the Wick expansion and then the Taylor expansion of the well-defined normal-ordered operator products prove (7.59) in the free scalar field theory.

(*b*) In the free quark model show that

$$TJ^a_\mu(x)J^b_\nu(0) \underset{x_\mu\to 0}{\sim} \frac{3\delta_{ab}(g_{\mu\nu}x^2 - 2x_\mu x_\nu)}{(2\pi)^4(x^2 - \mathrm{i}\varepsilon)}I + \frac{d_{abc}}{2\pi^2}\frac{\varepsilon_{\mu\nu\alpha\beta}x^\alpha A^\beta_c(0)}{(x^2 - \mathrm{i}\varepsilon)^2}$$
$$+ \frac{c_{abc}}{2\pi^2}\frac{x^\rho S_{\mu\rho\nu\sigma}J^\sigma_c(0)}{(x^2 - \mathrm{i}\varepsilon)^2} + \cdots$$

$$TJ^a_\mu(x)A^b_\nu(0) \underset{x_\mu\to 0}{\sim} \frac{d_{abc}}{2\pi^2}\frac{\varepsilon_{\mu\nu\alpha\beta}x^\alpha J^\beta_c(0)}{(x^2 - \mathrm{i}\varepsilon)^2} + \frac{c_{abc}}{2\pi^2}\frac{x^\rho S_{\mu\rho\nu\sigma}A^\sigma_c(0)}{(x^2 - \mathrm{i}\varepsilon)^2} + \cdots$$

where

$$J^a_\mu(x) = :\bar{\Psi}_c(x)\gamma_\mu \tfrac{1}{2}\lambda_a \Psi_c(x):$$
$$A^a_\mu(x) = :\bar{\Psi}_c(x)\gamma_\mu\gamma_5 \tfrac{1}{2}\lambda_a \Psi_c(x):$$

the repeated index c implies a sum over colours; λ_a are $SU(3)$ flavour matrices

$$\tfrac{1}{4}\lambda_a\lambda_b = \tfrac{1}{2}(d_{abc} + \mathrm{i}c_{abc})\lambda_c$$
$$\gamma_\mu\gamma_\rho\gamma_\nu = (S_{\mu\nu\rho\sigma} + \mathrm{i}\varepsilon_{\mu\rho\nu\sigma}\gamma_5)\gamma^\sigma$$
$$S_{\mu\rho\nu\sigma} = (g_{\mu\rho}g_{\nu\sigma} + g_{\rho\nu}g_{\mu\sigma} - g_{\mu\nu}g_{\rho\sigma})$$

Also derive the light-cone expansion for commutators of these currents (e.g. Ellis 1977).

7.5 Obtain the expansion

$$\mathrm{i}\int \mathrm{d}^4x \exp(\mathrm{i}qx)\, Tj_\mu(x)j_\nu(0) = (q_\mu q_\nu - q^2 g_{\mu\nu})\left(-\frac{1}{4\pi^2}\ln\frac{-q^2}{\mu^2} + \frac{2m_\Psi}{q^4}\bar{\Psi}\Psi + \cdots\right)$$

where $j_\mu = \bar{\Psi}\gamma_\mu\Psi$. The first term is given by the diagram

$\langle 0|j_\mu j_\nu|0\rangle =$ 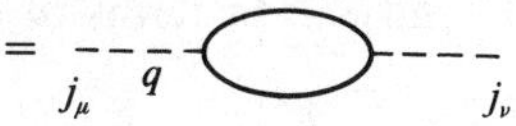 $= C^{(0)}_I\langle 0|I|0\rangle$

the second by

$\langle \Psi|j_\mu j_\nu|\Psi\rangle =$ 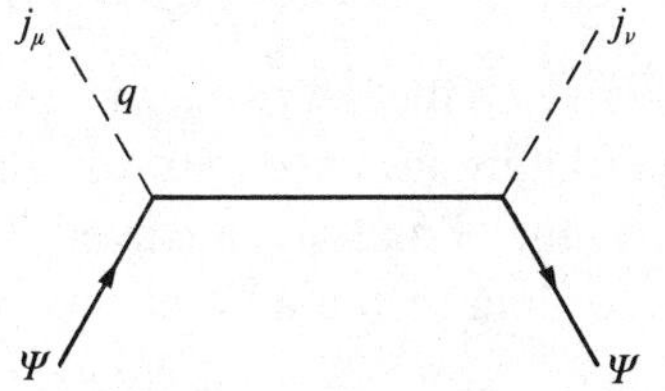 $= C^{(0)}_{\bar{\Psi}\Psi}\langle \Psi|\bar{\Psi}\Psi|\Psi\rangle$

$= C^{(0)}_{\bar{\Psi}\Psi}\cdot$  $= C^{(0)}_{\bar{\Psi}\Psi}$

Observe that to this order in α taking the appropriate matrix element singles out one particular contribution to the OPE. Extend the calculation to first order in α; in addition to $C^{(1)}_I$ and $C^{(1)}_{\bar{\Psi}\Psi}$ given by the same matrix elements (now taken to the first order in α) calculate also the two-gluon matrix element

$$\langle g|j_\mu j_\nu|g\rangle = C_{\bar{\Psi}\Psi}\langle g|\bar{\Psi}\Psi|g\rangle + C_G\langle g|G^\alpha_{\mu\nu}G^{\mu\nu}_\alpha|g\rangle$$

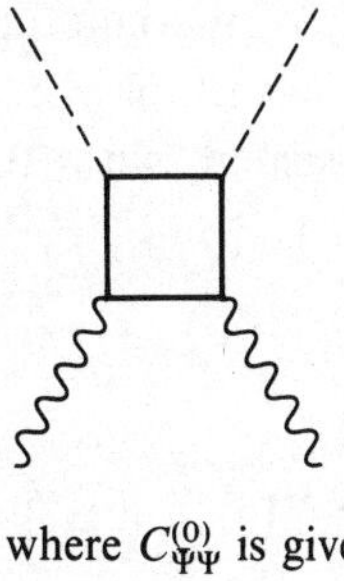 $+\,\mathrm{perm} = C^{(0)}_{\bar{\Psi}\Psi}\cdot$ 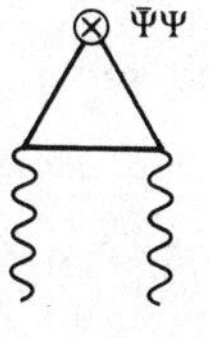 $+\, C^{(1)}_G\cdot$ 

where $C^{(0)}_{\bar{\Psi}\Psi}$ is given by the zeroth order calculation, and find $C^{(1)}_G$.

8
Quantum chromodynamics

8.1 General introduction

Renormalization and BRS invariance; counterterms

QCD is a theory of interactions of quarks and gluons. Its lagrangian density has been discussed in Sections 1.3 and 3.2. In terms of bare quantities it reads

$$\mathscr{L} = -\tfrac{1}{4}G^{\alpha}_{\mu\nu}G^{\alpha\mu\nu} + \sum_f \bar{\Psi}_f(\mathrm{i}\not{D} - m_f)\Psi_f$$
$$+ \text{gauge-fixing term} + \text{Faddeev–Popov term} \tag{8.1}$$

where the sum is over all quark flavours and the last two terms are given in (3.46) for the class of covariant gauges. The quarks are assumed to transform according to the fundamental representation: each flavour of quark is a triplet of the colour group $SU(3)$. Gauge bosons transform according to the adjoint representation, so that there are eight gluons.

In this Section we discuss the theory formulation in covariant gauges. The use of the axial gauge $n\cdot A^{\alpha} = 0$, $n^2 < 0$, or light-like gauge, $n^2 = 0$, where n is in each case some fixed four-vector, is also often very convenient in perturbative QCD calculations. QCD formulation in the axial gauge $n^2 < 0$ can be given in a similarly precise form to that in covariant gauges (e.g. Bassetto *et al.* 1985). In light-like gauges the general proof of renormalizability is still lacking but order-by-order calculations have been consistently performed.

Let us concentrate on the renormalization programme. As in QED, the crucial isssue is the gauge invariance or strictly speaking BRS invariance of the theory. It can be proved (e.g. Collins 1984) that the lagrangian is BRS invariant after renormalization. Let us therefore rewrite lagrangian (8.1) in terms of the renormalized quantities assuming the most general form consistent with BRS invariance.† We get (for one flavour)

$$\begin{aligned}\mathscr{L} = &-\tfrac{1}{4}Z_3[\partial_\mu A^{\alpha}_{\nu} - \partial_\nu A^{\alpha}_{\mu} + g(Z_{1\mathrm{YM}}/Z_3)c^{\alpha\beta\gamma}A^{\beta}_{\mu}A^{\gamma}_{\nu}]^2 \\ &+ Z_2\bar{\Psi}\mathrm{i}\gamma_\mu[\partial^\mu - \mathrm{i}\tilde{g}(Z_1/Z_2)A^{\alpha\mu}T^{\alpha}]\Psi - mZ_0\bar{\Psi}\Psi \\ &+ \tilde{Z}_2\eta^{*\alpha}[\delta^{\alpha\beta}\partial^2 - \tilde{\tilde{g}}(\tilde{Z}_1/\tilde{Z}_2)c^{\alpha\beta\gamma}A^{\gamma}_{\mu}\partial^\mu]\eta^\beta - (1/2a)(\partial_\mu A^{\beta\mu})^2\end{aligned} \tag{8.2}$$

† See also the beginning of Chapter 5

where

$$gZ_{1\mathrm{YM}}/Z_3 = \tilde{g}Z_1/Z_2 = \tilde{\tilde{g}}\tilde{Z}_1/\tilde{Z}_2 \tag{8.3}$$

If we use a renormalization scheme such as the minimal subtraction scheme in which the renormalized couplings satisfy

$$g = \tilde{g} = \tilde{\tilde{g}}$$

then (8.3) implies the corresponding equality for the ratios of the renormalization constants. As in QED there is no need for a gauge-fixing counterterm.

Relation (8.3) follows from the requirement of the BRS invariance for the lagrangian (8.2). Formally it can be derived from the Slavnov–Taylor identities (Ward identities for a non-abelian gauge theory) which themselves follow from the BRS invariance.

The correspondence between the renormalized quantities and the bare quantities of lagrangian (8.1) is the following

$$\left.\begin{array}{lll} A^{\alpha}_{\mathrm{B}\mu} = Z_3^{1/2}A^{\alpha}_{\mu} & \Psi_{\mathrm{B}} = Z_2^{1/2}\Psi & a_{\mathrm{B}} = Z_3 a \\ \eta_{\mathrm{B}} = \tilde{Z}_2^{1/2}\eta & \eta^*_{\mathrm{B}} = \tilde{Z}_2^{1/2}\eta^* & m_{\mathrm{B}} = mZ_0/Z_2 \end{array}\right\} \tag{8.4}$$

$$g_{\mathrm{B}} = gZ_{1\mathrm{YM}}/Z_3^{3/2} = \tilde{g}Z_1/Z_2Z_3^{1/2} = \tilde{\tilde{g}}\tilde{Z}_1/\tilde{Z}_2Z_3^{1/2} \tag{8.5}$$

Lagrangian (8.1) is invariant under the BRS transformation (3.76)

$$\left.\begin{array}{l} \delta_{\mathrm{BRS}}\Psi_{\mathrm{B}} = \mathrm{i}g_{\mathrm{B}}T^{\alpha}\Theta_{\mathrm{B}}\eta^{\alpha}_{\mathrm{B}}\Psi_{\mathrm{B}} \equiv \Theta_{\mathrm{B}}(\mathscr{s}\Psi_{\mathrm{B}}) \\ \delta_{\mathrm{BRS}}\bar{\Psi}_{\mathrm{B}} = \mathrm{i}g_{\mathrm{B}}\bar{\Psi}_{\mathrm{B}}T^{\alpha}\Theta_{\mathrm{B}}\eta^{\alpha}_{\mathrm{B}} \equiv \Theta_{\mathrm{B}}(\mathscr{s}\bar{\Psi}_{\mathrm{B}}) \\ \delta_{\mathrm{BRS}}A^{\alpha}_{\mathrm{B}\mu} = \Theta_{\mathrm{B}}(\partial_{\mu}\eta^{\alpha}_{\mathrm{B}} - g_{\mathrm{B}}c^{\alpha\beta\gamma}\eta^{\beta}_{\mathrm{B}}A^{\gamma}_{\mathrm{B}\mu}) \equiv \Theta_{\mathrm{B}}(\mathscr{s}A^{\alpha}_{\mathrm{B}\mu}) \\ \delta_{\mathrm{BRS}}\eta^{\alpha}_{\mathrm{B}} = -\tfrac{1}{2}g_{\mathrm{B}}c^{\alpha\beta\gamma}\eta^{\beta}_{\mathrm{B}}\eta^{\gamma}_{\mathrm{B}}\Theta_{\mathrm{B}} \equiv \Theta_{\mathrm{B}}(\mathscr{s}\eta^{\alpha}_{\mathrm{B}}) \\ \delta_{\mathrm{BRS}}\eta^{*\alpha}_{\mathrm{B}} = -(1/a_{\mathrm{B}})\Theta_{\mathrm{B}}\partial_{\mu}A^{\mu\alpha}_{\mathrm{B}} \equiv \Theta_{\mathrm{B}}(\mathscr{s}\eta^{*\alpha}_{\mathrm{B}}) \end{array}\right\} \tag{8.6}$$

and lagrangian (8.2) under

$$\left.\begin{array}{l} \delta_{\mathrm{BRS}}\Psi = \mathrm{i}g\tilde{Z}_1T^{\alpha}\Theta\eta^{\alpha}\Psi \\ \delta_{\mathrm{BRS}}\bar{\Psi} = \mathrm{i}g\tilde{Z}_1\bar{\Psi}T^{\alpha}\Theta\eta^{\alpha} \\ \delta_{\mathrm{BRS}}A^{\alpha}_{\mu} = \Theta(\partial_{\mu}\eta^{\alpha}\tilde{Z}_2 - g\tilde{Z}_1c^{\alpha\beta\gamma}\eta^{\beta}A^{\gamma}_{\mu}) \\ \delta_{\mathrm{BRS}}\eta^{\alpha} = -\tfrac{1}{2}g\tilde{Z}_1c^{\alpha\beta\gamma}\eta^{\beta}\eta^{\gamma}\Theta \\ \delta_{\mathrm{BRS}}\eta^{*\alpha} = -(1/a)\Theta\partial_{\mu}A^{\mu\alpha} \end{array}\right\} \tag{8.7}$$

Invariance of the lagrangian (8.2) under (8.7) can be checked explicitly but it also follows immediately from the invariance of bare lagrangian (8.1) under (8.6) if relation (8.3), (8.4) and (8.5) are inserted into (8.6) and if Θ is identified as

$$\Theta = (1/Z_3^{1/2}\tilde{Z}_2^{1/2})\Theta_{\mathrm{B}}$$

We note the presence of the divergent renormalization constants $\tilde{Z}_i$ in the BRS transformation (8.7). This is actually what is necessary for finiteness of the composite operators $\delta\Psi$, $\delta\bar{\Psi}$, δA_μ^α and $\delta\eta^\alpha$ (Collins 1984).

Lagrangian (8.2) can be rewritten introducing counterterms (we take $g = \tilde{g} = \tilde{\tilde{g}}$)

$$\begin{aligned}\mathscr{L} = & -\tfrac{1}{4}G_{\mu\nu}^\alpha G_\alpha^{\mu\nu} + \bar{\Psi}\mathrm{i}\not{D}\Psi - m\bar{\Psi}\Psi - (1/2a)(\partial_\mu A_\beta^\mu)^2 + \eta^{*\alpha}(\delta^{\alpha\beta}\partial^2 - gc^{\alpha\beta\gamma}A_\gamma^\mu\partial_\mu)\eta^\beta \\ & + (Z_2 - 1)\bar{\Psi}\mathrm{i}\not{\partial}\Psi + (Z_1 - 1)g\bar{\Psi}\gamma^\mu A_\mu^\alpha T^\alpha\Psi - m(Z_0 - 1)\bar{\Psi}\Psi \\ & + (\tilde{Z}_2 - 1)\eta^{*\alpha}\partial^{\alpha\beta}\partial^2\eta^\beta - (\tilde{Z}_1 - 1)g\eta^{*\alpha}c^{\alpha\beta\gamma}A_\gamma^\mu\partial_\mu\eta^\beta \\ & - \tfrac{1}{4}(Z_3 - 1)(\partial_\mu A_\nu^\alpha - \partial_\nu A_\mu^\alpha)^2 - \tfrac{1}{2}(Z_{1\mathrm{YM}} - 1)g(\partial_\mu A_\nu^\alpha - \partial_\nu A_\mu^\alpha)c^{\alpha\beta\gamma}A_\beta^\mu A_\gamma^\nu \\ & - \tfrac{1}{4}g^2(Z_{1\mathrm{YM}}^2 Z_3^{-1} - 1)c^{\alpha\beta\gamma}A_\mu^\beta A_\nu^\gamma c^{\alpha\rho\sigma}A^{\rho\mu}A^{\sigma\nu}\end{aligned} \tag{8.8}$$

The Feynman rules for the vertices of (8.8) have been derived in Section 3.2. They are collected in Appendix A.

Asymptotic freedom of QCD

The RGE, the notion of the effective coupling constant and its role in determining the behaviour of a quantum field theory at different momentum scales have been extensively discussed in Chapter 6. In this context the behaviour of observable cross sections has been discussed in Chapter 7 in the framework of the OPE which provides a systematic way of factorizing effects at different mass scales.

QCD is an asymptotically free theory: the coupling constant decreases as the scale at which it is defined is increased. Thus, the behaviour of the Green's functions when all the momenta are scaled up with a common factor is governed by a theory where $g \to 0$. This can explain the success of the parton model in describing the scaling phenomena in the deep inelastic processes. Only theories with a non-abelian gauge symmetry can be asymptotically free in four space-time dimensions (Coleman & Gross 1973).

The β-function in QCD defined as

$$\beta(\alpha) = \mu(\mathrm{d}/\mathrm{d}\mu)\alpha(\mu^2)|_{\alpha_\mathrm{B}\,\mathrm{fixed}}, \quad \alpha = g^2/4\pi$$

is given by expansion (6.45)

$$\beta(\alpha) = \alpha\left[\frac{\alpha}{\pi}b_1 + \left(\frac{\alpha}{\pi}\right)^2 b_2 + \left(\frac{\alpha}{\pi}\right)^3 b_3 + \cdots\right] \tag{6.45}$$

and the coefficients b_i are calculable perturbatively. Using dimensional regularization and a mass-independent renormalization scheme we have (6.33) and (6.35): $\beta(\alpha)$ is totally determined by the residue of the simple UV pole of Z_α (beyond the one-loop order we need to calculate a non-leading UV divergence) where according to (8.5) one can use one of the relations

$$Z_\alpha = Z_{1\mathrm{YM}}^2/Z_3^3 = Z_1^2/Z_2^2 Z_3 = \tilde{Z}_1^2/\tilde{Z}_2^2 Z_3 \tag{8.9}$$

Diagrams contributing to different renormalization constants are determined by the structure of counterterms in lagrangian (8.8). For instance at the one-loop level we have the contributions shown in Fig. 8.1 (sums over flavours and over different

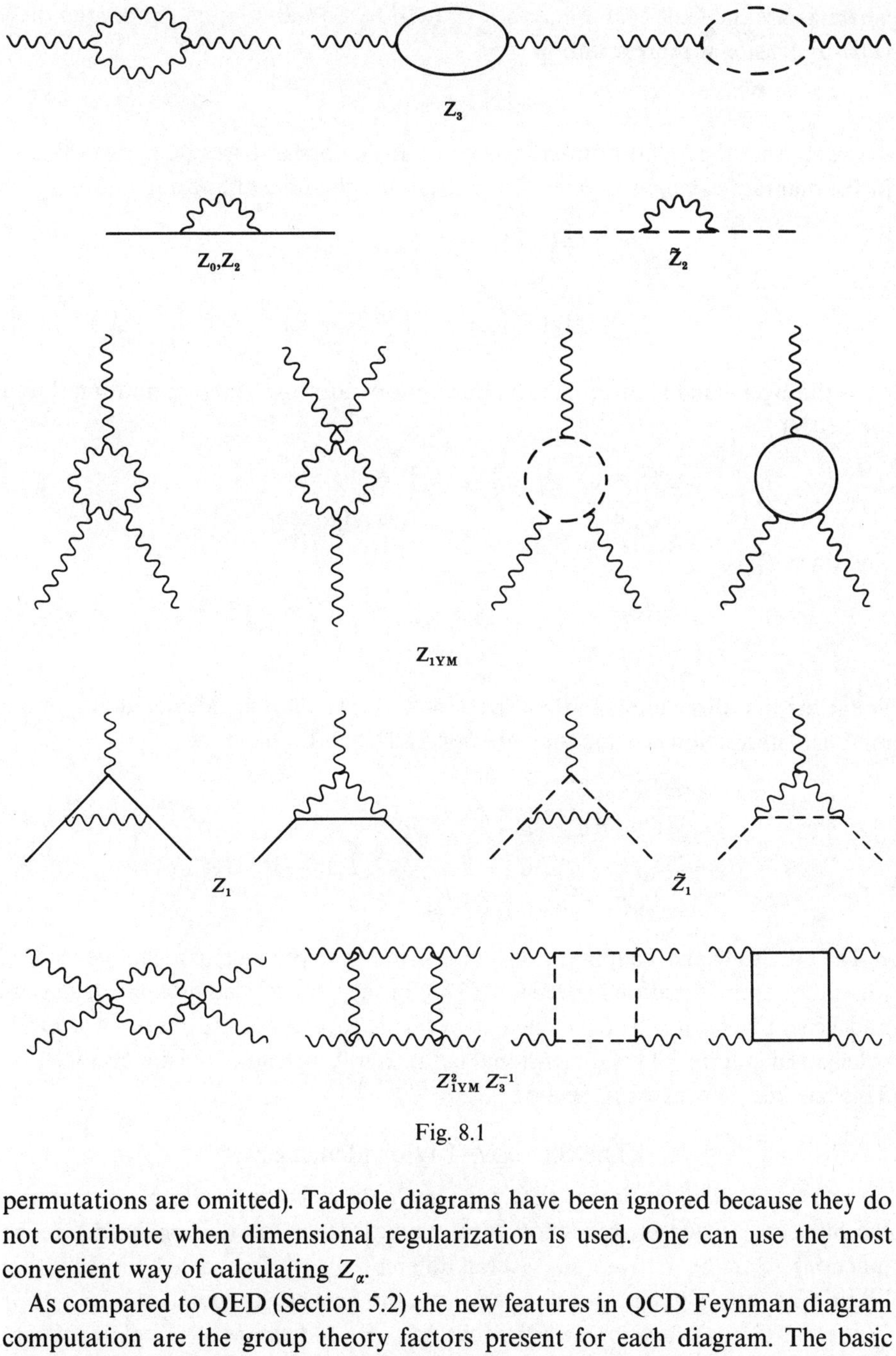

Fig. 8.1

permutations are omitted). Tadpole diagrams have been ignored because they do not contribute when dimensional regularization is used. One can use the most convenient way of calculating Z_α.

As compared to QED (Section 5.2) the new features in QCD Feynman diagram computation are the group theory factors present for each diagram. The basic ingredients are $SU(3)$ generators in the fundamental three-dimensional representation conventionally taken as

$$(T^a)_{bc} = \tfrac{1}{2}(\lambda^a)_{bc}$$

where λs are the Gell-Mann matrices, $\mathrm{Tr}[\lambda^a \lambda^b] = 2\delta^{ab}$ and in the regular (adjoint) *eight*-dimensional representation

$$(T^a)_{bc} = -\mathrm{i}c_{abc}$$

where c_{abc}s are the $SU(3)$ structure constants. Both appear in the QCD vertices, and in the diagram calculation one often encounters the following combinations

$$\sum_{a,c} (T^a_{\mathrm{R}})_{bc}(T^a_{\mathrm{R}})_{cd} = (T^2_{\mathrm{R}})_{bd} = C_{\mathrm{R}}\delta_{bd} \tag{8.10}$$

and

$$\sum_{c,d} (T^a_{\mathrm{R}})_{dc}(T^b_{\mathrm{R}})_{cd} = \mathrm{Tr}[T^a_{\mathrm{R}} T^b_{\mathrm{R}}] = T_{\mathrm{R}}\delta^{ab} \tag{8.11}$$

C_{R} is the eigenvalue of the quadratic Casimir operator in the representation R and for $SU(N)$

$$C_{\mathrm{F}} = (N^2 - 1)/2N \qquad \mathrm{F} \equiv \text{fundamental}$$

$$C_{\mathrm{A}} = N \qquad \mathrm{A} \equiv \text{adjoint}$$

The trace T_{R} is

$$T_{\mathrm{F}} = \tfrac{1}{2}$$

$$T_{\mathrm{A}} = N$$

So far the first three terms in the expansion (6.45) have been calculated and in the minimal subtraction scheme they are (for $SU(N)$ and n flavours)

$$\left.\begin{aligned}
-b_1 &= \tfrac{11}{6}C_{\mathrm{A}} - \tfrac{2}{3}T_{\mathrm{F}}n \\
-b_2 &= \tfrac{17}{12}C_{\mathrm{A}}^2 - \tfrac{1}{2}C_{\mathrm{F}}T_{\mathrm{F}}n - \tfrac{5}{6}C_{\mathrm{A}}T_{\mathrm{F}}n \\
-b_3 &= \tfrac{2857}{1728}C_{\mathrm{A}}^3 - \tfrac{1415}{864}C_{\mathrm{A}}^2 T_{\mathrm{F}}n + \tfrac{79}{432}C_{\mathrm{A}}T_{\mathrm{F}}^2 n^2 - \tfrac{205}{288}C_{\mathrm{A}}C_{\mathrm{F}}T_{\mathrm{F}}n \\
&\quad + \tfrac{44}{288}C_{\mathrm{F}}T_{\mathrm{F}}^2 n^2 + \tfrac{1}{16}C_{\mathrm{F}}^2 T_{\mathrm{F}}n
\end{aligned}\right\} \tag{8.12†}$$

The QCD effective coupling constant is given by expression (6.31) in the leading logarithm approximation expression, (6.53) in the next-to-leading approximation, and generally by (6.52). It is often convenient to express the result of a QCD calculation in terms of the dimensional parameter Λ instead of the dimensionless α. This has been discussed in Section 7.3.

The Slavnov–Taylor identities

As with the Ward identities in QED, their analogue in a non-abelian gauge theory – the Slavnov–Taylor identities for bare regularized or for renormalized Green's functions – can be derived in several different ways. One can proceed as in Section 5.1: apply gauge transformation to the generating functional for full or connected Green's functions which is written as a functional of the physical field

† The coefficient b_1 was first calculated by Politzer (1973) and Gross & Wilczek (1973), the coefficient b_2 by Caswell (1974) and Jones (1974) and the coefficient b_3 by Tarasov, Vladimirov & Zharkov (1980).

sources only. Then one derives the non-abelian analogue of (5.11) (Abers & Lee 1973; Marciano & Pagels 1978). This form generates all the identities for the connected Green's functions but, because of the presence of ghost fields it is very complicated to obtain the analogous relation for the 1PI generating functional $\Gamma(A, \bar{\Psi}, \Psi)$. A more elegant method is based on the BRS transformation (Lee 1976). In addition to sources $J, \alpha, \bar{\alpha}$ for the fields $A, \bar{\Psi}$ and Ψ, respectively, one introduces sources c^{α} and $c^{*\alpha}$ for the ghost field $\eta^{*\alpha}$ and η^{α} as well as sources $K, L, \bar{L}$ and M for the composite operators $\mathscr{s}A$, $\mathscr{s}\bar{\Psi}$, $\mathscr{s}\Psi$ and $\mathscr{s}\eta$ that appear in the BRS transformation (8.6). This is to obtain identities that are linear in derivatives with respect to the sources. Next one defines the generating functional

$$W[J, \alpha, \bar{\alpha}, c, c^*, K, L, \bar{L}, M] = \int \mathscr{D}(A\Psi\bar{\Psi}\eta\eta^*) \exp\left[iS_{\text{eff}} + i\int d^4x(JA + \bar{\alpha}\Psi + \bar{\Psi}\alpha + c^*\eta + \eta^*c + K\mathscr{s}A + \mathscr{s}\bar{\Psi}L + \bar{L}\mathscr{s}\Psi + M\mathscr{s}\eta)\right] \tag{8.13}$$

Changing variables in the functional integral according to the BRS transformation and using the invariance

$$\delta S_{\text{eff}}/\delta\Theta = 0$$

as well as the nilpotent relations

$$\mathscr{s}^2 A = \mathscr{s}^2\eta = \mathscr{s}^2\Psi = \mathscr{s}^2\bar{\Psi} = 0$$

one obtains the desired general relation generating the Slavnov–Taylor identities. Since the ghost fields are treated on equal footing with the physical fields it is now straightforward to obtain similar relation generating identities for the 1PI Green's functions. As usual this is achieved by means of the Legendre transform

$$\Gamma[A, \Psi, \bar{\Psi}, \eta, \eta^*, K, L, \bar{L}, M] = Z[J, \alpha, \bar{\alpha}, c, c^*, K, L, \bar{L}, M] - \int d^4x(JA + \bar{\alpha}\Psi + \bar{\Psi}\alpha + c^*\eta + \eta^*c) \tag{8.14}$$

where $\eta^{\alpha} = \delta Z/\delta c^{*\alpha}, c^{*\alpha} = -\delta\Gamma/\delta\eta^{\alpha}$, etc. One can convince oneself that in the presence of additional sources $K, L, \bar{L}$ and M the functional Γ still generates 1PI Green's functions.

The general relations whose derivation has just been briefly described contain all the information about BRS invariance and are useful in formal manipulations. The simplest way, however, to get specific identities for the Green's functions of interest is based on the observation that since BRS invariance is an exact symmetry of the theory all the Green's functions are BRS invariant (e.g. Llewellyn Smith 1980). For instance let us consider the regularized Green's function $\langle 0|TA^{\alpha}_{\mu}(x)\eta^{*\beta}(y)|0\rangle$. Using (8.6) we have

$$\delta_{\text{BRS}}\langle 0|TA^{\alpha}_{\mu}(x)\eta^{*\beta}(y)|0\rangle = \langle 0|T(\partial_{\mu}\eta^{\alpha}(x) - gc^{\alpha\beta\gamma}\eta^{\beta}(x)A^{\gamma}_{\mu}(x))\eta^{*\beta}(y)|0\rangle - (1/a)\langle 0|TA^{\alpha}_{\mu}(x)\partial_{\nu}A^{\nu\beta}(y)|0\rangle = 0 \tag{8.15}$$

Differentiating (8.15) with respect to x_μ and using the equation of motion for the Green's function $\langle 0|T\eta^\alpha(x)\eta^{*\beta}(y)|0\rangle$ which can be derived analogously to (2.56a) we conclude that

$$(1/a)\partial^\mu_{(x)}\partial^\nu_{(y)}\langle 0|TA^\alpha_\mu(x)A^\beta_\nu(y)|0\rangle = -\mathrm{i}\delta(x-y)\delta^{\alpha\beta} \tag{8.16}$$

Result (8.16) implies that as in QED (see (5.14)) the longitudinal part of the propagator is unchanged by the higher order corrections.

8.2 The background field method

This method is based on the idea of quantizing field theory in presence of a background field chosen so that certain symmetries of the generating functionals are restored. Let us consider gauge theories. We start with the gauge-fixed functional integral (the dependence on other fields and on the ghost sources is not indicated explicitly)

$$W[J] = \int \mathscr{D}A'_\mu \exp(\mathrm{i}S_{\mathrm{eff}} + \mathrm{i}J\cdot A'), \quad J\cdot A' = \int \mathrm{d}^4x J^\alpha_\mu A'^\mu_\alpha \tag{8.17}$$

and define, as usual,

$$Z[J] = -\mathrm{i}\ln W[J] \tag{8.18}$$

and the effective action

$$\Gamma[\bar{A}] = Z[J] - J\cdot\bar{A}, \quad \bar{A}^\mu_\alpha = \delta Z/\delta J^\alpha_\mu \tag{8.19}$$

Manifest gauge invariance is lost because of the gauge-fixing procedure. However, we can restore it by introducing a background field with properly chosen transformation properties. Let us consider first the gauge-invariant part of the action and introduce the background field by the replacement

$$A'_\mu = A_\mu + \mathbf{A}_\mu \tag{8.20}$$

where $\mathbf{A}^\alpha_\mu$ is the background field and A^α_μ is the quantum field (we recall that $A_\mu(x) = -\mathrm{i}gA^\alpha_\mu(x)T^\alpha$, $\Theta(x) = -\mathrm{i}\Theta^\alpha(x)T^\alpha$ and we define $\mathbf{A}_\mu(x) = -\mathrm{i}g\mathbf{A}^\alpha_\mu(x)T^\alpha$). Invariance of the action under the transformation (1.39)

$$\delta A'_\mu = -D_\mu\Theta(x) = -\partial_\mu\Theta(x) - [A'_\mu(x), \Theta(x)] \tag{8.21}$$

implies invariance under two kinds of transformations on A_μ and $\mathbf{A}_\mu$ which give the same $\delta A'_\mu$

quantum

$$\left.\begin{aligned}&\delta\mathbf{A}_\mu = 0\\&\delta A_\mu = -D_\mu\Theta(x) - [\mathbf{A}_\mu, \Theta(x)], \quad D_\mu = \partial_\mu + [A_\mu, \ldots]\end{aligned}\right\} \tag{8.22}$$

background

$$\left.\begin{aligned}&\delta\mathbf{A}_\mu = -\mathbf{D}_\mu\Theta(x), \quad \mathbf{D}_\mu = \partial_\mu + [\mathbf{A}_\mu, \ldots]\\&\delta A_\mu = [\Theta(x), A_\mu(x)]\end{aligned}\right\} \tag{8.23}$$

We can maintain the manifest gauge invariance of S_{eff} with respect to the *background* transformations if the gauge-fixing function $\tilde{F}^{\alpha}[A, \mathbf{A}]$ transforms covariantly under these transformations. For instance, choose

$$(\tilde{F}^{\alpha}(A_{\mu}, \mathbf{A}_{\mu}))^2 = -(1/2a)(\mathbf{D}_{\mu}A^{\mu})_{\alpha}^2 = -(1/2a)(\partial^{\mu}A_{\mu}^{\alpha} + gc^{\alpha\beta\gamma}\mathbf{A}_{\beta}^{\mu}A_{\mu}^{\gamma})^2 \qquad (8.24)$$

The Faddeev–Popov ghost term then takes account of

$$\det M_{\alpha\beta} = \det(\delta\tilde{F}^{\alpha}/\delta\Theta^{\beta})$$

where $\delta\tilde{F}^{\alpha}$ corresponds to the quantum gauge transformation

$$\delta A_{\mu}^{\alpha} = -(1/g)\partial_{\mu}\Theta^{\alpha} + c^{\alpha\beta\gamma}\Theta^{\beta}(A_{\mu}^{\gamma} + \mathbf{A}_{\mu}^{\gamma})$$

and it also transforms covariantly under the background gauge transformation.

The generating functional

$$\tilde{W}[J, \mathbf{A}_{\mu}] = \int \mathscr{D}A_{\mu} \exp\{\mathrm{i}S_{\text{eff}}[A_{\mu} + \mathbf{A}_{\mu}] + \mathrm{i}J\cdot A\} \qquad (8.25)$$

is manifestly invariant under the background gauge transformations if we complete them with the transformation

$$\delta J_{\mu}^{\alpha} = c^{\alpha\beta\gamma}\Theta^{\beta}J_{\mu}^{\gamma} \qquad (8.26)$$

The same is true for

$$\tilde{Z}[J, \mathbf{A}_{\mu}] = -\mathrm{i}\ln\tilde{W}[J, \mathbf{A}_{\mu}] \qquad (8.27)$$

and for

$$\tilde{\Gamma}[\bar{A}_{\mu}, \mathbf{A}_{\mu}] = \tilde{Z}[J, \mathbf{A}_{\mu}] - J\cdot\bar{A} \qquad (8.28)$$

$$\bar{A}_{\mu}^{\alpha} = \delta\tilde{Z}[J, \mathbf{A}_{\mu}]/\delta J_{\mu}^{\alpha}$$

if

$$\delta\bar{A}_{\mu}^{\alpha} = c^{\alpha\beta\gamma}\Theta^{\beta}\bar{A}_{\mu}^{\gamma}$$

The next step is to find the relationship between the original $\Gamma[\bar{A}]$ and the $\tilde{\Gamma}[\bar{A}_{\mu}, \mathbf{A}_{\mu}]$ in the presence of the background field. Changing the integration variables in (8.25) for $\tilde{W}[J, \mathbf{A}_{\mu}]$: $A_{\mu} \to A_{\mu} + \mathbf{A}_{\mu}$ one gets

$$\tilde{W}[J, \mathbf{A}_{\mu}] = W[J]\exp(-\mathrm{i}J\cdot\mathbf{A}_{\mu}) \qquad (8.29)$$

and correspondingly

$$\tilde{Z}[J, \mathbf{A}_{\mu}] = Z[J] - J\cdot\mathbf{A}_{\mu} \qquad (8.30)$$

Therefore

$$\tilde{\Gamma}[\bar{A}_{\mu}, \mathbf{A}_{\mu}] = Z[J] - J\cdot\mathbf{A}_{\mu} - J\cdot\bar{A}_{\mu} = \Gamma[\mathbf{A}_{\mu} + \bar{A}_{\mu}] \qquad (8.31)$$

and we conclude that the effective action $\tilde{\Gamma}[\bar{A}_{\mu}, \mathbf{A}_{\mu}]$ is the usual effective action $\Gamma[\bar{A}_{\mu}]$ evaluated at $\bar{A}_{\mu} = \mathbf{A}_{\mu} + \bar{A}_{\mu}$. It should also be remembered that this $\Gamma[\bar{A}_{\mu}]$ corresponds to the gauge-fixing term $F^{\alpha}[A'] = \tilde{F}^{\alpha}[A' - \mathbf{A}, \mathbf{A}]$ which may be an unusual gauge in the standard approach.

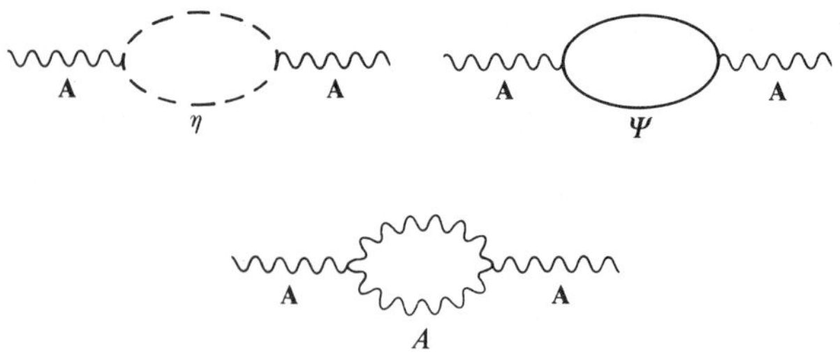

Fig. 8.2

In particular it is convenient to calculate $\tilde{\Gamma}[\tilde{A}_\mu, \mathbf{A}_\mu]$ for $\tilde{A}_\mu = 0$ i.e. restricting ourselves to diagrams with no external $\tilde{A}_\mu$ lines. These '$\tilde{A}_\mu$-vacuum' diagrams are the 1PI subset of diagrams obtained from $\tilde{W}[0, \mathbf{A}_\mu]$ given by (8.25) i.e. diagrams with only internal A_μ lines and external $\mathbf{A}_\mu$, ghost and other non-gauge field lines. We get

$$\Gamma[\mathbf{A}_\mu] = \tilde{\Gamma}[0, \mathbf{A}_\mu] \tag{8.32}$$

A very important fact is that $\tilde{\Gamma}[0, \mathbf{A}_\mu]$ is invariant under the transformation $\delta\mathbf{A}_\mu = -\mathbf{D}_\mu\Theta(x)$, i.e. it is a gauge-invariant functional of $\mathbf{A}_\mu$. This property is maintained in a perturbative loop-by-loop calculation of $\tilde{\Gamma}[0, \mathbf{A}_\mu]$ (check it). It implies, in particular, that the divergent terms and the counterterms are manifestly invariant with respect to the background gauge transformation. Thus the infinities appearing in $\tilde{\Gamma}[0, \mathbf{A}_\mu]$ must take the form of a divergent constant times $(\mathbf{G}^\alpha_{\mu\nu})^2$ where

$$\mathbf{G}^\alpha_{\mu\nu} = \partial_\mu \mathbf{A}^\alpha_\nu - \partial_\nu \mathbf{A}^\alpha_\mu + gc^{\alpha\beta\gamma}\mathbf{A}^\beta_\mu \mathbf{A}^\gamma_\nu$$

It is then straightforward to show (Problem 8.3) that the renormalization constants

$$(\mathbf{A}_\mu)_\mathrm{B} = \mathbf{Z}_A^{1/2}\mathbf{A}_\mu \tag{8.33}$$
$$g_\mathrm{B} = Z_g g$$

satisfy

$$Z_g = \mathbf{Z}_A^{-1/2} \tag{8.34}$$

and, for instance, the calculation of the β-function in the one-loop approximation simplifies to the calculation of the diagrams in Fig. 8.2.

It is recommended that the reader derives the Feynman rules necessary for the calculation of $\tilde{\Gamma}[0, \mathbf{A}_\mu]$ and completes the evaluation of the β-function in the one-loop approximation (Problem 8.3). The background field method is extensively used in supergravity theories (Gates, Grisaru, Roček & Siegel 1983).

8.3 The structure of the vacuum in non-abelian gauge theories

Homotopy classes and topological vacua

For the moment we shall consider the pure Yang–Mills gauge theory with $SU(2)$ as the gauge group. As we know, the so-called gauge-fixing procedure used in perturbation theory usually leaves some residual gauge freedom. For instance we can use the gauge $A_0=0$ (for the present discussion this is convenient but not necessary; see e.g. Coleman 1979) which obviously leaves room for further gauge transformations $U(\mathbf{r})$ depending on spatial variables only. Thus the vacuum defined by $G_{\mu\nu}=0$ restricts the potentials to 'pure gauge' configurations

$$A_i(\mathbf{r}) = -U^{-1}(\mathbf{r})\partial_i U(\mathbf{r}) \tag{8.35}$$

Further discussion relies on an additional restriction that

$$\lim_{r\to\infty} U(\mathbf{r}) = U_\infty \tag{8.36}$$

where U_∞ is a global (position-independent) gauge transformation. Equivalently, we assume that all vector potentials fall faster than $1/r$ at large distances. In pure Yang–Mills theory one can argue (Jackiw 1980), but not prove, that physically admissible gauge transformations satisfy (8.36) which assures that the non-abelian charge is well defined (see Problem 1.4).

As far as the gauge functions $U(\mathbf{r})$ are concerned the three-dimensional spatial manifold with points at infinity identified is topologically equivalent to S^3 – the surface of a four-dimensional Euclidean sphere labelled by three angles. The matrix functions U provide a mapping of S^3 into the manifold of the $SU(2)$ gauge group. Since any element M in the $SU(2)$ group can be written as

$$M = a + \mathrm{i}\mathbf{b}\cdot\boldsymbol{\tau}$$

where $\boldsymbol{\tau}$s are Pauli matrices and where real a and $\mathbf{b}$ satisfy

$$a^2 + \mathbf{b}^2 = 1$$

we conclude that the manifold of the $SU(2)$ group elements is topologically equivalent to S^3. Thus, gauge transformations $U(\mathbf{r})$ provide mappings $S^3 \to S^3$. They can be classified into homotopy classes.

Let X and Y be two topological spaces and f_0, f_1 be two continuous mappings from X into Y. They are said to be homotopic if they are continuously deformable into each other i.e. if, and only if, there exists a continuous deformation of maps $F(x,t)$, $0 \leqslant t \leqslant 1$, such that $F(x,0)=f_0(x)$ and $F(x,1)=f_1(x)$. The function $F(x,t)$ is called the homotopy. We can divide all mappings of X into Y into homotopic classes: two mappings are in the same class if they are homotopic. A group structure can be defined on the set of homotopy classes.

Let us consider a simple example. Let S^1 be a unit circle parametrized by the angle θ, with θ and $\theta+2\pi$ identified. We are interested in mappings from S^1 into the manifold of a Lie group G. The group of homotopy classes of mappings from S^1 into the manifold of G is called the first homotopy group of G and is denoted by $\Pi_1(G)$. Take the mappings from S^1 into $G \equiv U(1)$ represented by a set of unimodular complex numbers $u = \exp(\mathrm{i}\alpha)$ which itself is topologically equivalent to S^1. Thus $S^1 \to S^1$ and the continuous functions

$$\alpha(\theta) = \exp[\mathrm{i}(N\theta + a)] \tag{8.37}$$

form a homotopic class for different values of a and a fixed integer N. One can think of $\alpha(\theta)$ as a mapping of a circle into another circle such that N points of the first circle are mapped into one point of the second circle (winding N times around the second circle). Hence the integer N is called the winding number and each homotopy class is characterized by its winding number. For this example

$$\Pi_1(U(1)) = \Pi_1(S^1) = Z$$

where Z denotes an additive group of integers. We observe that in our example the winding number N can be written as

$$N = -\mathrm{i}\int_0^{2\pi} \frac{\mathrm{d}\theta}{2\pi}\left[\frac{1}{\alpha(\theta)}\frac{\mathrm{d}\alpha}{\mathrm{d}\theta}\right] \tag{8.38}$$

The mappings of any winding number can be obtained by taking powers of

$$\alpha^{(1)}(\theta) = \exp(\mathrm{i}\theta) \tag{8.39}$$

Instead of the unit circle S^1 we can consider the whole real axis $-\infty < x < \infty$ with the points $x = \pm\infty$ identified which is topologically equivalent to S^1 (see Problem 8.1).

This discussion can be generalized by taking $X = S^n$ (the n-dimensional sphere) or the topologically equivalent n-dimensional cube with all its boundaries identified and equivalent to some x_0 of S^n. The classes of mappings $S^n \to S^m$ with one fixed point $f(x_0) = y_0$ form a group called the nth homotopy group of S^m and designated by $\Pi_n(S^m)$. We have

$$\Pi_n(S^n) = Z \tag{8.40}$$

i.e. mappings $S^n \to S^n$ are classified by the number of times one n-sphere covers the other. Ordinary space R^3 with all points at ∞ identified is equivalent to S^3. As we have already mentioned the group $SU(2)$ is also equivalent to S^3. Thus

$$\Pi_3(SU(2)) = \Pi_3(S^3) = Z \tag{8.41}$$

and this together with relations (8.35) and (8.36) implies an infinity of topologically distinct vacua for the $SU(2)$ Yang–Mills theory. The same result holds for an arbitrary simple Lie group (for $SU(N)$ in particular) due to a theorem (Bott 1956) which says that any continuous mapping of S^3 into G can be continuously deformed into a mapping into an $SU(2)$ subgroup of G. Only if the group is $U(1)$, is every mapping of S^3 into $U(1)$ continuously deformable into the constant map $f(x) = y_0$ corresponding to $N = 0$.

It can be shown that analogously to (8.38) and (I) in Problem 8.1 the winding number for gauge transformations providing the mapping $S^3 \to G$ is given by

$$N = -(1/24\pi^2)\int \mathrm{d}^3x\, \mathrm{Tr}\,[\varepsilon_{ijk}A_iA_jA_k] \tag{8.42}$$

where A_is are given by (8.35). For prototype mapping with $N = 1$ see Problem 8.1.

Physical vacuum

The inequivalence of topologically distinct vacua can be understood as a consequence of Gauss' law

$$D_i^{ab} G_{0i}^b = 0 \tag{8.43}$$

To see this let us first recall the canonical formalism. In the $A_0^a = 0$ gauge the canonical variables are A_i^as and their conjugate momenta

$$\Pi_i^a = \frac{\delta \mathscr{L}}{\delta(\partial_0 A_a^i)} = G_{0i}^a = \partial_0 A_i^a = -E_i^a \tag{8.44}$$

The hamiltonian

$$H = \int \mathrm{d}^3 x \mathscr{H}(x), \quad \mathscr{H} = \Pi_i^a \partial_0 A_i^a - \mathscr{L} \tag{8.45}$$

is the energy

$$H = \frac{1}{2}\int \mathrm{d}^3 x (E_a^2 + B_a^2) = \int \mathrm{d}^3 x \Theta^{00} \tag{8.46}$$

$$E_i^a = G_{i0}^a = -\Pi_i^a, \quad B_i^a = -\tfrac{1}{2}\varepsilon_{ijk} G_{jk}^a \tag{8.47}$$

When the canonical commutation relations are imposed

$$\left.\begin{aligned} &[A_i^a(\mathbf{x},t), A_j^b(\mathbf{y},t)] = [E_i^a(\mathbf{x},t), E_j^b(\mathbf{y},t)] = 0 \\ &[E_i^a(\mathbf{x},t), A_j^b(\mathbf{y},t)] = \mathrm{i}\delta^{ab}\delta_{ij}\delta(\mathbf{x}-\mathbf{y}) \end{aligned}\right\} \tag{8.48}$$

the hamiltonian equations give

$$\partial_0 A_i^a = \mathrm{i}[H, A_i^a] = -E_i^a \tag{8.49}$$

$$\partial_0 E_i^a = \mathrm{i}[H, E_i^a] = \varepsilon_{ijk} D_j^{ab} B_k^b \tag{8.50}$$

but Gauss' law (8.43) is not found. We cannot set $D_i^{ab} E_i^b$ to zero as this operator does not commute with the canonical variables. Actually as seen from (8.53) it generates infinitesimal space-dependent gauge transformations. Thus, although we obtain a consistent quantum theory, it is different from the desired gauge theory. Gauss' law can be incorporated into the theory by demanding that the physical states are only those which are annihilated by $D_i^{ab} G_{0i}^b$

$$D_i^{ab} G_{0i}^b |\Psi\rangle = 0 \tag{8.51}$$

Having (8.51) we can return to our main question and recognize, as the next step, that only gauge transformations which belong to the homotopy class $N = 0$ can be built up by iterating the infinitesimal gauge transformations (see Problem 8.2). The generator of these gauge transformations is (see Problem 1.4)

$$Q_G = \int \mathrm{d}^3 x G_{0i}^a D_i^{ab} \Theta^b(\mathbf{x}) \tag{8.52}$$

and $\hat{U} = \exp(\mathrm{i}Q_G)$. Integrating by parts and noting that $\Theta^a(\mathbf{x})$ vanishes at $\mathbf{x} \to \infty$ for the $N = 0$ class one gets

$$Q_G = \int \mathrm{d}^3x[-D_i^{ab}G_{0i}^a\Theta^b(\mathbf{x})] \tag{8.53}$$

Using (8.51) and (8.53) we conclude that such transformations leave the vacuum state $|N\rangle$ invariant:

$$T^{(0)}|N\rangle = |N\rangle \tag{8.54}$$

where $T^{(0)}$ is the unitary transformation corresponding to $\hat{U} = \exp(\mathrm{i}Q_G)$ in the class $N = 0$. However, for a gauge transformation in the class $N = 1$ we have

$$T^{(1)}|N\rangle = |N+1\rangle \tag{8.55}$$

Since the states $|N\rangle$ are functionals of A_μ, (8.55) follows from the transformation properties of the winding number $N(A)$ given by (8.42)

$$N(U_n^{-1}\mathbf{A}U_n - U_n^{-1}\partial_i U_n) = N(\mathbf{A}) + n \tag{8.56}$$

where U_n is in the class $N = n$ and A_μ is pure gauge: $\mathbf{A} = H^{-1}\partial_\mu H$. Now the true vacuum must be at least phase invariant under all gauge transformations because they commute with observables. Because T is unitary its eigenvalues are $\exp(-\mathrm{i}\Theta)$, $0 \leqslant \Theta \leqslant 2\pi$, and the physical vacuum is given by

$$|\Theta\rangle = \sum_{N=-\infty}^{\infty} \exp(\mathrm{i}N\Theta)|N\rangle \tag{8.57}$$

so that

$$T^{(M)}|\Theta\rangle = \exp(-\mathrm{i}M\Theta)|\Theta\rangle \tag{8.58}$$

The physical vacuum is characterized by Θ which is a constant of motion and each $|\Theta\rangle$ vacuum is the ground state of an independent sector of the Hilbert space. Indeed, if $\hat{O}$ is any gauge invariant operator

$$[\hat{O}, \hat{U}_N] = 0$$

then

$$0 = \langle\Theta|[\hat{O}, \hat{U}_N]|\Theta'\rangle = [\exp(-\mathrm{i}N\Theta') - \exp(-\mathrm{i}N\Theta)]\langle\Theta|\hat{O}|\Theta'\rangle$$

and

$$\langle\Theta|\hat{O}|\Theta'\rangle = 0 \quad \text{if} \quad \Theta' \neq \Theta \tag{8.59}$$

We want to stress that we have been working in a fixed gauge $A_0 = 0$. Thus the definitions of the winding number as well as of the topological vacua are gauge dependent. In our gauge, pure gauge configurations for which **E** and **B** are zero are analogous to infinitely degenerate zero-energy configurations in a quantum mechanical problem with periodic potential. Gauge invariance led us to the notion of the physical $|\Theta\rangle$ vacua as superpositions of the topological vacua (see also Callan, Dashen & Gross 1976). As in the quantum mechanical problem, the degeneracy between different Θ-vacua is lifted if there exists tunnelling between topological vacua. In pure Yang–Mills theories such tunnelling is due to instantons. In the absence of tunnelling, e.g. when there are massless fermions in the theory, the Θ-vacua remain degenerate and the angle Θ has no physical meaning.

This interpretation of the $|\Theta\rangle$-vacuum is certainly gauge dependent (Bernard & Weinberg 1977). For instance one could choose a physical gauge where A_μ is uniquely determined in terms of $G_{\mu\nu}$ with no residual gauge freedom at all. The classical vacuum will be unique. However, the physical $|\Theta\rangle$-vacuum is, of course, a gauge-invariant concept. This will become clear in the next subsection where we discuss functional integral representation for vacuum-to-vacuum transitions.

Θ-vacuum and the functional integral formalism

Let us consider a vacuum-to-vacuum transition given by the functional integral

$$\langle 0|\exp(-\mathrm{i}Ht)|0\rangle \sim \int \mathscr{D}A_\mu \exp\left[\mathrm{i}\int \mathrm{d}^4x(\mathscr{L}_{\mathrm{YM}} + \mathscr{L}_{\mathrm{G}} + \mathscr{L}_{\mathrm{FP}})\right] \tag{8.60}$$

It remains to be understood what one means by 'vacuum' for the transition given by (8.60). Firstly, we must pay some attention to the possible boundary conditions on the fields in the integral (8.60). We are interested in transitions from one classical vacuum to another so we require that the initial $t \to -\infty$ and final $t \to +\infty$ field configurations are of the pure gauge form. Actually, we shall require that gauge potentials tend to a pure gauge at large distances in all four directions

$$A_\mu \xrightarrow[|x_\mu| \to \infty]{} U^{-1}\partial_\mu U \tag{8.61}$$

For the time being we leave the gauge unspecified.

Let us now consider the integral

$$Q = -(1/16\pi^2)\int \mathrm{d}^4x \, \mathrm{Tr}\,[\tilde{G}^{\mu\nu}G_{\mu\nu}] \tag{8.62}$$

where

$$\tilde{G}^{\mu\nu} = \tfrac{1}{2}\varepsilon^{\mu\nu\rho\sigma}G_{\rho\sigma}$$

Q is gauge invariant and is called the topological charge or the Pontryagin index. It is stationary with respect to any local variation δA_μ whether or not the equation of motion is satisfied

$$\begin{aligned}
\delta Q &\sim \frac{1}{2}\int \mathrm{d}^4x \, \mathrm{Tr}\,[\tilde{G}_{\mu\nu}(D^\mu \delta A^\nu - D^\nu \delta A^\mu)] \\
&= \int \mathrm{d}^4x \, \mathrm{Tr}\,[\tilde{G}_{\mu\nu}D^\mu \delta A^\nu] \\
&= -\int \mathrm{d}^4x \, \mathrm{Tr}\,[D_\mu \tilde{G}^{\mu\nu}\delta A_\nu] + \int \mathrm{d}^4x \partial^\mu \, \mathrm{Tr}\,[\tilde{G}_{\mu\nu}\delta A^\nu]
\end{aligned} \tag{8.63}$$

The first term vanishes from the Bianchi identities (Problem 1.3) and no surface terms arise from the second term; Q is a topological invariant. Since (see Problem 1.2)

$$-(1/16\pi^2)\,\mathrm{Tr}\,[G_{\mu\nu}\tilde{G}^{\mu\nu}] = \partial_\mu K^\mu \tag{8.64}$$

where

$$K_\mu = -(1/16\pi^2)\varepsilon^{\mu\alpha\beta\gamma}\mathrm{Tr}\,[G_{\alpha\beta}A_\gamma + \tfrac{2}{3}A_\alpha A_\beta A_\gamma]$$

we can write Q as an integral over the surface S at infinity

$$Q = \int \mathrm{d}S_\mu K^\mu \tag{8.65}$$

For potentials A_μ satisfying (8.61) Q may be written in terms of U by inserting the asymptotic form of K_μ in (8.65)

$$Q = -(1/24\pi^2)\int \mathrm{d}S_\mu \varepsilon^{\mu\alpha\beta\gamma}\mathrm{Tr}\,[U^{-1}\partial_\alpha U U^{-1}\partial_\beta U U^{-1}\partial_\gamma U] \tag{8.66}$$

To make contact with the discussion earlier in this Section let us take the gauge $A_0 = 0$. Then from (8.42) and (8.66)

$$Q = \int \mathrm{d}^3x K^0|_{-\infty}^{\infty} = N(+\infty) - N(-\infty) \tag{8.67}$$

Using the remaining gauge freedom we can set $N(-\infty) = 0$. Therefore in this gauge the topological charge Q can be interpreted as the winding number of the pure gauge configuration to which A_i tends (we always also assume (8.36) which is necessary for the integral in (8.67) to be finite) and the functional integral (8.60), with the integral restricted to the Q sector defined by (8.62), as the transition amplitude between topological vacua $|N=0\rangle$ and $|N=Q\rangle$ or more generally between $|N\rangle$ and $|N+Q\rangle$. In the gauge $A_0 = 0$ we are now ready to construct the functional integral corresponding to the $|\Theta\rangle \to |\Theta\rangle$ transition. The result turns out to have the gauge-invariant form for which we are looking. We have

$$\begin{aligned}
\langle\Theta|\exp(-\mathrm{i}Ht)|\Theta'\rangle &= \sum_{m,n}\exp[\mathrm{i}(n-m)\Theta]\exp[\mathrm{i}n(\Theta'-\Theta)]\langle m|\exp(-\mathrm{i}Ht)|n\rangle \\
&= 2\pi\delta(\Theta-\Theta')\sum_Q \exp(-\mathrm{i}Q\Theta)\langle N+Q|\exp(-\mathrm{i}Ht)|N\rangle \\
&\sim \sum_Q \exp(-\mathrm{i}Q\Theta)\int \mathscr{D}A_Q \exp\left[\mathrm{i}\int \mathrm{d}^4x(\mathscr{L}_{\mathrm{YM}} + \mathscr{L}_{\mathrm{G}} + \mathscr{L}_{\mathrm{FP}})\right] \\
&= \int \mathscr{D}A_\mu \exp\Big(\mathrm{i}\int \mathrm{d}^4x\{\mathscr{L}_{\mathrm{YM}} + \mathscr{L}_{\mathrm{G}} + \mathscr{L}_{\mathrm{FP}} \\
&\qquad + (1/16\pi^2)\Theta\,\mathrm{Tr}\,[G_{\mu\nu}\tilde{G}^{\mu\nu}]\}\Big)
\end{aligned} \tag{8.68}$$

where we have used (8.62). As promised the final result is gauge independent and it defines the theory having a well-defined vacuum state $|\Theta\rangle$. The $|\Theta\rangle$-vacuum is defined by the vacuum expectation values of operators which in turn are determined by the functional integral with a given value of Θ. The gauge-dependent notion of the topological vacua does not appear in this formulation. However using the path integral representation (8.68) we can always write the physical vacuum as a superposition

$$|\Theta\rangle = \sum\!\!\!\!\!\!\int \mathrm{d}\nu \exp(\mathrm{i}\nu\Theta)|\nu\rangle$$

where $|\nu\rangle$ is an eigenstate of the operator $\mathrm{Tr}\,[G_{\mu\nu}\tilde{G}^{\mu\nu}]$ with the eigenvalue ν. The new term in the lagrangian is a total divergence so it does not influence the classical equations of motion. Also it does not contribute to the energy–momentum tensor. However it modifies quantum dynamics which depend on the action. Explicit solutions for field configurations giving $Q = n \neq 0$ are known (instantons). It is convenient to work in Euclidean space and to look at Minkowski Green's functions as the analytic continuation of the Euclidean ones. The instantons are classical $E = 0$ trajectories in Euclidean space-time.†

Fermions must be included in a realistic theory. New interesting aspects then appear which cannot be considered in detail before discussion of chiral symmetry and anomalies. Let us summarize briefly only the most important facts (see also Chapter 12). Firstly, take fermions to be massless. Then the Θ-vacua turn out to be degenerate: they are the degenerate vacua of spontaneously broken axial $U_{\mathrm{A}}(1)$ global symmetry of a theory with massless fermions. The spontaneous breakdown of the $U_{\mathrm{A}}(1)$ is the combined effect of the anomaly in the divergence of the axial current and of the existence of field configurations with a non-zero topological charge Q. This mechanism does not imply the existence of the Goldstone boson coupled to the gauge-invariant $U_{\mathrm{A}}(1)$ current and it offers a prototype solution to the so-called $U_{\mathrm{A}}(1)$ problem.

With massless fermions Θ changes with chiral rotations and can be rotated away. This is no longer true when fermions are massive since then chiral symmetry is also broken explicitly. We face a situation where on one hand the field configurations with $Q \neq 0$ are welcome to solve the $U_{\mathrm{A}}(1)$ problem and on the other hand the experimentally allowed value of Θ is surprisingly small as for a strong interaction parameter: from the upper limit on the neutron dipole moment $\Theta < 10^{-9}$. The Θ-term violates P and CP conservation so one could argue that we should set $\Theta = 0$ in the strong interaction theory. However, this does not help since CP-violating weak interactions renormalize Θ to a non-zero value. So far no convincing explanation exists for the smallness of Θ (Peccei & Quinn 1977, Weinberg 1978, Wilczek 1978, Dine, Fischler & Srednicki 1981). Finally in the presence of fermions the spontaneous breakdown of the $U_{\mathrm{A}}(1)$ and of the chiral $SU(n) \times SU(n)$ are phenomena which are connected to each other by the chiral Ward identities based

† By a gauge transformation which removes A_4 (we are in the gauge $A_0 = 0$, hence in Euclidean space $A_4 = 0$) one can cast the instanton solution into a form which tends to pure gauge configurations satisfying (8.36) and having different winding numbers as $x_4 \to +\infty$ and $x_4 \to -\infty$. Thus tunneling takes place and the Θ-vacua are not degenerate.

on the spontaneous breaking of the $SU(n) \times SU(n)$ and on the anomalous non-conservation of the $U_A(1)$ current.†

The structure of the QCD vacuum briefly discussed in this Section is of fundamental physical importance. However, non-perturbative in nature, it is usually neglected in applications of perturbative QCD to short distance phenomena. One then simply sets $\Theta = 0$.

8.4 Perturbative QCD and hard collisions

Historically, the light-cone expansion discussed in Chapter 7 has been the original foundation for perturbative QCD applications. However this OPE combined with the renormalization group techniques are applicable to a limited number of problems and a more general approach based on direct summation of Feynman diagrams has been widely developed. It can be used for any hard scattering process, not necessarily light-cone dominated, in which there is a large momentum scale q^2. If in a given process there are several large momenta $O(q^2)$ then we assume the ratio of these variables to be fixed as $q^2 \to \infty$. This is, for instance, the Bjorken limit in the deep inelastic lepton–hadron scattering discussed in Section 7.5.

Parton picture

There are two basic ingredients in applications of perturbative QCD to hadronic processes. The first one is to assume the parton picture which provides a connection between QCD and hadrons. No such connection has so far been derived from QCD itself. By the parton picture one means, speaking most generally, the assumption that in processes with large momentum transfer q hadrons can be pictured as ensembles of partons (quarks and gluons) and that there is no interference between long time and distance physics responsible for the binding of quarks and gluons into hadrons and short distance effects relevant for the large momentum transfer process under consideration and occurring at the parton level. The usual argument to support this conjecture is as follows: view the scattering in a frame in which the hadron (or hadrons) is fast: $P \to \infty$. The time scale of the bound state effects is dilated as P rises. The life time of a virtual state with quasi-free constituents is $P/(\Delta E)^2$ and it becomes long compared to the collision interaction time which is $O(1/q_0)$ (here ΔE is the scale of binding effects: $\Delta E \sim O(1/r)$ where r is the radius of the hadron).

The parton picture is implicit in the OPE analysis of the deep inelastic lepton–hadron scattering in Section 7.5 and e.g. in the QCD calculation of the total cross section for the e^-e^+ annihilation into hadrons which is identified with the total

† This point has been particularly emphasized by Crewther (1979).

cross section $e^-e^+ \to$ quarks and gluons. In most applications however, this parton model assumption takes a more concrete form. We assume that the hard scattering cross section can be written as a convolution of soft hadronic wave functions, which are process independent, with the cross section for the hard subprocess which involves only partons and can be studied perturbatively. To be specific let us consider the deep inelastic electron–proton scattering depicted in Fig. 7.2. We call P and p the proton and parton momenta, respectively, and use the so-called light-cone parametrization of the four-momenta

$$\left.\begin{aligned} P &= (P + m^2/4P, \mathbf{O}, P - m^2/4P) \\ p &= (zP + (p_\perp^2 + p^2)/4zP, \mathbf{p}_\perp, zP - (p_\perp^2 + p^2)/4zP) \\ d^4p &= \tfrac{1}{2}(dz/|z|)d^2p_\perp\, dp^2 \end{aligned}\right\} \tag{8.69}$$

which defines z as $z = (p_0 + p_3)/2P$ and is very convenient for this discussion. We assume that in the Bjorken limit

$$d\sigma^{\text{had}}(P, q) = \sum_i \int dz\, d^2p_\perp\, dp^2 f_i(z, \mathbf{p}_\perp, p^2)\sigma_i^{\text{QCD}}(zP, \mathbf{p}_\perp, p^2, q) \tag{8.70}$$

where σ_i^{QCD} is the cross section for deep inelastic scattering on the parton 'i' as the target and f_i is the four-momentum distribution of partons of type 'i' in the proton i.e. it is related to the square of the soft wave function. We expect that both $p_\perp$ and p^2 are determined by soft binding effects. In the limit $P \to \infty$, $p_\perp$ and p^2 finite (infinite momentum frame) we see that $p = zP$ up to corrections $O(p_\perp/zP)$. In this limit we envisage the proton as a collection of massless, parallelly moving partons and $f_i(z)\,dz$ is the average number of partons of type 'i' having momentum fraction $z, z + dz$ of the total proton momentum. We shall see shortly that in the infinite momentum frame relation (8.70) has a very interesting physical interpretation (see (8.129) and the discussion following it).

Factorization theorem

The second central issue for perturbative QCD calculations is the factorization theorem (Ellis *et al.* 1979; Sachrajda 1983*b*). It is a generalization of the factorization encountered in the OPE approach and applies to the $d\sigma^{\text{QCD}}$ which under the parton model assumption is to be calculated perturbatively as the sum of appropriate quark–gluon Feynman diagrams. The problem is that, as we expect from our QED examples, the perturbative expression for $d\sigma^{\text{QCD}}$ contains singular contributions coming from certain regions of phase space in which the quark and gluon momenta are small or parallel. A typical example, but not the only one (Sachrajda 1983*b*), is large logarithms like $\alpha \ln(q^2/p^2)$ coming from collinear parton radiation where q^2 is the large scale of the process and p^2 is the parton mass. The idea of factorization can be illustrated by the identity

$$[1+\alpha\ln(q^2/p^2)+\cdots]=[1+\alpha\ln(M^2/p^2)+\cdots][1+\alpha\ln(q^2/M^2)+\cdots] \quad (8.71)$$

and the theorem which has been proved to all orders in perturbation theory up to power corrections $O(p^2/q^2)$ states that e.g. again for the lepton–parton deep inelastic scattering

$$\mathrm{d}\sigma_i^{\mathrm{QCD}}(p,q)=\int \mathrm{d}y\, F_i(y,p^2/M^2)\,\mathrm{d}\hat{\sigma}_i(yp,q^2/M^2)+O(p^2/M^2) \quad (8.72)$$

where $0<y<1$ and the F_i are universal process-independent functions containing all the logarithms singular in the limit $p^2\to 0$ (mass singularities or more generally all long distance singularities). The cross sections $\mathrm{d}\hat{\sigma}_i$ are finite in the limit $p^2\to 0$ and M^2 is an arbitrary mass introduced to factorize out the long distance effects of perturbation theory. Using (8.70) and (8.72) the hadronic cross section can be written as a convolution

$$\mathrm{d}\sigma^{\mathrm{had}}=\sum_i\int f_i * F_i\,\mathrm{d}\hat{\sigma}_i=\sum_i\int \tilde{F}_i\,\mathrm{d}\hat{\sigma}_i \quad (8.73)$$

In the following we explicitly get (8.72) and (8.73) in the first order perturbation theory. It is, however, absolutely crucial for any relevance of perturbative QCD to hard hadronic processes that the factorization theorem holds asymptotically to all orders in perturbation theory. Then, although we cannot calculate the cross sections $\mathrm{d}\sigma^{\mathrm{had}}$ or even $\mathrm{d}\sigma^{\mathrm{QCD}}$ from first principles, the theory has a clear predictive power by relating one cross section to another. This may concern the same process at different momentum transfers e.g. scaling violation in the deep inelastic lepton–hadron scattering or two different processes that involve the same distribution functions $\tilde{F}_i$. In this second case it is usually convenient to let M^2 be the scale q^2 itself and simultaneously be the renormalization scale. For instance one can compute the cross section for lepton pair production in $p\bar{p}$ collisions using distribution functions defined and measured in the inclusive lepto-production.

In the following we present a sample of lowest order QCD calculations and then briefly mention possible strategies for all order extensions. For more details both technical and phenomenological the reader should consult some of the excellent review articles (e.g. Buras 1980; Sachrajda 1983*a*; Reya 1981; Dokshitzer, Dyakonov & Troyan 1980) on this subject.

8.5 Deep inelastic electron–nucleon scattering in first order QCD (Feynman gauge)

Structure functions and Born approximation

We consider this process in the one-photon exchange approximation. The kinematical variables are defined in Fig. 7.2. The unpolarized cross section for the process

$eN \to eX$ has the form

$$d\sigma = \frac{e^4}{q^4}\frac{1}{\text{Flux}}\frac{1}{2k_0}\frac{m}{P_0}\frac{d^3\mathbf{k}'}{(2\pi)^3 2k_0'}$$

$$*\sum_N \prod_{n=1}^{N} \frac{d^3\mathbf{p}_n}{(2\pi)^3 2p_n^0}(2\pi)^4\delta\left(k+P-k'-\sum_n p_n\right)\frac{1}{4}\sum_{\text{pol}}|M_X|^2 \tag{8.74}$$

where p_n are four-momenta of the final particles and X includes all the necessary spin indices for a given final state (we neglect all particle masses with the exception of the initial nucleon mass). Writing

$$\tfrac{1}{4}\sum_{\text{pol}}|M_X|^2 = (1/2m)L_{\mu\nu}\tilde{W}^{\mu\nu} \tag{8.75}$$

where the tensor

$$L_{\mu\nu} = \tfrac{1}{2}\text{Tr}\,[\not k\gamma_\mu \not k'\gamma_\nu] = 2(k_\mu k'_\nu + k_\nu k'_\mu - k\cdot k' g_{\mu\nu}) \tag{8.76}$$

comes from the lepton vertex, and introducing the identity

$$1 = \int d^4q\,\delta(q+k'-k)$$

we can rewrite the cross section as follows

$$d\sigma = \frac{1}{\text{Flux}}\frac{e^4}{q^4}\frac{1}{2k_0}\frac{m}{P_0}\frac{d^3\mathbf{k}'}{(2\pi)^3 2k_0'}\frac{4\pi}{2m}L_{\mu\nu}W^{\mu\nu} \tag{8.77}$$

where we define

$$4\pi W_{\mu\nu} = \sum_N \int \prod_{n=1}^{N}\frac{d^3\mathbf{p}_n}{(2\pi)^3 2p_n^0}\delta(P+q-\sum p_n)(2\pi)^4\tilde{W}_{\mu\nu}$$

$$= m\sum_{\text{pol}}\int d^4x\,\exp(iqx)\langle P|J_\mu(x)J_\nu(0)|P\rangle \tag{8.78}$$

The most general Lorentz and parity-invariant form of the tensor $W_{\mu\nu}$ reads

$$W_{\mu\nu} = -(g_{\mu\nu} - q_\mu q_\nu/q^2)W_1 + (1/m^2)(P_\mu - q_\mu(P\cdot q/q^2))(P_\nu - q_\nu(P\cdot q/q^2))W_2 \tag{8.79}$$

where $W_i = W_i(q^2, \nu)$ i.e. they are Lorentz scalars. The cross section in the laboratory frame then reads

$$\frac{d\sigma}{d\Omega\, dE'} = \frac{mEE'}{\pi}\frac{d\sigma}{d|q^2|d\nu} = \frac{\alpha^2}{q^4}\frac{16E'^2}{4m}[\cos^2(\tfrac{1}{2}\theta)W_2 + 2\sin^2(\tfrac{1}{2}\theta)W_1] \tag{8.80}$$

This result follows directly from relations (8.74)–(8.79). From the expression (8.80) we see that the structure functions W_1 and W_2 defined by (8.75) and (8.78) are real dimensionless functions which are measurable experimentally.

We introduce also another often used notation

$$F_1 = W_1, \qquad F_2 = P\cdot qW_2/m^2 \tag{8.81}$$

and express the structure functions F_1 and F_2 in terms of the tensor $W_{\mu\nu}$ as follows

$$\left.\begin{aligned} g^{\mu\nu}W_{\mu\nu} &= -(n-1)F_1 + \frac{1}{2x}F_2 \\ 2x\frac{P_\mu P_\nu}{P\cdot q}W_{\mu\nu} &= -F_1 + \frac{1}{2x}F_2 \end{aligned}\right\} \tag{8.82}$$

where $x = -q^2/2P\cdot q$ is the Bjorken variable. Relations (8.82) are written for the general case of n space-time dimensions. Calculating the structure functions F_1 and F_2 we also neglect the nucleon mass as we are interested in the kinematical region $|q^2| \sim s \gg m^2$. Structure functions F_i defined by (8.75), (8.79) and (8.81) are free of kinematical singularities when $m \to 0$. We get finally

$$\left.\begin{aligned} F_1 &= \frac{1}{n-2}\left(-g^{\mu\nu} + 2x\frac{P^\mu P^\nu}{P\cdot q}\right)W_{\mu\nu} \\ \frac{1}{x}F_2 &= 2F_1 + 4x\frac{P^\mu P^\nu}{P\cdot q}W_{\mu\nu} \end{aligned}\right\} \tag{8.83}$$

We want to calculate the structure functions F_i in perturbative QCD under the parton model assumption (8.70).

We assume that partons in a nucleon have some momentum distributions $f_i(z)\,\mathrm{d}z$ where $p = zP$. According to (8.70) the lepton–hadron scattering is an incoherent mixture of elementary scatterings

$$\mathrm{d}\sigma^{\mathrm{had}} = \int_0^1 \mathrm{d}z \sum_i f_i(z)\,\mathrm{d}\sigma_i^{\mathrm{QCD}}(p = zP, k, k') \tag{8.84}$$

where $\mathrm{d}\sigma_i^{\mathrm{QCD}}$ is the cross section for the lepton scattering on a parton of type 'i'.

In the actual physical applications of (8.84) we must, of course, remember that hadrons consist of quarks with different flavours and of gluons. Equation (8.84) is a short-hand notation for:

in the case of the proton as a target

$$\sum_i f_i(z)\,\mathrm{d}\sigma_i^{\mathrm{QCD}} = [\tfrac{4}{9}(u_\mathrm{p} + \bar{u}_\mathrm{p}) + \tfrac{1}{9}(d_\mathrm{p} + \bar{d}_\mathrm{p}) + \tfrac{1}{9}(s_\mathrm{p} + \bar{s}_\mathrm{p})]\,\mathrm{d}\sigma^q + G_\mathrm{p}(z)\,\mathrm{d}\sigma^G \tag{8.85}$$

in the case of the neutron as a target

$$\sum_i f_i(z)\,\mathrm{d}\sigma_i^{\mathrm{QCD}} = [\tfrac{4}{9}(u_\mathrm{n} + \bar{u}_\mathrm{n}) + \tfrac{1}{9}(d_\mathrm{n} + \bar{d}_\mathrm{n}) + \tfrac{1}{9}(s_\mathrm{n} + \bar{s}_\mathrm{n})]\,\mathrm{d}\sigma^q + G_\mathrm{n}(z)\,\mathrm{d}\sigma^G \tag{8.86}$$

where the numerical coefficients are the squared electric charges of quarks, $q_N = q_N(z)$ denotes the momentum distribution of quark q in hadron N and $G_N(z)$ denotes the gluon distribution in hadron N. Once the squared electric charges of quarks are factorized out the $\mathrm{d}\sigma^q$ is flavour independent (and averaged over colours of the initial and summed over colours of the final partons). The $\mathrm{d}\sigma^G$ vanishes in the zeroth order in the strong coupling constant α but contributes in higher orders.

There are several obvious constraints on the $q_N(z)$ determined by the quantum numbers of the nucleon. For instance,

$$\int_0^1 [s_p(z) - \bar{s}_p(z)]\,dz = 0 \qquad \text{(strangeness)} \tag{8.87}$$

$$\int_0^1 \{\tfrac{2}{3}[u_p(z) - \bar{u}_p(z)] - \tfrac{1}{3}[d_p(z) - \bar{d}_p(z)]\}\,dz = 1 \qquad \text{(charge)} \tag{8.88}$$

etc. Isospin symmetry (p↔n, $u \leftrightarrow d$) gives

$$\left.\begin{aligned} u_p(z) &= d_n(z) \equiv u(z) \\ d_p(z) &= u_n(z) \equiv d(z) \\ s_p(z) &= s_n(z) \equiv s(z) \\ G_p(z) &= G_n(z) \equiv G(z) \end{aligned}\right\} \tag{8.89}$$

and then for protons

$$\sum_i f_i(z)\,d\sigma_i^{QCD} = [\tfrac{4}{9}(u + \bar{u}) + \tfrac{1}{9}(d + \bar{d}) + \tfrac{1}{9}(s + \bar{s})]\,d\sigma^q + G(z)\,d\sigma^G \tag{8.90}$$

and for neutrons

$$\sum_i f_i(z)\,d\sigma_i^{QCD} = [\tfrac{4}{9}(d + \bar{d}) + \tfrac{1}{9}(u + \bar{u}) + \tfrac{1}{9}(s + \bar{s})]\,d\sigma^q + G(z)\,d\sigma^G \tag{8.91}$$

It is convenient to decompose the quark distribution functions into valence quark and sea quark distributions

$$q(z) = q_v(z) + q_s(z) \tag{8.92}$$

where we assume

$$u_s = \bar{u}_s = d_s = \bar{d}_s = s_s = \bar{s}_s \equiv S(z) \tag{8.93}$$

Then for protons

$$\sum_i f_i(z)\,d\sigma_i^{QCD} = [\tfrac{4}{9}u_v + \tfrac{1}{9}d_v + \tfrac{4}{3}S(z)]\,d\sigma^q + G(z)\,d\sigma^G \tag{8.94}$$

and for neutrons

$$\sum_i f_i(z)\,d\sigma_i^{QCD} = [\tfrac{4}{9}d_v + \tfrac{1}{9}u_v + \tfrac{4}{3}S(z)]\,d\sigma^q + G(z)\,d\sigma^G \tag{8.95}$$

One also often talks about flavour non-singlet cross sections sensitive only to some combinations of the valence quark distributions e.g.

$$d\sigma^{ep}(x, q^2) - d\sigma^{en}(x, q^2) = \int dz\,\tfrac{1}{3}[u_v(z) - d_v(z)]\,d\sigma^q(zP, q^2) \tag{8.96}$$

and flavour singlet cross sections sensitive to the distributions $\frac{1}{2}(q_s + \bar{q}_s) = S(z)$ and $G(z)$. In the following we calculate $d\sigma^q$ in the zeroth and first orders in the

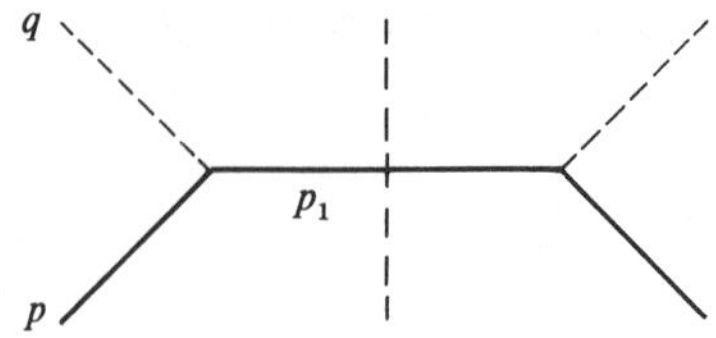

Fig. 8.3

strong coupling α and study flavour non-singlet hadron structure functions.

We can define the quark structure functions for the elementary subprocess again by (8.79) and (8.81)

$$W^q_{\mu\nu} = -\left(g_{\mu\nu} - \frac{q_\mu q_\nu}{q^2}\right)F^q_1(y, q^2) + \frac{1}{p\cdot q}\left(p_\mu - q_\mu\frac{p\cdot q}{q^2}\right)\left(p_\nu - q_\nu\frac{p\cdot q}{q^2}\right)F^q_2(y, q^2) \tag{8.97}$$

where

$$y = -q^2/2p\cdot q = x/z \tag{8.98}$$

is the Bjorken variable defined for electron–quark scattering. Using (8.84), (8.77) and remembering about the normalizing factor $1/zP_0$ in $d\sigma^q$ we can express the hadron structure functions in terms of $F^q_{1,2}(y, q^2)$ as follows

$$F^{\text{had}}(x, q^2) = \int_x^1 (\mathrm{d}z/z)\sum_i f_i(z)F^q(x/z, q^2) \tag{8.99}$$

where $F(y, q^2)$ stands for $F_1(y, q^2)$ or $(1/y)F_2(y, q^2)$.

In the Born approximation the quark structure functions are determined by elastic electron–quark scattering. We have

$$4\pi W^q_{\mu\nu} = \int \frac{\mathrm{d}^{n-1}\mathbf{p}_1}{(2\pi)^{n-1}2p_1^0}(2\pi)^n\delta^{(n)}(p + q - p_1)\tfrac{1}{2}\,\mathrm{Tr}\,[\not{p}\gamma_\mu\not{p}_1\gamma_\nu] \tag{8.100}$$

We can also use the relation (8.105) and calculate the cut diagram shown in Fig. 8.3. Then

$$W^q_{\mu\nu} = -\frac{1}{4\pi}\int\frac{\mathrm{d}^n p_1}{(2\pi)^n}(2\pi)^n\delta^{(n)}(p_1 - p - q)(-2\pi\mathrm{i})\delta_+(p_1^2)(-\mathrm{i})^2\tfrac{1}{2}\mathrm{Tr}\,[\gamma_\mu \mathrm{i}\not{p}_1\gamma_\nu\not{p}] \tag{8.101}$$

(we have used the identity $1 = \int[(\mathrm{d}^n p_1)/(2\pi)^n](2\pi)^n\delta^{(n)}(p_1 - p - q)$) which indeed coincides with (8.100). Using (8.83) we easily get the quark structure functions F_i in the Born approximation

$$(1/y)F^q_2 = 2F^q_1 = 2p\cdot q\delta(2pq + q^2) = \delta(1 - y) \tag{8.102}$$

where $y = -q^2/2p\cdot q$. Indeed, we know already that $y = 1$ for elastic scattering. We encounter here the so-called Bjorken scaling law: structure functions, which in general can depend on y and q^2, depend on y only. This will no longer be the case when we include radiative corrections. In terms of the Bjorken variable x defined for the hadronic process we have

$$\frac{1}{x/z}F_2^q\left(\frac{x}{z}\right)=2F_1^q\left(\frac{x}{z}\right)=\delta\left(1-\frac{x}{z}\right) \tag{8.103}$$

where $x=-q^2/2P\cdot q$ and $p=zP$. For the hadronic structure functions we get finally

$$(1/x)F_2^{\text{had}}(x,q^2)=2F_1^{\text{had}}(x,q^2)=\int(\mathrm{d}z/z)\sum_i f_i(z)\delta(1-x/z)=\sum_i f_i(x) \tag{8.104}$$

The Bjorken variable x in (8.104) is defined, as before, in terms of the total nucleon momentum P, which together with $q=k'-k$, is the kinematic parameter of the reaction. The Bjorken parameter is therefore determined by the specific kinematic configuration in our experiment. We arrive at a very interesting conclusion: it follows from (8.104) that in measuring the hadron structure functions as a function of x we effectively study the momentum distribution of the hadron constituents. In particular we have learned from experiment this way that half of the nucleon momentum is carried by gluons. It is clear from calculation of the elastic cross section (see (8.100)) that our result (8.104) holds up to corrections $O(p_\perp/zP)$ where $p_\perp$ is the typical transverse momentum of partons in the hadron. So the corrections are frame dependent and the transparent physical interpretation (8.104) of the hadron structure functions and of the Bjorken variable x (for $x\neq 0$) is valid in a so-called infinite momentum frame in which $P\to\infty$, $p_\perp$ finite.

Deep inelastic quark structure functions in the first order in the strong coupling constant

We now calculate the $\mathrm{d}\sigma^q$ in the first order in α^{strong}. Fig. 8.4 and Fig. 8.5 show the diagrams for the virtual and real corrections. We use the notation of cut diagrams where

$$W_{\mu\nu}=-\frac{1}{4\pi}$$ 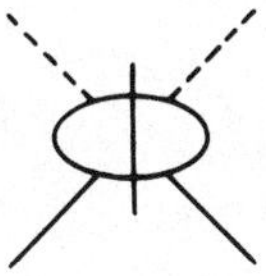 (8.105)

The calculation is performed in the Feynman gauge. The presence for each diagram of the group theoretical factor $C_F=\frac{4}{3}$ should be kept in mind. It is omitted in most of the following expressions. All the quantities are for lepton–quark scattering but the superscript 'q' will also be omitted.

Here and in the following we shall be mainly concerned with the quantity

$$\sigma=-\frac{1}{1+\frac{1}{2}\varepsilon}g_{\mu\nu}W^{\mu\nu}\quad(n=4+\varepsilon) \tag{8.106}$$

The term $p^\mu p^\nu W_{\mu\nu}$ in (8.83) is much simpler to calculate.

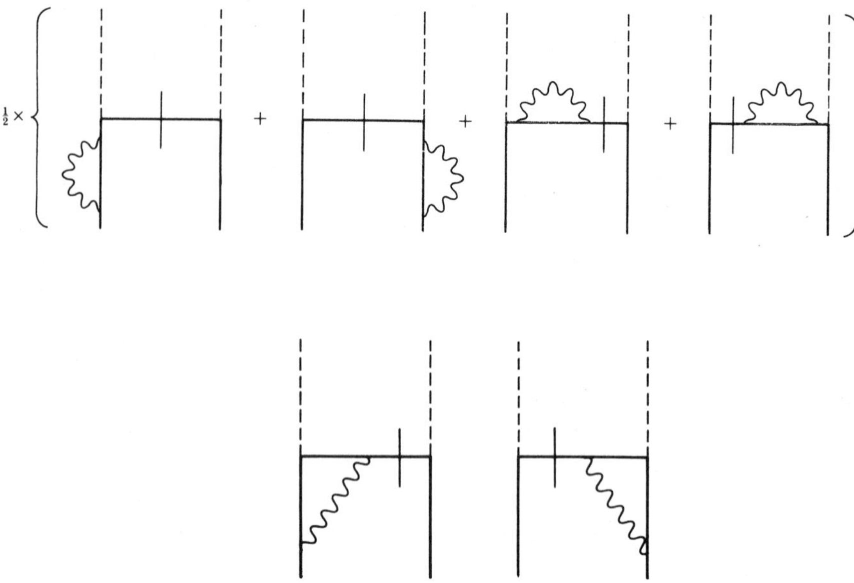

Fig. 8.4

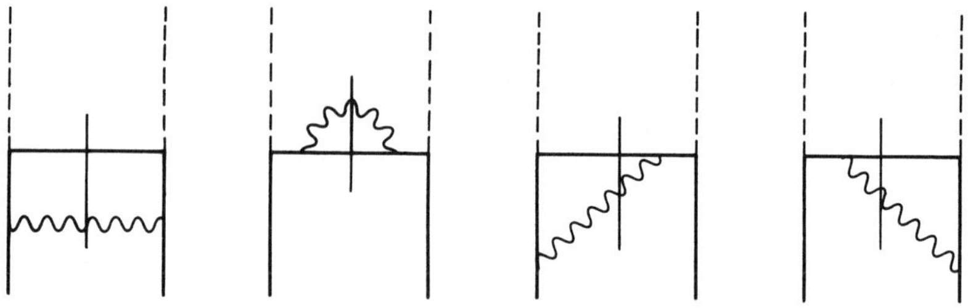

Fig. 8.5

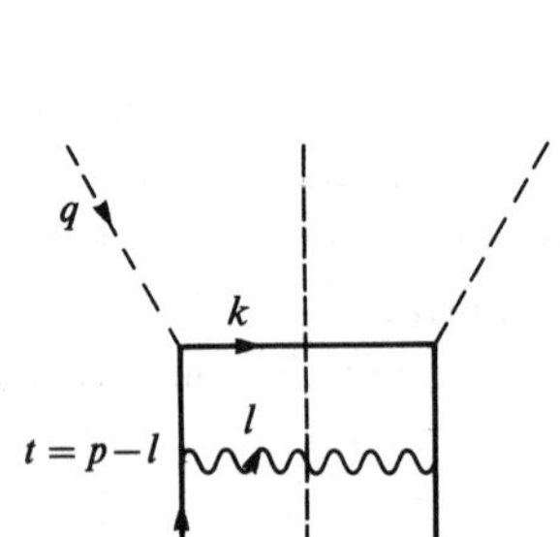

$x = -q^2/2p\cdot q$

Fig. 8.6

The ladder diagram in Fig. 8.6 gives the following contribution to the cross section σ

$$\sigma_L = \frac{1}{1+\frac{1}{2}\varepsilon}\alpha\mu^{-\varepsilon}\int d\Phi_2 \tfrac{1}{2}\mathrm{Tr}\,[\not p\gamma_\alpha(\not p - \not l)\gamma_\mu \not k\gamma^\mu(\not p - \not l)\gamma^\alpha]\left[\frac{1}{(p-l)^2}\right]^2 \tag{8.107}$$

with

$$d\Phi_2 = \frac{d^n l}{(2\pi)^n}2\pi\delta_+(l^2)\frac{d^n k}{(2\pi)^n}2\pi\delta_+(k^2)(2\pi)^n\delta^{(n)}(q+p-l-k)$$

(contrary to the calculation of Section 5.5 we do not assume the gluon to be soft). The trace gives

$$\tfrac{1}{2}\mathrm{Tr}[\cdots] = 4(1+\tfrac{1}{2}\varepsilon)^2 W^2[-(p-l)^2] = 4(1+\tfrac{1}{2}\varepsilon)^2\frac{1-x}{x}Q^2[-(p-l)^2] \tag{8.108}$$

where $W = q+p, Q^2 = -q^2$ and $x = -q^2/2p\cdot q$ is the Bjorken variable for the lepton–quark scattering. Using expression (5.90) for the two-body phase space in the (p,q) cms we get the following

$$\sigma_L = \alpha\frac{1}{8\pi}\frac{1}{\Gamma(1+\frac{1}{2}\varepsilon)}(1+\tfrac{1}{2}\varepsilon)4\left(\frac{Q^2}{4\pi\mu^2}\frac{1-x}{x}\right)^{\varepsilon/2}$$
$$*\int dy[y(1-y)]^{\varepsilon/2}\frac{1-x}{x}Q^2\left[\frac{-1}{(p-l)^2}\right] \tag{8.109}$$

with $y = \frac{1}{2}(1-z)$ and $z = \cos\theta$, where θ is the polar angle between the three-vectors $\mathbf{l}$ and $\mathbf{p}$. We observe that in the (p,q) cms $2p\cdot l = 2p_0 l_0(1-z) = p\cdot W(1-z) = (Q^2/x)y$ and the denominator in (8.107) gives a singularity at $y = 0$, i.e. for $\mathbf{l}\parallel\mathbf{p}$ and any value of the momentum $|\mathbf{l}|$ allowed by the conservation law $q+p = k+l$ ($4|\mathbf{l}|^2 = -q^2(1-x)/x$ in cms). This is the so-called mass singularity. We finally get

$$\sigma_L = \left(\frac{\alpha}{2\pi}\right)\left(\frac{Q^2}{4\pi\mu^2}\right)^{\varepsilon/2}\frac{(1+\frac{1}{2}\varepsilon)}{\Gamma(1+\frac{1}{2}\varepsilon)}\frac{\Gamma(\frac{1}{2}\varepsilon)\Gamma(1+\frac{1}{2}\varepsilon)}{\Gamma(1+\varepsilon)}\left[(1-x)\left(\frac{1-x}{x}\right)^{\varepsilon/2}\right] \tag{8.110}$$

The pole in ε reflects the mass singularity regularized by working in $n = 4+\varepsilon$ dimensions. The behaviour near $x = 1$ is IR behaviour ($|\mathbf{l}|\to 0$ when $x\to 1$). The cut ladder diagram in the Feynman gauge exhibits no IR singularity.† However, we expect that the full cross section for a single photon production is IR divergent. As we know from Section 5.5 in the properly defined physical cross section the IR divergence is cancelled by virtual corrections to the elastic scattering i.e. $\int A\delta(1-x)dx + \int\sigma(x,Q^2)dx$ is finite. In order to demonstrate this cancellation explicitly we write the cross section σ in terms of the so-called regularized

† Note the difference with Section 5.5: now $m = 0$.

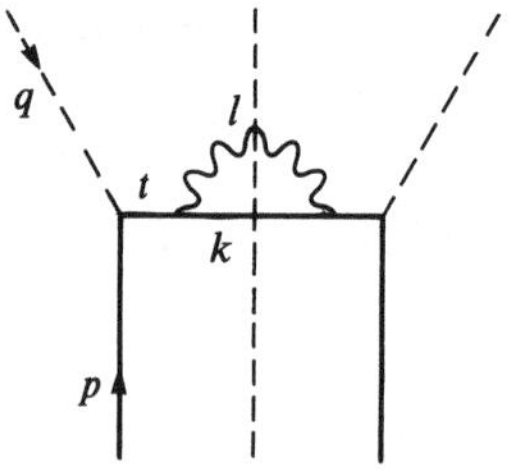

Fig. 8.7

distribution. For any function $f(x)$ we define the distribution $f_+(x)$ as follows

$$f_+(x) = f(x) - \delta(1-x)\int_0^1 \mathrm{d}y f(y) \tag{8.111}$$

i.e. we explicitly single out a possibly singular behaviour of the function $f(x)$ as $x \to 1$. We can then write

$$\sigma_{\mathrm{L}} = \left(\frac{\alpha}{2\pi}\right)\left(\frac{Q^2}{4\pi\mu^2}\right)^{\varepsilon/2}\left\{\left(\frac{1}{\varepsilon} + 1 + \tfrac{1}{2}\gamma\right)\delta(1-x) + \left(\frac{2}{\varepsilon} + 1 + \gamma\right)\left[(1-x)\left(\frac{1-x}{x}\right)^{\varepsilon/2}\right]_+\right\} \tag{8.112}$$

The contribution of the longitudinal term $p_\mu p_\nu W^{\mu\nu}$ can also be easily included. The ladder diagram gives

$$\left(\frac{1}{x}F_2\right)^{\mathrm{long}} = \frac{\alpha}{2\pi}3x = \frac{\alpha}{2\pi}[\tfrac{3}{2}\delta(1-x) + (3x)_+] \tag{8.113}$$

and the contribution of the rest of the diagrams to $p_\mu p_\nu W^{\mu\nu}$ vanishes.

We proceed to calculate cut self-energy corrections. The contribution of the diagram in Fig. 8.7 reads

$$\sigma_{\Sigma_c} = \frac{1}{1+\frac{1}{2}\varepsilon}\alpha\mu^{-\varepsilon}\int \mathrm{d}\Phi_2\left(\frac{1}{t^2}\right)^2 \tfrac{1}{2}\mathrm{Tr}\,[\not{p}\gamma^\mu \not{t}\gamma^\alpha \not{k}\gamma_\alpha \not{t}\gamma_\mu] \tag{8.114}$$

As before, $\mathrm{d}\Phi_2$ is the two-body (l, k) phase space and we work in $n = 4 + \varepsilon$ dimensions. The trace gives $2(1+\frac{1}{2}\varepsilon)^2\, 2(2p\cdot l)2k\cdot l = 2(1+\frac{1}{2}\varepsilon)^2 2(t^2)^2 y/(1-x)$ where $y = \frac{1}{2}(1+\cos\theta)$ and θ is the angle between vectors $\mathbf{p}$ and $\mathbf{k}$ in the (k, l) cm frame. We see that the propagators are cancelled and we get finally:

$$\begin{aligned}\sigma_{\Sigma_c} &= \alpha\frac{1}{8\pi}\left(\frac{Q^2}{4\pi\mu^2}\frac{1-x}{x}\right)^{\varepsilon/2} 4(1+\tfrac{1}{2}\varepsilon)\frac{1}{\Gamma(1+\frac{1}{2}\varepsilon)}\int_0^1 \mathrm{d}y[y(1-y)]^{\varepsilon/2}\frac{y}{1-x} \\ &= \frac{\alpha}{2\pi}\left(\frac{Q^2}{4\pi\mu^2}\right)^{\varepsilon/2}(1+\tfrac{1}{2}\varepsilon)\left[\left(\frac{1-x}{x}\right)^{\varepsilon/2}\frac{1}{1-x}\right]\frac{\Gamma(2+\frac{1}{2}\varepsilon)}{\Gamma(3+\frac{1}{2}\varepsilon)}\end{aligned} \tag{8.115}$$

In the present case the integration over y is finite: there is no mass singularity.

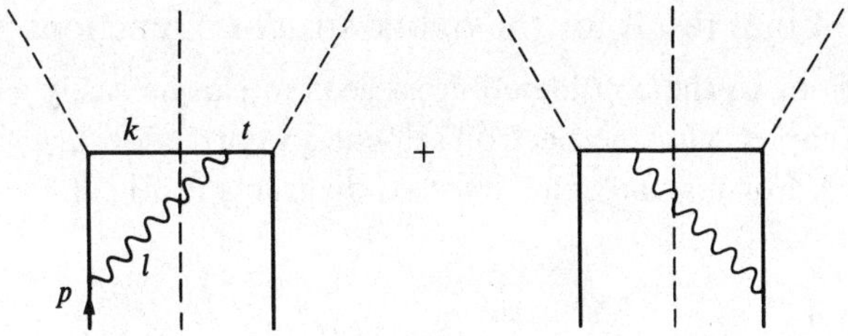

Fig. 8.8

The result is, however, divergent for $x \to 1$ (IR divergence). Using the regularized distribution we get:

$$\sigma_{\Sigma_c} = \frac{\alpha}{2\pi}\left(\frac{Q^2}{4\pi\mu^2}\right)^{\varepsilon/2}\left\{\left[\frac{1}{\varepsilon}+\tfrac{1}{2}(\gamma-1)\right]\delta(1-x)+\frac{1}{2}\left(\frac{1}{1-x}\right)_+\right\} \tag{8.116}$$

Finally we consider the diagrams in Fig. 8.8 and the corresponding cross section

$$\sigma_{\Gamma_c} = 2\alpha\mu^{-\varepsilon}\frac{1}{1+\frac{1}{2}\varepsilon}\int \mathrm{d}\Phi_2 \frac{1}{t^2(p-l)^2}\frac{1}{2}\mathrm{Tr}\,[\not{p}\gamma^\mu \not{t}\gamma^\alpha \not{k}\gamma_\mu(\not{p}-\not{l})\gamma_\alpha] \tag{8.117}$$

where $\mathrm{d}\Phi_2$ is as before. Using $\gamma^\mu \not{a}\not{b}\not{c}\gamma_\mu = -2\not{c}\not{b}\not{a} - \varepsilon\not{a}\not{b}\not{c}$ and $\gamma^\mu\not{a}\not{b}\gamma_\mu = 4ab + \varepsilon\not{a}\not{b}$ we evaluate the trace and get

$$\begin{aligned}\tfrac{1}{2}\mathrm{Tr}[\cdots] = -\{&4(2p\cdot k)(2p\cdot t - 2t\cdot l) + \tfrac{1}{2}\varepsilon 4(2p\cdot t)(2k\cdot p - 2k\cdot l)\\ &+ \varepsilon\,\mathrm{Tr}\,[\not{p}\not{k}\not{t}(\not{p}-\not{l})] + \tfrac{1}{2}\varepsilon^2\,\mathrm{Tr}\,[\not{p}\not{t}\not{k}(\not{p}-\not{l})]\}\end{aligned} \tag{8.118}$$

The last two terms do not contribute (one is exactly zero and the second gives $\varepsilon^2(-8)(p\cdot l)(k\cdot l)$ i.e. both propagators are cancelled and there is no divergence to cancel ε^2). In the (k, l) cm frame we have

$$2p\cdot t = Q^2/x, \quad 2p\cdot k = 2p\cdot(t-l) = Q^2/x + (p-l)^2, \quad 2t\cdot l = 2k\cdot l = (Q^2/x)(1-x)$$

and

$$\sigma_{\Gamma_c} = 2\alpha\frac{1}{1+\frac{1}{2}\varepsilon}\int \mathrm{d}\Phi_2 4\left[\frac{x}{1-x}\frac{1}{y}(1+\tfrac{1}{2}\varepsilon) - \frac{x}{1-x}\left(1+\frac{\varepsilon}{2x}\right)\right] \tag{8.119}$$

We see that the cut vertex has (in the Feynman gauge) both IR and mass singularity. Performing the integration over $\mathrm{d}\Phi_2$ we obtain the following

$$\begin{aligned}\sigma_{\Gamma_c} = 2\left(\frac{\alpha}{2\pi}\right)\left(\frac{Q^2}{4\pi\mu^2}\right)^{\varepsilon/2}\frac{\Gamma(1+\frac{1}{2}\varepsilon)}{\Gamma(1+\varepsilon)}&\left\{\frac{1}{1+\varepsilon}\left[\left(\frac{2}{\varepsilon}\right)^2 + \frac{\pi^2}{6} - 1\right]\delta(1-x)\right.\\ &\left.+\left(\frac{2}{\varepsilon}-1\right)\left[\frac{x}{1-x}\left(\frac{1-x}{x}\right)^{\varepsilon/2}\right]_+\right\}\end{aligned} \tag{8.120}$$

Final result for the quark structure functions

Virtual corrections to the considered cross section can be easily calculated using the results of Chapter 5 for the electron self-energy correction $\Sigma(p^2)$ and the vertex correction $\Lambda(q^2)$. For instance, the first two diagrams in Fig. 8.4 give together (in $n=4+\varepsilon$ dimensions)

$$\sigma = -\frac{g_{\mu\nu}W^{\mu\nu}}{1+\frac{1}{2}\varepsilon} = \frac{-1}{1+\frac{1}{2}\varepsilon}\frac{1}{4\pi}\int\frac{d^nk}{(2\pi)^n}(2\pi)^n\delta^{(n)}(k-p-q)2\pi\delta_+(k^2)(-i)^2\tfrac{1}{2}\mathrm{Tr}$$

$$*\left[\not p\gamma^\mu \not k\gamma_\mu\frac{i}{\not p}[-i\Sigma(p^2=0)\not p]\right] = \Sigma(p^2=0)\delta(1-x) = -\frac{\alpha}{2\pi}\frac{1}{\varepsilon}\delta(1-x) \quad (8.121)$$

The same result holds for the next two diagrams of Fig. 8.4. Also, with $\Sigma(p^2)$ changed into $\Lambda(q^2)$, each of the vertex correction diagrams can be calculated in a similar way.

We are now ready to collect all the results ($n=4+\varepsilon$ everywhere) Real

$$\left.\begin{aligned}
\sigma_{\mathrm{L}} &= \left(\frac{\alpha}{2\pi}\right)\left(\frac{Q^2}{4\pi\mu^2}\right)^{\varepsilon/2}\left\{\left(\frac{1}{\varepsilon}+1+\tfrac{1}{2}\gamma\right)\delta(1-x)+\left(\frac{2}{\varepsilon}+1+\gamma\right)\right.\\
&\quad\left. *\left[(1-x)\left(\frac{1-x}{x}\right)^{\varepsilon/2}\right]_+\right\}\\
\sigma_{\Sigma_c} &= \left(\frac{\alpha}{2\pi}\right)\left(\frac{Q^2}{4\pi\mu^2}\right)^{\varepsilon/2}\left[\left(\frac{1}{\varepsilon}+\tfrac{1}{2}\gamma-\tfrac{1}{2}\right)\delta(1-x)+\tfrac{1}{2}\left(\frac{1}{1-x}\right)_+\right]\\
\sigma_{\Gamma_c} &= 2\left(\frac{\alpha}{2\pi}\right)\left(\frac{Q^2}{4\pi\mu^2}\right)^{\varepsilon/2}\frac{\Gamma(1+\frac{1}{2}\varepsilon)}{\Gamma(1+\varepsilon)}\left\{\left[\left(\frac{2}{\varepsilon}\right)^2+\left(\frac{\pi^2}{6}-1\right)\right]\frac{1}{1+\varepsilon}\delta(1-x)\right.\\
&\quad\left.+\left(\frac{2}{\varepsilon}-1\right)\left[\frac{x}{1-x}\left(\frac{1-x}{x}\right)^{\varepsilon/2}\right]_+\right\}\\
\left(\frac{1}{x}F_2\right)^{\mathrm{long}} &= \frac{\alpha}{2\pi}\left[\tfrac{3}{2}\delta(1-x)+(3x)_+\right]
\end{aligned}\right\} \quad (8.122)$$

Virtual

$$\left.\begin{aligned}
\sigma_\Sigma &= 2\left(-\frac{\alpha}{2\pi}\right)\frac{1}{\varepsilon}\delta(1-x)\\
\sigma_\Gamma &= 2\frac{\alpha}{2\pi}\left\{\frac{1}{\varepsilon}-\frac{1}{\varepsilon}\frac{\Gamma(1+\frac{1}{2}\varepsilon)}{\Gamma(1+\varepsilon)}\left(\frac{Q^2}{4\pi\mu^2}\right)^{\varepsilon/2}+\left(\frac{Q^2}{4\pi\mu^2}\right)^{\varepsilon/2}\frac{\Gamma(1+\frac{1}{2}\varepsilon)}{\Gamma(1+\varepsilon)}\frac{(-1)}{1+\varepsilon}\right.\\
&\quad\left.*\left[\left(\frac{2}{\varepsilon}\right)^2+\frac{\pi^2}{6}\right]\right\}\delta(1-x)
\end{aligned}\right\} \quad (8.123)$$

The first two terms in σ_Γ correspond to the renormalized Λ^{UV} (Section 5.3) written in $n=4+\varepsilon$ dimensions. The final result for $(1/x)F_2$ in the lepton–quark scattering reads

$$\frac{1}{x}F_2 = C_F\left(\frac{\alpha}{2\pi}\right)\left[\left(\frac{2}{\varepsilon}+\gamma\right)\left(\frac{1+x^2}{1-x}\right)+\frac{1+x^2}{1-x}\left(\ln\frac{1-x}{x}-1\right)+2(1-x)\right.$$
$$\left.+\frac{1}{2}\left(\frac{1}{1-x}\right)+3x\right]_+\cdot\left(\frac{Q^2}{4\pi\mu^2}\right)^{\varepsilon/2}+\delta(1-x)+O(\alpha^2) \tag{8.124}$$

We now add several important comments on the result obtained to clarify its physical sense. As we already know from Section 5.5 the IR singularities cancel between real and virtual corrections and they are not present in (8.124). However, this is not the case with mass singularities (due to the decay of a massless particle into two parallelly moving massless particles) which explicitly reflect themselves in the final result as poles in ε. We certainly have to cut them off by some physical parameter before inserting result (8.124) into (8.99) to compare our calculation with experiment. The trick is to 'factorize out' the mass singularities by writing the result (8.124) in the limit $\varepsilon\to 0$ as (see (8.71))

$$\frac{1}{x}F_2(x,Q^2) = \int_x^1 \frac{dz}{z}\Phi\left(\frac{x}{z},Q^2,Q_0^2\right)\frac{1}{z}F_2(z,Q_0^2)+O(\alpha^2) \tag{8.125}$$

where Q_0^2 is an arbitrary parameter and from (8.124)

$$\Phi(t,Q^2,Q_0^2) = \delta(1-t)+C_F\frac{\alpha}{2\pi}\left(\ln\frac{Q^2}{Q_0^2}\right)\left(\frac{1+t^2}{1-t}\right)_+ + O(\alpha) \tag{8.126}$$

where the $O(\alpha)$ terms are finite when $Q^2\to\infty$. The meaning of (8.125) is clear: the structure function for any Q^2 is expressed in terms of the structure function for Q_0^2 and of the theoretically calculated function $\Phi(x/z,Q^2,Q_0^2)$ free of mass singularity.

Hadron structure functions; probabilistic interpretation

To calculate hadronic structure functions we use the convolution (8.99) and get

$$\frac{1}{x}F_2^{\text{had}}(x,Q^2) = \int_x^1 \frac{dz}{z}f(z)\frac{z}{x}F_2\left(\frac{x}{z},Q^2\right)$$
$$= \frac{1}{x}\int_x^1 dt\,\Phi(t,Q^2,Q_0^2)\int_{x/t}^1 dz f(z)F_2\left(\frac{x}{zt},Q_0^2\right) \tag{8.127}$$

where $x = Q^2/2Pq$ is the Bjorken variable for the lepton–hadron scattering. Consequently

$$\frac{1}{x}F_2^{\text{had}}(x,Q^2) = \int_x^1 \frac{dz}{z}\Phi\left(\frac{x}{z},Q^2,Q_0^2\right)\frac{1}{z}F_2^{\text{had}}(z,Q_0^2) \tag{8.128}$$

and interestingly enough we again recover the result (8.125). The effect of the intrinsic quark distribution in the nucleon is included in $F_2^{\text{had}}(z, Q_0^2)$. We do not have to know it theoretically to compare the theory with experiment as far as Q^2 evolution of the structure functions is concerned.

By introducing an intermediate scale Q_0^2 we escape two problems: the intrinsic parton distribution which originates from non-perturbative effects need not be known theoretically and the perturbative mass singularities are not present in our final formula. They have been absorbed into $F_2^{\text{had}}(z, Q_0^2)$ which we take as the experimentally measured structure function. Our theoretical prediction is the evolution function $\Phi(x/z, Q^2, Q_0^2)$ and we do not worry that $F_2^{\text{had}}(z, Q_0^2)$ is actually singular in our calculation since this is due to our ignorance of how to include correctly the soft binding effects.

It is also worth observing that the leading Q^2 dependence of the result (8.124) for the quark structure functions and therefore also for the hadron structure functions (8.128) can be interpreted in a very transparent probabilistic parton-like way. To this end let us formally rewrite (8.124) as follows

$$\frac{1}{x}F_2(x, Q^2) = \int_0^1 \frac{\mathrm{d}z}{z}\frac{1}{z}F_2(z, Q^2)\delta\left(1 - \frac{x}{z}\right) \tag{8.129}$$

Formally, we obtain an expression for F_2 which is the same as (8.104) for the structure function of a composite object consisting of a set of partons which scatter elastically at large Q^2. We are tempted to interpret the quantity $(1/z)F(z, Q^2)$ as the average number of partons, with momentum in the interval $zp, (z + \mathrm{d}z)p$, seen in the initial quark with momentum p by a probe (electron) in the process of deep inelastic scattering with momentum transfer Q^2. The theory predicts its evolution with Q^2. Let us assume for the moment that our interpretation is sensible. The hadronic structure function $(1/x)F_2^{\text{had}}(x, Q^2)$ is then a convolution of the intrinsic parton momentum distribution in a hadron (Q^2-independent) and the parton momentum distribution in a parton when probed with large Q^2. Technically, the second effect is present in perturbation theory whereas the first one has its origin in non-perturbative binding effects.

Is our interpretation of (8.129) sensible? It is for the leading Q^2 dependence. This point will become clearer in the next Section when we repeat our calculation in a so-called axial gauge. The reason is that interference diagrams with a gluon emitted or absorbed by the final quark contribute to the final result in the present calculation. It turns out, however, that with the proper choice of gauge the leading effect comes from the ladder diagram and so our interpretation is possible. It applies, however, only to the leading Q^2 dependence because to interpret (8.129) in terms of the elastic scattering of a set of partons with momentum distribution $(1/z)F_2(z, Q^2)$ we have to worry, as before, about corrections $O(p_\perp/zp)$. But now the transverse momentum of a parton is limited only by kinematics and can be as large as $O((Q^2)^{1/2})$. For the probabilistic interpretation we have not only to stay in an infinite momentum frame but also in a limited region of phase space ($p_\perp$ small). This we can do by writing $\ln(Q^2/Q_0^2) = \ln(\varepsilon Q^2/Q_0^2) +$

$\ln(Q^2/\varepsilon Q^2)$ where ε is some fixed arbitrarily small number. The first logarithm comes from the region where $p_\perp \lesssim (\varepsilon Q^2)^{1/2}$ is small and at the same time it gives the leading Q^2 dependence correctly (for $Q^2 \to \infty$).

The function

$$P(z) = \left(\frac{1+z^2}{1-z}\right)_+ \tag{8.130}$$

is called the Altarelli–Parisi function. One often considers the Q^2 evolution of moments of the structure function e.g.

$$M_N(Q^2) = \int_0^1 \mathrm{d}x\, x^{N-1} F(x, Q^2) \tag{8.131}$$

where again $F(x) = F_1(x)$ or $F(x) = (1/x)F_2(x)$. From (8.128) and (8.126) one gets

$$M_N = \int_0^1 \mathrm{d}x\, x^{N-1} \int_0^1 \mathrm{d}z \int_0^1 \mathrm{d}t\, \Phi(t, Q^2, Q_0^2)\delta(x - zt)(1/z)F_2(z, Q_0^2) \tag{8.132}$$

or in the leading Q^2 dependence approximation

$$M_N(Q^2) = M_N(Q_0^2)\left[C_\mathrm{F}\frac{\alpha}{2\pi}P_N \ln\left(\frac{Q^2}{Q_0^2}\right) + 1\right] \tag{8.133}$$

where P_N is the Nth moment of the Altarelli–Parisi function:

$$P_N = \int_0^1 \mathrm{d}x\, x^{N-1}\left(\frac{1+x^2}{1-x}\right)_+ = \int_0^1 \mathrm{d}x \frac{1+x^2}{1-x}(x^{N-1} - 1) \tag{8.134}$$

8.6 Light-cone variables, light-like gauge

In this Section we describe briefly the calculation of $\mathrm{d}\sigma^q$ given by the diagrams in Fig. 8.4 and Fig. 8.5 using the light-cone parametrization of the momenta and the light-like gauge. To be specific let us consider first the ladder diagram in Fig. 8.6. The gluon propagator in the $n^2 = 0$ gauge is

$$-\mathrm{i}\left(g_{\mu\nu} - \frac{l_\mu n_\nu + l_\nu n_\mu}{l \cdot n}\right)\delta_{\alpha\beta} \tag{8.135}$$

and we parametrize the momenta as follows

$$\left.\begin{aligned}
&p = (P + p^2/4P, \mathbf{O}_\perp, P - p^2/4P)\\
&l = (zP + l_\perp^2/4zP, \mathbf{l}_\perp, zP - l_\perp^2/4zP)\\
&t = (yP + (t^2 + t_\perp^2)/4yP, \mathbf{t}_\perp, yP - (t^2 + t_\perp^2)/4yP)\\
&q = (\eta P + q^2/4\eta P, \mathbf{O}_\perp, \eta P - q^2/4\eta P)\\
&y = 1 - z, t^2 z = -l_\perp^2
\end{aligned}\right\} \tag{8.136}$$

For the gauge four-vector n we take

$$n = (p\cdot n/2P, \mathbf{O}, -p\cdot n/2P) \tag{8.137}$$

We can interpret (8.136) and (8.137) noticing that P is an arbitrary boost parameter defining the frame (invariant quantities are P independent) so that we may start, e.g. from the p rest frame ($4P^2 = p^2$) and make an infinite boost $P \to \infty$ in the direction oposite to n. We observe that

(i) $l\cdot n/p\cdot n = z$
(ii) $\eta = -x$ for $p^2 = 0$ because $2p\cdot q = q^2/\eta$
(iii) $1 - z = x + O(l_\perp^2/(-q^2))$ because $(p-l)^2 = -l_\perp^2/z$ and $(p - l + q)^2 = 0$.

The two-body phase space

$$\mathrm{d}\Phi_2 = \frac{\mathrm{d}^n l}{(2\pi)^n} 2\pi\delta_+(l^2) \frac{\mathrm{d}^n k}{(2\pi)^n} 2\pi\delta_+(k^2)(2\pi)^n \delta^{(n)}(q + p - l - k) \tag{8.138}$$

can be written as

$$\mathrm{d}\Phi_2 = \frac{\mathrm{d}^n t}{(2\pi)^{n-2}} \delta_+((p-t)^2)\delta_+((q+t)^2) \tag{8.139}$$

From

$$\left.\begin{aligned} \delta_+((p-t)^2) &= \delta(-(1-y)t^2/y - t_\perp^2/y), & (p-t)_0 > 0 \\ \delta_+((q+t)^2) &= \delta((y-x)(t^2/y - q^2/x) - xt_\perp^2/y), & (q+t)_0 > 0 \end{aligned}\right\} \tag{8.140}$$

one concludes that

$$\begin{gathered} x < y < 1 \\ t^2 = \frac{y-x}{1-x}\frac{q^2}{x}, \quad \frac{q^2}{x} < t^2 < 0 \end{gathered} \tag{8.141}$$

Furthermore

$$\frac{\mathrm{d}^n t}{(2\pi)^{n-2}} = \frac{1}{(2\pi)^{n-2}} \mathrm{d}t^2 \frac{\mathrm{d}y}{2|y|} \mathrm{d}^{n-2}\mathbf{t}_\perp \tag{8.142}$$

and

$$\int \mathrm{d}^{n-2}\mathbf{t}_\perp = \int t_\perp^{n-3}\,\mathrm{d}t_\perp\,\mathrm{d}\Omega_{n-2} = \frac{\pi^{(n-2)/2}}{\Gamma(\frac{1}{2}(n-2))} \int \mathrm{d}t_\perp^2 (t_\perp^2)^{(n-4)/2} \tag{8.143}$$

So finally, integrating over $\mathrm{d}(t_\perp^2/y)$ by means of the $\delta((p-t)^2)$, we get ($n = 4 + \varepsilon$)

$$\Phi_2 = \frac{1}{8\pi}\frac{1}{\Gamma(1+\frac{1}{2}\varepsilon)} \int_{q^2/x}^{0} \mathrm{d}t^2 \int_x^1 \frac{\mathrm{d}y}{y} y \left[\frac{(1-y)|t^2|}{4\pi}\right]^{\varepsilon/2} \delta\left(t^2(1-x) - \frac{(y-x)q^2}{x}\right) \tag{8.144}$$

$$= \frac{1}{8\pi}\frac{1}{\Gamma(1+\frac{1}{2}\varepsilon)}\frac{1}{1-x} \int_x^1 \mathrm{d}y \left[\frac{(1-y)(y-x)}{x(1-x)}\frac{-q^2}{4\pi}\right]^{\varepsilon/2} \tag{8.145}$$

For the Feynman part of the ladder diagram, using (8.107) and (8.108) together with (8.144) we recover the previous result (8.110).

The axial part of the ladder diagram

$$\sigma_{\rm L}^{\rm A} = C_{\rm F}\frac{1}{1+\frac{1}{2}\varepsilon}\alpha\mu^{-\varepsilon}\int \mathrm{d}\Phi_2\tfrac{1}{2}\mathrm{Tr}\left[\not{p}\not{l}\frac{1}{\not{t}}\gamma_\mu(\not{t}+\not{q})\gamma^\mu\frac{1}{\not{t}}\not{n}+\not{p}\not{n}\frac{1}{\not{t}}\gamma_\mu(\not{t}+\not{q})\gamma^\mu\frac{1}{\not{t}}\not{l}\right]\left(-\frac{1}{l\cdot n}\right) \tag{8.146}$$

can be calculated using the expansion

$$\left.\begin{aligned}\not{t} &= y\not{p}+\frac{t^2+t_\perp^2}{2y}\frac{\not{n}}{p\cdot n}+\not{t}_\perp\\ \not{q} &= -x\not{p}+\frac{-q^2}{x}\frac{\not{n}}{2p\cdot n}\end{aligned}\right\} \tag{8.147}$$

and (8.144). After some effort one gets

$$\begin{aligned}\sigma_{\rm L}^{\rm A} = C_{\rm F}\left(\frac{\alpha}{2\pi}\right)\left(\frac{-q^2}{4\pi\mu^2}\right)^{\varepsilon/2}&\left\{2\frac{\Gamma(1+\frac{1}{2}\varepsilon)}{\Gamma(1+\varepsilon)}\left[\left(\frac{2}{\varepsilon}\right)^2-\frac{2}{\varepsilon}+\frac{\pi^2}{6}\right]\right.\\ &\left.*\,\delta(1-x)+\left(\frac{2}{\varepsilon}+\gamma\right)\left[\frac{2x}{1-x}\left(\frac{1-x}{x}\right)^{\varepsilon/2}\right]_+\right\}\end{aligned} \tag{8.148}$$

We see that

$$\sigma_{\rm L}^{\rm F}+\sigma_{\rm L}^{\rm A} = C_{\rm F}\left(\frac{\alpha}{2\pi}\right)\left[\left(\frac{2}{\varepsilon}+\gamma\right)\left(\frac{1+x^2}{1-x}\right)_+ + A(\varepsilon)\delta(1-x)+\cdots\right]\left(\frac{-q^2}{4\pi\mu^2}\right)^{\varepsilon/2} \tag{8.149}$$

where the neglected terms do not have a pole in ε and are polynomials in x and logarithms in x and $(1-x)$. Thus in the axial gauge the ladder diagram by itself gives the leading logarithmic contribution (8.126) to the quark structure function. (Remember that we are interested in the finite $x \neq 0$ region so $\ln x$ and $\ln(1-x)$ can be dropped.) We expect therefore that the contribution from the remaining diagrams with real emissions is non-leading. By explicit evaluation of the axial parts of these amplitudes one gets:

$$\sigma_{\Sigma_c}^{\rm A} = \left(\frac{1}{x}F_2\right)^{\rm long} = 0, \quad \sigma_{\Gamma_c}^{\rm A} = -\sigma_{\rm L}^{\rm A}$$

so that

$$\sigma_{\rm L}^{\rm F}+\sigma_{\rm L}^{\rm A}+\sigma_{\Gamma_c}^{\rm F}+\sigma_{\Gamma_c}^{\rm A}+\sigma_{\Sigma_c}^{\rm F}+\sigma_{\Sigma_c}^{\rm A} = \sigma_{\rm L}^{\rm F}+\sigma_{\Gamma_c}^{\rm F}+\sigma_{\Sigma_c}^{\rm F} \tag{8.150}$$

and $(\sigma_{\Gamma_c}^{\rm F}+\sigma_{\Gamma_c}^{\rm A})$ does not contribute to the leading logarithm, i.e. it is free of mass singularities. Actually, one can support this conclusion by the following general argument. The leading logarithm reflecting the presence of mass singularity obviously comes from the integration over t^2 in the region $\varepsilon q^2 < t^2 < 0$ i.e. from the region where t and l are almost collinear with p: $t \cong yp$. Part of the Dirac algebra involves

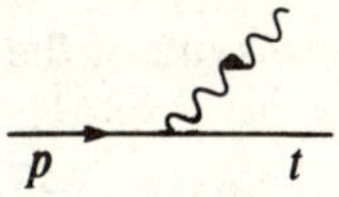

$$\not{t}\gamma_\mu\not{p} \cong y\not{p}\gamma_\mu\not{p} = y2p_\mu\not{p} = \frac{2y}{1-y}l_\mu\not{p} \tag{8.151}$$

and since in the axial gauge the on-shell gluons have only physical polarization the contraction with the gluon tensor $d^{\mu\nu}(l) = \sum_{\lambda=1,2}\varepsilon^{*\mu}(\lambda, l)\varepsilon^{\nu}(\lambda, l)$, $l^2 = 0$ gives

$$l_\mu d^{\mu\nu}(l) = 0 \tag{8.152}$$

Actually, one can check that the vertex vanishes linearly in θ where θ is the gluon emission angle. So the ladder diagram gives

$$\int d\Phi_2 (1/t^2)^2 \theta^2 \sim \int d\theta\, \theta\theta^2/(\theta^2)^2 \sim \ln\theta \tag{8.153}$$

whereas the cut-vertex diagram gives

$$\int d\Phi_2 (1/t^2)\theta \sim \int d\theta\, \theta(\theta/\theta^2) \tag{8.154}$$

and hence no logarithm.

The poles in ε in the coefficient of the $\delta(1-x)$ are cancelled by the virtual corrections. Correspondingly the situation with real emissions, in the axial gauge mass singularity is present only in the quark self-energy correction. It is instructive to calculate the virtual corrections in the light-like gauge and using the light-cone parametrization of the momenta. This we leave as an exercise for the reader limiting ourselves to the following two remarks. Take, for instance the self-energy correction

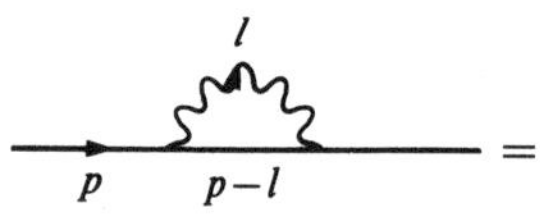

$= -i\Sigma$

where Σ can be decomposed as follows

$$\Sigma = A\not{p} + B\frac{p^2}{2p\cdot n}\not{n} \tag{8.155}$$

with

$$A = \tfrac{1}{2}\mathrm{Tr}\left[\frac{\not{n}}{2p\cdot n}\Sigma\right] \tag{8.156}$$

$$A + B = \tfrac{1}{2}\mathrm{Tr}\left[\frac{\not{p}\not{n}\not{p}}{2p\cdot n}\Sigma\right]\frac{1}{p^2} \tag{8.157}$$

The full propagator reads

$$G(p^2) = \frac{i}{\not{p}-\Sigma} = \frac{i}{1-(A+B)}\left[\frac{\not{p}}{p^2} - \frac{B}{1-A}\frac{\not{n}}{2p\cdot n}\right] = [1 + (A+B) + \cdots]\left(\frac{i\not{p}}{p^2} + \cdots\right) \tag{8.158}$$

Apart from IR and mass singularities Σ is also UV divergent, i.e. contains poles in ε_{UV} where $n = 4 - \varepsilon_{UV}$. The necessary counterterms are of the form

$$a\not{p} + b\not{n}$$

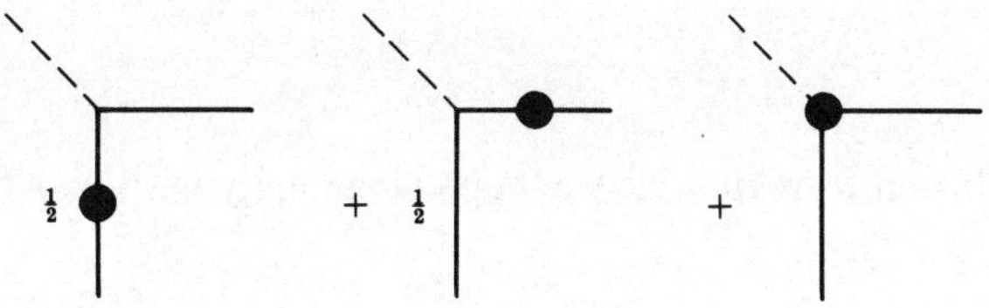

Fig. 8.9

The term proportional to $\not{n}$ reflects the pathology of the $n^2 = 0$ gauge because

$$\int d^4 l \frac{1}{(n \cdot l) l^2 (p - l)^2}$$

should be convergent on dimensional grounds,† and it makes the general proof of the renormalizability in this gauge difficult. However, in order-by-order calculation one can check explicitly that such terms cancel out in the full vertex function, see Fig. 8.9.

Our second remark is that an explicit calculation of the virtual corrections using the light-cone parametrization of the momenta is called the Sudakov technique. It consists of noting that the integral

$$\int \frac{d^n l}{(2\pi)^n} \frac{1}{l^2 + i\varepsilon} \frac{1}{(p-l)^2 + i\varepsilon} \sim \int \frac{dl^2}{l^2} \frac{dz}{2|z|} dl_\perp^2 (l_\perp^2)^{(n-4)/2} \frac{z}{z(1-z)(p^2 - l^2/z) - l_\perp^2 + i\varepsilon} \tag{8.159}$$

can be evaluated by integrating over $dl_\perp^2$ first. Going to the complex $l_\perp^2$-plane, since there is a pole at

$$l_\perp^2 = z(1-z)(p^2 - l^2/z) + i\varepsilon \tag{8.160}$$

and the integral over the semi-circle, which is at infinity, vanishes, we can replace the denominator by $\delta(z(1-z)(p^2 - l^2/z) - l_\perp^2)$.

Since in the leading logarithmic approximation, only the ladder diagram contributes to the final result, its probabilistic interpretation is quite natural. In fact this result has the following structure

$$\sigma_L(p, q^2) = \frac{\alpha}{2\pi} C_F \int_{p^2}^{q^2} \frac{dt^2}{t^2} \int_x^1 \frac{dy}{y} \left(\frac{1 + y^2}{1 - y} \right)_+ \sigma_L(yp, q) \tag{8.161}$$

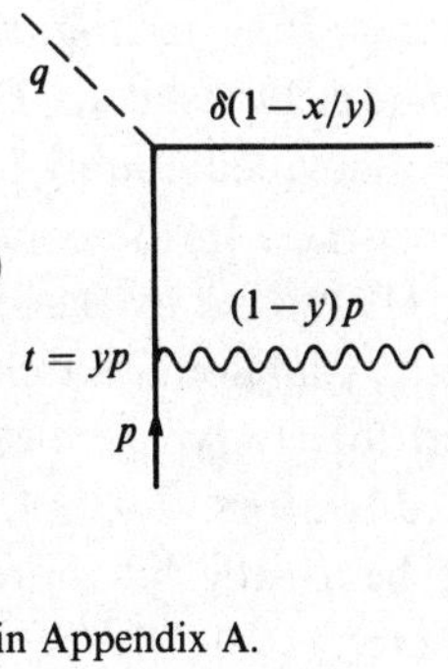

† In the Feynman parametrization such integrals are discussed in Appendix A.

where

$$\sigma_L(yp, q) = \delta(1 - x/y)$$

and we have chosen to regularize the mass singularity with $p^2 \neq 0$. This coincides with (8.129) and

$$P_{q\to q}(y) = \left(\frac{1+y^2}{1-y}\right)_+$$

can be interpreted as the variation of the probability density per unit t^2 of finding a quark in a quark with fraction y of its momentum. Of course, our previous discussion concerning the hadron structure functions, (8.127)–(8.129), is still valid.

8.7 Beyond the one-loop approximation

Since $\alpha_{\text{strong}} \sim O(1)$ the relevance of perturbative QCD to the description of the actual experimental data often relies on our ability to go beyond the one-loop approximation. In addition we must also include in considerations the singlet and the gluon structure and fragmentation functions. This vast area of technical and phenomenological activity goes far beyond the scope of the present book and can by itself be a subject of an extended monograph. Let us only briefly indicate the techniques that have been developed to include higher order effects in α^{strong}. One approach is based on the OPE and on the RGE for the Wilson coefficient functions (Chapter 7). Another one, applicable to a wider class of problems, consists of summing infinite sets of Feynman diagrams by cleverly picking out the leading contribution, the next-to-leading one and so on (Gribov & Lipatov 1972; Llewellyn Smith 1978; Dokshitzer, Dyakonov & Troyan 1980). For the regions $x \neq 0$ and $x \neq 1$ i.e. neglecting $\ln x$ and $\ln(1-x)$ terms as compared to $\ln(q^2/p^2)$ terms the summation of the so-called leading logarithms $\alpha^n \ln^n(q^2/p^2)$ and also of the next-to-leading $\alpha^n \ln^{n-1}(q^2/p^2)$ terms has been explicitly performed and compared, whenever possible, with the OPE. It is very interesting that these results can also be given a probabilistic interpretation. In the leading logarithm approximation it is concisely formulated by means of the Altarelli–Parisi equations (Altarelli & Parisi 1977).

Further very interesting and more recent progress consists of extending the QCD perturbative techniques to processes where there are two types of large logarithm (e.g. Dokshitzer, Dyakonov & Troyan 1980; Parisi & Petronzio 1979) and to the so-called doubly-logarithmic region of the structure and fragmentation functions ($x \cong 0$). In the context of QCD pioneered by Furmański, Petronzio & Pokorski (1979) and by Bassetto, Ciafaloni & Marchesini (1980) and fully clarified by Bassetto, Ciafaloni, Marchesini & Mueller (1982), this extension is, for instance, important for studying new particle production in hard collisions in very high energy accelerators.

Let us be a little bit more explicit concerning the summation of the leading logarithm terms $\alpha^n \ln^n(q^2/p^2)$ generated by mass singularities in the quark structure

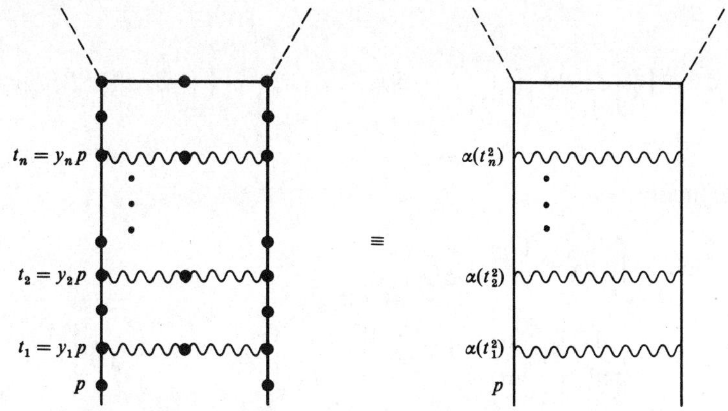

Fig. 8.10

function. It turns out that the dominant diagrams in the axial gauges are the generalized ladder diagrams, Fig. 8.10, i.e. ladder diagrams with vertex and self-energy insertions. The dominant region of integration is where

$$|p^2| \ll |t_1^2| \ll |t_2^2| \ll \cdots \ll |t_n^2| \ll |q^2|$$

Furthermore it can be shown that the generalized ladder can be replaced by a ladder with the coupling constant $\alpha(t_i^2)$ at each vertex. Thus in the leading logarithm approximation the cross section is a generalization of (8.161) and reads

$$\sigma_{\mathrm{L}}^n = \int_{p^2}^{|q^2|} \frac{\mathrm{d}t_n^2}{t_n^2} C_{\mathrm{F}} \frac{\alpha(t_n^2)}{2\pi} \int_{p^2}^{|t_n^2|} \frac{\mathrm{d}t_{n-1}^2}{t_{n-1}^2} C_{\mathrm{F}} \frac{\alpha(t_{n-1}^2)}{2\pi} \cdots$$
$$* \int_{p^2}^{|t_2^2|} \frac{\mathrm{d}t_1^2}{t_1^2} C_{\mathrm{F}} \frac{\alpha(t_1^2)}{2\pi} \int_x^1 \mathrm{d}y_1 P_{q\to q}(y_1) \int_x^{y_1} \frac{\mathrm{d}y_2}{y_1} P_{q\to q}\left(\frac{y_2}{y_1}\right) \cdots$$
$$* \int_x^{y_{n-2}} \frac{\mathrm{d}y_{n-1}}{y_{n-2}} P_{q\to q}\left(\frac{y_{n-1}}{y_{n-2}}\right) \int_x^{y_{n-1}} \frac{\mathrm{d}y_n}{y_{n-1}} P_{q\to q}\left(\frac{y_n}{y_{n-1}}\right) \delta(y_n - x) \frac{1}{y_n} \tag{8.162}$$

Using (8.162) we can easily calculate the moments of $(1/x)F_2$. First we note that

$$\left(\frac{C_{\mathrm{F}}}{2\pi}\right)^n \int_{p^2}^{|q^2|} \frac{\mathrm{d}t_n^2}{t_n^2} \alpha(t_n^2) \int_{p^2}^{|t_n^2|} \frac{\mathrm{d}t_{n-1}^2}{t_{n-1}^2} \alpha(t_{n-1}^2) \cdots \to \frac{1}{n!} (C_{\mathrm{F}}/2\pi)^n$$
$$* \int_{p^2}^{|q^2|} \frac{\mathrm{d}t_n^2}{t_n^2} \alpha(t_n^2) \int_{p^2}^{|q^2|} \frac{\mathrm{d}t_{n-1}^2}{t_{n-1}^2} \alpha(t_{n-1}^2) \cdots = \frac{1}{n!} \left[\ln \frac{\ln(-q^2/\Lambda^2)}{\ln(-p^2/\Lambda^2)} \right]^n \left(-\frac{C_{\mathrm{F}}}{b_1} \right)^n \tag{8.163}$$

where $\alpha(t^2) = -2\pi/[b_1 \ln(t^2/\Lambda^2)]$, $-b_1 = \frac{11}{2} - \frac{1}{3} n_f$. Changing variables

$$z_1 = y_1, \quad z_2 = y_2/y_1, \quad \ldots, \quad z_n = y_n/y_{n-1}$$
$$z_1 z_2 \ldots z_n = y_n$$

we get

$$\int_0^1 \mathrm{d}x\, x^N \prod_i \int \mathrm{d}z_i P_{q\to q}(z_i) \frac{1}{z_1 \ldots z_n} \delta(x - z_1 \ldots z_n) = \prod_i \int_0^1 \mathrm{d}z_i z_i^{N-1} P_{q\to q}(z_i) \tag{8.164}$$

Therefore, finally

$$\begin{aligned} M_N &= \int_0^1 \mathrm{d}x\, x^N \frac{1}{x} F_2(x, q^2) \\ &= \sum_n \frac{1}{n!} \left\{ \left[\ln \frac{\ln(-q^2/\Lambda^2)}{\ln(-p^2/\Lambda^2)} \right] \left[-\frac{C_\mathrm{F}}{b_1} \int_0^1 \mathrm{d}z\, z^{N-1} P_{q\to q}(z) \right] \right\}^n \\ &= \left[\frac{\ln(-q^2/\Lambda^2)}{\ln(-p^2/\Lambda^2)} \right]^{-(C_\mathrm{F}/b_1)\int_0^1 \mathrm{d}z\, z^{N-1} P_{q\to q}(z)} \end{aligned} \tag{8.165}$$

and

$$\frac{M_N(Q^2)}{M_N(Q_0^2)} = \left[\frac{\ln(Q^2/\Lambda^2)}{\ln(Q_0^2/\Lambda^2)} \right]^{\gamma_N^0/b_1}, \quad Q^2 = -q^2 \tag{8.166}$$

where

$$-\gamma_N^0 = C_\mathrm{F} \int_0^1 \mathrm{d}z\, z^{N-1} \left(\frac{1+z^2}{1-z} \right)_+ = -\frac{2}{3} \left[1 - \frac{2}{N(N+1)} + 4 \sum_{j=2}^{N} \frac{1}{j} \right] \tag{8.167}$$

The same result holds, of course, for the moments of the hadron structure function and it coincides with (7.124) obtained using the OPE. We see that γ_N^0 are the g^2-coefficients in the expansion of the anomalous dimensions of the operators in (7.99) in the coupling constant.

Comments on the IR problem in QCD

The Bloch–Nordsieck theorem in QED states that IR divergences cancel for physical processes i.e. for processes with an arbitrary number of undetectable soft photons. In QCD it has been demonstrated that IR divergences cancel for processes with none or one coloured particle in the initial state (for a review see e.g. Sachrajda 1983*b*). However, it has been shown by means of an explicit example that such a cancellation does not occur for processes with two coloured particles in the initial state even if an average over different colour states is taken. For instance, in the process $\mathrm{qq} \to \mathrm{l}^+\mathrm{l}^-\mathrm{X}$ these divergences arise at the two-loop level. Of course, in physical processes coloured partons never appear as free on-shell particles and these divergences are regulated by a mass-scale characteristic of the hadronic wave function which is of the order of the inverse hadronic radius. However, residual large logarithms remain and could in principle spoil the factorization property discussed in Section 8.4 which states that all long distance effects can be absorbed into the quark and gluon distribution functions and is crucial for the sensible use of perturbative QCD. Fortunately enough it has been shown recently that such large IR logarithms are suppressed by at least one power of Q^2 and asymptotically the factorization property is still maintained.

In the context of the IR problem in QCD it is worth recalling the Kinoshita–Lee–Nauenberg (KLN) theorem (Kinoshita 1962, Lee & Nauenberg 1964). The KLN theorem is a mathematical theorem saying that, as a consequence of unitarity, transition probabilities are finite when the sum over all degenerate states (final and initial) is taken. This is true order-by-order in perturbation theory in bare quantities or if the minimal subtraction renormalization is used (to avoid IR or mass singularities in the renormalization constants). The physical meaning of this theorem seems, however, to be unclear. In QED the cancellation of IR singularities occurs separately in the final state and there is no need to invoke the KNL theorem. In non-abelian gauge theories this does not happen but the initial state is determined by the non-perturbative confinement effects and the relevance of the KLN theorem is doubtful.

Problems

8.1 (*a*) Consider mappings of the real axis with the end-points identified into a set of unimodular complex numbers $u = \exp[i\alpha(x)]$. Check that the winding number can be written as

$$N = \frac{1}{2\pi}\int_{-\infty}^{\infty} dx\left[\frac{-i}{\alpha(x)}\frac{d\alpha(x)}{dx}\right] \tag{I}$$

Check that $N = 1$ for $\alpha(x) = \exp[i\pi x/(x^2 + a^2)^{1/2}]$.

(*b*) For $S^3 \to SU(2)$ the transformation

$$A_\mu \to \exp(-\tfrac{1}{2}i\Theta^a\tau^a)(A_\mu + \partial_\mu)\exp(\tfrac{1}{2}i\Theta^a\tau^a)$$

with $\Theta^a(\mathbf{x}) \equiv 0$ or $\Theta^a(\mathbf{x})$ continuously deformable to $\Theta^a(\mathbf{x}) \equiv 0$ belongs to $N = 0$. Clearly, then $\Theta^a(\infty) = 0$. Check that with

$$\Theta^a(\mathbf{x}) = 2\pi x^a/(\mathbf{x}^2 + a^2)^{1/2}$$

or with Θ^a continuously deformable to it, $U(\mathbf{x})$ belongs to $N = 1$ class. What is U_∞? Can such a transformation be built up from infinitesimal transformations $\Theta^a(\mathbf{x}) \to \delta\Theta^a(\mathbf{x})$ for all $\mathbf{x}$?

8.2 Show that gauge transformations which belong to the homotopy class $N \neq 0$ cannot be built up by iterating the infinitesimal gauge transformations. For this assume that a 'large' ($N \neq 0$) gauge transformation

$$U = \exp(-\tfrac{1}{2}i\Theta^a\tau^a)$$

where $\Theta^a(\mathbf{x})$ approaches a non-zero limit at spatial infinity is given by the operator

$$\exp(iQ_G)\mathbf{A}\exp(-iQ_G) = U^{-1}\mathbf{A}U - U^{-1}\nabla U$$

where

$$\exp(iQ_G) = \exp\left(-i\int d^3x D^i_{ab}\Theta_b\cdot\Pi^i_a\right)$$

Applying it to

$$W(\mathbf{A}) = \int d^3x K^0$$

where K^0 is given by (8.64), show that

$$\exp(iQ_G)W(\mathbf{A})\exp(-iQ_G) = W(\mathbf{A})$$

contradicting the correct result

$$\exp(iQ_G)W(\mathbf{A})\exp(-iQ_G) = W(U^{-1}\mathbf{A}U - U^{-1}\nabla U) = W(\mathbf{A}) + N$$

(Use $\delta W(\mathbf{A})/\delta A^i_a = (1/8\pi^2)B^a_i$)

8.3 Derive the Feynman rules for calculating the 1PI Green's functions generated by $\tilde{\Gamma}[0, \mathbf{A}_\mu]$ defined in Section 8.2. Use the shifted action $S[A_\mu + \mathbf{A}_\mu]$, the gauge-fixing term (8.24) and the Faddeev–Popov term obtained from (3.15) and (3.39) with $F^\alpha \to \tilde{F}^\alpha[A_\mu, \mathbf{A}_\mu]$ and δA_μ given by (8.22). Prove (8.34) and complete the calculation of the β-function in Yang–Mills theories in the one-loop approximation (Abbott 1982).

9

Chiral symmetry; spontaneous symmetry breaking

Chiral symmetry plays an important role in the dynamics of fermions coupled to gauge fields. Historically, its importance for particle physics originates in the idea that $SU(2)_L \times SU(2)_R$ is an approximate symmetry of the strong interactions, realized as spontaneously broken symmetry with pions being Goldstone bosons in the symmetry limit. Much of the progress of particle physics since 1960 is related to understanding the phenomenological aspects of spontaneous and explicit chiral symmetry breaking. The successes of gauge theories in describing the fundamental interactions stimulate further our interest in chiral symmetry and this is for at least two reasons: one may hope to understand the underlying mechanism of spontaneous chiral symmetry breaking in the strong interactions and in gauge theories in general and, secondly, to find the dynamical theory of the fermion mass matrix which might eventually lead to the true theory of fundamental interactions. This chapter serves as an introduction to the subject of chiral symmetry and to techniques used in its exploration.

9.1 Chiral symmetry of the QCD lagrangian

We begin our discussion by considering the QCD lagrangian with two quarks u and d in the theoretical limit in which the quark mass parameters in the lagrangian are put equal to zero. In fact, the masses of the light quarks are small in a sense which will be specified later on and we shall assume in the following that the theory with the quark masses neglected is a correct first approximation.

Consider the general massless Dirac lagrangian of fermion fields Ψ_{ri} coupled to gauge fields A^α_μ

$$\mathscr{L} = \sum_r \sum_i^{n(r)} \bar{\Psi}_{ri} \mathrm{i}\gamma^\mu D_\mu \Psi_{ri}, \quad \Psi = \Psi(x) \tag{9.1}$$

where

$$D_\mu = \partial_\mu - \mathrm{i}gA^\alpha_\mu t^\alpha_r$$

and the index α runs over the generators of the gauge group, the matrices t^α_r

represent these generators in the representation r of the gauge group to which the fermions are assigned and the index i runs over all $n(r)$ flavours belonging to the same representation r. For QCD with two flavours $r \equiv 3$, $i = 1,2$ and Ψ_i (we omit now the subscript r) denotes a vector in the colour space. In this case lagrangian (9.1) is invariant under the general unitary transformation in the two-dimensional flavour space. Introducing $\Psi = \begin{pmatrix} \Psi_1 \\ \Psi_2 \end{pmatrix}$ we can write

$$\mathscr{L} = \bar{\Psi}\begin{pmatrix} i\not{D} & \\ & i\not{D} \end{pmatrix}\Psi \tag{9.2}$$

which is invariant under $SU(2) \times U(1)$ transformations

$$\left.\begin{aligned} \Psi &\to \exp(-i\alpha^a \tfrac{1}{2}\tau^a)\Psi \\ \bar{\Psi} &= \bar{\Psi}\exp(+i\alpha^a \tfrac{1}{2}\tau^a) \end{aligned}\right\} \tag{9.3}$$

where $a = 0, 1, 2, 3$ and τ^a are Pauli matrices for $a = 1, 2, 3$ and $\tau^0 = 1$. The corresponding conserved currents are

$$\mathscr{V}^a_\mu = \bar{\Psi}\gamma_\mu \tfrac{1}{2}\tau^a \Psi \tag{9.4}$$

and the $SU(2) \times U(1)$ invariance gives the isospin and the baryon number

$$Q^a = \int d^3x \bar{\Psi}(x)\gamma_0 \tfrac{1}{2}\tau^a \Psi(x), \quad a = 0, 1, 2, 3 \tag{9.5}$$

conservation.

However, the massless theory (9.2) has also another global symmetry, involving γ_5

$$\left.\begin{aligned} \Psi &\to \exp(-i\alpha^a \tfrac{1}{2}\tau^a \gamma_5)\Psi \\ \bar{\Psi} &\to \bar{\Psi}\exp(-i\alpha^a \tfrac{1}{2}\tau^a \gamma_5) \end{aligned}\right\} \tag{9.6}$$

where $a = 0, 1, 2, 3$. Indeed, since $\{\gamma_5, \gamma_\mu\} = 0$ the exponentials again cancel when (9.6) is introduced into lagrangian (9.2). The axial currents read

$$\mathscr{A}^a_\mu = \bar{\Psi}\gamma_\mu \gamma_5 \tfrac{1}{2}\tau^a \Psi \tag{9.7}$$

and the axial charges†

$$Q^a_5 = \int d^3x \bar{\Psi}(x)\gamma_0 \gamma_5 \tfrac{1}{2}\tau^a \Psi(x) \tag{9.8}$$

together with the charges (9.5) form the chiral $SU_L(2) \times SU_R(2) \times U(1) \times U_A(1)$ algebra (as can be seen using (9.5) and (9.8) and the equal-time canonical

† Charges may be formal in the sense that the integral over all space may not exist. Even so, their commutators exist provided we first commute under the space integrals and then do the integrations. This remark is relevant for the case of spontaneous symmetry breakdown.

commutation relations between the fields)

$$\left.\begin{aligned} Q^a_{\rm L} = \tfrac{1}{2}(Q^a - Q^a_5), \quad Q^a_{\rm R} = \tfrac{1}{2}(Q^a + Q^a_5), \quad a = 0, 1, 2, 3 \\ [Q^a_{\rm L,R}, Q^b_{\rm L,R}] = {\rm i}\varepsilon^{abc} Q^c_{\rm L,R}, \quad a, b, c = 1, 2, 3 \\ [Q^a_{\rm L}, Q^b_{\rm R}] = 0, \quad a, b = 0, 1, 2, 3 \end{aligned}\right\} \tag{9.9}$$

(for completeness: $[Q^a, Q^b] = [Q^a_5, Q^b_5] = {\rm i}\varepsilon^{abc} Q^c$; $[Q^a_5, Q^b] = {\rm i}\varepsilon^{abc} Q^c_5$). An important property of the generators $Q^a_{\rm L}$ and $Q^a_{\rm R}$ is that they transform into each other under the parity operation: $PQ^a_{\rm R,L}P^{-1} = Q^a_{\rm L,R}$. Hence the name chiral transformations.

Before we proceed to explore the chiral symmetry of the massless QCD lagrangian (9.2) we introduce some useful notation. Any spinor field Ψ can be decomposed into 'chiral' fields† as follows

$$\Psi = \Psi_{\rm L} + \Psi_{\rm R} = \tfrac{1}{2}(1 - \gamma_5)\Psi + \tfrac{1}{2}(1 + \gamma_5)\Psi = P_{\rm L}\Psi + P_{\rm R}\Psi \tag{9.10}$$

where

$$P^2_{\rm L} + P^2_{\rm R} = 1, \quad P_{\rm L}P_{\rm R} = 0$$

and the lagrangian (9.2) can, then, be rewritten as

$$\mathscr{L} = \bar{\Psi}_{\rm L}{\rm i}\not{D}\mathbb{1}\Psi_{\rm L} + \bar{\Psi}_{\rm R}{\rm i}\not{D}\mathbb{1}\Psi_{\rm R} \tag{9.11}$$

($\mathbb{1}$ is unit matrix in the flavour space).

Chiral symmetry of the lagrangian (9.11) means invariance under the transformations

$$\left.\begin{aligned} \Psi^i_{\rm L} \to \Psi'^i_{\rm L} = (U_{\rm L})^i_j \Psi^j_{\rm L} \\ \Psi^i_{\rm R} \to \Psi'^i_{\rm R} = (U_{\rm R})^i_j \Psi^j_{\rm R} \end{aligned}\right\} \tag{9.12}$$

where indices i, j are flavour indices and the unitary transformations $U_{\rm L}$ and $U_{\rm R}$ read

$$\left.\begin{aligned} U_{\rm L} = \exp(-{\rm i}\alpha^a_{\rm L} T^a_{\rm L}) \\ U_{\rm R} = \exp(-{\rm i}\alpha^a_{\rm R} T^a_{\rm R}) \end{aligned}\right\} \tag{9.13}$$

with $T^a_{\rm L} = T^a \otimes P_{\rm L}$ and $T^a_{\rm R} = T^a \otimes P_{\rm R}$. The matrices T form a representation of the corresponding generators Q in the space of fermion multiplets Ψ. We see that multiplet $\Psi_{\rm L}(\Psi_{\rm R})$ is a singlet with respect to $SU_{\rm R}(2)$ $(SU_{\rm L}(2))$ transformations. At this level we should also conclude that because of the $U(1) \times U_{\rm A}(1)$ invariance the fermion numbers of $\Psi_{\rm R}$ and $\Psi_{\rm L}$ (corresponding to $U(1) \pm U_{\rm A}(1)$ charges) are separately conserved. The problem of the $U_{\rm A}(1)$ symmetry will be briefly discussed in Chapter 12. For the time being we will be concerned with the chiral $SU(2) \times SU(2)$ only. Some useful properties of the chiral fields $\Psi_{\rm L}$ and $\Psi_{\rm R}$ are collected in the Appendix C.

† Decomposition (9.10) has a clear interpretation in terms of the free particle solutions of the massless Dirac equation. This is discussed in the Appendix C.

In many problems we work with one type of chiral fields Ψ_R or Ψ_L only. For instance in 'chiral' gauge theories such that Ψ_R and Ψ_L transform differently under the gauge group (e.g. in grand unification schemes) only Ψ_L (or only Ψ_R) fields can be placed in irreducible multiples of the gauge group. Indeed, $\bar{\Psi}_R\gamma_\mu\Psi_L$ currents vanish and we would not be able to introduce symmetry operations between Ψ_R and Ψ_L. The description in terms of the e.g. only left-handed fields can be obtained if we replace Ψ_R by their charge conjugate partners

$$(\Psi_R)^C = (\Psi^C)_L \underset{\text{df}}{=} \tilde{\Psi}_L \tag{9.14}$$

Then we can formulate our theory in terms of Ψ_L and $\tilde{\Psi}_L$. Another notation worth mentioning is the one using the two-component Weyl spinors. We refer the reader to Appendix C for further details on both subjects.

9.2 Hypothesis of spontaneous chiral symmetry breaking in strong interactions

We have assumed that the massless QCD with its chiral $SU_L(2) \times SU_R(2)$ global symmetry is a good approximation to the real world. Now let us imagine in addition that the physical vacuum defined by the minimum of the expectation values of the hamiltonian $\langle 0|H|0\rangle = \langle H\rangle_{\text{min}}$ is invariant under the chiral transformations i.e.

$$Q^a_R|0\rangle = Q^a_L|0\rangle = 0 \tag{9.15}$$

We can invoke Coleman's theorem (see e.g. Itzykson & Zuber 1980, p. 513): 'a symmetry of the vacuum is a symmetry of the world' or

$$Q_i|0\rangle = 0 \Rightarrow [Q_i, H] = 0 \tag{9.16}$$

to conclude that the physical states in the spectrum of H can be classified according to the irreducible representations of the chiral group generated by $Q^a_{R,L}$ (as can be seen by constructing them from the vacuum by means of the field operators and then using the properties (9.16) and the transformation properties of the field operators under the chiral group). It is easy then to see that all isospin multiplets would have to have at least one degenerate partner of opposite parity. Let the state $|\Psi\rangle$ be an energy and parity eigenstate:

$$H|\Psi\rangle = E|\Psi\rangle \quad \text{and} \quad P|\Psi\rangle = |\Psi\rangle$$

Since Q_L and Q_R commute with the Hamiltonian we also have

$$HQ_L|\Psi\rangle = EQ_L|\Psi\rangle \quad \text{and} \quad HQ_R|\Psi\rangle = EQ_R|\Psi\rangle$$

In addition

$$PQ_{\substack{L\\R}}|\Psi\rangle = PQ_{\substack{L\\R}}P^+P|\Psi\rangle = Q_{\substack{R\\L}}|\Psi\rangle$$

and therefore the state

$$|\Psi'\rangle = \frac{1}{\sqrt{2}}(Q_R - Q_L)|\Psi\rangle$$

degenerate with $|\Psi\rangle$ is also a parity eigenstate

$$P|\Psi'\rangle = -|\Psi'\rangle$$

Presence of parity degenerate states is not a general feature of the known hadron spectrum. Therefore, if we wish to maintain our assumption about massless QCD being a good approximation to the strong interaction we must give up the assumption (9.15). Instead we assume that the physical vacuum is not invariant (but the Hamiltonian remains invariant) under the full chiral group and

$$Q_5^a|0\rangle \neq 0, \quad Q^a|0\rangle = 0 \tag{9.17}$$

One then talks about spontaneous breakdown, or Nambu–Goldstone realization (Nambu 1960, Goldstone 1961), of the chiral symmetry $SU_L(2) \times SU_R(2)$ into isospin symmetry $SU(2)$ generated by Q^a. Of course, given a theory like massless QCD the choice between (9.15) and (9.17) is no longer free for us. The theory itself should, in principle, tell us whether the symmetry or part of it is spontaneously broken or not. The standard example of a physical theory with spontaneous symmetry breakdown is the Heisenberg ferromagnet, an infinite array of spin one-half magnetic dipoles. The spin–spin interactions between neighbouring dipoles cause them to align. The Hamiltonian is rotationally invariant but the ground state is not: it is a state in which all the dipoles are aligned in some arbitrary direction. So for an infinite ferromagnet there is an infinite degeneracy of the vacuum. Unfortunately, the problem of calculating the spontaneous chiral symmetry breaking in QCD has not yet been satisfactorily solved. It is then interesting to begin as we do: assume the spontaneous breakdown (9.17) and explore its consequences in the hope of finding support for the assumption.

The most important consequence of the spontaneous symmetry breakdown is summarized in Goldstone's theorem: if a theory has a global symmetry of the lagrangian which is not a symmetry of the vacuum there must be a massless (Goldstone) boson, scalar or pseudoscalar, corresponding to each generator (with its quantum numbers) which does not leave the vacuum invariant. The proof of the theorem will be given later on. A very simple heuristic argument is as follows: if $Q_5^a|0\rangle \neq 0$ then there is a state $Q_5^a|0\rangle$ degenerate with the vacuum† because $[Q_5^a, H] = 0$. By successive application of Q_5^a one can construct an infinite number of degenerate states. A massless boson with internal quantum numbers, spin and parity of Q_5^a must exist to account for the degeneracy.

† This interpretation requires some caution (Bernstein 1974) since $Q_5^a|0\rangle$ is not a normalizable state. It means that $U = \exp(-\mathrm{i}\alpha^a Q_5^a)$ is not a unitary operator. Nevertheless, expressions like UAU^+ can be meaningful. If the exponentials are expanded, the resulting commutators may be well defined if we first commute currents and then integrate over space.

Very plausible candidates for Goldstone bosons in the strong interactions with spontaneously broken chiral symmetry are pseudoscalar mesons. Restricting ourselves to the $SU(2) \times SU(2)$ case we see that the triplet of pions has the right quantum numbers for such an interpretation. That they are not exactly massless (but nevertheless much lighter than other hadrons) is attributed to an explicit chiral symmetry breaking by the quark mass terms in the lagrangian.

Another consequence of spontaneous chiral symmetry breaking which agrees with the experimental data is the Goldberger–Treiman relation. It is based on the identification of the axial current of the two-flavour QCD with the axial current of the weak interactions for the first generation of quarks. This identification is explicit in the $SU(2) \times U(1)$ Glashow–Salam–Weinberg theory of weak interactions. The derivation of the Goldberger–Treiman relation in the limit of exact chiral symmetry (with quark masses and pion mass neglected) which is spontaneously broken, goes then as follows: the pion decay amplitude reads (we neglect the Cabbibo angle on both sides of (9.18))

$$\langle 0|\mathscr{A}^a_\mu(x)|\pi^b(q)\rangle = \exp(-\mathrm{i}qx)\langle 0|\mathscr{A}^a_\mu(0)|\pi^b(q)\rangle \tag{9.18}$$

where

$$\langle 0|\mathscr{A}^a_\mu(0)|\pi^b(q)\rangle = \mathrm{i} f_\pi q_\mu \delta^{ab} \tag{9.19}$$

and $\pi^{(1)} = (1/\sqrt{2})(\pi^+ + \pi^-)$, $\pi^{(2)} = -(\mathrm{i}/\sqrt{2})(\pi^+ - \pi^-)$, $\pi^3 = \pi^0$ (therefore, the $\pi^\pm$ decay constant is $\sqrt{2} f_\pi$). We assume the axial charge does not annihilate the vacuum, therefore $f_\pi \neq 0$. In our approximation the axial current is conserved: $\partial^\mu \mathscr{A}^a_\mu(x) = 0$ and from (9.18) and (9.19) we get

$$\langle 0|\partial^\mu \mathscr{A}^a_\mu(x)|\pi^b(q)\rangle = \exp(-\mathrm{i}qx) f_\pi m_\pi^2 \delta^{ab} \tag{9.20}$$

This is Goldstone's theorem: $f_\pi \neq 0$ therefore $m_\pi^2 = 0$. Next consider nucleon matrix elements of the axial current (β-decay). Assuming C invariance and isospin invariance one has

$$\langle \mathrm{N}(p')|\mathscr{A}^-_\mu(x)|\mathrm{P}(p)\rangle = \exp(\mathrm{i}qx)\bar{u}(p')\tfrac{1}{2}\tau^- [\gamma_\mu \gamma_5 g_\mathrm{A}(q^2) + q_\mu \gamma_5 h(q^2)]u(p) \tag{9.21}$$

where

$$\tau^- = (1/\sqrt{2})(\tau^1 - \mathrm{i}\tau^2), \quad q = p' - p$$

Current conservation implies

$$2m_\mathrm{N} g_\mathrm{A}(q^2) + q^2 h(q^2) = 0 \tag{9.22}$$

where m_N is the nucleon mass. For $q^2 = 0$ this relation can be satisfied either for $g_\mathrm{A}(0) = 0$ or if $h(q^2)$ has a pole for $q^2 = 0$. If pions are Goldstone bosons they contribute such a pole as can be seen from (9.23)

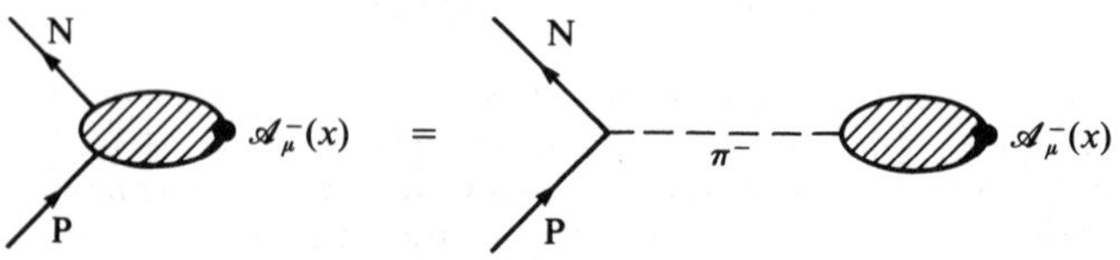

$$= \sqrt{2} g_{\pi\mathrm{N}}(\mathrm{i}/q^2)\mathrm{i}\sqrt{2} f_\pi q_\mu \bar{u}\gamma_5 u \exp(\mathrm{i}qx) \tag{9.23}$$

where $g_{\pi N}$ is the pion–nucleon coupling constant with the same normalization convention as for f_π. Hence

$$m_N g_A(0) = g_{\pi N} f_\pi \tag{9.24}$$

which is the Goldberger–Treiman relation derived from the assumption of spontaneously broken chiral symmetry of strong interactions. One can now ask whether the relation (9.24) is a good approximation to the real world where $m_\pi^2 \neq 0$. Experimentally, from πN scattering $g_{\pi N}^2/4\pi = 14.6$, from (physical) $\pi \to l\nu |f_\pi| = 93$ MeV, from β-decay $g_A(0) = 1.24$ and the relation (9.24) happens to be satisfied within 7%. This agreement supports our main assumption: massless QCD with spontaneously broken chiral symmetry is a good approximation for the strong interactions.

Relating physical quantities to the results obtained in the exact symmetry limit $m_\pi^2 = 0$ is called the PCAC (partially conserved axial vector current) approximation. Originally, the PCAC approximation had been formulated in another, equivalent, way as an operator equation

$$\partial^\mu \mathscr{A}_\mu^a(x) = m_\pi^2 f_\pi \pi^a(x) \tag{9.25}$$

m_π^2 is the physical pion mass, $\pi^a(x)$ is the pion field normalized as follows: $\langle \pi^a(p)|\pi^b(x)|0\rangle = \delta^{ab}\exp(ipx)$. The relation (9.25) is an identity on the pion mass shell, as can be seen from (9.20), and it is assumed that the pion field so defined is a good interpolating field off-shell at $p^2 = 0$. For a more extensive discussion of the derivation of the Goldberger–Treiman relation based on the relation (9.25) and its connection to the chiral derivation see e.g. Pagels (1975).

9.3 Phenomenological chirally symmetric model of the strong interactions (σ-model)

As we have already mentioned, calculating spontaneous chiral symmetry breaking in QCD is a difficult problem not solved yet. It has, therefore, proved instructive in several aspects to study a phenomenological chirally symmetric field theory model of the strong interactions, the so-called σ-model. In this model the scalar fields are introduced as elementary fields and their interactions are arranged to produce the spontaneous breakdown of the symmetry.

The elementary fields in the σ-model are: the nucleon doublet, massless to start with so in fact we have two doublets N_R and N_L transforming like (1, 2) and (2, 1) under $SU_L(2) \times SU_R(2)$; the triplet pion field $\boldsymbol{\pi}(0^-)$ and σ' – a scalar 0^+ meson.

The postulated lagrangian is as follows

$$\begin{aligned}\mathscr{L} = \bar{N}i\not{\partial}N + g\bar{N}(\sigma' + i\boldsymbol{\tau}\cdot\boldsymbol{\pi}\gamma_5)N + \tfrac{1}{2}[(\partial_\mu\boldsymbol{\pi})^2 + (\partial_\mu\sigma')^2] \\ - \tfrac{1}{2}\mu^2(\sigma'^2 + \boldsymbol{\pi}^2) - \tfrac{1}{4}\lambda(\sigma'^2 + \boldsymbol{\pi}^2)^2\end{aligned} \tag{9.26}$$

where

$$\boldsymbol{\pi} = (\pi^1, \pi^2, \pi^3), \quad \boldsymbol{\tau} = (\tau^1, \tau^2, \tau^3)$$

This is the usual pseudoscalar pion–nucleon coupling. The fourth field σ' has been introduced to get chirally symmetric theory. We rewrite

$$\left.\begin{aligned}\bar{N}\mathrm{i}\not{\partial}N &= \bar{N}_{\mathrm{R}}\mathrm{i}\not{\partial}N_{\mathrm{R}} + \bar{N}_{\mathrm{L}}\mathrm{i}\not{\partial}N_{\mathrm{L}}\\ \bar{N}(\sigma' + \mathrm{i}\boldsymbol{\pi}\cdot\boldsymbol{\tau}\gamma_5)N &= \bar{N}_{\mathrm{L}}(\sigma' + \mathrm{i}\boldsymbol{\pi}\cdot\boldsymbol{\tau})N_{\mathrm{R}} + \bar{N}_{\mathrm{R}}(\sigma' - \mathrm{i}\boldsymbol{\pi}\cdot\boldsymbol{\tau})N_{\mathrm{L}}\end{aligned}\right\} \tag{9.27}$$

to see that for chiral invariance of the lagrangian the $(\sigma' + \mathrm{i}\boldsymbol{\pi}\cdot\boldsymbol{\tau})$ must transform as the (2,2) representation of the $SU(2) \times SU(2)$ (remember that $2^* \equiv 2$). This corresponds to the following transformations (see Appendix B)

under vector $SU(2)$ i.e. isotopic spin group

$$\boldsymbol{\pi} \to \boldsymbol{\pi} + \delta\boldsymbol{\alpha} \times \boldsymbol{\pi}$$
$$\sigma' \to \sigma'$$

under axial vector rotations (9.28)

$$\boldsymbol{\pi} \to \boldsymbol{\pi} + \delta\boldsymbol{\alpha}\sigma'$$
$$\sigma' \to \sigma' - \delta\boldsymbol{\alpha}\cdot\boldsymbol{\pi}$$

($\delta\boldsymbol{\alpha}$ is an infinitesimal vector in the isospin space). Chiral symmetry implies conserved axial and vector currents. They read

$$\left.\begin{aligned}\mathscr{V}^a_\mu(x) &= \bar{N}(x)\gamma_\mu\tfrac{1}{2}\tau^a N(x) + [\boldsymbol{\pi}(x) \times \partial_\mu\boldsymbol{\pi}(x)]^a\\ \mathscr{A}^a_\mu(x) &= \bar{N}(x)\gamma_\mu\gamma_5\tfrac{1}{2}\tau^a N(x) + \sigma'(x)\partial_\mu\pi^a(x) - \pi^a(x)\partial_\mu\sigma'(x)\end{aligned}\right\} \tag{9.29}$$

Constructing then charges $Q^a, Q^a_5, Q_{\mathrm{L}}, Q_{\mathrm{R}}$ and using canonical commutation relations for fields one gets

$$\left.\begin{aligned}[Q^a_{\substack{\mathrm{L}\\ \mathrm{R}}}, \sigma'] &= \mp\tfrac{1}{2}\mathrm{i}\pi^a\\ [Q^a_{\substack{\mathrm{L}\\ \mathrm{R}}}, \pi^b] &= \tfrac{1}{2}\mathrm{i}\varepsilon^{abc}\pi^c \pm \tfrac{1}{2}\mathrm{i}\delta^{ab}\sigma'\end{aligned}\right\} \tag{9.30}$$

Note (Appendix B) that in general if a field multiplet Φ transforms as $\Phi_i \to \Phi'_i = (\delta_{ij} - \mathrm{i}\delta\alpha^a T^a_{ij})\Phi_j$ under linear transformations generated by generators Q^a then, since on the other hand the operator relation

$$\begin{aligned}\Phi_i \to \Phi'_i &= \exp(\mathrm{i}\alpha^a Q^a)\Phi_i\exp(-\mathrm{i}\alpha^a Q^a)\\ &\cong \Phi_i + \mathrm{i}\delta\alpha^a[Q^a, \Phi_i]\end{aligned} \tag{9.31}$$

holds, it must be that

$$[Q^a, \Phi_i(x)] = -T^a_{ij}\Phi_j(x) \tag{9.32}$$

(matrices T^a form a representation of generators Q^a).

It is now necessary to establish the correct ground state of the model. The vacuum is defined by the minimum of the expectation values of the hamiltonian. As we know e.g. from the effective potential formalism, in the tree approximation the vacuum expectation values of the fields are equal to the values of the classical fields in the minimum of the potential $V(\sigma', \boldsymbol{\pi})$. We can rewrite the potential as

$$V(\sigma', \boldsymbol{\pi}) = -\tfrac{1}{4}\lambda(\sigma'^2 + \boldsymbol{\pi}^2 + \mu^2/\lambda)^2 \tag{9.33}$$

to see that for $\mu^2/\lambda > 0$ the minimum occurs for $\sigma' = \boldsymbol{\pi} = 0$ whereas for $\mu^2/\lambda < 0$ it exists for

$$\sigma'^2 + \boldsymbol{\pi}^2 = |\mu^2/\lambda| \tag{9.34}$$

With this second choice the model accounts for the spontaneous breakdown of chiral symmetry. The physical vacuum is chosen (defined) from the degenerate ground states given by the condition (9.34). In particular we would like to preserve unbroken isospin symmetry. The physical vacuum must then satisfy (9.34) and also

$$Q^a|0\rangle = 0, \quad Q_5^a|0\rangle \neq 0 \tag{9.35}$$

These conditions are satisfied with the choice

$$\langle 0|\boldsymbol{\pi}|0\rangle = 0, \quad \langle 0|\sigma'|0\rangle = -(|\mu^2/\lambda|)^{1/2} \equiv v \tag{9.36}$$

Indeed the vacuum must be an isospin singlet (hence $\langle 0|\boldsymbol{\pi}|0\rangle = 0$) but not a chiral singlet (hence $\langle 0|\sigma'|0\rangle \neq 0$ where σ' belongs to the chiral doublet). Formally, one can see that $\langle 0|\sigma'|0\rangle \neq 0$ implies $Q_5^a|0\rangle \neq 0$ (and vice versa) by taking the vacuum expectation value of the commutator $[Q_5^a, \pi^b] = -\mathrm{i}\delta^{ab}\sigma'$. Another way is to consider $\langle 0|\exp(\mathrm{i}\alpha^a Q_5^a)\sigma' \exp(-\mathrm{i}\alpha^a Q_5^a)|0\rangle$ to conclude that generators of the transformations which do not leave $\langle 0|\sigma'|0\rangle$ invariant are spontaneously broken. It is worth stressing at this point that in our discussion of the σ-model we have fixed at the very begining the physical interpretation of the group generators and of the fields. In the basis so defined the choice of the physical vacuum among the degenerate ground states given by the condition (9.34) is no longer arbitrary if we want to keep the isospin unbroken. However, we could also proceed in another way: define any of the *a priori* equivalent degenerate ground states as the physical vacuum, find the broken and unbroken generators of the theory, give to them and to the fields the desired physical interpretation. The pattern of the spontaneous symmetry breaking from such a general viewpoint will be discussed in Section 9.5.

In view of (9.36) we must, to assure the orthogonality of the vacuum to the one particle state, define the physical σ field as

$$\sigma = \sigma' - v \tag{9.37}$$

The lagrangian (9.26) can be now rewritten in terms of the field σ

$$\begin{aligned}\mathscr{L} = \bar{N}\mathrm{i}\not{\partial}N + g\bar{N}(\sigma + \mathrm{i}\boldsymbol{\tau}\cdot\boldsymbol{\pi}\gamma_5)N + gv\bar{N}N + \tfrac{1}{2}[(\partial_\mu\boldsymbol{\pi})^2 + (\partial_\mu\sigma)^2] \\ - |\mu^2|\sigma^2 - \tfrac{1}{4}\lambda(\sigma^2 + \boldsymbol{\pi}^2)^2 - \lambda v(\sigma^3 + \sigma\boldsymbol{\pi}^2) + \text{const.}\end{aligned} \tag{9.38}$$

It describes nucleons of mass $m_{\mathrm{N}} = g(|\mu^2/\lambda|)^{1/2} = -gv$, a σ meson of mass $m_\sigma = 2(-\mu^2/2)^{1/2} = (2v^2\lambda)^{1/2}$ and the isospin triplet of massless pions. In addition we observe that in the tree approximation, from (9.29)

$$\begin{aligned}\langle 0|\mathscr{A}_\mu^a(x)|\pi^b(q)\rangle &= \langle 0|\sigma'|0\rangle\langle 0|\partial_\mu\pi^a(x)|\pi^b(q)\rangle \\ &= -\langle 0|\sigma'|0\rangle \mathrm{i}q_\mu\delta^{ab}\exp(-\mathrm{i}qx)\end{aligned}$$

In the tree approximation the other terms in the axial current (9.29) give a vanishing contribution to this matrix element. Comparing with (9.19) one gets

$$f_\pi = -\langle 0|\sigma'|0\rangle$$

The v or f_π is the only mass scale of the model.

Our last remark in this section is as follows: the model can be extended to account for an explicit breakdown of the chiral symmetry. To do it in an isotopically invariant manner one adds to $\mathscr{L}$ a term $-\varepsilon\sigma'$. Then minimizing the potential (9.33) plus the extra term one finds instead of (9.36) the following equation for the vacuum expectation value v:

$$-\varepsilon - \lambda v^3 - \mu^2 v = 0$$

With the shift (9.37) one now gets

$$m_\pi^2 = \mu^2 + \lambda v^2 = -\varepsilon/v = \varepsilon/f_\pi$$

The axial current is no longer conserved

$$\partial^\mu \mathscr{A}_\mu^a(x) = \varepsilon\pi^a(x) \qquad (9.39)$$

and finally we get

$$\partial^\mu \mathscr{A}_\mu^a(x) = m_\pi^2 f_\pi \pi^a(x) \qquad (9.40)$$

which is the operator PCAC relation (9.25) introduced earlier. Another effect of the $\varepsilon\sigma'$ term is that it defines the unique physical vacuum. This vacuum alignment problem will be discussed in more detail later on.

9.4 Goldstone bosons as eigenvectors of the mass matrix and poles of Green's functions in theories with elementary scalars

We have learned in the previous Section that spontaneous breakdown of the chiral symmetry in the σ-model gives massless Goldstone bosons in the physical spectrum. Here we study the problem of Goldstone bosons in theories with elementary scalars (QCD-like theories are discussed in Section 9.6) in a more general context.

Goldstone bosons as eigenvectors of the mass matrix

First we generalize the discussion of Section 9.3 and show in the tree approximation that in a theory with spontaneously broken global symmetry the mass matrix of the scalar sector has eigenvalues zero. The corresponding eigenvectors are Goldstone boson fields. Next we give the general proof of Goldstone's theorem (Goldstone, Salam & Weinberg 1962) which does not rely on perturbation theory. This proof can also be easily extended to theories without elementary scalars.

Consider a theory described by a lagrangian $\mathscr{L}$ which has some global symmetry G generated by charges Q^a. Among others, the theory contains elementary scalar fields Φ_i, $i = 1, \ldots, n$, which we take as real; any complex representation can always be turned into a real one by doubling the number of the basis vectors. The piece $\mathscr{L}_s$ of the lagrangian containing the scalar fields Φ_i is

$$\mathscr{L}_s = \tfrac{1}{2}\partial_\mu \Phi_i \partial^\mu \Phi_i - V(\Phi_i) + \cdots \tag{9.41}$$

where the unspecified terms, irrelevant for our discussion in the tree approximation, are possible couplings of the scalar fields to other fields of the theory. The $V(\Phi_i)$ is a real polynomial in fields Φ_i, of the fourth order at most, if the theory is to be renormalizable. We assume furthermore that the set Φ_i form a multiplet Φ transforming under some irreducible representation of the group G of the global symmetry of the lagrangian $\mathscr{L}$ (if the original fields form a reducible multiplet we change the basis to write it as a direct sum of irreducible multiplets)

$$\Phi_i \to \Phi_i' \approx \Phi_i - \mathrm{i}\Theta^a T^a_{ij}\Phi_i \tag{9.42}$$

Matrices T^a form the representation of charges Q^a in the n-dimensional space of scalar fields Φ_i. Because $\mathrm{i}T^a$ must be real, T^a is purely imaginary and being hermitean it is antisymmetric. The transformation rule (9.42) is equivalent to the following commutation relation between charges Q^a and fields Φ_i (see (9.32))

$$[Q^a(t), \Phi_i(\mathbf{x}, t)] = -\,T^a_{ij}\Phi_j(\mathbf{x}, t) \tag{9.43}$$

We are interested in theories where the global symmetry G is spontaneously broken by the vacuum which is such that some of the fields Φ_i have non-vanishing vacuum expectation values

$$\langle 0|\Phi_i|0\rangle = v_i \neq 0 \tag{9.44}$$

In the tree or classical approximation, when the energy density of the vacuum state is just given by the minimum of the potential $V(\Phi_i)$, one then concludes that for the spontaneous breakdown of the symmetry G the condition for the minimum of $V(\Phi_i)$

$$0 = \delta V(\Phi_i) = \frac{\partial V}{\partial \Phi_i}\delta\Phi_i = -\mathrm{i}\frac{\partial V}{\partial \Phi_i}\Theta^a T^a_{ij}\Phi_j \tag{9.45}$$

must have solutions with $\Phi_i = v_i \neq 0$ for some of the Φ_is. Notice that if the vector $v = (v_1 \ldots v_n)^{\mathrm{T}}$ is a solution of (9.45) then it follows from the G invariance of the potential V that the vector

$$v' = \exp(-\mathrm{i}\Theta^a T^a)v \tag{9.46}$$

is another solution to this equation. Thus, for a given potential V which gives the spontaneous breakdown of the symmetry G the vacuum is infinitely degenerate and any choice of it is physically equivalent (choice of a solution v is a choice of the vacuum). Of course different potentials may in principle give physically different, non-degenerate, vacua. Possible patterns of spontaneous symmetry breaking are discussed in more detail in the next Section.

Choosing some solution v of (9.45) as the physical vacuum of our theory it is convenient to define the physical fields Φ'

$$\Phi'_i = \Phi_i - v_i \tag{9.47}$$

and to rewrite in terms of them the potential V

$$V(\Phi_i) = \text{const} + \tfrac{1}{2}M^2_{ik}(\Phi - v)_i(\Phi - v)_k + \cdots \tag{9.48}$$

where

$$M^2_{ik} = \left(\frac{\partial^2 V}{\partial\Phi_i\partial\Phi_k}\right)_{\Phi = v} \tag{9.49}$$

is the mass matrix for the physical scalar fields. The term linear in fields does not appear in (9.48) because of (9.45). Differentiating (9.45) one gets the following result

$$M^2_{ik}\mathrm{i}T^a_{ij}v_j = 0 \tag{9.50}$$

which suggests that the mass matrix has eigenvectors $T^a v$ corresponding to zero eigenvalues. To discuss this point more precisely we divide the generators of G into two groups. The first one consists of those which remain unbroken with our choice of the vacuum. Denoting their matrix representation by $Y^i, i = 1,\ldots,n$, one has $Y^i v = 0$; the vacuum is invariant with respect to Y^i and (9.50) does not give any new information. To the second group belong generators which are spontaneously broken. Denoting their matrix representation by $X^i, i = n+1,\ldots,N$ (note the values of the index i) we have $X^i v \neq 0$ and (9.50) implies the existence, for each broken generator of G, of the eigenvectors $X^i v$ corresponding to zero eigenvalues of the mass matrix. The massless boson fields (normal coordinates) are then the following

$$\Pi^l = \mathrm{i}(X^l v)_i\Phi_i \tag{9.51}$$

We recall at this point that our discussion is for spontaneously broken global symmetries. For spontaneously broken gauge symmetries the Goldstone boson fields can be gauged away and do not appear in the physical spectrum. These degrees of freedom are used for longitudinal components of the massive gauge boson. This so-called Higgs mechanism is discussed in Section 11.1.

We conclude these considerations with several useful remarks. Obviously, if some of the fields have non-vanishing vacuum expectation values, some generators of G must be broken. The generators which remain unbroken must form the algebra of some subgroup H of group G (it may, of course, be that $H \equiv 1$). Indeed, if it is not the case then

$$[Y^i, Y^j] = \mathrm{i}c_{ijk}Y^k + \mathrm{i}c_{ijl}X^l, \quad i,j,k \leqslant n, \quad l > n \tag{9.52}$$

with some of the c_{ijl} different from zero. Acting with both sides of (9.52) on the vector v we get a contradiction. Therefore

$$c_{ijl} = 0, \quad i,j \leqslant n, \quad l > n \tag{9.53}$$

and the unbroken generators form the algebra of a group. Furthermore it is easy to see that broken generators transform under some representation of the subgroup H

$$[Y^i, X^j] = \mathrm{i}c_{ijk}X^k, \quad i \leqslant n, \quad j,k > n \tag{9.54}$$

This follows from the property (9.53) and from the antisymmetry of the structure constants in indices i, j, k. Thus, the r.h.s. of (9.54) cannot contain a term $c_{ijl}Y^l$ where $i, l \leqslant n, j > n$. Equations (9.54) and (9.51) taken together tell us that Goldstone bosons transform under some representation of the unbroken subgroup H. Finally, in many physical problems one can define a parity operation P which leaves the Lie algebra of the group G invariant and such that

$$PY^iP^{-1} = Y^i, \quad PX^iP^{-1} = -X^i \tag{9.55}$$

This is the case e.g. for chiral symmetry; see Section 9.1. Then there is one more useful and obvious relation

$$[X^i, X^j] = \mathrm{i}c_{ijk}Y^k, \quad i,j > n, \quad k \leqslant n \tag{9.56}$$

General proof of Goldstone's theorem

We turn now to the general proof of Goldstone's theorem (Goldstone, Salam & Weinberg 1962) in the class of theories specified in the begining of this Section. One can show that spontaneous breakdown of some global symmetry G of the lagrangian implies poles at $p^2 = 0$ in certain Green's functions and consequently implies the existence of massless bosons in the physical spectrum.

Let us consider the Green's function

$$G^a_{\mu,k}(x - y) = \langle 0|Tj^a_\mu(x)\Phi_k(y)|0\rangle \tag{9.57}$$

where $j^a_\mu(x)$ is the current corresponding to a generator Q^a of the symmetry G and $\Phi_k(y)$ belongs to an irreducible multiplet of real scalar fields. The Green's function $G^a_{\mu,k}(x - y)$ satisfies a Ward identity obtained by differentiating (9.57) with proper account of the Θ-functions involved in the definition of the T-product

$$\partial^\mu_{(x)}G^a_{\mu,k}(x - y) = \delta(x^0 - y^0)\langle 0|[j^a_0(x), \Phi_k(y)]|0\rangle \tag{9.58}$$

This is an example of the so-called non-anomalous Ward identity (see Section 10.1); for the discussion of anomalies in theories with fermions see Chapter 12. Using the relation

$$[j^a_0(\mathbf{x}, t), \Phi_k(\mathbf{y}, t)] = -T^a_{kj}\Phi_j(\mathbf{y}, t)\delta(\mathbf{x} - \mathbf{y}) \tag{9.59}$$

which follows from the assumed transformation properties, see (9.43), of fields Φ_i under the group G, one gets the following Ward identity (translational invariance is assumed)

$$\partial^\mu_{(x)}G^a_{\mu,k}(x - y) = -\delta(x - y)T^a_{kj}\langle 0|\Phi_j(0)|0\rangle \tag{9.60}$$

A few words are worthwhile here about renormalization properties of (9.60) (see also the next Chapter). The Ward identity is, in principle, derived for bare currents and fields for which transformation properties (9.59), are assumed. However, the conserved currents are not subject to renormalization and the field renormalization constants $\Phi_i^{\mathrm{B}} = Z_i^{1/2}\Phi_i^{\mathrm{R}} = Z^{1/2}\Phi_i^{\mathrm{R}}$ do not carry the group index because the lagrangian is G invariant. Thus, the renormalized fields transform under the group G in the same way as the bare fields and the same Ward identity (9.60) holds both for bare and renormalized Green's functions. Let us take it as a relation for renormalized quantities. Introducing the Fourier transform $\tilde{G}^a_{\mu,k}(p)$

$$G^a_{\mu,k}(x-y) = \int \frac{\mathrm{d}^4 p}{(2\pi)^4} \exp[-\mathrm{i}p(x-y)]\tilde{G}^a_{\mu,k}(p) \tag{9.61}$$

one gets from the relation (9.60) the following

$$\mathrm{i}p^\mu \tilde{G}^a_{\mu,k}(p) = T^a_{kj}\langle 0|\Phi_j(0)|0\rangle \tag{9.62}$$

This last equation implies that if some of the vacuum expectation values $\langle 0|\Phi_j|0\rangle$ do not vanish then there are poles at $p^2 = 0$ in the Green's functions $\tilde{G}^a_{\mu,k}(p)$, related to these $\langle 0|\Phi_j|0\rangle$ by (9.62). Indeed, from Lorentz invariance the general structure of $\tilde{G}^a_{\mu,k}(p)$ can be written as

$$\tilde{G}^a_{\mu,k}(p) = p_\mu F^a_k(p^2) \tag{9.63}$$

and therefore

$$F^a_k(p^2) = -\mathrm{i}T^a_{kj}\langle 0|\Phi_j(0)|0\rangle(1/p^2) \tag{9.64}$$

Our last step is to show that poles at $p^2 = 0$ imply the existence of massless bosons in the physical spectrum. To this end we consider the matrix element

$$\langle 0|\, j^a_\mu(x)|\pi^k(p)\rangle = \mathrm{i}f^a_k p_\mu \exp(-\mathrm{i}px) \tag{9.65}$$

where the vector $|\pi^k(p)\rangle$ describes a particle of mass m_k which is a quantum of the field Φ_k. Using the reduction formula (see Section 2.7) one can relate the matrix element $\langle 0|\, j^a_\mu(x)|\pi^k(p)\rangle$ to the Green's function $\tilde{G}^a_{\mu,k}(p)$

$$\langle 0|\, j^a_\mu(x)|\pi^k(p)\rangle = \int \mathrm{d}^4 y\, \mathrm{d}^4 z f_p(z) G^a_{\mu,k'}(x-y)\mathrm{i}G^{-1}_{k'k}(y-z) \tag{9.66}$$

where

$$f_p(z) = \exp(-\mathrm{i}pz)$$

and

$$\left.\begin{aligned} G_{k'k}(y-z) &= \int \frac{\mathrm{d}^4 q}{(2\pi)^4} \frac{-\delta_{k'k}}{q^2 - m_k^2 + \mathrm{i}\varepsilon} \exp[-\mathrm{i}q(y-z)] \\ &\int \mathrm{d}^4 y G^{-1}(x-y)G(y-z) = \delta(x-z) \end{aligned}\right\} \tag{9.67}$$

Writing $\exp(-\mathrm{i}pz)$ as $\exp(-\mathrm{i}px)\exp[\mathrm{i}p(x-y)]\exp[\mathrm{i}p(y-z)]$ we get from (9.66) the following result

$$\langle 0|\, j^a_\mu(x)|\pi^k(p)\rangle = \lim_{p^2\to m_k^2} \exp(-ipx)\tilde{G}^a_{\mu,k'}(p)\mathrm{i}\widetilde{G^{-1}_{k'k}}(p)$$
$$= \lim_{p^2\to m_k^2} -\mathrm{i}\exp(-ipx)\tilde{G}^a_{\mu,k}(p)(p^2-m_k^2) \qquad (9.68)$$

Comparison with (9.65) gives us our final equation

$$\lim_{p^2\to m_k^2} \tilde{G}^a_{\mu,k}(p)(p^2-m_k^2) = -f^a_k p_\mu \qquad (9.69)$$

If $\tilde{G}^a_{\mu,k}(p)$ is one of those Green's functions which (according to (9.64)) have for spontaneously broken symmetry a pole at $p^2=0$ then (9.69) implies $m_k^2=0$ and $f^a_k \neq 0$

$$f^a_k = \mathrm{i}T^a_{kj}\langle 0|\Phi_j(0)|0\rangle \qquad (9.70)$$

Thus there must be massless bosons $|\Pi^a(p)\rangle = \mathrm{i}T^a_{kj}\langle 0|\Phi_j(0)|0\rangle|\pi^k(p)\rangle$ in the physical spectrum corresponding to each broken generator which does not leave the vacuum invariant $T^a_{kj}\langle 0|\Phi_j(0)|0\rangle \neq 0$. If the broken generators and the corresponding currents form a single real irreducible representation of the unbroken subgroup H which leaves the vacuum invariant then from Schur's *Lemma*:

$$\langle 0|\, j^a_\mu|\pi^k(p)\rangle \sim \delta^{ak} \qquad (9.71)$$

and

$$f^a_k = f^a\delta_{ak}$$

For instance, for the σ-model with $\langle 0|\sigma'|0\rangle \neq 0$ we get from (9.70), with matrices T^a_{kj} from Appendix B,

$$f_\pi = -\langle 0|\sigma'|0\rangle \qquad (9.72)$$

as the generalization of the result obtained in the tree approximation.

Goldstone's theorem does not hold for spontaneously broken *gauge* symmetries (see Section 11.1). The breakdown of the presented proof is most easily seen if we work in a 'physical' gauge in which the gauge field has only physical degrees of freedom. Gauge fixing then requires a specification of some four-vector n_μ and the general form of the matrix element involving the gauge source current is no longer given by (9.63): a term proportional to n_μ can be present. For a discussion in a covariant gauge see e.g. Bernstein (1974).

9.5 Patterns of spontaneous symmetry breaking

For a theory with some global symmetry G and a given multiplet content of fields, patterns of spontaneous symmetry breaking are determined by the properties of the vacuum defined by the non-vanishing of the vacuum expectation values of some fields. In theories with elementary scalars the non-vanishing vacuum expectation values may already appear in the tree approximation as a solution minimizing the assumed potential $V(\Phi)$ or may be generated by radiative corrections (Section 11.2). In theories without elementary scalars the symmetry is spontaneously broken by some condensates (e.g. $\langle 0|\bar{\Psi}\Psi|0\rangle$) which, in the absence of reliable methods for calculating non-perturbative effects, are usually introduced by assumption. A choice of the vacuum specifies the pattern of the spontaneous symmetry breaking which may, in general, be such that some subgroup H of G remains unbroken. As we have already mentioned in the previous Section for any choice of the vacuum specified by the non-vanishing vacuum expectation values there exists a set of physically equivalent vacua, which also minimize the same potential $V(\Phi)$, obtained from the original one by the set of unitary transformations belonging to G. Different choices of the physical vacuum among the degenerate vacua give the same pattern of symmetry breaking; they simply correspond to different orientations of the unbroken subgroup H in the group G which can be transformed one into the other by a redefinition of the basis.

The question we would like to discuss here is what are the possible patterns of spontaneous breaking of a given symmetry G. Although different patterns, if they exist, must, in general, correspond to different potentials $V(\Phi)$, no explicit reference to the potential is necessary for our discussion.

Our first example is a group of orthogonal transformations, $O(3)$ for definiteness, and the scalar fields are assumed to form a real multiplet $\Phi = (\Phi_1, \Phi_2, \Phi_3)^T$ transforming under the vector representation. The matrix representation of generators Q_i in the space of Φs is

$$T_x = \begin{pmatrix} 0 & 0 & 0 \\ 0 & 0 & -\mathrm{i} \\ 0 & \mathrm{i} & 0 \end{pmatrix} \quad T_y = \begin{pmatrix} 0 & 0 & \mathrm{i} \\ 0 & 0 & 0 \\ -\mathrm{i} & 0 & 0 \end{pmatrix} \quad T_z = \begin{pmatrix} 0 & -\mathrm{i} & 0 \\ \mathrm{i} & 0 & 0 \\ 0 & 0 & 0 \end{pmatrix} \tag{9.73}$$

A possible choice for the vacuum expectation value of the multiplet Φ is, for instance

$$\langle 0|\Phi|0\rangle = \begin{pmatrix} 0 \\ 0 \\ v \end{pmatrix} \equiv \Phi_0$$

The vacuum defined this way is invariant under the transformations generated by Q_z: taking the transformed vacuum $\exp(-\mathrm{i}\alpha Q_z)|0\rangle$ and using (9.32) we get:

$$\langle 0|\exp(\mathrm{i}\alpha Q_z)\Phi\exp(-\mathrm{i}\alpha Q_z)|0\rangle \cong \Phi_0 + \mathrm{i}\alpha\langle 0|[Q_z, \Phi]|0\rangle = \Phi_0 - \mathrm{i}\alpha T_z\Phi_0 = \Phi_0 \tag{9.74}$$

However the vacuum defined by (9.74) breaks spontaneously the other two generators:

$$\mathrm{i}T_x\Phi_0 = \begin{pmatrix} 0 \\ v \\ 0 \end{pmatrix}, \quad \mathrm{i}T_y\Phi_0 = \begin{pmatrix} -v \\ 0 \\ 0 \end{pmatrix} \tag{9.75}$$

Thus the group $O(3)$ is broken to the group $O(2)$ of rotations around the z axis. There are two Goldstone bosons corresponding to two linearly independent vectors $\mathrm{i}T_x\Phi_0$ and $\mathrm{i}T_y\Phi_0$. Any other choice of the vacuum expectation value

$$\langle 0|\Phi|0\rangle = \begin{pmatrix} v_1 \\ v_2 \\ v_3 \end{pmatrix} \tag{9.76}$$

which can be obtained from the original one by unitary transformations belonging to G gives the same pattern of symmetry breaking but this time with the combination $v_1 T_x + v_2 T_y + v_3 T_z$ unbroken. Both theories are physically equivalent and transform one into the other by a redefinition of the basis. This would not be the case if there was, in space, a physically distinguished direction, that is if the $O(3)$ symmetry was also broken explicitly. This point will be discussed in detail in the next Chapter.

Consider now unitary groups with scalars in a real representation. Any complex field Φ can be written in a real basis by defining two real fields

$$\left.\begin{aligned} \varphi_i &= (1/\sqrt{2})(\Phi_i + \Phi_i^*) \\ \chi_i &= (1/\sqrt{2}\mathrm{i})(\Phi_i - \Phi_i^*) \end{aligned}\right\} \tag{9.77}$$

For the group $U(1)$ of the unitary transformations

$$\Phi \to \Phi' = \exp(-\mathrm{i}\Theta)\Phi$$

the real representation is

$$\Phi^{\mathrm{r}} \to \Phi'^{\mathrm{r}} = \exp(-\mathrm{i}\Theta T)\Phi^{\mathrm{r}}$$

where

$$\Phi^{\mathrm{r}} = \begin{pmatrix} \varphi \\ \chi \end{pmatrix} \tag{9.78}$$

and T is purely imaginary

$$T = \begin{pmatrix} 0 & \mathrm{i} \\ -\mathrm{i} & 0 \end{pmatrix}$$

because Φ^{r} must be real. Obviously, in this case, there is just one pattern of spontaneous symmetry breaking and any non-zero choice of the vacuum expectation value $\langle 0|\Phi^{\mathrm{r}}|0\rangle = (v_1 v_2)^{\mathrm{T}}$ is equally good.

Our next example is the group $SU(2)$ with a doublet of complex fields Φ_1 and Φ_2. In the real representation

$$\Phi^{\mathrm{r}} = \begin{pmatrix} \varphi_1 \\ \chi_1 \\ \varphi_2 \\ \chi_2 \end{pmatrix}$$

the generators of the $SU(2)$ transformations are

$$T^1 = \begin{pmatrix} & & & \mathrm{i} \\ & & -\mathrm{i} & \\ & \mathrm{i} & & \\ -\mathrm{i} & & & \end{pmatrix} \qquad T^2 = \left(\begin{array}{cc|cc} \multicolumn{2}{c|}{0} & -\mathrm{i} & \\ & & & -\mathrm{i} \\ \hline \mathrm{i} & & \multicolumn{2}{c}{0} \\ & \mathrm{i} & & \end{array}\right)$$

$$T^3 = \left(\begin{array}{cc|cc} & \mathrm{i} & \multicolumn{2}{c}{0} \\ -\mathrm{i} & & & \\ \hline \multicolumn{2}{c|}{0} & & -\mathrm{i} \\ & & \mathrm{i} & \end{array}\right) \tag{9.79}$$

Let us take, for instance, $\langle 0|\Phi^{\mathrm{r}}|0\rangle = (0, 0, v, 0)^{\mathrm{T}}$. Following the discussion of our first example we see that the group is broken completely because the three vectors $\mathrm{i}T^a\langle 0|\Phi^{\mathrm{r}}|0\rangle \neq 0$ are linearly independent. All other possible non-vanishing vacuum expectation values can be obtained from the one above by unitary $SU(2)$ transformations and, thus, give the same pattern of symmetry breaking.

It is easy to extend the analysis to the $SU(2) \times U(1)$ group with one complex doublet of scalar fields (Glashow–Salam–Weinberg model). In this case we have one more generator

$$T = \left(\begin{array}{cc|cc} & \mathrm{i} & \multicolumn{2}{c}{0} \\ -\mathrm{i} & & & \\ \hline \multicolumn{2}{c|}{0} & & \mathrm{i} \\ & & -\mathrm{i} & \end{array}\right)$$

but the vector $\mathrm{i}T\langle 0|\Phi^{\mathrm{r}}|0\rangle$ is not linearly independent from the other three vectors $\mathrm{i}T^a\langle 0|\Phi^{\mathrm{r}}|0\rangle$. For instance, for $\langle 0|\Phi^{\mathrm{r}}|0\rangle = (0, 0, v, 0)^{\mathrm{T}}$ we have $T\langle 0|\Phi^{\mathrm{r}}|0\rangle = -T^3\langle 0|\Phi^{\mathrm{r}}|0\rangle$ and the combination $T + T^3$ remains unbroken. Other choices of the vacuum give only different unbroken combinations of the generators T and T^a but a redefinition of the basis transforms one case into another. Thus, again only one pattern of symmetry breaking is possible.

The spontaneous breaking of chiral symmetry in the σ-model can be discussed in a similar manner. Using the real basis $\Phi^{\mathrm{r}} = (\pi_1\pi_2\pi_3\sigma')^{\mathrm{T}}$ and the matrices T^a and ${}_5T^a$ collected in Appendix B we see that e.g. for $\langle 0|\Phi^{\mathrm{r}}|0\rangle = (0, 0, 0, v)^{\mathrm{T}}$ the generators ${}_5T^a$ are broken and the T^as remain unbroken: $SU_{\mathrm{L}}(2) \times SU_{\mathrm{R}}(2) \to SU_{\mathrm{L+R}}(2)$. This corresponds to the standard choice $\langle 0|\sigma'|0\rangle \neq 0$ and to the πs forming a triplet under the $SU(2)$. However any other choice of vacuum among the $SU(2) \times SU(2)$ degenerate set gives the same pattern of symmetry breaking. Take e.g. $\langle 0|\Phi^{\mathrm{r}}|0\rangle = (v, 0, 0, 0)^{\mathrm{T}}$. Then the generators ${}_5T^2$, ${}_5T^3$ and T^1 remain unbroken and we see from the commutation relations (9.9) that they also form an $SU(2)$ algebra. We can call this group the isospin group and interpret as physical pions those combinations of the fields π^i and σ which transform as a triplet under this new $SU(2)$, thus obtaining the same theory as before.

So far we have been discussing examples with only one possible pattern of spontaneous symmetry breaking. This is, of course, not always the case. As our last example we consider the group $SU(3)$ with a real multiplet Φ of scalar fields transforming under the adjoint representation of the $SU(3)$

$$[Q^a, \Phi_b] = -T^a_{bc}\Phi_c$$

where

$$T^a_{bc} \equiv -\mathrm{i}c_{abc} \tag{9.80}$$

It is convenient to write the scalar multiplet in the matrix form

$$\hat{\Phi} = \tfrac{1}{2}\Phi^b\lambda^b, \qquad b = 1,\ldots,8 \tag{9.81}$$

and λ^b are the Gell-Mann matrices. The matrix $\hat{\Phi}$ is hermitean and $\hat{\Phi}_{ii} = 0$ so it has eight independent elements. The transformation rule for the matrix $\hat{\Phi}$ is

$$\hat{\Phi} \to \hat{\Phi}' = \exp(-\mathrm{i}\Theta^a\tfrac{1}{2}\lambda^a)\hat{\Phi}\exp(\mathrm{i}\Theta^a\tfrac{1}{2}\lambda^a) \cong \hat{\Phi} - \mathrm{i}\Theta^a[\tfrac{1}{2}\lambda^a, \hat{\Phi}] \tag{9.82}$$

and because of (9.81) it is equivalent to

$$\Phi \to \Phi' = \exp(-\mathrm{i}\Theta^a T^a)\Phi$$

with T^a given by (9.80), for the multiplet Φ. Because the matrix $\hat{\Phi}$ is hermitean it can be diagonalized by unitary transformations and written in terms of commuting generators of $SU(3)$. Taking them as λ^3 and λ^8 we can write the most general form of the vacuum expectation values of the scalar fields as follows

$$\langle 0|\hat{\Phi}|0\rangle = \tfrac{1}{2}v_3\lambda^3 + \tfrac{1}{2}v_8\lambda^8 \tag{9.83}$$

First of all we notice that in our example the maximal possible breaking is $SU(3) \to U_3(1) \times U_8(1)$ because it is clear from (9.83) and (9.82) that λ^3 and λ^8 remain unbroken for any choice of v_3 and v_8. In this case we have six Goldstone bosons. However, another pattern of symmetry breaking is also possible, namely $SU(3) \to SU(2) \times U(1)$. If we assume $v_3 = 0$, $v_8 \neq 0$ the generators $\lambda^1, \lambda^2, \lambda^3$ ($SU(2)$) and λ^8($U(1)$) remain unbroken; if $v_3 = +\sqrt{3}v_8$ the λ^6, λ^7, $\tfrac{1}{2}(\lambda^3 - \sqrt{3}\lambda^8)$ ($SU(2)$) and $\tfrac{1}{2}(\sqrt{3}\lambda^3 + \lambda^8)$ ($U(1)$) are unbroken; if $v_3 = -\sqrt{3}v_8$ the λ^4, λ^5, $\tfrac{1}{2}(\lambda^3 + \sqrt{3}\lambda^8)$ ($SU(2)$) and $\tfrac{1}{2}(\sqrt{3}\lambda^3 - \lambda^8)$ ($U(1)$) are unbroken. It is also worth noticing that a vacuum expectation value giving $SU(3) \to SU(2) \times U(1)$ cannot be transformed by unitary transformations belonging to $SU(3)$ to one which gives $SU(3) \to U(1) \times U(1)$ and they must correspond to different potentials $V(\Phi)$. However, for each pattern there is an infinitely degenerate set of possible vacua obtained from (9.83) (with specified v_3 and v_8) by unitary $SU(3)$ transformations.

For a given pattern of spontaneous symmetry breaking one usually works from the begining in a convenient basis with a specified physical interpretation of the group generators and of the fields, as we do for the σ-model in Section 9.3 and for QCD in Section 9.6.

9.6 Goldstone bosons in QCD

The purpose of this Section is to extend the proof of Goldstone's theorem to theories like QCD, with no elementary scalars. Chiral symmetry of the massless QCD lagrangian has been discussed in Section 9.1. Considerable phenomenological evidence has then been summarized that the group $SU(2) \times SU(2)$ is an approximate symmetry of the strong interactions which is spontaneously broken by the vacuum. One should be aware, however, of the fact, that our understanding of the underlying mechanism of spontaneous symmetry breaking in theories without elementary scalars is not yet satisfactory. It has at least been realized that only asymptotically free theories allow spontaneous chiral symmetry breaking at reasonable momentum scales (Lane 1974; Gross & Neveu 1974). It is not our aim here to review partial results suggesting that this indeed happens; we shall assume the spontaneous breakdown of chiral symmetry and then show that it implies the existence of massless bosons (composite states) in the physical spectrum. It is, nevertheless, useful to give at least an intuitive argument in favour of the spontaneous breaking of chiral symmetries in asymptotically free theories. It relies on the fact that the gauge coupling in QCD becomes strong at large distances. Assuming that it becomes arbitrarily strong we expect that the ground state of the theory has an indefinite number of massless fermion pairs which can be created and annihilated by the strong coupling. We still expect the ground state to be invariant under Lorentz transformations so these pairs must have zero total momentum and angular momentum. Thus we find that the vacuum $|0\rangle$ has the property that operators which destroy or create such a fermion pair have non-zero vacuum expectation values $\langle 0|\bar{\Psi}_{\mathrm{R}}\Psi_{\mathrm{L}}|0\rangle \neq 0$, $\langle 0|\bar{\Psi}_{\mathrm{L}}\Psi_{\mathrm{R}}|0\rangle \neq 0$; we omit here indices. In the following we assume that the chiral symmetry of QCD is indeed spontaneously broken by non-vanishing vacuum expectation values of some fermion condensates.

We consider QCD with two flavours and the chiral symmetry $SU_{\mathrm{L}}(2) \times SU_{\mathrm{R}}(2)$. Let us introduce scalar operators Φ and $\Phi^\dagger$

$$\Phi_{ij} = \bar{\Psi}_{\mathrm{L}j}\Psi_{\mathrm{R}i}, \quad (\Phi^\dagger)_{ij} = \bar{\Psi}_{\mathrm{R}j}\Psi_{\mathrm{L}i} \tag{9.84}$$

($i, j = 1, 2$ are flavour indices). From the transformation properties of the fields $\Psi_{\mathrm{L}i}$ and $\Psi_{\mathrm{R}j}$ under the chiral group

$$\Psi'_{\mathrm{R}i} = (U_{\mathrm{R}})_{ik}\Psi_{\mathrm{R}k}$$
$$\Psi'^\dagger_{\mathrm{L}j} = \Psi^\dagger_{\mathrm{L}k}(U^\dagger_{\mathrm{L}})_{kj}$$

it follows that

$$\left.\begin{aligned} \Phi' &= U_{\mathrm{R}}\Phi U_{\mathrm{L}}^{\dagger} \\ (\Phi^{\dagger})' &= U_{\mathrm{L}}\Phi^{\dagger}U_{\mathrm{R}}^{\dagger} \end{aligned}\right\} \tag{9.85}$$

We assume

$$\langle 0|\Phi_{ij}|0\rangle \neq 0, \qquad \langle 0|\Phi_{ij}^{\dagger}|0\rangle \neq 0$$

and in addition

$$\langle 0|\Phi_{ij}|0\rangle = \langle 0|\Phi_{ij}^{\dagger}|0\rangle$$

The latter equality assures that parity is not spontaneously broken since then

$$\langle 0|\bar{\Psi}_i\gamma_5\Psi_j|0\rangle = \langle 0|(\Phi - \Phi^{\dagger})_{ji}|0\rangle = 0 \tag{9.86}$$

(we always work in a specified basis of chiral fields).

The vacuum expectation value $\langle 0|\Phi|0\rangle$ being a hermitean 2×2 matrix can be most generally written as (see the discussion preceeding (9.83))

$$\langle 0|\Phi|0\rangle = v\mathbb{1} + v_3\tau^3 \tag{9.87}$$

where τ^3 is the diagonal Pauli matrix. We want the chiral $SU_{\mathrm{L}}(2) \times SU_{\mathrm{R}}(2)$ symmetry to be broken to the $SU_{\mathrm{L+R}}(2)$ symmetry which is the isospin symmetry in our basis. Therefore we assume $v \neq 0$, $v_3 = 0$ in (9.87): it is clear from (9.85) that the vacuum expectation value $v\mathbb{1}$ is invariant under the simultaneous $U_{\mathrm{L}} = U_{\mathrm{R}}$ transformation. Notice the difference with the σ-model with one complex doublet, or four real fields, where the $SU(2) \times SU(2) \rightarrow SU(2)$ is the only possible pattern of symmetry breaking; now the $\langle 0|\Phi|0\rangle$ is a 2×2 complex hermitean matrix. We can also write

$$\langle 0|\Phi_{ij}|0\rangle = \tfrac{1}{2}\langle 0|\bar{\Psi}\Psi|0\rangle\delta_{ij} \tag{9.87a}$$

where

$$\bar{\Psi}\Psi = \bar{\Psi}_1\Psi_1 + \bar{\Psi}_2\Psi_2$$

We are ready to prove Goldstone's theorem. The proof is similar to that in Section 9.4 and it is based on Ward identities for certain appropriate Green's functions. First we collect the useful equal-time commutation relations following from (9.59) and (9.85)

$$\left.\begin{aligned} [j_{0\mathrm{L}}^a(\mathbf{x}, t), \Phi(\mathbf{y}, t)] &= \Phi(\mathbf{x}, t)T^a\delta(\mathbf{x} - \mathbf{y}) \\ [j_{0\mathrm{R}}^a(\mathbf{x}, t), \Phi(\mathbf{y}, t)] &= -\,T^a\Phi(\mathbf{x}, t)\delta(\mathbf{x} - \mathbf{y}) \\ [j_{0\mathrm{L}}^a(\mathbf{x}, t), \Phi^{\dagger}(\mathbf{y}, t)] &= -\,T^a\Phi^{\dagger}(\mathbf{x}, t)\delta(\mathbf{x} - \mathbf{y}) \\ [j_{0\mathrm{R}}^a(\mathbf{x}, t), \Phi^{\dagger}(\mathbf{y}, t)] &= \Phi^{\dagger}(\mathbf{x}, t)T^a\delta(\mathbf{x} - \mathbf{y}) \end{aligned}\right\} \tag{9.88}$$

where the $j_{0\mathrm{L}}^a(j_{0\mathrm{R}}^a)$ are the zeroth components of the currents of the $SU_{\mathrm{L}}(2)$ ($SU_{\mathrm{R}}(2)$) transformations and the matrices T^a represent the generators of the $SU_{\mathrm{L}}(2)$ (when to the right of the operator Φ) and of the $SU_{\mathrm{R}}(2)$ (when to the left of the

Φ) in the space of chiral doublets Ψ_L and Ψ_R, respectively. The above commutation relations are valid both for the bare and the renormalized quantities because chiral symmetry is an exact symmetry of the lagrangian (see also the next Chapter). From now on we always refer to renormalized quantities. For the proof of Goldstone's theorem we use a Ward identity for Green's functions involving composite operator $\Pi^a(x)$ defined as follows

$$\Pi^a(x) = \mathrm{Tr}\,[\Pi(x)\cdot T^a]$$

where

$$\Pi(x) = \mathrm{i}(\Phi - \Phi^\dagger), \qquad \Pi^\dagger = \Pi \tag{9.89}$$

and therefore

$$\Pi^a(x) = \mathrm{i}(\Phi_{ij} - \Phi^\dagger_{ij})T^a_{ji} = \mathrm{i}\bar{\Psi} T^a \gamma_5 \Psi$$

($T^a = \frac{1}{2}\tau^a$, τ^a are Pauli matrices). Using the commutation relations (9.88) we get the following result for the axial currents

$$[\mathscr{A}^a_0(\mathbf{x}, t), \Pi(\mathbf{y}, t)] = -\mathrm{i}\{T^a, \Phi(\mathbf{y}, t) + \Phi^\dagger(\mathbf{y}, t)\}\delta(\mathbf{x} - \mathbf{y}) \tag{9.90}$$

where the bracket on the r.h.s. means anticommutator. Using this last result we finally arrive at the desired Ward identity:

$$\partial^\mu_{(x)}\langle 0|T\mathscr{A}^a_\mu(x)\Pi^b(y)|0\rangle = -\mathrm{i}\delta(x-y)\langle 0|\mathrm{Tr}\,[\{T^a, \Phi(y) + \Phi^\dagger(y)\}T^b]|0\rangle \tag{9.91}$$

(it is unaltered by the chiral anomalies discussed in Chapter 12). Using $\mathrm{Tr}\,[T^aT^b] = \frac{1}{2}\delta^{ab}$ and (9.87) one gets

$$\partial^\mu_{(x)}\langle 0|T\mathscr{A}^a_\mu(x)\Pi^b(y)|0\rangle = -2v\mathrm{i}\delta(x-y)\delta^{ab} \tag{9.92}$$

This result is analogous to (9.60) and it implies a pole at $p^2 = 0$ in the Fourier transform $\tilde{G}^{ab}_{\mu,\Pi}(p)$ of the Green's function $G^{ab}_{\mu,\Pi}(x-y)$

$$G^{ab}_{\mu,\Pi}(x-y) = \langle 0|T\mathscr{A}^a_\mu(x)\Pi^b(y)|0\rangle = \int \frac{\mathrm{d}^4p}{(2\pi)^4} \exp[-\mathrm{i}p(x-y)]\tilde{G}^{ab}_{\mu,\Pi}(p) \tag{9.93}$$

Following the steps (9.62)–(9.64) we get

$$\tilde{G}^{ab}_{\mu,\Pi}(p) = p_\mu \frac{2v}{p^2}\delta^{ab} = p_\mu \frac{\langle\bar{\Psi}\Psi\rangle}{p^2}\delta^{ab} \tag{9.94}$$

The pole at $p^2 = 0$ in the Green's function $\tilde{G}^{ab}_{\mu,\Pi}(p)$ implies the existence in the physical spectrum of the massless Goldstone boson with the quantum numbers of the operators Π^a. The proof goes as in Section 9.4. The only modification is that some care must be taken about the interpretation of the operator $\Pi(x)$. Indeed, consider the Green's function

$$G^{ab}(x-y) = \langle 0|T\Pi^a(x)\Pi^b(y)|0\rangle \tag{9.95}$$

A single particle state with appropriate quantum numbers gives a pole at $p^2 = m^2$ in the Fourier transform of (9.95)

$$\int d^4x G^{ab}(x) \exp(ipx) \underset{p^2 \to m^2}{\approx} \frac{iZ^2}{p^2 - m^2} \delta^{ab} \tag{9.96}$$

where Z is a finite dimensionful constant (Π^a is a renormalized operator). It does not carry the group index because G is an exact symmetry of the lagrangian.

Introducing the physical, in the sense of Chapter 2, field $\pi^a(x)$:

$$\pi^a(z) = Z^{-1}\Pi^a(x) \tag{9.97}$$

we repeat the step (9.65)

$$\langle 0|\mathscr{A}^a_\mu(x)|\pi^b(p)\rangle = \mathrm{i} f_\pi \delta^{ab} p_\mu \exp(-\mathrm{i}px) \tag{9.98}$$

and the steps (9.66)–(9.68)

$$\langle 0|\mathscr{A}^a_\mu(x)|\pi^b(p)\rangle \underset{p^2 \to m^2}{\approx} -\mathrm{i}\exp(-\mathrm{i}px) Z^{-1} \tilde{G}^{ab}_{\mu,\Pi}(p)(p^2 - m^2) \tag{9.99}$$

to get

$$\lim_{p^2 \to m^2} \tilde{G}^{ab}_{\mu,\Pi}(p)(p^2 - m^2) = -Z f_\pi \delta^{ab} p_\mu \tag{9.100}$$

Since $\tilde{G}^{ab}_{\mu,\Pi}(p)$ has the pole (9.94) due to the spontaneous breakdown of chiral symmetry, relation (9.100) implies $m^2 = 0$ and

$$Z f_\pi = -\langle \bar{\Psi}\Psi \rangle \tag{9.101}$$

This completes the proof of Goldstone's theorem. In this case we have two low energy parameters: f_π and $\langle \bar{\Psi}\Psi \rangle$.

10

Spontaneous and explicit global symmetry breaking

So far we have been discussing theories described by lagrangians having some exact global symmetry G spontaneously broken by the vacuum. Now we would like to introduce the possibility that the symmetry G of the lagrangian is also broken explicitly by a small perturbation. Several interesting physical problems fall into this category. We begin with a systematic discussion of Ward identities between Green's functions involving global symmetry currents. This will also be useful as a supplement to our use of Ward identities in the earlier Chapters and as an introduction to Chapter 12 devoted to the problem of anomalies.

10.1 Internal symmetries and Ward identities

Preliminaries

We are interested in theories described by lagrangians $\mathscr{L}$ such that

$$\mathscr{L} = \mathscr{L}_0 + \mathscr{L}_1$$

(or $\mathscr{H} = \mathscr{H}_0 + \mathscr{H}'$) where $\mathscr{L}_0$ is invariant under global internal symmetry G and $\mathscr{L}_1$ explicitly breaks this symmetry. In addition the symmetry G may or may not be spontaneously broken. We assume that $\mathscr{L}_1$ can be treated as a perturbation whose effect can be calculated as a perturbative expansion in an appropriate small parameter. Furthermore let the bare fields in the lagrangian $\mathscr{L}$ form a basis for some definite representation R of the symmetry group G of the lagrangian $\mathscr{L}_0$

$$\delta\Phi_i = -\mathrm{i}\Theta^a T^a_{ij}\Phi_j \equiv \Theta^a \delta^a \Phi_i \tag{10.1}$$

(Φ stands for boson or fermion fields).

The bare currents $j^a_\mu(x)$ constructed according to Noether's theorem

$$j^a_\mu(x) = -\mathrm{i}\frac{\partial\mathscr{L}}{\partial(\partial_\mu\Phi_i)} T^a_{ij}\Phi_j \tag{10.2}$$

which are conserved currents of the G symmetric classical theory described by

$\mathscr{L}_0$, remain unchanged when $\mathscr{L}_0 \to \mathscr{L}$ if $\mathscr{L}_1$ does not contain derivatives of the fields present in the $j^a_\mu(x)$; we assume this to be the case.

Under an infinitesimal global symmetry transformation

$$\Phi'_i(x) = \Phi_i(x) + \Theta^a \delta^a \Phi_i(x) \tag{10.3}$$

the lagrangian $\mathscr{L}$ transforms as follows

$$\mathscr{L}'(x) = \mathscr{L}(x) + \Theta^a \delta^a \mathscr{L}(x) = \mathscr{L}(x) + \Theta^a \delta^a \mathscr{L}_1(x) \tag{10.4}$$

and using the classical equations of motion,

$$\partial^\mu j^a_\mu(x) = \delta^a \mathscr{L}_1(x) \tag{10.5}$$

In the following we shall also need the variation of the same lagrangian $\mathscr{L}$ under local transformations

$$\Phi'_i(x) = \Phi_i(x) + \Theta^a(x) \delta^a \Phi_i(x) \tag{10.6}$$

Remembering that $\mathscr{L}_0$ is invariant under global transformations (10.3) and that $\mathscr{L}_1$ does not contain derivatives of the fields we get

$$\delta \mathscr{L}(x) = \Theta^a(x) \delta^a \mathscr{L}(x) + j^a_\mu(x) \partial^\mu \Theta^a(x) \tag{10.7}$$

We want to derive Ward identities for bare regularized or for renormalized Green's functions following from the symmetries of the lagrangian. We assume in this Section that the UV regularization of the theory does preserve its classical symmetries. As we will see in Chapter 12 this is not true in theories with chiral fermions. In such theories some of the Ward identities derived here are afflicted with anomalies.

Let us first consider the case when the symmetry G is an exact symmetry of the lagrangian i.e. $\mathscr{L}_1 = 0$. If the regularization preserves this symmetry then the necessary counterterms are also G invariant. Both bare regularized and renormalized Green's functions are G symmetric. The renormalized fields transform under the same representation R of G as the bare fields: the renormalization constants are G invariant. The same discussion applies when the symmetry G is spontaneously broken since in this case also only symmetric counterterms are needed: this follows from (2.161). Whether G is spontaneously broken or not we are able to derive the same Ward identities.

With $\mathscr{L}_1$ present we consider the case when the symmetry breaking operators have dimension $d < 4$ ($\mathscr{L}_1$ then involves dimensionful constants). Then the counterterms required by $\mathscr{L}_1$, being of dimension lower or equal to d (see the discussion in Section 4.2) do not affect counterterms of dimension 4. Thus, for such a 'soft' breaking the wave-function renormalization constants, included in counterterms of dimension 4, remain symmetric under G (strictly speaking there exists a renormalization scheme such that they remain symmetric) and the renormalized fields of the theory with softly broken symmetry G transform under G in the same way as the bare fields. Our general remarks can be illustrated with the minimal subtraction renormalization scheme used for QCD with the mass term $m\bar{\Psi}\Psi$; the wave-function renormalization constants are mass independent and thus the same as in the chirally symmetric theory.

Ward identities from the path integral

Given all these preliminary constraints we can proceed to derive Ward identities using the functional integral formulation of quantum theory. Consider the generating functional $W[J]$ e.g. for bare regularized Green's functions

$$W[J] = N\int \mathscr{D}\Phi_i \exp\left(\mathrm{i}\left\{\int \mathrm{d}^4x[\mathscr{L}(\Phi_i) + J^i\Phi_i]\right\}\right) \tag{10.8}$$

The integral is invariant under the change of integration variables defined by (10.6). If the integration measure is invariant,† using (10.7) we get

$$0 = \int \mathrm{d}^4x \int \mathscr{D}\Phi_i \exp\left(\mathrm{i}S + \mathrm{i}\int \mathrm{d}^4x J^i\Phi_i\right)(\Theta^a(x)\delta^a\mathscr{L} + j^a_\mu(x)\partial^\mu\Theta^a(x) + J^i(x)\Theta^a(x)\delta^a\Phi_i(x)) \tag{10.9}$$

The crucial point in deriving (10.9) is the invariance of the integration measure under the transformation (10.6). It breaks down for chiral transformations in theories with chiral fermions (see Chapter 12). In other words, in this case there is no regularization of the path integral and no renormalization prescription for the Green's functions which preserve the full chiral symmetry and anomalies are present. We now define the Green's functions

$$\langle 0|TA(x)\prod_i \Phi_i(y_i)|0\rangle \sim \int \mathscr{D}\Phi_i A(x)\prod_i \Phi_i(y_i)\exp(\mathrm{i}S) \tag{10.10}$$

where $A(x)$ denotes one of the $j^a_\mu(x)$, $\delta^a\mathscr{L}(x)$ and $\delta^a\Phi_i(x)$. One should remember that the quantity on the l.h.s. of (10.10) is just a short-hand notation for its r.h.s. The notation has been chosen, however, to make easy contact with the operator language. Integrating (10.9) by parts (remember the remarks following (2.55)) differentiating it functionally with respect to the sources J_i and then setting them to zero we get the following general Ward identity

$$(\partial/\partial x_\mu)\langle 0|Tj^a_\mu(x)\prod_i \Phi_i(y_i)|0\rangle = \langle 0|T\delta^a\mathscr{L}(x)\prod_i \Phi_i(y_i)|0\rangle - \mathrm{i}\sum_i \delta(x-y_i)\langle 0|T\delta^a\Phi_i(y_i)\prod_{j\neq i}\Phi_j(y_j)|0\rangle \tag{10.11}$$

Ward identity (10.11) is valid irrespective of whether the symmetry is spontaneously broken or not. If the symmetry G is an exact global symmetry of the lagrangian the first term on the r.h.s. of (10.11) vanishes. Under the constraint of explicit breaking being only soft the Ward identity is valid both for bare regularized and for the renormalized Green's functions.

† We consider cases when $\det(\mathscr{D}\Phi'_i/\mathscr{D}\Phi_j) = 1 + O(\Theta^2)$. This is true when generators T^a in (10.1) are traceless. We also assume the absence of anomalies.

We may also be interested in Ward identities for Green's functions involving several symmetry currents j^a_μ and/or the symmetry breaking operator $\mathscr{L}_1$. We take the latter to have well-defined transformation properties under the symmetry group. The most compact way to derive such Ward identities is based on the general idea (see Chapter 8) of using background fields to get, by construction, an action which is exactly invariant under certain symmetry transformations. In our case the background fields are the sources J^i, A^a_μ and K for the operators Φ^i, j^a_μ and $\mathscr{L}_1$, respectively. We want to introduce them in such a way that the action $S[\Phi, J, A, K]$ is invariant under the *local* transformation (10.6). This is achieved if

$$S[\Phi, J, A, K] = S_0[\Phi, A] + \int d^4x(J^i\Phi_i + K\cdot\mathscr{L}_1) \tag{10.12}$$

where $S_0[\Phi]$ is the part of the action symmetric under global transformation (10.1) and the background fields A^a_μ have been introduced as gauge fields by changing the normal derivatives in S_0 into covariant derivatives $\partial_\mu \to \partial_\mu - iA^a_\mu T^a$. Action (10.12) is invariant under (10.6) if simultaneous transformations on A, J and K are defined as

$$\left.\begin{aligned}\delta A^a_\mu &= -\partial_\mu\Theta^a(x) + c^{abc}\Theta^b A^c\\ \delta J^i\Phi_i &= -J^i\delta\Phi_i, \quad \delta K\cdot\mathscr{L}_1 = -K\cdot\delta\mathscr{L}_1\end{aligned}\right\} \tag{10.13}$$

The generating functional

$$W[A, J, K] = N\int \mathscr{D}\Phi_i \exp\{iS[\Phi, J, A, K]\}$$

is then invariant under transformations (10.13) of the background fields

$$W[A, J, K] = W[A', J', K']$$

The last statement contains all the information equivalent to Ward identities which follow from the global symmetry of the original theory. For infinitesimal transformations the variation of the functional $W[A, J, K]$ reads

$$0 = \int d^4x\left[\frac{\delta W}{\delta J_i(x)}\delta J^i(x) + \frac{\delta W}{\partial A^a_\mu(x)}\delta A^a_\mu(x) + \frac{\delta W}{\delta K(x)}\delta K(x)\right] \tag{10.14}$$

or specifying j^a_μ, for instance, to flavour fermionic currents,

$$0 = \int d^4x\int \mathscr{D}\Phi_i \exp\{iS[A, J, K, \Phi]\}(J^i\delta^a\Phi_i + K\cdot\delta^a\mathscr{L}_1 - \partial_\mu j^{\mu a} - c^{abc}A^b_\mu j^{\mu c})\Theta^a(x) \tag{10.15}$$

Equation (10.15) generates the desired Ward identities valid to any order in $\mathscr{L}_1$ by differentiating with respect to sources J^i, K, and A^a_μ and setting $J^i = A^a_\mu = 0$, $K = 1$. The method can be generalized to generate Ward identities involving other composite operators with well-defined transformation properties under the global symmetry group if we introduce sources for such operators.

We stress again that the validity of the symmetry relations for the Green's functions relies on the existence of the regularization and renormalization procedures which preserve this symmetry. In particular renormalization of Green's functions involving products of composite operators, in addition to multiplicative renormalization, may also require subtractions in the form of polynomials in momenta in momentum space or of δ-functions and their derivatives in position space.

Finally we observe that in the special cases of no massless Goldstone bosons coupled to the current j^a_μ Ward identities like (10.11) can be integrated over space-time and the surface terms neglected. Thus in the case of exact symmetry which is not spontaneously broken one gets

$$\delta^a \langle 0 | T \prod_i \Phi_i(y_i) | 0 \rangle = 0 \tag{10.16}$$

which simply means an exact invariance of the Green's functions under the symmetry transformations. When the symmetry is explicitly broken by perturbation $\mathscr{L}_1$ we have

$$\begin{aligned} \delta^a \langle 0 | T \prod_i \Phi_i(y_i) | 0 \rangle &= -\mathrm{i} \int \mathrm{d}^4x \langle 0 | T \delta^a \mathscr{L} \prod_i \Phi_i(y_i) | 0 \rangle \\ &\equiv -\mathrm{i} \langle 0 | T \delta^a S \prod_i \Phi_i(y_i) | 0 \rangle \end{aligned} \tag{10.17}$$

Note that in the latter case the symmetry may also be spontaneously broken since in presence of perturbation $\mathscr{L}_1$ the massless Goldstone bosons acquire masses. This can for instance, already be seen at the level of the tree approximation i.e. by analysing the classical lagrangian.

Comparison with the operator language

The equal-time bare current commutation relations following from the equal-time canonical commutation relations for the bare field operators are

$$[j^a_0(\mathbf{x}, t), j^b_0(\mathbf{y}, t)] = \mathrm{i}c^{abc} j^c_0(\mathbf{x}, t) \delta(\mathbf{x} - \mathbf{y}) \tag{10.18}$$

The relation (10.18) is true in the symmetric theory and also with the symmetry breaking term $\mathscr{L}_1$ included as long as such terms do not contain derivatives of the fields. However, in presence of the symmetry breaking term $\mathscr{L}_1$ the currents are no longer conserved and the corresponding charges are time dependent. Defining

$$Q^a(t) = \int \mathrm{d}^3x j^a_0(\mathbf{x}, t)$$

we get from (10.18)

$$[Q^a(t), Q^b(t)] = \mathrm{i}c^{abc} Q^c(t) \tag{10.19}$$

In accordance with the transformation properties (10.1) we have furthermore

$$[Q^a(t), \Phi_i(\mathbf{x}, t)] = -T^a_{ij}\Phi_j(\mathbf{x}, t) \tag{10.20}$$

or

$$[j^a_0(\mathbf{x}, t), \Phi_i(\mathbf{y}, t)] = -T^a_{ij}\Phi_j(\mathbf{y}, t)\delta(\mathbf{x}-\mathbf{y}) \tag{10.21}$$

If the symmetry breaking is soft the relations (10.18)–(10.21) hold for the renormalized quantities as well because wave-function renormalization constants can remain invariant (for properly chosen renormalization schemes) and as will be seen shortly the currents are not subject to renormalization. The time dependence of the charges is given by

$$\frac{\mathrm{d}Q^a}{\mathrm{d}t} = \int \mathrm{d}^3x \frac{\mathrm{d}}{\mathrm{d}t} j^a_0(\mathbf{x}, t) = \mathrm{i}\int \mathrm{d}^3x [\mathscr{H}'(\mathbf{x}, t), Q^a(t)] \tag{10.22}$$

where $\mathscr{H}'$ is the symmetry breaking part of the hamiltonian, and therefore

$$\partial_\mu j^{a\mu}(\mathbf{x}, t) = \mathrm{i}[\mathscr{H}'(\mathbf{x}, t), Q^a(t)] \tag{10.23}$$

(the surface terms at infinity can be neglected because due to perturbation $\mathscr{L}_1$ the would-be Goldstone bosons acquire masses).

The Ward identity (10.11) is the same as one derived for the time-ordered product of operators defined as

$$Tj^a_\mu(x)\Phi(y) = j^a_\mu(x)\Phi(y)\Theta(x^0-y^0) + \Phi(y)j^a_\mu(x)\Theta(y^0-x^0) \tag{10.24}$$

(with a straightforward generalization to products of more than two operators) by direct differentiation of the Green's function and use of the commutation relations (10.21). For instance

$$(\partial/\partial x_\mu)\langle 0|Tj^a_\mu(x)\Phi(y)|0\rangle = \langle 0|T\partial^\mu j^a_\mu(x)\Phi(y)|0\rangle + \delta(x^0-y^0)\langle 0|[j^a_0(x), \Phi(y)]|0\rangle \tag{10.25}$$

and using (10.21) together with (10.5) for the operator $\partial^\mu j^a_\mu$ we recover the Ward identity (10.11). Similarly, one can derive Ward identities for Green's functions involving products of currents and they are the same as the relations following from (10.13) or (10.15) if we neglect in the current commutators the presence of the derivatives of the δ-functions (the so-called Schwinger terms). However, in many cases these terms certainly do not vanish (see Jackiw 1972 for a systematic discussion) and one may wonder about the compatibility of the two methods. The answer is that the Green's functions defined by the path integral do not always coincide with the vacuum expectation values of the T-products defined by (10.24). The former are always understood to be regularized or renormalized covariant quantities whereas the latter are not and they may differ by local terms, the so-called seagull terms, which cancel the Schwinger terms. The path integral approach tells us that it is always possible to arrange such cancellation by a suitable renormalization prescription if there exists one which preserves the symmetry of the classical lagrangian. Otherwise anomalies appear. We shall come back to these problems in some detail in the following.

Ward identities and short distance singularities of the operator products

Consider, for instance, the Ward identity (10.25) with $\Phi(x)$ replaced by any operator $O(x)$. This relation is obtained by differentiating the T-product defined by (10.24). However, generally speaking, the meaning of time-ordering as given by (10.24) is clear everywhere except at coinciding points i.e. at $x_0 = y_0$. Therefore there is an apparent ambiguity

$$TA(x)B(0) \to TA(x)B(0) + c_1\delta(x) + c_2\partial_x\delta(x) + \cdots \tag{10.26}$$

(Lorentz and other symmetry group indices must, of course, be properly taken into account). The question thus arises of the role of such terms which in momentum space are polynomials in momentum.

To discuss this point we recall first that equations like (10.25) are for distributions, that is, they hold in the sense that both sides are integrated over $\int d^4x f(x)$ with $f(x)$ being any 'smooth' test function vanishing at $x_\mu \to \pm\infty$. In particular, the derivative $\partial_\mu T(x)$ of the distribution $T(x)$ is defined as follows

$$\int d^4x f(x)\partial_\mu T(x) \underset{\mathrm{df}}{\equiv} \int d^4x[-\partial_\mu f(x)]T(x) \tag{10.27}$$

We also define the Fourier transform of a distribution

$$(2\pi)^{-4}\int d^4q \tilde{f}(-q)\tilde{T}(q) = \int d^4x f(x)T(x) \tag{10.28}$$

where $\tilde{f}(q)$ is the Fourier transform of $f(x)$ and use the notation

$$\tilde{T}(q) = \int d^4x \exp(iqx)T(x)$$

The Fourier transform $(\widetilde{\partial_\mu T})$ of $\partial_\mu T(x)$ is

$$\begin{aligned}(2\pi)^{-4}\int d^4q \tilde{f}(-q)(\widetilde{\partial_\mu T}) &= \int d^4x f(x)\partial_\mu T(x)\\ &= -\int d^4x \partial_\mu f(x)T(x)\end{aligned} \tag{10.29}$$

i.e.

$$(\widetilde{\partial_\mu T}) = -iq_\mu \tilde{T}(q) \tag{10.30}$$

In the other notation

$$\int d^4x \exp(iqx)\partial_\mu T(x) = -iq_\mu \int d^4x \exp(iqx)T(x) \tag{10.31}$$

We expect the meaningfulness of the Ward identity (10.25) to be related to the short-distance singularity structure of the distributions on both sides of (10.25).

To see it clearly, let us rederive (10.25) paying explicit attention to the singularities at $x \sim 0$. Using the definition (10.27) the l.h.s. of (10.25) can be written as

$$-\lim_{\varepsilon \to 0} \int \mathrm{d}^4 x\, \partial_\mu f(x) \langle 0 | T j_\mu(x) O(0) | 0 \rangle \Theta(|x_0| - \varepsilon) \tag{10.32}$$

Performing Wick's rotation we can work in the Euclidean space, so that the light-cone singularities of the operator product collapse to the origin. Then the integrand in (10.32) is finite and unambigous for $\varepsilon > 0$. Integrating by parts we get the following

$$\begin{aligned} -\lim_{\varepsilon \to 0} &\left(\int_{-\infty}^{-\varepsilon} + \int_{\varepsilon}^{\infty} \right) \mathrm{d}x_0 \int \mathrm{d}^3 x \partial_\mu f(x) \langle 0 | T j^\mu O | 0 \rangle \\ &= -\lim_{\varepsilon \to 0} \int \mathrm{d}^3 x f(x) \langle 0 | O(0) j_0(\mathbf{x}, -\varepsilon) - j_0(\mathbf{x}, \varepsilon) O(0) | 0 \rangle \\ &\quad + \lim_{\varepsilon \to 0} \int \mathrm{d}^4 x f(x) \langle 0 | T \partial^\mu j_\mu(x) O(0) | 0 \rangle \Theta(|x_0| - \varepsilon) \end{aligned} \tag{10.33}$$

In the last term the operator ∂_μ has been interchanged with T because, after excluding $x = 0$, $T j_\mu(x) O(0)$ is regular. We have also assumed the absence of massless particles; then the surface terms at $\pm\infty$ do not contribute.

Similarly, we may consider the Fourier transform

$$\begin{aligned} &\int \mathrm{d}^4 x \exp(ipx) \partial^\mu \langle 0 | T j_\mu(x) O(0) | 0 \rangle \\ &\quad = \lim_{\varepsilon \to 0} \int \mathrm{d}^4 x [-\partial^\mu \exp(ipx)] \langle 0 | T j_\mu O | 0 \rangle \Theta(|x_0| - \varepsilon) \\ &\quad = -\lim_{\varepsilon \to 0} \int \mathrm{d}^4 x \exp(ipx) \langle 0 | T j_0(x) O(0) | 0 \rangle [\delta(x_0 + \varepsilon) - \delta(x_0 - \varepsilon)] \\ &\quad\quad + \lim_{\varepsilon \to 0} \int \mathrm{d}^4 x \exp(ipx) \langle 0 | T \partial_\mu j^\mu(x) O(0) | 0 \rangle \Theta(|x_0| - \varepsilon) \end{aligned} \tag{10.34}$$

If the limit $\varepsilon \to 0$ exists, we have

$$-\mathrm{i} p_\mu \tilde{T}(p) = \Delta(p) + \int \mathrm{d}^3 x \exp(\mathrm{i}\mathbf{p}\mathbf{x}) \langle 0 | [j_0(\mathbf{x}, 0), O(0)] | 0 \rangle \tag{10.35}$$

where $\Delta(p)$ is the Fourier transform of $\langle 0 | T \partial^\mu j_\mu(x) O(0) | 0 \rangle$

Let us discuss the limit $\varepsilon \to 0$ in (10.33) and (10.34) using the OPE. One can immediately convince oneself that the limit exists if

$$T j_\mu(x) O(0) \sim C^1_\mu(x) O_1(0) + \text{less singular terms} \tag{10.36}$$

where $C^1_\mu(x) \sim O(1/x^3)$, and similarly for $\partial^\mu j_\mu(x)$. Indeed, with $O(1/x^3)$ singularity the integrals in (10.33) and (10.34) are convergent. From (10.36) we also get (see Section 7.4)

$$[j_\mu(\mathbf{x}, 0), O(0)] = a^1 O_1(\mathbf{x}, 0)\delta(\mathbf{x}) \tag{10.37}$$

(no Schwinger terms, a^1 finite). Thus, with no additional subtractions the Ward identity (10.33) or (10.35) with the equal-time commutator given by (10.37) is a well-defined relation.

Often, $Tj_\mu(x)O(0)$ is more singular than $O(1/x^3)$ (e.g. $Tj_\mu^{\text{em}}(x)j_\nu^{\text{em}}(0) \sim (1/x^6)\cdot\mathbb{1}$). Stronger than $O(1/x^3)$ singularity in the OPE of the T-product of operators translates into terms additional to the $\delta(\mathbf{x})$ term with a finite coefficient in their equal-time commutators (see Section 7.4). These are the so-called Schwinger terms. However at the same time subtractions at $x = 0$ are necessary to have a well-defined limit when $\varepsilon \to 0$ in the integrals in (10.33) and (10.34). Depending on the nature of the singularity, without subtractions these integrals may depend on the order of integration over x_0 and $\mathbf{x}$ (and be then non-covariant) or may be divergent. Subtractions are of the form

$$c_1\delta(x) + c_2\partial_{(x)}\delta(x) + \cdots$$

possibly with divergent c_i. For instance, let

$$TA(x)B(0) \sim \left(\frac{1}{x^2 - \mathrm{i}\varepsilon}\right)^2 \mathbb{1} + \cdots$$

Performing Wick's rotation and integrating in the Euclidean space we conclude that for a well-defined $\varepsilon \to 0$ limit in the l.h.s. integral in (10.33) a subtraction of the form

$$\mathrm{i}\pi^2 \ln \varepsilon\delta(0) + c\delta(0)$$

is necessary. The finite constant c is to be fixed by some renormalization conditions. This subtractive renormalization of the Green's functions comes in addition to the multiplicative renormalization of operators. The path integral approach teaches us that when regularization and renormalization schemes exist which preserve the classical symmetry of the lagrangian it is possible to define the renormalized T-products so that they satisfy Ward identities in a form like (10.11) i.e. as if the Schwinger terms were absent. In other words by suitable renormalization the Schwinger terms can be cancelled by subtraction terms. Our last remark is that in applications of Ward identities to physical problems one should pay attention to the fact that part of their content may be subtraction dependent.

Renormalization of currents

As the first application of (10.11) we discuss the renormalization properties of currents and their divergences. Let us consider, for instance, the following specific case of (10.11)

$$\partial^\mu_{(x)}\langle 0|Tj^a_\mu(x)\Psi^b(y)\bar{\Psi}^c(z)|0\rangle = \langle 0|T\partial^\mu j^a_\mu(x)\Psi^b(y)\bar{\Psi}^c(z)|0\rangle - \delta(x-y)T^a_{bd}\langle 0|T\Psi^d(y)\bar{\Psi}^c(z)|0\rangle + \delta(x-z)T^a_{dc}\langle 0|T\Psi^b(y)\bar{\Psi}^d(z)|0\rangle \quad (10.38)$$

We take (10.38) to be for bare quantities. In the symmetry limit

$$\partial^\mu j^a_\mu = 0 \quad (10.39)$$

$$\Psi^a_{\mathrm{B}} = Z_2^{1/2}\Psi^a_{\mathrm{R}} \quad (10.40)$$

where Z_2 is invariant under symmetry transformations. Expressing the bare fermion fields in (10.38) by the renormalized fields and dividing both sides by Z_2 we immediately conclude from the finiteness of the r.h.s. of (10.38) that

$$j^a_{\mu\mathrm{B}}(x) = j^a_{\mu\mathrm{R}}(x)$$

up to a finite multiplicative constant which is fixed as one by the current commutation relations. Thus the conserved currents are not subject to renormalization, e.g.

$$j^a_{\mu\mathrm{B}} = \bar{\Psi}_{\mathrm{B}}T^a\gamma_\mu\Psi_{\mathrm{B}} = Z_2\bar{\Psi}_{\mathrm{R}}T^a\gamma_\mu\Psi_{\mathrm{R}} = j^a_{\mu\mathrm{R}} \quad (10.41)$$

In the case of explicit symmetry breaking by soft ($d < 4$) terms one has $\partial^\mu j_\mu(x) \neq 0$ but, as discussed before, (10.40) can still be assumed. Writing in general $j^a_{\mu\mathrm{B}} = Z_j j^a_{\mu\mathrm{R}}$ and $\partial^\mu j^a_{\mu\mathrm{B}} = Z_{\mathrm{D}}\partial^\mu j^a_{\mu\mathrm{R}}$ we first conclude, arguing as before, that $Z_j = Z_{\mathrm{D}}$. Next, Fourier transforming (10.38) and taking the limit $p = 0$ (p is conjugate to x) we get $Z_{\mathrm{D}} = 1$ since in the limit $p = 0$ the Fourier transform of the l.h.s vanishes (explicit symmetry breaking accompanies the spontaneous symmetry breaking and therefore there are no massless bosons in the spectrum coupled to j^a_μ currents). In conclusion, even if the symmetry is broken explicitly but only softly the currents and their divergences are not subject to renormalization. For instance, for chiral symmetry broken by a mass term

$$\mathscr{L}_1 = -m\bar{\Psi}\Psi \qquad (d = 3) \quad (10.42)$$

for the divergence of the axial currents $j^a_{5\mu}$ we have (from the equations of motion)

$$\partial^\mu j^a_{5\mu} = 2mi\bar{\Psi}T^a\gamma_5\Psi \quad (10.43)$$

and the non-renormalization theorem implies

$$Z_m = Z^{-1}_{\bar{\Psi}\gamma_5\Psi} \quad (10.44)$$

where

$$(\bar{\Psi}\gamma_5 T^a\Psi)_{\mathrm{B}} = Z_{\bar{\Psi}\gamma_5\Psi}(\bar{\Psi}\gamma_5 T^a\Psi)_{\mathrm{R}} \quad (10.45)$$

Absence of wave-function renormalization for conserved or partially conserved currents does not, of course, exclude subtractive renormalizations for time-ordered products of such currents, as discussed earlier.

10.2 Quark masses and chiral perturbation theory

Simple approach

Using the Ward identity for the Green's function involving the axial current $\mathscr{A}^a_\mu(x)$ and the pseudoscalar density $\Pi^a(x)$ we have proved in Section 9.5 that in chirally symmetric QCD with chiral symmetry spontaneously broken there exist massless Goldstone particles in the physical spectrum. The same Ward identity gave us the relation (9.101) between the parameters f_π and $\langle\bar{\Psi}\Psi\rangle$ of spontaneous breaking of chiral symmetry

$$Zf_\pi = -\langle\bar{\Psi}\Psi\rangle \tag{9.101}$$

where the constants f_π and Z are defined by (9.97) and (9.98)

$$\langle 0|\Pi^a(x)|\pi^b(p)\rangle = Z\delta^{ab}\exp(-ipx)$$

and

$$\langle 0|\mathscr{A}^a_\mu(x)|\pi^b(p)\rangle = \mathrm{i}f_\pi\delta^{ab}p_\mu\exp(-ipx)$$

respectively. (In view of our discussion in the previous Section check, using the OPE for $T\mathscr{A}^a_\mu(x)\Pi^b(y)$, that the Ward identity (9.91) does not need subtractions and therefore in the chiral limit the relation (9.101) is free of any subtraction dependent parameters.)

Imagine now a realistic case when the chiral symmetry is also broken explicitly by the mass term (10.42). Then the proof of Goldstone's theorem breaks down because

$$\partial^\mu\mathscr{A}^a_\mu(x) = 2m\mathrm{i}\bar{\Psi}T^a\gamma_5\Psi \neq 0 \tag{10.43a}$$

Instead, one can now calculate masses of the would-be (pseudo-) Goldstone bosons in terms of the explicit symmetry breaking parameter: quark mass m. Such calculation is based on perturbative treatment of the symmetry breaking interaction (chiral perturbation theory) but takes account of all orders of the interactions described by the chirally symmetric part $\mathscr{L}_0$ of the lagrangian.

It is worth mentioning at this point that the subject of chiral perturbation theory is not limited to the problem of the pseudoscalar meson masses but covers the whole low energy physics of pions and kaons. Several techniques have been developed to expand the bound state masses, Green's functions and any other quantity in the low energy pseudoscalar meson sector (which in the limit of exact chiral symmetry depend on only two parameters f_π and $\langle\bar{\Psi}\Psi\rangle$) in powers of the light quark masses m_u, m_d and m_s (Dashen 1969, Dashen & Weinstein 1969, Pagels 1975). The most systematic approach proposed recently in a very interesting series of papers (Leutwyler 1984, Gasser & Leutwyler 1984, 1985*a*, *b*) relies on the effective lagrangian technique (see Chapter 13).

Here we shall only be concerned with the pseudoscalar meson masses calculated in the first order in the quark mass m. The answer is almost immediate. Equation (10.43a) sandwiched between the vacuum and the one pion state together with the definitions (9.97) and (9.98) imply the exact relation

$$f_\pi m_\pi^2 = 2mZ \tag{10.46}$$

Working to the first order in m we can now use (9.101), valid in the chiral limit, to get

$$m_\pi^2 = -(1/f_\pi^2)2m\langle 0|\bar{\Psi}\Psi|0\rangle \tag{10.47}$$

which is the desired result valid to the lowest order in chiral perturbation theory. It is the pion mass e.g. in $SU(2) \times SU(2)$ symmetric QCD with $\Psi = (u, d)$ and $m_u = m_d = m$. The pion mass has been expressed in terms of the parameters of chiral symmetry breaking: spontaneous $\langle 0|\bar{\Psi}\Psi|0\rangle$ and explicit m.

Our discussion is always in terms of renormalized quantities. We recall that interactions described by the chirally invariant part $\mathscr{L}_0$ of the lagrangian are included to all orders in our discussion. Therefore $m = m(\mu)$, where μ is some renormalization point, and operators $\bar{\Psi}\gamma_5 T^a\Psi$ and $\bar{\Psi}\Psi$ are also renormalized at the same point. However, as discussed in Section 10.1 the products $m\bar{\Psi}\gamma_5 T^a\Psi$, $m\bar{\Psi}\Psi$ are renormalization scheme independent.

The result (10.47) can be extended to the case of three flavours with $m_u \neq m_d \neq m_s$ (for heavy flavours the breaking of chiral symmetry by the mass terms is too strong to be reliably considered as a perturbation). We get the following

$$\begin{aligned} m_{\pi^0}^2 = m_{\pi^\pm}^2 &= (m_u + m_d)C + O(m_q^2) \\ m_{K^+}^2 &= (m_u + m_s)C + \cdots \\ m_{K^0}^2 = m_{\bar{K}^0}^2 &= (m_d + m_s)C + \cdots \\ C = -(2/f_\pi^2)\langle 0|\bar{u}u|0\rangle, &\ \langle 0|\bar{u}u|0\rangle = \langle 0|\bar{d}d|0\rangle = \langle 0|\bar{s}s|0\rangle \end{aligned}$$

To extract reliable values for the quark mass ratios we need, however, to include the electromagnetic mass corrections which will be considered in Section 10.4.

Finally we would like to stress that the quark masses we talk about are the renormalized quark mass parameters of the lagrangian (the so-called current quark masses). For a mass-independent renormalization procedure ratios of the renormalized masses do not depend on μ.

Approach based on use of the Ward identity

The result (10.47) can also be derived using a Ward identity directly for the realistic case with the chiral symmetry broken explicitly. This approach can be used in other problems too and its generalizations are used to calculate higher order corrections in the chiral perturbation theory.

The starting point is the Ward identity for the Green's function $\langle 0|T\mathscr{A}_\mu^a(x)\partial^\nu\mathscr{A}_\nu^b(0)|0\rangle$ where $\mathscr{A}_\mu^a(x)$ is the axial current and $\partial^\nu\mathscr{A}_\nu^b$ stands for $2m\mathrm{i}\bar{\Psi}\gamma_5 T^b\Psi$; similar considerations hold for any other current corresponding to a generator broken both spontaneously and explicitly when according to (10.23)

$$\partial^\mu j_\mu^a(\mathbf{x}, t) = \mathrm{i}[\mathscr{H}'(\mathbf{x}, t), Q^a(t)] \tag{10.23}$$

The Ward identity reads

$$\partial^{\mu}_{(x)}\langle 0|T\mathscr{A}^{a}_{\mu}(x)\partial^{\nu}\mathscr{A}^{b}_{\nu}(0)|0\rangle = D^{ab}_{5}(x) + \delta(x_0)\langle 0|[\mathscr{A}^{a}_{0}(x), \partial^{\nu}\mathscr{A}^{b}_{\nu}(0)]|0\rangle \tag{10.49}$$

where

$$D^{ab}_{5}(x) = \langle 0|T\partial^{\mu}\mathscr{A}^{a}_{\mu}(x)\partial^{\nu}\mathscr{A}^{b}_{\nu}(0)|0\rangle$$

The divergence $\partial^{\mu}\mathscr{A}_{\mu}$ is soft: the identity (10.49) can then be regarded as an identity for the renormalized Green's functions.

Defining the Fourier transform $\Delta^{ab}_{5}(q)$ of the $D^{ab}_{5}(x)$

$$\Delta^{ab}_{5}(q) = \mathrm{i}\int \mathrm{d}^4x \exp(\mathrm{i}qx) D^{ab}_{5}(x) \tag{10.50}$$

Fourier transforming both sides of (10.49) and taking the limit $q \to 0$ one gets the following relation

$$0 = \Delta^{ab}_{5}(q=0) + \mathrm{i}\langle 0|[Q^{a}_{5}(t=0), \partial^{\nu}\mathscr{A}^{b}_{\nu}(0)]|0\rangle \tag{10.51}$$

which in a general case, using (10.23), can also be written as

$$\Delta^{ab}_{5}(q=0) = -\langle 0|[Q^{a}_{5}(t=0), [Q^{b}_{5}(t=0), \mathscr{H}'(0)]]|0\rangle \tag{10.52}$$

For us $\mathscr{H}' = m\bar{\Psi}\Psi$. In the limit $q \to 0$ the l.h.s. of (10.49) is zero because in the present case (explicitly broken symmetry) there are no massless bosons in the physical spectrum and therefore no poles at $q^2 = 0$. The relation (10.52) can be used to calculate the would-be Goldstone boson (pion) masses in the first order in chiral perturbation theory if we approximate the Fourier transform $\Delta^{ab}_{5}(q)$ by the pion pole. In general $\Delta^{ab}_{5}(q)$ is given by the pole and the continuum contribution starting at $s = 9m^2_{\pi}$

$$\Delta^{ab}_{5}(q^2) = \frac{f^2_{\pi}m^4_{\pi}\delta^{ab}}{m^2_{\pi} - q^2} + \int_{(3m_{\pi})^2}^{\infty} \mathrm{d}s \frac{\rho(s, m^2_{\pi})}{s - q^2} \tag{10.53}$$

We have used $\langle 0|\partial^{\mu}\mathscr{A}^{a}_{\mu}(0)|\pi^b\rangle = f_{\pi}m^2_{\pi}\delta^{ab}$ and have written (10.50) as the spectral integral using the standard technique (see e.g. Bjorken & Drell 1965). For $q^2 = 0$ one gets

$$\Delta^{ab}_{5}(q^2 = 0) = f^2_{\pi}m^2_{\pi}\delta^{ab} + O(m^4_{\pi}) \tag{10.54}$$

The leading term in m^2_{π} is given by the pion pole because this is the only term where in the chiral symmetry limit the zero of the numerator in (10.53) partially cancels with the denominator. The pion mass measures the departure from chiral symmetry induced by perturbation $\mathscr{H}'(x)$, so expansion in m^2_{π} is the chiral perturbation theory. In the order $O(m^2_{\pi})$ we get the so-called Dashen's relation

$$\delta^{ab}m^2_{\pi} = -\frac{1}{f^2_{\pi}}\langle 0|[Q^{a}_{5}(t=0), [Q^{b}_{5}(t=0), \mathscr{H}'(0)]]|0\rangle \tag{10.55}$$

where f_π and $|0\rangle$ should be taken, for consistency, as in the chiral limit. In our specific problem, from commutation relations (9.90), we easily get (10.47).

It is also worth observing that the Green's function $D_5^{ab}(x)$ or equivalently the integral in the spectral representation (10.53) do require subtractions. For instance, it can be seen from the OPE (see Chapter 7) that the product $T\partial^\mu \mathscr{A}_\mu^a(x)\partial^\nu \mathscr{A}_\nu^b(0)$ has a singularity $O(m^2x^{-6})$. The subtraction ambiguity is $O(m^2)$ or according to our discussion following (10.54) it is $O(m_\pi^4)$ and does not affect the first order results. However in higher orders the subtraction constants appear as additional free parameters.

10.3 Dashen's theorems

Formulation of Dashen's theorems

We have learned in Section 9.4 that if a global symmetry G of a theory is spontaneously broken by the vacuum state $|0\rangle$ to a subgroup H then there is a set of degenerate vacua corresponding to different orientations of H in G. This set of vacua is generated by the action of the G transformations on the state $|0\rangle$

$$|\Omega(g)\rangle = U(g)|0\rangle \tag{10.56}$$

where $U(g)$ is a unitary operator corresponding to a group element g. If the vacuum $|0\rangle$ is left invariant by H then $|\Omega(g)\rangle$ is left invariant by the equivalent G-rotated subgroup gHg^{-1}. The set of degenerate vacua is isomorphic to the coset G/H space. If we introduce generators X^a and Y^a which are respectively broken and unbroken by the $|0\rangle$ then the transformation (10.56) can be written as

$$|\Omega(\Theta)\rangle = \exp(-\mathrm{i}\Theta^a X^a)|0\rangle \tag{10.57}$$

Transformations in G/H space correspond to variations of the vacuum expectation values for which the effective action is level at its minimum. Zero mass particles are quantized excitations – one for each orthogonal direction in G/H (see (9.51) and Chapter 13).

A small symmetry breaking perturbation $\mathscr{H}'(x)$ may change the above picture in a qualitative way: it may lift the degeneracy of the vacuum. The criteria for identifying the correct vacuum in the presence of a small perturbation $\mathscr{H}'(x)$ are called Dashen's theorems, and generally speaking, assure the vacuum stability. To the leading order in the perturbation $\mathscr{H}'(x)$ the energy density in each of the degenerate vacua is given by

$$\Delta E(\Theta) = \langle \Omega(\Theta)|\mathscr{H}'(0)|\Omega(\Theta)\rangle = \langle 0|\exp(\mathrm{i}\Theta^a X^a)\mathscr{H}'(0)\exp(-\mathrm{i}\Theta^a X^a)|0\rangle \tag{10.58}$$

To identify $|0\rangle$ as the true vacuum in the presence of the perturbation $\mathscr{H}'(x)$, $\Delta E(\Theta)$ should have a minimum (at least a local one) for $\Theta = 0$. We therefore get the following two conditions

$$\frac{\partial}{\partial \Theta^a}\Delta E(\Theta)|_{\Theta=0} = \mathrm{i}\langle 0|[X^a, \mathscr{H}'(0)]|0\rangle = 0 \tag{10.59}$$

$$\frac{\partial^2}{\partial\Theta^a\partial\Theta^b}\Delta E(\Theta)|_{\Theta=0} = -\langle 0|[X^b,[X^a,\mathscr{H}'(0)]]|0\rangle \geqslant 0 \tag{10.60}$$

In view of the relation $\partial^\mu j^a_\mu(x) = \mathrm{i}[\mathscr{H}'(x), Q^a]$, the condition (10.59) also follows from the requirement of the translational invariance of the vacuum: the vacuum value of a divergence then vanishes. Since charges X^a which do not annihilate the vacuum couple to Goldstone bosons $\langle 0|X^a|\pi^a\rangle \neq 0$, (10.59) says that $\langle \pi^a|\mathscr{H}'|0\rangle = 0$ i.e. the Goldstone boson tadpoles

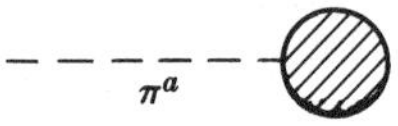

vanish. Equation (10.60) has a straightforward interpretation too: up to the renormalization factor the second derivative of the potential with respect to Θ is the Goldstone boson mass matrix (see also Chapter 13) and (10.60) assures that in the first order perturbation in $\mathscr{H}'(x)$ the pseudo-Goldstone boson masses m^2_{ab} are positive. So we rederive the Dashen's formula (10.55) referring however to the previous derivation for fixing the normalization factor.

Let us first see what the conditions (10.59) and (10.60) give for the σ-model with chiral symmetry explicitly broken by $\mathscr{H}'(x) = +\varepsilon\sigma'$ term. From the commutation relations (9.30) we get

$$\langle 0|\pi|0\rangle = 0 \qquad -\varepsilon\langle 0|\sigma'|0\rangle \geqslant 0 \tag{10.61}$$

for the correct vacuum, in agreement with (9.36). Using the normalization given in (10.55) we also rederive the result (9.39) for the pseudo-Goldstone boson mass.

For the $SU_{\mathrm{L}}(2)\times SU_{\mathrm{R}}(2)$ symmetric QCD lagrangian with chiral symmetry broken spontaneously to $SU(2)$ by a condensate $U_{\mathrm{R}}\langle\Phi\rangle U^+_{\mathrm{L}}$ where $\langle\Phi\rangle = v\mathbb{1}$ as in (9.87) and explicitly by the mass term $\mathscr{H}' = m\bar{\Psi}\Psi$ one gets

$$\langle 0|\bar{\Psi}\gamma_5\Psi|0\rangle = 0 \qquad -\langle 0|\bar{\Psi}\Psi|0\rangle > 0 \tag{10.62}$$

This follows from the commutation relation (9.88) and from the relations $\bar{\Psi}\Psi = \mathrm{Tr}[\Phi+\Phi^\dagger]$ and $\bar{\Psi}\gamma_5\Psi = \mathrm{Tr}[\Phi-\Phi^\dagger]$ where Φ is defined by (9.84). In particular we recover (9.86) which we have taken as the definitions of the vacuum in the chiral limit and in a specified basis for the chiral fields. Explicit symmetry breaking lifts the degeneracy and leaves (10.61) and (10.62) as the only possibilities.

Dashen's conditions find very interesting applications in models of the dynamical breaking of the gauge symmetry of weak interactions (technicolour and extended technicolour models). Some of those ideas will be discussed in Section 11.3. Here we give a general technical introduction to such problems which will also be useful in the next Section where we calculate the $\pi^+ - \pi^0$ mass difference.

Dashen's conditions and global symmetry broken by weak gauge interactions

A physically very interesting case of an explicit global symmetry breaking by a small perturbation is the following one: consider a strongly interacting theory (a gauge theory) with some exact global symmetry G e.g. the chiral symmetry which is spontaneously broken. Imagine then that some of the G-symmetry currents couple to some of the gauge bosons of the group G_W of weak interactions (in the rest of this Section all sums over the group indices are explicitly written down)

$$\mathscr{L}_1 = \sum_\alpha g^\alpha A_\mu^\alpha(x) j_\alpha^\mu(x) \tag{10.63}$$

where $j_\alpha^\mu(x)$ are linear combinations of some of the G currents; operators corresponding to quantum numbers commuting with G are also possible but irrelevant for our discussion. In the first non-vanishing order of perturbation theory in g^α the G-symmetry breaking term of such a theory is given by the following effective hamiltonian

$$\mathscr{H}'(0) = -\sum_{\alpha,\beta} \tfrac{1}{2} g^\alpha g^\beta \int \mathrm{d}^4x \Delta_{\alpha\beta}^{\mu\nu}(x) T j_\mu^\alpha(x) j_\nu^\beta(0) \tag{10.64}$$

where $\Delta_{\alpha\beta}^{\mu\nu}(x)$ is the G_W gauge boson propagator

$$\Delta_{\alpha\beta}^{\mu\nu}(x) = \mathrm{i}\langle 0|TA_\alpha^\mu(x) A_\beta^\nu(0)|0\rangle \equiv \delta^{\alpha\beta}\Delta^{\mu\nu}(x) \tag{10.65}$$

The question is whether our considerations in Section 10.1 based on the assumed softness of $\mathscr{H}'$ apply to this case when the symmetry breaking term $\mathscr{H}'(0)$ is not soft ($d=6$), and at first glance it could affect the wave-function renormalization of fields. It turns out, however, that this does not necessarily happen. The problem can be studied by means of the operator expansion of the product of the two currents in (10.64). First of all the counterterms affected by $\mathscr{H}'(0)$ are easily enumerated (by assumption $\mathscr{H}'(0)$ is used only in the first order perturbation): they involve only those operators of the expansion which are accompanied by coefficients that produce non-integrable singularities. It is clear from (10.64) and (10.65) (remember that $\Delta^{\mu\nu}(x) \sim 1/x^2$ for $x \to 0$) that these are operators of dimension $\leqslant 4$. The higher dimension operators in the expansion of $j_\mu^\alpha j_\nu^\beta$ do not cause a non-integrable singularity at $x=0$, do not require renormalization and are irrelevant for our discussion; notice the difference in the case if $\mathscr{H}'(0)$ was considered in any order of perturbation theory. Among the operators of dimension 4 there may be singlets under G which are again irrelevant. The non-singlets with $d=4$ would affect our discussion but in several interesting cases they do not appear (see e.g. Section 10.4 for the breaking of the chiral group by $\mathscr{H}'$ given by the product of two electromagnetic currents). We assume here that the expression (10.64) converges without subtractions which are non-singlets under the group G.

With perturbation $\mathscr{H}'(0)$ given by (10.64) Dashen's conditions (10.59) and (10.60) can be rewritten in a more useful form if we express the vacuum energy density $\Delta E(\Theta)$ given by (10.58) in terms of the G currents (Peskin 1981, Preskill 1981). We recall that (10.58) and (10.64) together give the following expression

$$\Delta E(\Theta) = -\frac{1}{2}\int \mathrm{d}^4x \Delta^{\mu\nu}(x) \sum_\alpha g_\alpha^2 \langle 0|TU^\dagger j_\mu^\alpha(x) U U^\dagger j_\nu^\alpha(0) U|0\rangle \tag{10.66}$$

where U is any unitary operator representing the group G. The G_W currents are linear combinations of the G currents

$$g_\alpha j^\mu_\alpha(x) = \sum_a g^a_\alpha j^\mu_a(x) \tag{10.67}$$

(α is the G_W index, the small a is the G index and there is no sum over α on the l.h.s.) and consequently

$$g_\alpha U^\dagger j^\mu_\alpha(x) U = \sum_a g^a_\alpha U^\dagger j^\mu_a(x) U = \sum_{a,c} g^a_\alpha R^{ac}(\Theta) j^\mu_c(x) \tag{10.68}$$

where $R(\Theta)$ is the adjoint representation of G. Because the vacuum is H invariant ($G \to H$ spontaneously) only the H invariant part of the product $j^c_\mu j^d_\nu$ contributes to (10.66) expressed in terms of (10.68). For any given vacuum from the degenerate set the generators of G split into two groups: the generators Y^a such that $Y^a|0\rangle = 0$ and X^a where $X^a|0\rangle \neq 0$. Correspondingly, each G current carries index Y or X. Since Y^a span a single real representation of H (the adjoint representation), by Schur's *Lemma* the only H invariant term in the product ${}^Y j^c_\mu {}^Y j^d_\nu$ is proportional to δ^{cd}

$$\langle 0|T\,{}^Y j^c_\mu(x)\,{}^Y j^d_\nu(0)|0\rangle = \delta^{cd}\langle 0|T\,{}^Y j_\mu\,{}^Y j_\nu|0\rangle = \mathrm{Tr}[Y^c Y^d]\langle 0|T\,{}^Y j_\mu\,{}^Y j_\nu|0\rangle \tag{10.69}$$

(here we normalize the broken and unbroken generators to $\mathrm{Tr}[Y^a Y^b] = \mathrm{Tr}[X^a X^b] = \delta^{ab}$, $\mathrm{Tr}[Y^a X^b] = 0$) where ${}^Y j_\mu$ denotes any single current ${}^Y j^a_\mu$; there is no summation in $T\,{}^Y j_\mu\,{}^Y j_\nu$. We will limit our discussion to cases when in the product $X^a X^b$ also the only H invariant term is δ^{ab}. It can be shown (Peskin 1981) that this restriction implies the existence of a parity operation P which preserves the Lie algebra of G, such that

$$PY^aP^{-1} = Y^a, \quad PX^aP^{-1} = -X^a \tag{10.70}$$

For chiral symmetry (10.70) are, of course, true since $PQ^a_{\mathrm{R,L}}P^{-1} = Q^a_{\mathrm{L,R}}$. Equations (10.70) also imply that the joint expectation values of one ${}^Y j^a_\mu$ and one ${}^X j^b_\nu$ are zero by parity conservation. The decomposition in (10.68) can be split into the Y part and the X part

$$\begin{aligned} U^\dagger g_\alpha j^\mu_\alpha(x) U &= U^\dagger g_\alpha j^\mu_\alpha(x) U|_Y + U^\dagger g_\alpha j^\mu_\alpha(x) U|_X \\ &\equiv {}^Y j^\mu_{\alpha U}(x) + {}^X j^\mu_{\alpha U}(x) \end{aligned} \tag{10.71}$$

(the last line defines our notation) such that only $Y(X)$ currents from the sum $\sum_{a,c} g^a_\alpha R^{ac} j^c_\mu$ contribute to the first (second) component. Taking all this into account we obtain for $\Delta E(\Theta)$ the following

$$\begin{aligned} \Delta E(\Theta) = -\frac{1}{2}\int d^4x \Delta^{\mu\nu}(x) \sum_\alpha \{ &\mathrm{Tr}[{}^Y\hat{J}_{\alpha U}\,{}^Y\hat{J}_{\alpha U}]\langle 0|T\,{}^Y j_\mu(x)\,{}^Y j_\nu(0)|0\rangle \\ &+ \mathrm{Tr}[{}^X\hat{J}_{\alpha U}\,{}^X\hat{J}_{\alpha U}]\langle 0|T\,{}^X j_\mu(x)\,{}^X j_\nu(0)|0\rangle\} \end{aligned} \tag{10.72}$$

where $\hat{J}_{\alpha U}$ is the linear combination of the G generators corresponding to the current $U^\dagger g_\alpha j_\mu^\alpha(x)U$ with g_α included. From the orthogonality of X^a and Y^b ($\mathrm{Tr}[X^a Y^b]=0$)

$$\left.\begin{aligned}\mathrm{Tr}[{}^Y\hat{J}_{\alpha U}{}^Y\hat{J}_{\alpha U}]=\mathrm{Tr}[{}^Y\hat{J}_{\alpha U}\hat{J}_{\alpha U}]\\ \mathrm{Tr}[{}^X\hat{J}_{\alpha U}{}^X\hat{J}_{\alpha U}]=\mathrm{Tr}[{}^X\hat{J}_{\alpha U}\hat{J}_{\alpha U}]\end{aligned}\right\} \tag{10.73}$$

Using in addition an obvious relation

$$\mathrm{Tr}[{}^Y\hat{J}_{\alpha U}\hat{J}_{\alpha U}]=\mathrm{Tr}[\hat{J}_{\alpha U}\hat{J}_{\alpha U}]-\mathrm{Tr}[{}^X\hat{J}_{\alpha U}\hat{J}_{\alpha U}] \tag{10.74}$$

we finally get

$$\begin{aligned}\Delta E(\Theta)=&-\frac{1}{2}\int \mathrm{d}^4x\Delta^{\mu\nu}(x)\sum_\alpha\{\mathrm{Tr}[\hat{J}_{\alpha U}\hat{J}_{\alpha U}]\langle 0|T\,{}^Yj_\mu(x){}^Yj_\nu(0)|0\rangle\}\\ &-\frac{1}{2}\left\{\int \mathrm{d}^4x\Delta^{\mu\nu}(x)\langle 0|T[{}^Xj_\mu(x){}^Xj_\nu(0)-{}^Yj_\mu(x){}^Yj_\nu(0)]|0\rangle\right\}\\ &*\sum_\alpha \mathrm{Tr}[{}^X\hat{J}_{\alpha U}{}^X\hat{J}_{\alpha U}]\end{aligned} \tag{10.75}$$

The first term in the sum is Θ-independent (without subtraction it may even be infinite) and in the second term Θ dependence has been factored out from the current matrix elements. Assuming that the coefficient of the last trace in (10.75) is negative as can be argued in some cases one gets the preferred vacuum simply by minimizing the quantity

$$\sum_\alpha \mathrm{Tr}[{}^X\hat{J}_{\alpha U}{}^X\hat{J}_{\alpha U}]$$

i.e. by minimizing the trace for projections of currents $\hat{J}^\alpha$ on the subspace of possible equivalent representations of the X^a and Y^a or, regarding the vacuum $|0\rangle$ as fixed so that the rotation is applied to the perturbation $\mathscr{H}'$, for projections of transformed currents on the subspace of the original broken generators. One sees that the preferred vacuum orientation corresponds to the minimal number of broken generators in the decomposition of $\hat{J}^\alpha$ into the generators of G.

We can rewrite Dashen's conditions (10.59) and (10.60) in the present notation. Writing $U=\exp(-\mathrm{i}\Theta^a X^a)$ one has

$$\frac{\partial}{\partial\Theta^a}\sum_\alpha \mathrm{Tr}({}^X\hat{J}_{\alpha U})^2|_{U=1}=2\mathrm{i}\sum_\alpha \mathrm{Tr}[{}^X[X^a,\hat{J}_\alpha]{}^X\hat{J}_\alpha]=2\mathrm{i}\sum_\alpha \mathrm{Tr}[X^a[{}^Y\hat{J}_\alpha,{}^X\hat{J}_\alpha]] \tag{10.76}$$

where the last result follows from the commutation relations $[X,X]=\mathrm{i}Y$ following from (10.70), orthogonality of Y and X and the trace property $\mathrm{Tr}[[A,B]C]=\mathrm{Tr}[A[B,C]]$. Thus, the vacuum orientation is a stationary point of ΔE if $[{}^Y\hat{J}_\alpha,{}^X\hat{J}_\alpha]=0$ for all αs.

The second Dashen's condition reads

$$m_{ab}^2=\frac{1}{2}\frac{\partial^2}{\partial\Theta^a\partial\Theta^b}\sum_\alpha \mathrm{Tr}({}^X\hat{J}_{\alpha U})^2|_{U=1}\times M^2\geqslant 0 \tag{10.77}$$

where, from (10.75)

$$M^2 = (1/f_\pi^2)\int d^4x\Delta^{\mu\nu}(x)\langle 0|T[{}^Yj_\mu(x){}^Yj_\nu(0) - {}^Xj_\mu(x){}^Xj_\nu(0)]|0\rangle \qquad (10.78)$$

and ${}^Yj_\mu$ (${}^Xj_\mu$) denotes, as before, any single current corresponding to an unbroken (broken) generator of G. The Θ dependence is only in the trace which can be expressed as follows

$$\frac{\partial^2}{\partial\Theta^a\partial\Theta^b}\sum_\alpha \mathrm{Tr}({}^X\hat{J}_{\alpha U})^2|_{U=1}$$

$$= -2\sum_\alpha\{\mathrm{Tr}[{}^X[X^a,[X^b,\hat{J}_\alpha]]{}^X\hat{J}_\alpha] + \mathrm{Tr}[{}^X[X^a,\hat{J}_\alpha]{}^X[X^b,\hat{J}_\alpha]]\}$$

$$= 2\sum_\alpha \mathrm{Tr}[[{}^Y\hat{J}_\alpha,[{}^Y\hat{J}_\alpha,X^a]]X^b - [{}^X\hat{J}_\alpha,[{}^X\hat{J}_\alpha,X^a]]X^b] \qquad (10.79)$$

Dashen's condition (10.60) or equivalently the positivity of the pseudo-Goldstone bosons mass matrix depends on the signs of the trace (10.79) and of the integral (10.78). The latter can often be studied by means of the spectral function sum rules. Several applications of the results of this Section will be discussed in the next Section and in Section 11.3.

10.4 Electromagnetic $\pi^+-\pi^0$ mass difference and spectral function sum rules

Electromagnetic $\pi^+-\pi^0$ mass difference from Dashen's formula

In Section 10.2 we have studied an explicit chiral symmetry breaking in QCD caused by non-zero current quark masses. Another source of such a breaking is the electromagnetic interaction of quarks which couples the T^3 generator of the isospin group to the electromagnetic gauge field: $Q = T^3 + \frac{1}{2}B$. The effective hamiltonian is given by (10.64) with $j_\mu(x)$ being the electromagnetic current and $\Delta_{\mu\nu}$ the photon propagator. Following our discussion in Section 10.3 we notice that the expansion of the operator product $Tj_\mu(x)j^\mu(x)$ reads

$$Tj_\mu(x)j^\mu(x) = C_1(x)\mathbb{1} + \sum_q C_q(x)m_q\bar{q}q + C_G(x)G^\alpha_{\mu\nu}G^{\mu\nu}_\alpha$$
$$+ \text{operators with } d>4 + \text{non-scalar operators} \qquad (10.80)$$

where $G^\alpha_{\mu\nu}$ is the gluon field strength. Only scalar operators contribute to the integral in (10.64). Thus, the only $d=4$ operator ($G^\alpha_{\mu\nu}G^{\mu\nu}_\alpha$) is a singlet under the chiral group and our considerations of Section 10.1 and 10.3 apply. We also see that in the chiral limit $m_q = 0$ all the divergent terms in $\mathscr{H}_{\text{em}}$ are singlets under the chiral group and may be disregarded: they induce a universal energy shift for all states, including the vacuum.

This Section is devoted to a study of the electromagnetic contribution to the pion masses. The quark masses will be neglected in the calculation of the electromagnetic masses in the spirit of the first order perturbation in any explicit symmetry breaking term. As mentioned, in this case the electromagnetic pion mass difference is finite without renormalization. The calculation can be performed by means of Dashen's formula.

Using (10.77), (10.79) and the fact that $\hat{J}^\alpha_Y = eT^3$, $\hat{J}^\alpha_X = 0$ for the electromagnetic current (as in Section 9.6 we define the vacuum of the $SU_L(2) \times SU_R(2)$ symmetric QCD so that the $SU_{L+R}(2)$ remains unbroken; then $\hat{J}^\alpha_X = 0$ and Dashen's condition for the stationary vacuum is satisfied in the presence of the electromagnetic interactions) we immediately get the following result for the one-photon exchange contribution to the pion mass

$$(m^2_{\pi^+})^\gamma = e^2 M^2, \quad (m^2_{\pi^0})^\gamma = 0 \tag{10.81}$$

where M^2 can be explicitly rewritten in terms of the vector and axial currents

$$M^2 = (1/f^2_\pi) \int d^4x \Delta^{\mu\nu}(x) \langle 0|T[\mathscr{V}^3_\mu(x)\mathscr{V}^3_\nu(0) - \mathscr{A}^3_\mu(x)\mathscr{A}^3_\nu(0)]|0\rangle \tag{10.82}$$

We recall again that all, possibly infinite, singlet contributions to the electromagnetic masses have been disregarded in (10.81) and (10.82) (see the derivation of (10.77) in the previous Section). We also remember that both equations are valid only in the first order in the symmetry breaking i.e. for $m_q = 0$. Non-zero quark masses introduce second order corrections $O(\alpha m_q)$ which need renormalization and are therefore renormalization scale dependent. Formula (10.82) has been derived by Das *et al.* (1967). To calculate the integral (10.82) one can use the spectral function sum rules.

Spectral function sum rules

Spectral function sum rules, originally derived by Weinberg (1967*a*) for the $SU(2) \times SU(2)$ chiral symmetry can be discussed in a general way by invoking the OPE (Bernard, Duncan, Lo Secco & Weinberg 1975). Let us consider the Källen–Lehmann spectral representation (Bjorken & Drell 1965) for the time-ordered product of two currents

$$\langle 0|Tj^a_\mu(x)j^b_\nu(0)|0\rangle = \int_0^\infty d\mu^2 \int \frac{d^4k}{(2\pi)^4} \exp(-ikx) \frac{i}{k^2 - \mu^2 + i\varepsilon} \rho^{ab}_{\mu\nu}(k, k^2 = \mu^2) \tag{10.83}$$

where the spectral function ρ can be decomposed into the spin one and spin zero parts

$$(2\pi)^3 \sum_{\substack{n \\ \text{spin 1}}} \langle 0|j^a_\mu(0)|n\rangle\langle n|j^b_\nu(0)|0\rangle \delta(k - k_n) = -(g_{\mu\nu} - k_\mu k_\nu/\mu^2)\rho^{(1)ab}(\mu^2) \tag{10.84}$$

$$(2\pi)^3 \sum_{\substack{n \\ \text{spin 0}}} \langle 0|j^a_\mu(0)|n\rangle\langle n|j^b_\nu(0)|0\rangle \delta(k - k_n) = k_\mu k_\nu \rho^{(0)ab}(\mu^2) \tag{10.85}$$

Taking the Fourier transform of (10.83)

$$\text{F.T.} = \int \mathrm{d}^4x \exp(ikx)\langle 0|Tj^a_\mu(x)j^b_\nu(0)|0\rangle = \mathrm{i}\int_0^\infty \mathrm{d}\mu^2 \frac{\rho^{ab}_{\mu\nu}(k, k^2 = \mu^2)}{k^2 - \mu^2} \tag{10.86}$$

and expanding it formally in powers of $1/k^2$ we get

$$\begin{aligned}\text{F.T.} = {} & \mathrm{i}\frac{k_\mu k_\nu}{k^2}\int_0^\infty \mathrm{d}\mu^2\left[\frac{\rho^{(1)}_{ab}(\mu^2)}{\mu^2} + \rho^{(0)}_{ab}(\mu^2)\right] - \mathrm{i}\frac{g_{\mu\nu}}{k^2}\int_0^\infty \mathrm{d}\mu^2\,\rho^{(1)}_{ab}(\mu^2)\\ & + \mathrm{i}\frac{k_\mu k_\nu}{(k^2)^2}\int_0^\infty \mathrm{d}\mu^2[\rho^{(1)}_{ab}(\mu^2) + \mu^2\rho^{(0)}_{ab}(\mu^2)] + \cdots\end{aligned} \tag{10.87}$$

This formal expansion can be meaningful only when the Fourier transform $\int \mathrm{d}^4x \exp(ikx)\langle 0|Tj^a_\mu(x)j^b_\nu(0)|0\rangle$ behaves for large k^2 as $O(1)$ or softer. Otherwise the coefficients of the expansion must be divergent. If one constructs a linear combination of currents such that the Fourier transform (10.86) is for large k^2 softer than $O(1)$, then the first few terms in the expansion (10.87) must vanish, e.g. for softer than $(1/k^2)$ behaviour,

$$\left.\begin{aligned}\int_0^\infty \mathrm{d}s[\rho^{(1)}_{ab}(s)/s + \rho^{(0)}_{ab}(s)] = 0\\ \int_0^\infty \mathrm{d}s\,\rho^{(1)}_{ab}(s) = 0\\ \int_0^\infty \mathrm{d}s\,s\rho^{(0)}_{ab}(s) = 0\end{aligned}\right\} \tag{10.88}$$

and these are the spectral function sum rules.

The OPE is useful in finding appropriate combinations of current products with soft high energy behaviour. In looking for such products one can use the global symmetry of the lagrangian which is respected by the OPE coefficient functions even if the symmetry is spontaneously broken. In our problem we are interested in the group $G = SU_{\mathrm{L}}(2) \times SU_{\mathrm{R}}(2)$ broken to the $H = SU_{\mathrm{L+R}}(2)$. Spectral function sum rules can be derived by studying the high momentum limit of

$$G^{ab}_{\mu\nu}(k) = \int \mathrm{d}^4x \exp(ikx)\langle 0|Tj^a_{\mu\mathrm{L}}(x)j^b_{\nu\mathrm{R}}(0)|0\rangle$$

which transforms as the adjoint representation $(3, 3)$ under $SU(2) \times SU(2)$. The asymptotic behaviour of $G^{ab}_{\mu\nu}(k)$ is determined by the lowest-dimension operator in the expansion of $j^a_{\mathrm{L}\mu}j^b_{\mathrm{R}\nu}$ which has a non-zero vacuum expectation value. This operator must be Lorentz invariant, gauge invariant, H invariant i.e. a singlet under $SU_{\mathrm{L+R}}(2)$, and, because the expansion coefficient functions respect the G symmetry, it must transform as the adjoint representation $(3, 3)$ under $SU(2) \times SU(2)$. The lowest-dimension operator satisfying these requirements is a four-fermion operator of dimension (mass)^6 ($\bar{\Psi}\Psi$ transforms as $(2, \bar{2}) \oplus (\bar{2}, 2)$). Therefore, $G^{ab}_{\mu\nu} \sim (k^2)^{-2}$ up to logarithms and the sum rules (10.88) hold for the spectral functions ρ_{LR}^{ab}.

According to (10.82) we are interested in the combination $\mathscr{V}\mathscr{V} - \mathscr{A}\mathscr{A} = 4j_L j_R$. Taking into account that

$$\langle 0|\mathscr{A}_\mu^a|\pi^b\rangle = \mathrm{i}k_\mu f_\pi \delta^{ab}$$

and therefore

$$\rho_{\mathscr{A}ab}^{(0)}(\mu^2) = \delta_{ab} f_\pi^2 \delta(\mu^2) + \cdots$$

and neglecting any continuum contribution to $\rho_{\mathscr{A}}^{(0)}$ which should be small we get from (10.88) the following sum rules

$$\left.\begin{aligned} \int_0^\infty (\mathrm{d}s/s)[\rho_{\mathscr{V}ab}^{(1)}(s) - \rho_{\mathscr{A}ab}^{(1)}(s)] &= f_\pi^2 \delta_{ab} \\ \int_0^\infty \mathrm{d}s[\rho_{\mathscr{V}ab}^{(1)}(s) - \rho_{\mathscr{A}ab}^{(1)}(s)] &= 0 \end{aligned}\right\} \tag{10.89}$$

which are valid in the chirally symmetric theory. If the chiral symmetry is broken by non-zero quark masses the OPE for the difference $\mathscr{V}\mathscr{V} - \mathscr{A}\mathscr{A}$ contains a term proportional to $m\bar{q}q$. This can be seen by a spurion analysis (Wilson 1969*a*): the current product belongs to (3, 3) of the $SU(2) \times SU(2)$ whereas the operator $\bar{q}q$ belongs to $(2, \bar{2}) \oplus (\bar{2}, 2)$. However the symmetry is broken by the mass term $m\bar{q}q$. Representing the symmetry breaking interaction by a spurion which must also belong to $(2, \bar{2}) \oplus (\bar{2}, 2)$ we see that combining one spurion with the $\bar{q}q$ operator one can produce the (3, 3) representation. Thus, in the expansion of the two-current product we need one power of the symmetry breaking parameter m for the coefficient function in front of the operator $\bar{q}q$ to be non-zero. The second sum rule is then no longer valid: the coefficient of this $m\bar{q}q$ term scales as x^{-2}, so its Fourier transform behaves as k^{-2}. The corrections are, however, expected to be small because the masses m_u and m_d are very small in comparison to the scale of QCD.

Results

We are now ready to calculate M^2, (10.82). Introducing explicitly

$$\Delta^{\mu\nu}(x) = g_{\mu\nu} \int \frac{\mathrm{d}^4k'}{(2\pi)^4} \frac{\exp(-\mathrm{i}k'x)}{k'^2 + \mathrm{i}\varepsilon}$$

we get

$$M^2 = -\frac{\mathrm{i}}{f_\pi^2} \int_0^\infty \mathrm{d}\mu^2\, 3[\rho_{\mathscr{V}}^{(1)}(\mu^2) - \rho_{\mathscr{A}}^{(1)}(\mu^2)] \int \frac{\mathrm{d}^4k}{(2\pi)^4} \frac{1}{k^2(k^2 - \mu^2)} \tag{10.90}$$

The momentum integral is logarithmically divergent and gives, using dimensional regularization,

$$\frac{1}{(4\pi)^2}\frac{1}{\varepsilon} + \frac{1}{(4\pi)^2} \ln\frac{\mu_0^2}{\mu^2} + \text{const.} \tag{10.91}$$

However, it follows from the second sum rule (10.89) that the coefficient of the divergent term vanishes. This just reflects the already discussed fact that the first order electromagnetic contribution is finite after the singlet piece under $SU(2) \times SU(2)$ has been disregarded. The remaining finite term gives

$$(m_{\pi^+}^2)^\gamma = \frac{3\alpha}{4\pi(f_\pi)^2} \int_0^\infty \mathrm{d}s \left(\ln \frac{\mu_0^2}{s} \right) [\rho_{\mathscr{V}}^{(1)}(s) - \rho_{\mathscr{A}}^{(1)}(s)] \tag{10.92}$$

and on account of the second sum rule (10.89) it is renormalization point (μ_0^2) independent. Finally we can calculate (10.92) assuming that it is reasonable to saturate the integral (10.92) with the two lowest-lying narrow resonances, one (ρ) coupled to the vector current and the other (A_1) to the axial current. Taking

$$\rho_{\mathscr{V}}(s) = f_\rho^2 s\delta(s - m_\rho^2)$$
$$\rho_{\mathscr{A}}(s) = f_{A_1}^2 s\delta(s - m_{A_1}^2)$$

and using both sum rules (10.89) one gets

$$f_\rho^2 = f_{A_1}^2 + f_\pi^2$$

and

$$m_\rho^2 f_\rho^2 = f_{A_1}^2 m_{A_1}^2$$

Therefore we can eliminate the A_1 parameters to get the final answer in terms of the ρ parameters

$$(m_{\pi^+}^2)^\gamma = \frac{3\alpha}{4\pi} m_\rho^2 \left(\frac{f_\rho}{f_\pi} \right)^2 \ln \frac{f_\rho^2}{f_\rho^2 - f_\pi^2} \tag{10.93}$$

To calculate the π^+–π^0 mass difference we have to combine the results of Section 10.2 and of this Section. In the first order in both symmetry breaking perturbations, quark masses and electromagnetic interactions, we have, using the first equation (10.48),

$$m_{\pi^+}^2 = m_{\pi^0}^2 + (m_{\pi^+}^2)^\gamma \tag{10.94}$$

where

$$m_{\pi^0}^2 = C(m_u + m_d), \qquad C = -(2/f_\pi^2)\langle 0|\bar{u}u|0\rangle$$

and therefore

$$m_{\pi^+} - m_{\pi^0} \approx (m_{\pi^+}^2)^\gamma / 2m_{\pi^0}$$

Using $f_\pi = 93\,\mathrm{MeV}$ and $f_\rho = (145 \pm 8)\,\mathrm{MeV}$, one gets $m_{\pi^+} - m_{\pi^0} = (4.9 \pm 0.2)$. Although the error due to the resonance saturation is hard to estimate accurately this result should be regarded as at least in qualitative agreement with the experimental value 4.6 MeV.

Corrections to this result due to the π^0–η mixing caused by $m_u \neq m_d$ can be estimated to be negligibly small (Gasser & Leutwyler 1982). Finally, as already mentioned before, corrections of order αm_q require renormalization but are also expected to be small (Gasser & Leutwyler 1982).

Extending our considerations to the $SU(3) \times SU(3)$ case we get $(m^2_{K^+})^\gamma = (m^2_{\pi^+})^\gamma$ and $(m^2_{K^0})^\gamma = 0$. Therefore, again using (10.48),

$$\left.\begin{aligned} m^2_{K^+} &= C(m_u + m_s) + (m^2_{\pi^+})^\gamma \\ m^2_{K^0} &= C(m_d + m_s) \end{aligned}\right\} \tag{10.95}$$

and the K^+–K^0 mass difference calculated in the first order of the explicit chiral symmetry breaking is not purely electromagnetic but it also reflects a possible m_u–m_d difference. Equations (10.94) and (10.95) taken together can be used to estimate the quark mass ratios (Weinberg 1977). Completing our discussion given at the end of Section 10.2 we get

$$\frac{m_d}{m_u} = \frac{m^2_{K^0} - m^2_{K^+} + m^2_{\pi^+}}{2m^2_{\pi^0} + m^2_{K^+} - m^2_{K^0} - m^2_{\pi^+}} = 1.8$$

$$\frac{m_s}{m_d} = \frac{m^2_{K^0} + m^2_{K^+} - m^2_{\pi^+}}{m^2_{K^0} - m^2_{K^+} + m^2_{\pi^+}} = 20.1$$

We stress that these results are valid in the first order of the chiral symmetry breaking.

11

Spontaneous breaking of gauge symmetry

11.1 Higgs mechanism

The proof of Goldstone's theorem given in Chapter 9 for theories with spontaneously broken global symmetry requires Lorentz invariance and Hilbert space with positive-definite scalar products. Gauge theories do not obey both requirements simultaneously. In a covariant gauge, the theory contains states of negative norm. In a gauge in which the theory has only states of positive norm it is not manifestly covariant. In consequence, Goldstone's theorem does not hold and the so-called Higgs mechanism operates (see, for instance, Abers & Lee 1973 and references therein).

We shall consider only the simplest example of the $U(1)$ gauge theory with the gauge field coupled to a complex scalar field. This example illustrates all the essential points; we leave it to the reader to study the spontaneous symmetry breaking in the standard electroweak theory: $SU(2) \times U(1)$ gauge theory with one complex scalar doublet.

The lagrangian for a single self-interacting complex field Φ'

$$\mathcal{L} = \partial_\mu \Phi'^* \partial^\mu \Phi' - \lambda(\Phi'^*\Phi' + m^2/2\lambda)^2 \tag{11.1}$$

is invariant under global $U(1)$ transformation: ${}^{\mathrm{U}}\Phi' = \exp(-\mathrm{i}\Theta)\Phi'$. The $U(1)$ symmetry is spontaneously broken (see Section 9.5) for $m^2 < 0$. In the real basis $\Phi' = (\varphi' + \mathrm{i}\chi')/\sqrt{2}$ we can choose, for instance,

$$\langle 0|\varphi'|0\rangle = v = (-m^2/\lambda)^{1/2}, \quad \langle 0|\chi'|0\rangle = 0$$

and defining the physical fields $\varphi = \varphi' - v$, $\chi = \chi'$ we get

$$\mathcal{L} = \tfrac{1}{2}(\partial_\mu\varphi\partial^\mu\varphi + \partial_\mu\chi\partial^\mu\chi) - v^2\lambda\varphi^2 - \tfrac{1}{4}\lambda(\varphi^2 + \chi^2)^2 - \lambda\varphi v(\varphi^2 + \chi^2) \tag{11.2}$$

This lagrangian describes the interaction of the massive particle φ and the massless Goldstone boson χ.

It is also useful to introduce 'angular' variables and parametrize the field Φ' as follows (see (9.78))

$$\sqrt{2}\Phi'(x) = \rho'(x)\exp[-i\eta'(x)T]\begin{pmatrix}0\\1\end{pmatrix} = \begin{pmatrix}\rho'\sin\eta'\\ \rho'\cos\eta'\end{pmatrix}; \quad T = \begin{pmatrix}0 & i\\ -i & 0\end{pmatrix} \tag{11.3}$$

Using the freedom in our choice of the vacuum expectation value among the $U(1)$ degenerate set (Section 9.5) we can write

$$\Phi'(x) - \langle 0|\Phi'(x)|0\rangle = [\rho'(x) - v]\exp[-i\eta'(x)T]\begin{pmatrix}0\\1\end{pmatrix} \tag{11.4}$$

and in terms of the fields $\rho = \rho' - v$ and $\eta = v\eta'$ the lagrangian (11.1) reads

$$\mathscr{L} = \tfrac{1}{2}(\partial_\mu\rho\partial^\mu\rho + \partial_\mu\eta\partial^\mu\eta) - v^2\lambda\rho^2 + \text{cubic and higher order terms} \tag{11.5}$$

Thus, the particle interpretation of the fields ρ and η is the same as that of φ and χ.

Now let us introduce a gauge field A_μ; a lagrangian invariant under local gauge transformation is

$$\mathscr{L} = -\tfrac{1}{4}F^{\mu\nu}F_{\mu\nu} + (D_\mu\Phi')^*(D_\mu\Phi') - \lambda(\Phi'^*\Phi' + m^2/2\lambda)^2 \tag{11.6}$$

where

$$D_\mu = \partial_\mu - igA_\mu$$

The gauge freedom has profound implications for the mechanism of spontaneous symmetry breaking. This is easiest to see working with the angular variables. Choosing the gauge function to be $-\eta'(x) = -\eta(x)/v$ and using (11.4) we get (in the real basis)

$$\left.\begin{aligned} {}^{\mathrm{U}}\Phi'(x) - \langle 0|\Phi'(x)|0\rangle^{\mathrm{U}} &= [\rho'(x) - v]\begin{pmatrix}0\\1\end{pmatrix}\\ A_\mu^{\mathrm{U}}(x) &= A_\mu(x) - (1/gv)\partial_\mu\eta(x)\end{aligned}\right\} \tag{11.7}$$

Since the lagrangian (11.6) is gauge invariant we can write it in terms of fields ${}^{\mathrm{U}}\Phi$ and A_μ^{U} and then

$$\mathscr{L} = -\tfrac{1}{4}F^{\mathrm{U}\mu\nu}F_{\mu\nu}^{\mathrm{U}} + \tfrac{1}{2}\partial_\mu\rho\partial^\mu\rho + \tfrac{1}{2}g^2(\rho + v)^2 A_\mu^{\mathrm{U}}A^{\mathrm{U}\mu} - \tfrac{1}{4}\lambda(\rho^2 + 2\rho v)^2 \tag{11.8}$$

where

$$\rho = \rho' - v$$

In this gauge called the unitary gauge the particle spectrum is evident. There is a scalar meson ρ with the bare mass $(2\lambda v^2)^{1/2}$ and a massive vector meson with mass gv and no particle corresponding to the field η which has disappeared from the lagrangian. The η degree of freedom has been traded for the longitudinal component of the vector field, now massive. Thus, the gauge symmetry is spontaneously broken and no massless Goldstone boson remains in the physical spectrum.

Quantization in the unitary gauge gives a theory which is manifestly unitary order-by-order in perturbation theory but not manifestly renormalizable. Because of the massive vector meson propagator which for large k grows like $k_\mu k_\nu/m^2k^2$ the Green's functions contain divergences that cannot be removed by the renormalization counterterms. Such divergences vanish in the S-matrix elements but this we know from the proof of renormalizability of a spontaneously broken gauge theory given in covariant gauges (Abers & Lee 1973 and references therein). Hence it is useful to extend our discussion to covariant gauges.

Let us first take a gauge fixed by the standard gauge-fixing term $-(\partial_\mu A^\mu)^2/2a$. Returning to the variables φ and χ we get from (11.6) the following lagrangian

$$\begin{aligned}\mathscr{L} = &-\tfrac{1}{4}F^{\mu\nu}F_{\mu\nu} + \tfrac{1}{2}(\partial_\mu\varphi\partial^\mu\varphi + \partial_\mu\chi\partial^\mu\chi) - \lambda v^2\varphi^2 - (1/2a)(\partial_\mu A^\mu)^2 && (S_0)\\ &+\tfrac{1}{2}g^2v^2A_\mu A^\mu - gvA_\mu\partial^\mu\chi && (\tilde{S}_0)\\ &\left.\begin{aligned}&+gA_\mu\partial^\mu\varphi\chi - gA_\mu\partial^\mu\chi\varphi + \tfrac{1}{2}g^2A_\mu A^\mu(\varphi^2+2\varphi v)\\ &+\tfrac{1}{2}g^2A_\mu A^\mu\chi^2 - \tfrac{1}{4}\lambda(\varphi^2+\chi^2)^2 - \lambda(\varphi^2+\chi^2)\varphi v\end{aligned}\right\} && (S_\mathrm{I})\end{aligned} \tag{11.9}$$

One can now proceed in several different ways. One possibility is to take S_0 as the free lagrangian of the theory with the corresponding definitions of the free particle states (which may not be physical states). With the standard methods of Chapters 2 and 3 we get then the following free particle propagators (e.g. in the Landau gauge $a \to 0$)

$$\left.\begin{aligned}&A_\mu(\text{massless}): && -\mathrm{i}(g_{\mu\nu} - p_\mu p_\nu/p^2)/(p^2+\mathrm{i}\varepsilon) \equiv -\mathrm{i}D_{\mu\nu}\\ &\varphi\,(\text{massive}): && \mathrm{i}/(p^2 - 2\lambda v^2 + \mathrm{i}\varepsilon)\\ &\chi\,(\text{massless}): && \mathrm{i}/(p^2+\mathrm{i}\varepsilon)\end{aligned}\right\} \tag{11.10}$$

The full gauge boson propagator

$$\Delta_{\mu\nu}(p) = -\mathrm{i}D_{\mu\nu}\frac{1}{1+\Pi(p^2)} \tag{11.11}$$

where $\Pi(p^2)$ is defined by the gauge invariant vacuum polarization tensor

$$\equiv \mathrm{i}\Pi_{\mu\nu}(p^2) = -\mathrm{i}(g_{\mu\nu}p^2 - p_\mu p_\nu)\Pi(p^2) \tag{11.12}$$

gets a contribution from the remaining terms in the lagrangian and in particular from $\tilde{S}_0$ which, although quadratic in fields too, is counted as an interaction term. In the lowest order the $\tilde{S}_0$ contribution is

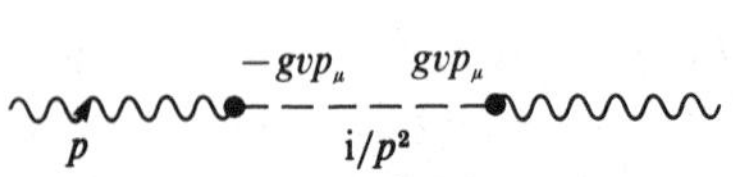

$$\equiv -\mathrm{i}(p_\mu p_\nu/p^2)g^2v^2 \tag{11.13a}$$

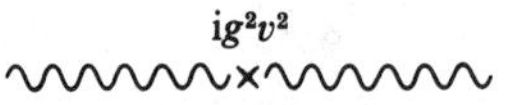

$$\equiv +\mathrm{i}g^2v^2g_{\mu\nu} \tag{11.13b}$$

Hence

$$\Pi(p^2) = -g^2v^2/p^2 \tag{11.14}$$

and consequently

$$\Delta_{\mu\nu}(p) = -\mathrm{i}(g_{\mu\nu} - p_\mu p_\nu/p^2)/(p^2 - g^2v^2) \tag{11.15}$$

which can be rewritten as a combination of the standard propagator for a massive vector boson and of a term corresponding to a scalar negative norm boson coupled to the source of the vector meson gradiently

$$\Delta_{\mu\nu}(p) = -\mathrm{i}\left(g_{\mu\nu} - \frac{p_\mu p_\nu}{g^2 v^2}\right)\frac{1}{p^2 - g^2 v^2 + \mathrm{i}\varepsilon} - \mathrm{i}\frac{p_\mu p_\nu}{g^2 v^2}\frac{1}{p^2 + \mathrm{i}\varepsilon} \tag{11.16}$$

Note also that the χ propagator remains massless in the Landau gauge.

One sometimes says that the coupling (11.13a) of a gauge boson to a massless Goldstone boson is the origin of the gauge boson mass and of the spontaneous breaking of gauge invariance. In this context one should observe that the massless Goldstone boson contributes only to the $p_\mu p_\nu$ term in the $\Pi_{\mu\nu}$ but by gauge invariance the $g_{\mu\nu}$ term must then be present. This remark will be particularly relevant in Section 11.3 where other mechanisms for spontaneous gauge symmetry breaking are discussed. Secondly, such an interpretation is certainly procedure and gauge dependent. Indeed, we can as well take $(S_0 + \tilde{S}_0)$ as our free particle lagrangian. The presence of the mixing term $gvA_\mu \partial^\mu \chi$ leads to a system of coupled differential equations for A_μ and χ propagators and in the limit $a \to 0$ we get (11.15) as the *free* boson propagator. The coupling  disappears from the theory as well as its previous interpretation.

Finally we can get rid of the coupling $gvA_\mu \partial^\mu \chi$ in the lagrangian by a clever choice of gauge. This so-called R_ξ gauge is in wide use. Take the gauge-fixing term as

$$f(A_\mu, \chi) = -\frac{1}{2\xi}(\partial^\mu A_\mu + \xi g v \chi)^2 \tag{11.17}$$

Then

$$\tilde{S}_0 + f(A_\mu, \chi) = \tfrac{1}{2}g^2 v^2 A_\mu A^\mu - \tfrac{1}{2}\xi g^2 v^2 \chi^2 - (1/2\xi)(\partial_\mu A^\mu)^2 \tag{11.18}$$

and the free propagators are

$$A_\mu\colon \frac{-\mathrm{i}}{p^2 - M^2 + \mathrm{i}\varepsilon}\left[g_{\mu\nu} - (1-\xi)\frac{p_\mu p_\nu}{p^2 - \xi M^2}\right] = -\mathrm{i}\frac{g_{\mu\nu} - p_\mu p_\nu/M^2}{p^2 - M^2 + \mathrm{i}\varepsilon} - \mathrm{i}\frac{p_\mu p_\nu/M^2}{p^2 - \xi M^2 + \mathrm{i}\varepsilon}$$

$$\chi\colon \mathrm{i}/(p^2 - \xi M^2 + \mathrm{i}\varepsilon)$$

$$\varphi\colon \mathrm{i}/(p^2 - 2\lambda v^2 + \mathrm{i}\varepsilon)$$

where $M = gv$. In the limit $\xi = 0$ we get the Landau gauge and for $\xi = 1$ the so-called 't Hooft–Feynman gauge. The unitary gauge is recovered in the limit $\xi \to \infty$.

For any finite value of ξ there are unphysical poles at $p^2 = \xi M^2$ in the gauge boson propagator and in the χ propagator. They cancel, however, in the S-matrix elements which are obtained from the Green's functions by removing external lines, setting external momenta on the mass-shell and contracting tensor indices with appropriate physical polarization vectors. This can be shown, for instance, by proving that the renormalized S-matrix is independent of ξ (Abers & Lee 1973).

11.2 Spontaneous gauge symmetry breaking by radiative corrections

In the model of the previous Section gauge symmetry has already been spontaneously broken at the tree level. Here we discuss spontaneous gauge symmetry breakdown by radiative corrections (Coleman & Weinberg 1973). This can occur when elementary scalar fields of zero mass are present in the theory (Georgi & Pais 1977). In the tree approximation there is then no spontaneous symmetry breaking because in this approximation a non-zero vacuum expectation value of the scalar fields can appear only as a result of interplay between the Φ^4 interaction terms and the scalar mass terms. Spontaneous symmetry breaking is, however, produced by non-zero vacuum expectation values induced by the higher order corrections. This is also an example of a dimensional transmutation: a dimensionful parameter $\langle 0|\Phi|0\rangle$ is generated in a quantum field theory whose classical limit is described by a scale-invariant lagrangian.

We consider first $\lambda\Phi^4$ theory with a massless meson field and discuss spontaneous breaking of global reflection symmetry $\Phi \to -\Phi$. The lagrangian density for this theory written in terms of the renormalized quantities

$$\mathscr{L} = \tfrac{1}{2}(\partial_\mu \Phi)(\partial^\mu \Phi) - (\lambda/4!)\Phi^4 + \tfrac{1}{2}A(\partial_\mu \Phi)(\partial^\mu \Phi) - \tfrac{1}{2}B\Phi^2 - 1/4!C\Phi^4 \tag{11.19}$$

contains the usual wave-function and coupling-constant renormalization counterterms. Also a mass renormalization counterterm is present, even though we are studying the massless theory, because the theory possesses no symmetry which would guarantee vanishing bare mass in the limit of vanishing renormalized mass (the scale invariance is broken by anomalies). To study spontaneous symmetry breaking in such a theory we will use the effective potential formalism developed in Section 2.6. To the one-loop approximation the effective potential is given by (2.165) and (2.166) (with $m^2 = 0$ and with the contributions from the mass and coupling-constant counterterms included)

$$V = \frac{\lambda}{4!}\varphi^4 + \tfrac{1}{2}B\varphi^2 + \frac{1}{4!}C\varphi^4 + \frac{1}{2}\int \frac{d^4k}{(2\pi)^4} \ln\left(1 + \frac{\lambda\varphi^2}{2k^2}\right) \tag{11.20}$$

In this expression we have performed the usual Wick rotation so the integral is in Euclidean space: $d^4k = \tfrac{1}{2}k^2 dk^2 d\Omega_4$, $\Omega_4 = 2\pi^2$. Observe also that, even though for $m^2 = 0$ the series in (2.166) is disastrously IR divergent, its sum exhibits only a logarithmic singularity at $\varphi = 0$. Cutting off the integral in (11.20) at $k^2 = \Lambda^2$ (exceptionally we use the UV cut-off rather than dimensional regularization method), performing the elementary integration and throwing away terms which vanish as Λ^2 goes to infinity we obtain

$$V = \frac{\lambda}{4!}\varphi^4 + \tfrac{1}{2}B\varphi^2 + \frac{1}{4!}C\varphi^4 + \frac{\lambda\Lambda^2}{64\pi^2}\varphi^2 + \frac{\lambda^2\varphi^4}{256\pi^2}\left(\ln\frac{\lambda\varphi^2}{2\Lambda^2} - \frac{1}{2}\right) \tag{11.21}$$

Imposing now the definitions of the renormalized mass and coupling constant we determine the value of the counterterms. Defining the renormalized squared mass of the field $\Phi(x)$ as the value of the inverse propagator at zero momentum and recalling the interpretation following (2.164) of the derivatives of the effective potential we get

$$\mathrm{d}^2V/\mathrm{d}\varphi^2|_{\varphi=0} = 0 \tag{11.22}$$

as the mass renormalization condition. We take the derivative at $\varphi = 0$ because we are interested in the inverse propagator of the original (not shifted) field $\Phi(x)$; note also that the perturbative calculation of $V(\varphi)$ makes use of the Feynman rules for the lagrangian written in terms of the field $\Phi(x)$ and therefore it is based on (2.164) even though the theory may be spontaneously broken. From (11.21) and (11.22) we get

$$B = -\lambda\Lambda^2/32\pi^2$$

A possible definition of the renormalized coupling constant would be to identify it with the value of the 1PI four-point function in the original fields $\Phi(x)$ at zero external momenta

$$\mathrm{d}^4V/\mathrm{d}\varphi^4|_{\varphi=0} = \lambda$$

Unfortunately, for a massless field we encounter the logarithmic IR singularities. In the standard renormalization procedure one defines in such a case the coupling constant at some off-mass-shell position in momentum space. Using the effective potential formalism it is much more convenient to adopt an alternative definition, namely

$$\mathrm{d}^4V/\mathrm{d}\varphi^4|_{\varphi=M} = \lambda(M) \tag{11.23}$$

where M is some arbitrary number with the dimension of mass. It is clear from our discussion in Section 2.6 (see (2.156)) that the fourth derivative of $V(\varphi)$ at $\varphi = M$ is the 1PI Green's function, at zero external momenta, for the shifted fields $\Phi'(x) = \Phi(x) - M$. In particular if we took $M = \langle 0|\Phi|0\rangle$ it was the four-point function taken at zero external momenta for the physical massive field $\bar{\Phi}(x)$:$\langle 0|\bar{\Phi}(x)|0\rangle = 0$.

Imposing condition (11.23) on the effective potential $V(\varphi)$ given by (11.21) we finally get

$$V = \frac{\lambda(M)}{4!}\varphi^4 + \frac{\lambda^2(M)\varphi^4}{256\pi^2}\left(\ln\frac{\varphi^2}{M^2} - \frac{25}{6}\right) \tag{11.24}$$

We see that the one-loop radiative corrections have turned the minimum at the origin $\varphi = 0$ into a maximum and caused a new minimum of V to appear at

$$\lambda(M)\ln(\langle 0|\Phi|0\rangle/M) = -\tfrac{32}{3}\pi^2 + O(\lambda) \tag{11.25}$$

Thus, the one-loop corrections have generated spontaneous breaking of reflection symmetry. However, in our example, the new minimum lies outside the range of validity of perturbation theory as higher orders will give higher powers of $\lambda\ln(\varphi^2/M^2)$ which according to (11.25) is large for $\varphi = \langle 0|\Phi|0\rangle$.

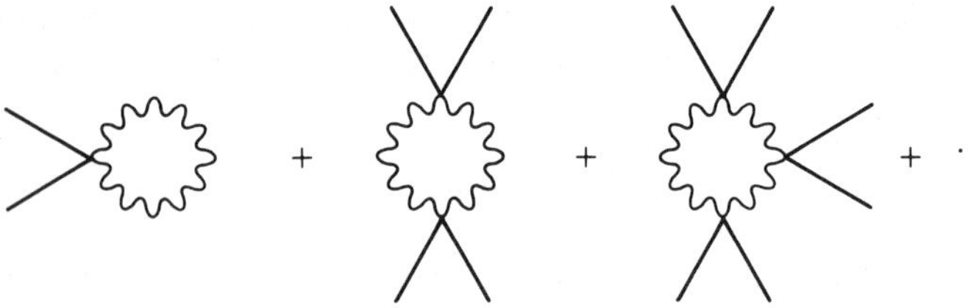

Fig. 11.1

Physically interesting theories in which spontaneous symmetry breaking can be studied in perturbation theory are gauge theories with elementary massless scalar fields and possibly fermion fields coupled to the gauge fields. Let us consider the case of a single gauge field coupled to a charged scalar to see qualitatively what happens. In the one-loop approximation the effective potential of the scalar sector gets a contribution from the scalar loops which, aside from irrelevant numerical factors (there are now two real scalar fields in the theory: $\Phi = (\Phi_1 + \mathrm{i}\Phi_2)/\sqrt{2}$; the scalar self-coupling is $\lambda(\Phi\Phi^*)^2$; the effective potential can only depend on $\varphi^2 = (\varphi_1^2 + \varphi_2^2)$ because the theory is $U(1)$ invariant), have the same structure as before, and from the gauge field loops in Fig. 11.1 arising from the minimal coupling $|(\partial_\mu - \mathrm{i}gA_\mu)\Phi|^2$. The trilinear terms lead to diagrams such as

which vanish if we work in the Landau gauge where the gauge boson propagator is

$$D_{\mu\nu} = -\mathrm{i}\frac{g_{\mu\nu} - k_\mu k_\nu/k^2}{k^2 + \mathrm{i}\varepsilon}$$

This is because we calculate diagrams with zero external momenta, and therefore the momentum of the internal scalar meson is the same as that of the internal gauge boson.

It is obvious now that the gauge boson loop contribution to the effective potential has the same structure as the scalar loop contribution, so we get

$$V = \lambda(M)\varphi^4 + [(1/16\pi^2)\lambda^2(M) + bg^4(M)]\varphi^4[\ln(\varphi^2/M^2) - \tfrac{25}{6}] \tag{11.26}$$

where b is some numerical factor. If it happens that $\lambda(M) \ll g^2(M) \ll 1$ we can neglect the $O(\lambda^2)$ term and then $\mathrm{d}V/\mathrm{d}\varphi = 0$ for

$$|\langle\varphi\rangle| = (\langle 0|\Phi_1|0\rangle^2 + \langle 0|\Phi_2|0\rangle^2)^{1/2} = M\exp\left[\frac{11}{6} - \frac{1}{2b}\frac{\lambda(M)}{g^4(M)}\right] \tag{11.27}$$

Thus, in the one-loop approximation we find in our theory spontaneous breaking of the $U(1)$ symmetry and, this time, the new minimum is in the range of validity of our approximation: due to our choice of the renormalization point such that

$\lambda(M) \ll g^2(M)$ (11.27) constrains only the value of $\ln(\langle\varphi\rangle^2/M^2)$ and for small λ and g the terms $\lambda \ln(\langle\varphi\rangle^2/M^2)$ and $g^2 \ln(\langle\varphi\rangle^2/M^2)$ are small. A non-vanishing field vacuum expectation value generated by radiative corrections is an example of the dimensional transmutation phenomenon. This can be seen more clearly if we choose $M = \langle\varphi\rangle$. Then, from (11.27)

$$\lambda(\langle\varphi\rangle) = \tfrac{11}{3} b g^4(\langle\varphi\rangle) \tag{11.28}$$

and we recall that $M = \langle\varphi\rangle$ corresponds to on-shell renormalization for the shifted fields. We can take dimensionless λ and g or g and dimensionful $\langle\varphi\rangle$ as independent parameters of our theory.

Spontaneous gauge symmetry breaking leads to the physical consequences discussed in Section 11.1: the would-be Goldstone boson combines with the gauge boson to make a massive vector meson $m^2(V) = g^2\langle\varphi\rangle^2$ and the remaining scalar field, after shifting, corresponds to a scalar meson of mass $m^2(S) = V''(\langle\varphi\rangle) \sim g^4\langle\varphi\rangle^2$.

Physics cannot depend on the arbitrary parameter M but one choice of M can be more convenient than the others, particularly since we are working in perturbation theory and have to control higher order corrections. We have already learned that the perturbative calculation of $\langle\varphi\rangle$ is reliable when the renormalization point M is such that $\lambda(M) \ll g^2(M) \ll 1$. It is also worth noting that from (11.27)

$$\langle\varphi\rangle = M_0 e^{11/6} \tag{11.29}$$

where $\lambda(M_0) = 0$. Usually, we suppose that $\lambda(M)$ and $g(M)$ are given at some scale $M = \tilde{M}$ e.g. at the grand unification scale and we can then use renormalization group arguments to look for a renormalization scale suitable for our calculation (e.g. M_0). In our renormalization procedure the scalar quartic coupling is defined in terms of the fourth derivative of V. Thus, to study $\lambda(M)$ one should derive the RGE for this fourth derivative (see Coleman & Weinberg 1973 and Gildener 1976). We know, however, that the function $\beta(\lambda)$ in the one-loop approximation does not depend on the renormalization conventions. Therefore to the lowest order we can use the results of calculations which adopt momentum space renormalization conventions. The one-loop RGEs for λ and g then read

$$\left.\begin{aligned} 16\pi^2 M(\mathrm{d}\lambda/\mathrm{d}M) &= A\lambda^2(M) + B'\lambda(M)g^2(M) + Cg^4(M) \\ 16\pi^2 M(\mathrm{d}g^2/\mathrm{d}M) &= -bg^4(M) \end{aligned}\right\} \tag{11.30}$$

where the dimensionless coefficients A, B' and C are given, respectively, by the diagrams shown in Fig. 11.2. To solve (11.30) we introduce $\alpha = \lambda/g^2$. Then we get $(t = \ln M)$

$$\frac{16\pi^2}{g^2(t)}\frac{\mathrm{d}\alpha}{\mathrm{d}t} = -bg^2(t)\frac{\mathrm{d}\alpha}{\mathrm{d}g^2} = A\alpha^2(t) + B\alpha(t) + C = A(\alpha - \alpha_1)(\alpha - \alpha_2), \quad \alpha_1 < \alpha_2 \tag{11.31}$$

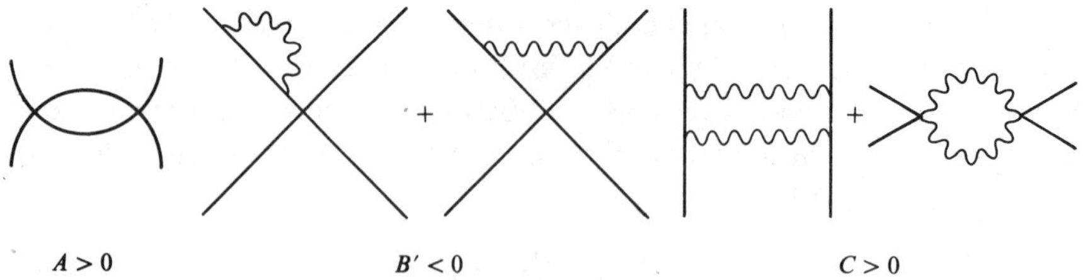

Fig. 11.2

where $B = B' + b$. For real roots there is a UV stable fixed point at $\alpha = \alpha_1 < \alpha_2$ and an IR stable fixed point at $\alpha = \alpha_2$. Thus, if $0 < \alpha(\tilde{M}) = \lambda(\tilde{M})/g^2(\tilde{M}) < \alpha_1$ we expect the RGEs (11.31) to have a solution $\alpha(M) = 0$ for $M \neq 0$. A direct integration gives

$$\exp\left(-b\int_0^{\alpha(\tilde{M})}\frac{d\alpha}{A\alpha^2 + B\alpha + C}\right) = \frac{g^2(\tilde{M})}{g^2(M)} = 1 + \frac{1}{16\pi^2}g^2(\tilde{M})b\ln\frac{M}{\tilde{M}} \quad (11.32)$$

or, more explicitly, $\alpha(M) = 0$ for

$$M = \tilde{M}\exp\left[-\frac{16\pi^2}{g^2(\tilde{M})}F(\alpha(\tilde{M}))\right] \quad (11.33)$$

where

$$F(\alpha(\tilde{M})) = \frac{1}{b}\left[1 - \exp\left(-b\int_0^{\alpha(\tilde{M})}\frac{d\alpha}{A\alpha^2 + B\alpha + C}\right)\right] \quad (11.34)$$

If $\alpha(\tilde{M})$ is of the order $O(1)$ (that is, if $\lambda(\tilde{M}) \sim g^2(\tilde{M})$) then $F(\alpha(\tilde{M}))$ is also of order unity and

$$\frac{M}{\tilde{M}} = \exp\left[-\frac{O(1)}{g^2(\tilde{M})}\right] \quad (11.35)$$

Thus, using (11.29) we conclude that for a reasonably small value of $g^2(\tilde{M})$, the dimensional transmutation can generate a scale M of spontaneous symmetry breaking which is enormously smaller than the scale $\tilde{M}$ at which as we suppose $\alpha(\tilde{M})$ is given. This small ratio of mass scales is not put into the theory by hand but arises automatically from the assumption of a vanishing scalar mass. Thus in order for this to happen some massless scalars which escape getting masses of order $\tilde{M}$ must be available in our theory. For instance, if $\tilde{M}$ is the grand unification scale they must remain massless after the spontaneous breakdown of the grand unification group. Such a situation can perhaps arise naturally from supersymmetries which are unbroken at $\tilde{M}$. The mechanism described here very much resembles the generation of the scale Λ at which QCD coupling becomes large (see (7.53a)). It can also be generalized to the case of several quartic scalar couplings present in the theory (Gildener & Weinberg 1976).

The generation of a very small ratio of mass scales by radiative corrections and the dimensional transmutation phenomenon is to be contrasted with making $M/\tilde{M}$ very small in the tree approximation. The latter is highly 'unnatural'. To be more specific, we have in mind the following: consider a gauge group G which we want to be spontaneously broken to G_1 at some scale $\tilde{M}$, and G_1 to be broken to G_2 at M with $M/\tilde{M}$ very small. It turns out that one can obtain such a pattern of spontaneous symmetry breaking in the tree approximation by a proper choice of the scalar potential, but only at the expense of tuning the parameters of the potential to an accuracy much higher than the neglected one-loop corrections. This is called the gauge hierarchy problem.

The formalism presented in this Section can be extended to non-abelian gauge theories with scalars and also fermions (see Coleman & Weinberg 1973 and Problem 11.6).

11.3 Dynamical breaking of gauge symmetries and vacuum alignment

Dynamical breaking of gauge symmetry

We consider now the possibility of spontaneous gauge symmetry breaking in the absence of elementary scalar fields. The mechanism presented here is often termed dynamical gauge symmetry breaking. Our discussion in this Section is closely connected with that in Section 10.3.

Imagine a spontaneously broken chiral *global* symmetry G of some strong gauge interactions with gauge group G_H (we shall call it the hypercolour group) also to be broken explicitly by weak interactions with a gauge symmetry group G_W; some of the chiral currents of G couple to gauge bosons of G_W. By assumption, it is a good approximation to treat G_W interaction to the lowest order (see 10.64)). In Section 10.3 we have learned how to find the preferred vacuum which minimizes the vacuum energy in the presence of the perturbation G_W. Assuming the positivity condition for certain current matrix elements the correct vacuum corresponds to the minimal value of the quantity (see (10.75))

$$\sum_{\alpha} \mathrm{Tr}\,[{}^X\hat{J}^{\alpha}_U\, {}^X\hat{J}^{\alpha}_U] \tag{11.36}$$

built of the transformed currents projected on the subspace of the broken generators in the original vacuum; we recall that $\hat{J}^{\alpha}$ is the linear combination of the G generators corresponding to the G_W current $g_{\alpha} j^{\alpha}_{\mu}(x)$ and ${}^X\hat{J}^{\alpha}$ is that part of $\hat{J}^{\alpha}$ which contains only broken generators X^{α}. Equivalently, the correct vacuum must satisfy Dashen's conditions (10.59) and (10.60). Assume that the correct vacuum has been found: it gives a specific orientation of the spontaneously unbroken subgroup H in the G, a set of broken generators X^{α} and the corresponding Goldstone bosons, some of which get masses induced by the perturbation G_W. In general one might also single out another subgroup of G: the maximal subgroup S of elements of G which commute with all the generators of G_W. Then the coupling of gauge bosons to the currents of G breaks G explicitly to $G_W \times S$ and we get the picture of the G symmetry breaking shown in Fig. 11.3 where the orientation of different subgroups is illustrated in Fig. 11.4.

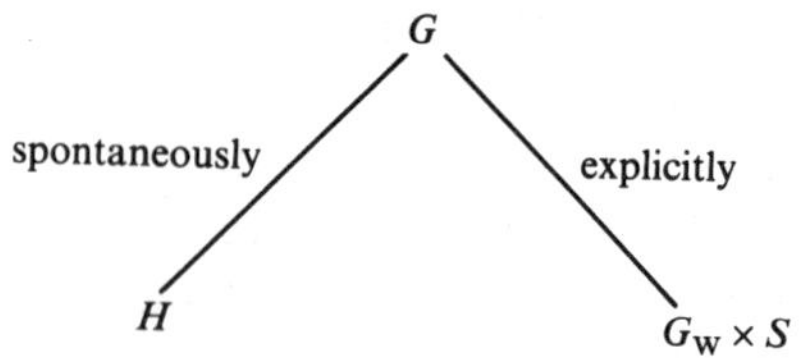

Fig. 11.3

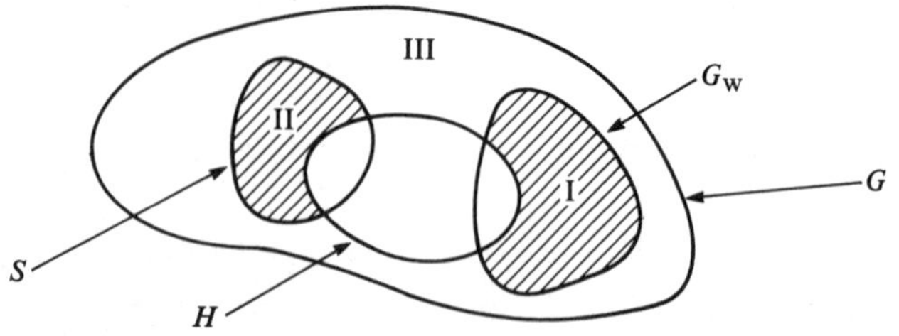

Fig. 11.4

In the zeroth order in the G_W couplings there are Goldstone bosons associated with each spontaneously broken generator (regions I + II + III). In presence of the G_W interactions the Goldstone bosons split into three different groups which are easy to classify. The $G_W \times S$ symmetry remains the exact symmetry of the lagrangian and therefore the Goldstone bosons in the regions I and II remain massless in any order in the G_W interactions, as can be proved following the arguments of Section 9.6. The generators in region III do not correspond to exact symmetries of the full lagrangian and the corresponding Goldstone bosons acquire, in general, masses which in the first order in G_W are given by (10.55) or (10.77) and (10.78). From the calculation of the π^+–π^0 mass difference we can see that the masses are of the order $g\Lambda$ ($m_\rho \sim \Lambda$), where g is the coupling constant of the weak interactions and Λ is the confinement scale of the hypercolour interactions. The Goldstone bosons in regions I and II still split into two groups: clearly, those in region II remain in the physical spectrum; however, those in region I are absorbed by the G_W gauge bosons which get masses through the Higgs mechanism. One then talks about the dynamical breaking of the G_W gauge symmetry. Let us discuss the mass generation for the G_W gauge bosons somewhat more explicitly. The Goldstone bosons in region I couple to the G_W currents j^α_μ

$$\langle 0|j^\alpha_\mu(0)|\pi^a\rangle = \mathrm{i}p_\mu f^{\alpha a} \tag{11.37}$$

and therefore the gauge-invariant vacuum polarization tensor $\Pi^{\alpha\beta}_{\mu\nu}$ defined by

$$\begin{aligned} -\mathrm{i}\Pi^{\alpha\beta}_{\mu\nu}(p^2) &= g^\alpha g^\beta \int \mathrm{d}^4x \exp(\mathrm{i}px)\langle 0|Tj^\alpha_\mu(x)j^\beta_\nu(0)|0\rangle \\ &= -\mathrm{i}(g_{\mu\nu}p^2 - p_\mu p_\nu)\Pi^{\alpha\beta}(p^2) \end{aligned} \tag{11.38}$$

has a pole at $p^2 = 0$ (see also Section 11.1)

$$\lim_{p^2 \to 0} \Pi_{\mu\nu}^{\alpha\beta}(p^2) = (g_{\mu\nu}p^2 - p_\mu p_\nu)\sum_a g^\alpha g^\beta f^{\alpha a} f^{\beta a}/p^2 \tag{11.39}$$

It is important to remember the role of gauge invariance in writing (11.39): the Goldstone bosons contribute only to the $p_\mu p_\nu$ term of the polarization tensor; the $g_{\mu\nu}$ term must be present by gauge invariance and we do not need to specify its origin more explicitly. The full gauge boson propagator $\Delta_{\mu\nu}^{\alpha\beta}$ reads†

$$\Delta_{\mu\nu}^{\alpha\beta} = -\mathrm{i}D_{\mu\nu}[\delta^{\alpha\beta} + \Pi^{\alpha\beta}(p^2)] = -\mathrm{i}D_{\mu\nu}(1-\Pi)^{-1\alpha\beta} + \text{higher order terms} \tag{11.40}$$

We are working in the Landau gauge and $-\mathrm{i}D_{\mu\nu\delta}{}^{\alpha\beta}$ is the free propagator in this gauge. Thus, the residue (11.39) gives the vector boson mass matrix

$$m_{\alpha\beta}^2 = \sum_a g_\alpha g_\beta f_{\alpha a} f_{\beta a}$$

The Goldstone boson decay constants $f^{\alpha a}$ can be expressed in terms of the appropriate traces. Under the assumption (10.70) we have

$$\langle 0 | j_\mu^a(0) | \pi^b \rangle = \mathrm{i}p_\mu f_\pi \delta^{ab} = \mathrm{i}p_\mu f_\pi \,\mathrm{Tr}\,[X^a X^b] \tag{11.41}$$

where j_μ^a is a G current and consequently

$$m_{\alpha\beta}^2 = f_\pi^2 \sum_a \mathrm{Tr}\,[\hat{J}^\alpha X^a]\,\mathrm{Tr}\,[\hat{J}^\beta X^a] = f_\pi^2\,\mathrm{Tr}\,[{}^X\hat{J}^\alpha\,{}^X\hat{J}^\beta] \tag{11.42}$$

(where no sum has been taken over α and β). This result has been obtained for the vacuum state $|0\rangle$. For any other vacuum state related by the unitary transformation $U|0\rangle$ we find

$$m_{\alpha\beta}^2 = f_\pi^2\,\mathrm{Tr}\,[{}^X\hat{J}_U^\alpha\,{}^X\hat{J}_U^\beta] \tag{11.43}$$

Comparing with (10.75) we see that

$$\Delta E(\Theta) \sim \mathrm{Tr}\, m^2 \tag{11.44}$$

The conclusion is that the energetically preferred vacuum alignment and the relative orientation of the groups G_W and H in the presence of the perturbation G_W are such that the trace of the vector boson mass matrix is the minimal one (the gauge symmetry is broken as little as possible). The weak interactions determine their own pattern of symmetry breaking by the choice of the vacuum orientation. We note also that the vector boson mass matrix obeys, in general, certain relations which are consequences of H invariance.

The above ideas may be relevant for the gauge symmetry breaking in the Glashow–Salam–Weinberg theory of the weak interactions (see for instance Fahri & Susskind 1981) however, no convincing model has been constructed yet. We will illustrate them with two simple examples.

† $\Delta_{\mu\nu}^{\alpha\beta} = (-\mathrm{i}D_{\mu\nu}\delta^{\alpha\beta}) + (-\mathrm{i}D_{\mu\lambda}\delta^{\alpha\beta'}\mathrm{i}\Pi^{\lambda\rho,\beta'\gamma'}(-\mathrm{i}D_{\rho\nu}\delta^{\gamma'\beta})$

Examples

Consider a strong hypercolour gauge group G_H with $2N$ left-handed fermions and $2N$ right-handed fermions transforming as the same representation of G_H. The theory has $G = SU_L(2N) \times SU_R(2N)$ chiral symmetry which, we assume, breaks down spontaneously to $H = SU(2N)$. The embedding of H in G is characterized by a fermion condensate

$$\langle 0|\bar{\Psi}_{Li}\Psi_{Rj}|0\rangle = \Phi_{ij} \tag{11.45}$$

As in Section 9.6, assuming $\Phi_{ij} = v\delta_{ij}$, the chiral group is broken to $SU_{L+R}(2N)$. Any equivalent orientation of the unbroken $SU(2N)$ corresponds to a condensate obtained from (11.45) by a unitary transformation

$$U_R \Phi U_L^\dagger = v U_R U_L^\dagger \tag{11.46}$$

(remember that $U^\dagger \Phi U = U_R \Phi U_L^\dagger$; U is an unitary operator, $U_{L(R)}$ are unitary matrices) i.e. it is specified by a unimodular unitary matrix $U_R U_L^\dagger$. We take the weak gauge group to be $G_W = SU_L(2) \times U(1)$ with fermions in N left-handed G_W doublets and $2N$ right-handed singlets, and with the $U(1)$ charge assignments as follows

$$\begin{array}{cccc} & \begin{pmatrix} \mathscr{U}_i \\ \mathscr{D}_i \end{pmatrix}_L & \mathscr{U}_{iR} & \mathscr{D}_{iR} \\ \tfrac{1}{2}Y = & q & q+\tfrac{1}{2} & q-\tfrac{1}{2} \end{array}$$

Consider first the case $N = 1$. Denoting the gauge bosons of the Weinberg–Salam model by W_μ^α, B_μ the weak coupling is

$$\Delta\mathscr{L} = \sum_\alpha g W_\mu^\alpha \bar{\Psi}_L \gamma^\mu T^\alpha \Psi_L + g' B_\mu (\bar{\Psi}_R \gamma^\mu T^3 \Psi_R + q\bar{\Psi}\gamma^\mu \Psi) \tag{11.47}$$

where T^α is an $SU(2)$ generator normalized to $\mathrm{Tr}\,[T^\alpha T^\beta] = \frac{1}{2}\delta^{\alpha\beta}$ and q is the mean electric charge of the doublet. Assuming the condensate (11.45) and expressing the G_W currents j_μ^α in terms of the G currents corresponding to the unbroken subgroup (vector currents) and to the broken generators (axial currents) $j_L = \frac{1}{2}(\mathscr{V} - \mathscr{A})$, $j_R = \frac{1}{2}(\mathscr{V} + \mathscr{A})$ one gets in the (Ψ_L, Ψ_R) basis the following expression:

$$\begin{aligned} \hat{J}^\alpha &= g\left(\begin{array}{c|c} T^\alpha & \\ \hline & 0 \end{array}\right) + g'\left(\begin{array}{c|c} q & \\ \hline & q+T^3 \end{array}\right) = \left[\tfrac{1}{2}g\left(\begin{array}{c|c} T^\alpha & \\ \hline & T^\alpha \end{array}\right) + \tfrac{1}{2}g'\left(\begin{array}{c|c} 2q+T^3 & \\ \hline & 2q+T^3 \end{array}\right)\right] \\ &\quad + \left[\tfrac{1}{2}g\left(\begin{array}{c|c} T^\alpha & \\ \hline & -T^\alpha \end{array}\right) + \tfrac{1}{2}g'\left(\begin{array}{c|c} -T^3 & \\ \hline & T^3 \end{array}\right)\right] \end{aligned} \tag{11.48}$$

Inserting (11.48) into (11.42) gives the familiar vector boson matrix

	W_1	W_2	W_3	B
W_1	$\frac{1}{4}g^2 f_\pi^2$			
W_2		$\frac{1}{4}g^2 f_\pi^2$		
W_3			$\frac{1}{4}g^2 f_\pi^2$	$-\frac{1}{4}gg' f_\pi^2$
B			$-\frac{1}{4}gg' f_\pi^2$	$\frac{1}{4}g'^2 f_\pi^2$

After diagonalization one gets:

$$\left.\begin{aligned} m_{W_{1,2}} &= \tfrac{1}{2} g f_\pi, \quad m_Z = \tfrac{1}{2} f_\pi (g'^2 + g^2)^{1/2} = m_W/\cos\Theta_W, \\ m_\gamma &= 0 \end{aligned}\right\} \tag{11.49}$$

where Θ_W, the Weinberg angle, is given by

$$\cos\Theta_W = \frac{g}{(g'^2 + g^2)^{1/2}}$$

and

$$\begin{aligned} \gamma &= \cos\Theta_W B + \sin\Theta_W W_3 \\ Z &= -\sin\Theta_W B + \cos\Theta_W W_3 \end{aligned}$$

In this model all the Goldstone bosons have been used to give masses to the weak gauge bosons. There is no problem of vacuum alignment since all the equivalent vacua (11.46) can be transformed into (11.45) by the gauge transformations in the weak sector.

For $N > 1$ left-handed doublets with the same $U(1)$ charge assignment the situation is more complex. If we introduce the $4N$-plet $\Psi = (\mathscr{U}_{1L}\mathscr{D}_{1L}\ldots \mathscr{D}_{NL}\mathscr{D}_{NL}\mathscr{U}_{1R}\mathscr{D}_{1R}\ldots\mathscr{U}_{NR}\mathscr{D}_{NR})$ then the Salam–Weinberg coupling is

$$\Delta\mathscr{L} = g W^\alpha_\mu \bar{\Psi}\gamma^\mu \begin{pmatrix} T^\alpha \otimes \mathbb{1} & \\ & 0 \end{pmatrix} \Psi + g' B_\mu \bar{\Psi}\gamma^\mu \begin{pmatrix} q & \\ & q + T^3 \otimes \mathbb{1} \end{pmatrix} \Psi \tag{11.50}$$

where the $2N \times 2N$ block $T^\alpha \otimes \mathbb{1}$ is a direct product of the 2×2 weak isospin and the $N \times N$ matrix on the space of N doublets. The weak coupling with the gauge symmetry $G_W = SU_L(2) \times U(1)$ breaks the original chiral symmetry $G = SU_L(2N) \times SU_R(2N)$ of the strong interaction sector into $G \to G_W \times S$ where $S = SU_L(N) \times SU_R(N) \times SU_R(N)$ with the first $SU(N)$ corresponding to horizontal transformations among N left-handed doublets, and the second (the third) $SU(N)$ corresponding to transformations among N right-handed singlets $\mathscr{U}_{iR}$ ($\mathscr{D}_{iR}$); $\mathscr{U}_{iR}$ and $\mathscr{D}_{iR}$ have different charges and therefore the original $SU_R(2N)$ is broken into $SU_R(N) \times SU_R(N)$. The spontaneous breakdown of G into $SU_{L+R}(2N)$ induced by (11.46) leaves the subgroup $SU_{L+R}(N)$ of S unbroken. The spontaneous breakdown of the original $SU(2N) \times SU(2N)$ chiral symmetry produces $(4N^2 - 1)$ Goldstone bosons. Three of them are used to give masses to the gauge bosons. Again, there is no vacuum alignment problem. The breaking of S into $SU(N)$ which remains the exact symmetry of the full theory leaves $2(N^2 - 1)$ exactly massless Goldstone bosons. They are electrically neutral because they correspond to generators of transformations among fermions with the same charge. There remain $2(N^2 - 1)$ charged Goldstone bosons which do not correspond to generators of exact symmetries of the full lagrangian and they should, therefore, receive masses from the G_W interactions. Splitting the weak currents into the unbroken ${}^Y\hat{J}^\alpha$ and the broken parts

Fig. 11.5

${}^{X}\hat{J}^{\alpha}$ as in (11.48)

$$\left.\begin{aligned} {}^{Y}\hat{J}^{\alpha} &\equiv \tfrac{1}{2}g\begin{pmatrix} T^{\alpha}\otimes 1 & \\ & T^{\alpha}\otimes 1 \end{pmatrix} + \tfrac{1}{2}g'\begin{pmatrix} 2q + T^{3}\otimes 1 & \\ & 2q + T^{3}\otimes 1 \end{pmatrix} \\ {}^{X}\hat{J}^{\alpha} &\equiv -\tfrac{1}{2}g\begin{pmatrix} -T^{\alpha}\otimes 1 & \\ & T^{\alpha}\otimes 1 \end{pmatrix} + \tfrac{1}{2}g'\begin{pmatrix} -T^{3}\otimes 1 & \\ & T^{3}\otimes 1 \end{pmatrix} \end{aligned}\right\} \tag{11.51}$$

it is evident from (10.79) that the Goldstone bosons receive no mass in the first order in G_{W}. The broken generators X_a are of the block-diagonal form

$$\begin{pmatrix} -A & \\ & A \end{pmatrix}$$

where the As are $(4N^2 - 1)$ matrices which can be easily constructed for any N, and therefore for ${}^{Y}\hat{J}$ and ${}^{X}\hat{J}$ given by (11.51) the two contributions to the trace in (10.79) cancel each other. Actually, one can see that, if G_{W} commutes with one of the chiral groups $SU_{\mathrm{L}}(2N)$ or $SU_{\mathrm{R}}(2N)$ then there will be pseudo-Goldstone bosons which remain massless to the lowest order in G_{W}. Indeed, one sees directly from (10.66), using H invariance as in (10.69), that the Θ-dependent (U-dependent) part of the vacuum energy $\Delta E(\Theta)$ relevant for the pseudo-Goldstone boson masses (10.79) comes from the term

$$\Delta E(\Theta) = -\tfrac{1}{2}\int \mathrm{d}^4x\,\Delta^{\mu\nu}(x)\sum_{\alpha} g_{\alpha}^2 \langle 0|T U_{\mathrm{L}}^{\dagger} j^{\alpha}_{\mu\mathrm{L}}(x) U_{\mathrm{L}} U_{\mathrm{R}}^{\dagger} j^{\alpha}_{\nu\mathrm{R}}(0) U_{\mathrm{R}}|0\rangle \tag{11.52}$$

whereas $j^{\alpha}_{\mu\mathrm{L}} j^{\alpha}_{\nu\mathrm{L}}$ and $j^{\alpha}_{\mu\mathrm{R}} j^{\alpha}_{\nu\mathrm{R}}$ terms contribute only the Θ-independent constants. In our example the $U(1)$ current also does not contribute to $\Delta E(\Theta)$ (although it has both the left-handed and the right-handed parts) because the left-handed $U(1)$ current is an $SU_{\mathrm{L}}(2N)$ singlet. The $U(1)$ contribution is therefore $SU_{\mathrm{L}}(2N)$ invariant and also $SU_{\mathrm{L+R}}(2N)$ invariant because the vacuum is, so it is $SU_{\mathrm{L}}(2N) \times SU_{\mathrm{R}}(2N)$ invariant and hence Θ-independent. The charged pseudo-Goldstone bosons receive masses in the next order in G_{W} due to the mass splitting of the γ and Z which has to be taken into account in that order, e.g. as in Fig. 11.5.

In the model discussed above there remain massless neutral Goldstone bosons in the physical spectrum because the $SU(N)$ group of transformations between doublets remains the exact symmetry of the theory. This can be avoided by giving the left-handed doublets different $U(1)$ charge. Such a model is very instructive because then the $U(1)$ boson exchange contributes the Θ-dependent term to the $\Delta E(\Theta)$ and the pattern of the G_{W} breakdown depends on the $U(1)$ charge assignments.

To be specific we consider a model with four left-handed and four right-handed fermions with the G_{W} representation content given by

$$\begin{array}{cccccc} \begin{pmatrix} \mathscr{U}_1 \\ \mathscr{D}_1 \end{pmatrix}_{\mathrm{L}} & \begin{pmatrix} \mathscr{U}_2 \\ \mathscr{D}_2 \end{pmatrix}_{\mathrm{L}} & \mathscr{U}_{1\mathrm{R}} & \mathscr{D}_{1\mathrm{R}} & \mathscr{U}_{2\mathrm{R}} & \mathscr{D}_{2\mathrm{R}} \\ \tfrac{1}{2}Y = \Delta, & -\Delta, & (\Delta + \tfrac{1}{2}), & (\Delta - \tfrac{1}{2}), & (-\Delta + \tfrac{1}{2}), & (-\Delta - \tfrac{1}{2}) \end{array} \tag{11.53}$$

Similarly to (11.50) the G_{W} coupling in the Ψ_{L}, Ψ_{R} basis ($\Psi = (\mathscr{U}_{1\mathrm{L}}\mathscr{D}_{1\mathrm{L}}\mathscr{U}_{2\mathrm{L}}\mathscr{D}_{2\mathrm{L}}$ $\mathscr{U}_{1\mathrm{R}}\mathscr{D}_{1\mathrm{R}}\mathscr{U}_{2\mathrm{R}}\mathscr{D}_{2\mathrm{R}})$) is

$$\Delta\mathscr{L} = gW^{\alpha}_{\mu}\bar{\Psi}\gamma^{\mu}\left(\begin{array}{c|c} T^{\alpha}\otimes\mathbb{1} & \\ \hline & 0 \end{array}\right)\Psi + g'B_{\mu}\bar{\Psi}\gamma^{\mu}\left[\left(\begin{array}{c|c} 0 & \\ \hline & T^3\otimes\mathbb{1} \end{array}\right) + \Delta\left(\begin{array}{c|c} \mathbb{1}\otimes T^3 & \\ \hline & \mathbb{1}\otimes T^3 \end{array}\right)\right]\Psi \tag{11.54}$$

where in the product $A\otimes B$ the first matrix is in the 2×2 weak space and the second one is on the space of doublets. Let us first assume that the $SU(4)\times SU(4)$ symmetry of the G_{H} lagrangian is broken by the

$$\langle 0|\bar{\mathscr{U}}_{\mathrm{L}i}\mathscr{U}_{\mathrm{R}j}|0\rangle = \langle 0|\bar{\mathscr{D}}_{\mathrm{L}i}\mathscr{D}_{\mathrm{R}j}|0\rangle = v\delta_{ij} \tag{11.55}$$

condensates. This defines the vacuum and the breaking $SU_{\mathrm{L}}(4)\times SU_{\mathrm{R}}(4)\to SU_{\mathrm{L+R}}(4)$ in the basis (11.53). We can now split the G_{W} currents into unbroken and broken parts in the vacuum (11.55). It is clear from our previous discussion that only the $U(1)$ current is interesting to us; as before the contribution of the $SU(2)$ current to the two terms under the trace in (10.79) cancels. For the $U(1)$ current we have

$$\hat{J} = \tfrac{1}{2}g'\left(\begin{array}{c|c} T^3\otimes\mathbb{1} + 2\Delta\mathbb{1}\otimes T^3 & \\ \hline & T^3\otimes\mathbb{1} + 2\Delta\mathbb{1}\otimes T^3 \end{array}\right) + \tfrac{1}{2}g'\left(\begin{array}{c|c} -T^3\otimes\mathbb{1} & \\ \hline & T^3\otimes\mathbb{1} \end{array}\right) \tag{11.56}$$

There are 15 Goldstone bosons corresponding to the broken generators

$$X = \frac{1}{2}\left(\begin{array}{c|c} -A & \\ \hline & A \end{array}\right)$$

where A represents $T^{a}\otimes\mathbb{1}$, $2T^{a}\otimes T^3$, $\mathbb{1}\otimes T^3$ and eight matrices of the form, e.g.

$$\frac{1}{\sqrt{2}}\left(\begin{array}{cc|cc} \multicolumn{2}{c|}{0} & 1 & 0 \\ & & 0 & 0 \\ \hline 1 & 0 & \multicolumn{2}{c}{0} \\ 0 & 0 & & \end{array}\right) \quad\text{or}\quad \frac{1}{\sqrt{2}}\left(\begin{array}{cc|cc} \multicolumn{2}{c|}{0} & -\mathrm{i} & 0 \\ & & 0 & 0 \\ \hline \mathrm{i} & 0 & \multicolumn{2}{c}{0} \\ 0 & 0 & & \end{array}\right) \tag{11.57}$$

linking one of $(\mathscr{U}_1, \mathscr{D}_1)$ with one of $(\mathscr{U}_2, \mathscr{D}_2)$. A look at (10.79) and (11.56) tells us that Goldstone bosons corresponding to broken generators $T^a \otimes \mathbb{1}$, $T^a \otimes T^3$ and $\mathbb{1} \otimes T^3$ receive no mass in the first order in G_{W}. An explicit calculation based on (10.79), (11.56) and (11.57) give the following masses for the remaining pseudo-Goldstone bosons which are denoted by the flavours linked by the corresponding generators: non-diagonal terms of the mass matrix vanish and bosons corresponding to the two generators like (11.57) are degenerate in mass and

$$\left.\begin{array}{l} m^2_{\mathscr{U}_1\mathscr{U}_2} = m^2_{\mathscr{D}_1\mathscr{D}_2} \sim g'^2\Delta^2 M^2 \\ m^2_{\mathscr{U}_1\mathscr{D}_2} \sim g'^2\Delta(\Delta+1)M^2 \\ m^2_{\mathscr{D}_1\mathscr{U}_2} \sim g'^2\Delta(\Delta-1)M^2 \end{array}\right\} \tag{11.58}$$

where M^2 is the integral given by (10.78). Assuming that the integral M^2 is dominated by the lowest-lying vector mesons of the correct quantum numbers we find, as for the π^+–π^0 mass difference in Section 10.4, $M^2 > 0$. Therefore, we find an interesting result: for $|\Delta| > 1$ Dashen's condition is satisfied by masses (11.58) and the vacuum (11.55) is the preferred vacuum in the presence of perturbation G_{W}. However, for the $U(1)$ charge assignments for doublets such that $|\Delta| < 1$ the vacuum (11.55) turns out to be unstable. We must, therefore, find the correct vacuum in the set of vacua obtained from (11.55) by all unitary transformations U of $SU(4) \times SU(4)$. According to (11.36) the right transformation is the one which minimizes the $\mathrm{Tr}\,{}^X\hat{J}^2_U$ or maximizes the $\mathrm{Tr}\,{}^Y\hat{J}^2_U$ of the projection of the transformed current on the subspace of the original broken (X^a) and unbroken (Y^a) generators, respectively. Before we find this transformation let us notice that in the original basis (11.54) the term in the lagrangian proportional to Δ is $SU_{\mathrm{L+R}}(4)$ invariant. It is then clear that for $|\Delta| > 1$ that basis is the right one: the $\Delta\mathscr{L}$ given by (11.54) is 'as much as possible' $SU_{\mathrm{L+R}}(4)$ invariant. However, for $|\Delta| < 1$ one can imagine that a transformation $U^\dagger \hat{J} U$, which breaks the $SU_{\mathrm{L+R}}(4)$ invariance of the Δ term, simultaneously transforms the other term of the $U(1)$ current to such a form that the full transformed current has 'better' $SU_{\mathrm{L+R}}(4)$ symmetry than the original one. The right transformation can be found by noting that the transformed $U(1)$ current, split into unbroken and broken parts, is as follows

$$U^\dagger \left(\begin{array}{cccc|cccc} \Delta &&&&&&& \\ & \Delta &&&&&& \\ && -\Delta &&&&& \\ &&& -\Delta &&&& \\ \hline &&&& 1+\Delta &&& \\ &&&&& -1+\Delta && \\ &&&&&& 1-\Delta & \\ &&&&&&& -1-\Delta \end{array}\right) U = \left(\begin{array}{c|c} A & \\ \hline & A \end{array}\right) + \left(\begin{array}{c|c} B & \\ \hline & -B \end{array}\right)$$

where the matrix on the l.h.s. is the $U(1)$ current in the original basis (11.54). Since the trace of B^2 should be the minimal one, the transformed current should have both 4×4 blocks as similar as possible ($A + B \approx A - B \approx A$). Hence, for $|\Delta| < 1$ the right transformation is the one which gives

$$U^\dagger \hat{J} U = \left(\begin{array}{cccc|cccc} \Delta & & & & & & & \\ & \Delta & & & & & & \\ & & -\Delta & & & & & \\ & & & -\Delta & & & & \\ \hline & & & & 1+\Delta & & & \\ & & & & & 1-\Delta & & \\ & & & & & & -1+\Delta & \\ & & & & & & & -1-\Delta \end{array}\right)$$

From (11.54) one sees that this is the transformation $\mathscr{D}_{1R} \leftrightarrow \mathscr{U}_{2R}$. The new vacuum is then defined by the condensate

$$U_R \Phi U_L^\dagger = v \begin{array}{c} \begin{array}{cccc} \bar{\mathscr{U}}_{1L} & \bar{\mathscr{D}}_{1L} & \bar{\mathscr{U}}_{2L} & \bar{\mathscr{D}}_{2L} \end{array} \\ \begin{pmatrix} 1 & 0 & 0 & 0 \\ 0 & 0 & 1 & 0 \\ 0 & 1 & 0 & 0 \\ 0 & 0 & 0 & 1 \end{pmatrix} \end{array} \begin{array}{c} \mathscr{U}_{1R} \\ \mathscr{D}_{1R} \\ \mathscr{U}_{2R} \\ \mathscr{D}_{2R} \end{array}$$

so the electrically charged composite field $\bar{\mathscr{U}}_2 \mathscr{D}_1$ acquires a non-zero vacuum expectation value. Thus, the $U(1)_{em}$ subgroup of the gauge group $SU(2) \times U(1)$ is broken and the photon gets a mass when $|\Delta| < 1$. Of course, one can easily find a set of broken and unbroken generators corresponding to the new vacuum as well as all the gauge boson masses and pseudo-Goldstone boson masses.

An interesting conclusion of the above exercise is that the coupling of a weak gauge group to the chiral currents of a spontaneously broken chiral symmetry of strong interactions, may, for dynamical reasons, lead to different theories depending on the specific assignment of the chiral multiplets to the weak group representations.

Problems

11.1 Derive the remaining Feynman rules for the $U(1)$ theory of Section 11.1 in the R_ξ gauge.

11.2 Check by explicit calculation that at the tree level the unphysical ξ-dependent poles cancel in the S-matrix for the process $\varphi_1 + \varphi_2 \to A_1 + A_2$ where φ and A are the physical scalar particle and the massive vector meson of the $U(1)$ theory in Section 11.1, respectively.

11.3 Study the spontaneous symmetry breaking in the $SU(2) \times U(1)$ theory with one complex scalar doublet.

11.4 Calculate the longitudinal vector boson scattering $W_L + W_L \to W_L + W_L$ in the tree approximation in the 't Hooft–Feynman gauge.

11.5 Prove that in the 't Hooft–Feynman gauge the high energy amplitude for the scattering of longitudinal vector bosons is up to terms $O(m_W/s^{1/2})$ equal to the amplitude for scattering of unphysical Goldstone bosons (Lee, Quigg & Thacker 1977).

11.6 Extend the discussion of Section 11.2 to non-abelian gauge theories including fermions. Calculate the Higgs particle (physical scalar) mass in terms of the W and Z masses in the $SU(2) \times U(1)$ electroweak theory with one massless scalar complex doublet and taking fermions as massless.

12
Chiral anomalies

12.1 Triangle diagram and different renormalization conditions

Introduction

Anomalies have already been mentioned in this book on several occasions. In this Chapter we systematically discuss the fermion anomaly in $(3+1)$-dimensional quantum field theory (Adler 1969, Bell & Jackiw 1969). Its existence can be traced back to the short-distance singularity structure of products of local operators.

To be specific let us consider QCD. As discussed in Chapter 9, apart from being invariant under gauge transformations in the colour space its lagrangian is also invariant under global $SU(N) \times SU(N) \times U(1) \times U_A(1)$ chiral group of transformations acting in the flavour space.† Fermions belong to a vector-like i.e. real representation of the gauge group and for $N > 2$ to a complex representation of the chiral group (see Appendix B). We know from Chapters 9 and 10 that the Noether currents corresponding to the global flavour symmetry, although external with respect to the strong interaction gauge group, acquire important dynamical sense. The axial non-abelian currents couple to Goldstone bosons (pseudoscalar mesons) and the left-handed chiral currents couple to the intermediate vector bosons i.e. they are gauge currents of the weak interaction gauge group. Moreover, the conservation of the $U(1)$ current corresponds to baryon number conservation whereas the conservation of the $U_A(1)$ current is a problem (the so-called $U_A(1)$ problem): it can be seen that the spontaneous breakdown of the $SU(N) \times SU(N)$ implies the same for the $U_A(1)$ but there is no good candidate for the corresponding Goldstone boson in the particle spectrum.

It is clear from the preceding discussion that Green's functions involving chiral or axial currents are of direct physical interest. In the framework of the renormalization programme in QCD the finiteness of the matrix elements of the T-ordered

† We take quarks as massless. For massive quarks the chiral group is softly broken but, as we shall see, this is irrelevant for the discussion of anomalies.

products of such local composite operators as chiral currents is an additional requirement which, in general, must be separately verified. They may require an additional renormalization. Anomalies reflect the absence of a renormalization procedure such that Green's functions respect all the symmetries of the lagrangian. In other words an anomaly means a deviation of the renormalized theory from the canonical behaviour.

Historically, the main tool for studying anomalies is the Ward identities in perturbation theory. Relations obtained this way between various renormalized Green's functions involving currents can then be summarized in operator language: certain currents have anomalous divergences. Ward identities, being a generalization of Gauss' theorem, also provide a beautiful manifestation of the UV–IR interrelation in the theory: short-distance singularities manifest themselves in low energy theorems. This way anomalies can be seen experimentally. However, if gauge fields are coupled to linear combinations of vector and axial-vector flavour currents, as in the standard $SU(2) \times U(1)$ model of weak interactions, these combinations must be anomaly free. This is an important constraint on any theory. Otherwise, the anomaly spoils the conservation of the currents coupled to gauge bosons and gauge theories with gauge fields coupled to non-conserved currents are inconsistent. In the next Section we will see explicitly that the anomaly constitutes a breakdown of gauge invariance.

Anomalies originate in the short-distance singularities of the free fermion propagator which generate renormalization ambiguities in the Green's functions containing chiral currents. It turns out that in four space-time dimensions the Green's function which cannot be renormalized preserving all the classical symmetries of the lagrangian is the one containing two vector and one axial currents or three chiral currents. The basic structures to study are the triangle diagrams depicted in Fig. 12.1. As we know, in perturbation theory, the requirement of unitarity in each channel makes any one-loop amplitude well-defined up to a polynomial in the external momenta. This freedom is used in the renormalization procedure when we get rid of all infinite pieces. The triangle diagram is anomalous because it is impossible to add a polynomial so that all the classical symmetries are simultaneously preserved.

In the following we calculate explicitly the triangle amplitude and study its Ward identities. All internal quantum numbers of fermions are suppressed in this calculation. We consider the $U(1)$ and $U_A(1)$ currents of one-fermion free field theory. All the necessary generalizations can be easily introduced at the end.

To complete the introductory remarks it is worth mentioning that anomalies are also of basic importance for most theories going beyond the standard model. In particular the problem of chiral anomalies in more than four space-time dimensions and of gravitational anomalies as well as new techniques for their calculation have been vigorously studied (e.g. Frampton & Kephart 1983, Alvarez-Gaumé & Witten 1983, Zumino, Wu & Zee 1984). We shall come back to some of these subjects later on.

Calculation of the triangle amplitude

We consider a vacuum matrix element of three chiral currents $j^{\mathrm{L}}_{\mu}(x)$, say, or of one axial vector current $j^5_{\mu}(x)$, and two vector currents $j_{\mu}(x)$. For the time being we treat them as abelian currents of one-fermion free-field theory which, at the classical level, are conserved. If we allow for a Dirac mass term in the lagrangian then the axial-vector current is partially conserved

$$\partial^{\mu}j^5_{\mu}(x) = -2\partial^{\mu}j^{\mathrm{L}}_{\mu}(x) = 2mi\bar{\Psi}\gamma_5\Psi$$

Using the techniques of Section 10.1 and the fact that the canonical equal-time commutators $[j_0(\mathbf{x},t),\ j_{\mu}(\mathbf{y},t)]$, $[j^5_0(\mathbf{x},t),\ j_{\mu}(\mathbf{y},t)]$ and $[j^5_0(\mathbf{x},t),\ j^5_{\mu}(\mathbf{y},t)]$ vanish one can derive the standard Ward identities for the matrix elements $\langle 0|Tj^{\mathrm{L}}_{\mu}(z)\, j^{\mathrm{L}}_{\nu}(y)j^{\mathrm{L}}_{\delta}(x)|0\rangle$ and $\langle 0|Tj^5_{\mu}(z)j_{\nu}(y)j_{\delta}(x)|0\rangle$. In momentum space they read e.g.

$$-(p+q)^{\mu}\Gamma^5_{\mu\nu\delta}(p,q) = p^{\delta}\Gamma^5_{\mu\nu\delta}(p,q) = q^{\nu}\Gamma^5_{\mu\nu\delta}(p,q) = 0 \tag{12.1}$$

for $m=0$ or, for $m\neq 0$

$$-(p+q)^{\mu}\Gamma^5_{\mu\nu\delta}(p,q) = 2m\Gamma^5_{\nu\delta}(p,q) \tag{12.2}$$

and

$$p^{\delta}\Gamma^5_{\mu\nu\delta}(p,q) = q^{\nu}\Gamma^5_{\mu\nu\delta}(p,q) = 0 \tag{12.3}$$

where p and q are momenta conjugate to x and y, respectively, $\Gamma^5_{\mu\nu\delta}(p,q)$ is the Fourier transform of $\langle 0|Tj^5_{\mu}(z)j_{\nu}(y)j_{\delta}(x)|0\rangle$

$$\begin{aligned}&\Gamma^5_{\mu\nu\delta}(p,q)(2\pi)^4\delta(r+p+q)\\ &\quad= \int \mathrm{d}^4x\mathrm{d}^4y\mathrm{d}^4z \exp[-\mathrm{i}(px+qy+rz)]\langle 0|Tj^5_{\mu}(z)j_{\nu}(y)j_{\delta}(x)|0\rangle\end{aligned} \tag{12.4}$$

and $\Gamma^5_{\nu\delta}(p,q)$ is the Fourier transform of $\langle 0|TP(z)j_{\nu}(y)j_{\delta}(x)|0\rangle$ with $P=\bar{\Psi}\gamma_5\Psi$.

In a quantized theory, the current matrix elements get, at the one-loop level, a contribution from the triangle diagrams shown in Fig. 12.1 and from crossed diagrams with the external lines interchanged, $p\leftrightarrow q$ and $\delta\leftrightarrow\nu$. As we shall see, these contributions destroy the naive Ward identities (12.1)–(12.3).

We shall first consider the diagram on the r.h.s. of Fig. 12.1. The amplitude corresponding to this diagram reads

$$F_{\mu\nu\delta}(p,q) = (-1)\int\frac{\mathrm{d}^4k}{(2\pi)^4}\mathrm{Tr}\left[\frac{\mathrm{i}}{\not{k}+\not{q}-m}\gamma_{\nu}\frac{\mathrm{i}}{\not{k}-m}\gamma_{\delta}\frac{\mathrm{i}}{\not{k}-\not{p}-m}\gamma_{\mu}\gamma_5\right] \tag{12.5}$$

and

$$\Gamma^5_{\mu\nu\delta}(p,q) = F_{\mu\nu\delta}(p,q) + F_{\mu\delta\nu}(q,p) \tag{12.6}$$

We take here a non-zero fermion mass but, as we shall see, the anomaly is mass independent. Thus the result we will obtain is valid for the $m=0$ case, too. We also keep in mind that, if currents are coupled to gauge fields with some coupling constant g, an extra factor $(-\mathrm{i}g)$ appears in each such vertex.

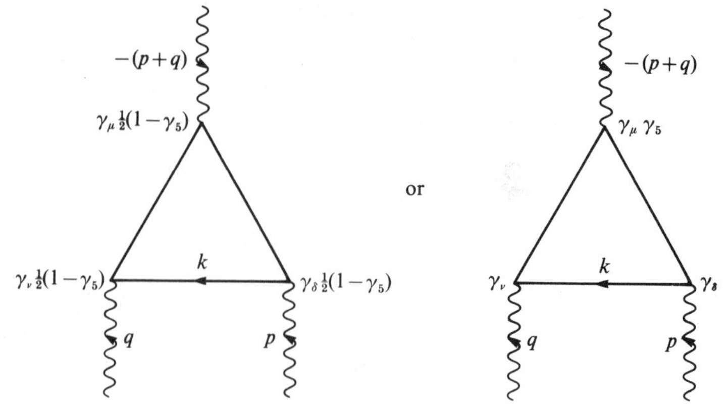

Fig. 12.1

The first observation worth making is the symmetry property

$$F_{\mu\nu\delta}(p,q) = F_{\mu\delta\nu}(q,p) \tag{12.7}$$

This follows from (12.5) by using the change of variables $k' = -k$, anticommuting $\gamma_\mu\gamma_5 = -\gamma_5\gamma_\mu$ and reversing the order of factors

$$\mathrm{Tr}\,[\gamma_\alpha\gamma_\nu\gamma_\beta\gamma_\delta\gamma_\gamma\gamma_\mu\gamma_5] = \mathrm{Tr}\,[\gamma_\mu\gamma_\gamma\gamma_\delta\gamma_\beta\gamma_\nu\gamma_\alpha\gamma_5] \tag{12.8}$$

We also notice that the sign of the mass factor is irrelevant because the only non-vanishing mass-dependent terms in the trace are proportional to m^2.

By power counting the amplitude $F_{\mu\nu\delta}(p,q)$ is superficially linearly divergent. It is impossible to renormalize this amplitude preserving simultaneously all three identities (12.1) or (12.2) and (12.3). It depends on our choice of the renormalization constraints, dictated by the physics of the problem, where the anomaly is actually placed.

For clarity of the following discussion it is useful to first write down the most general form of the amplitude $\Gamma^5_{\mu\nu\delta}(p,q) = 2F_{\mu\nu\delta}(p,q)$ consistent with the requirements of parity and Lorentz invariance. It reads

$$\begin{aligned}\Gamma^5_{\mu\nu\delta}(p,q) = {} & A_1(p,q)\varepsilon_{\mu\nu\delta\alpha}p^\alpha + A_2(p,q)\varepsilon_{\mu\nu\delta\alpha}q^\alpha + B_1(p,q)p_\nu\varepsilon_{\mu\delta\alpha\beta}p^\alpha q^\beta \\ & + B_2(p,q)q_\nu\varepsilon_{\mu\delta\alpha\beta}p^\alpha q^\beta + B_3(p,q)p_\delta\varepsilon_{\mu\nu\alpha\beta}p^\alpha q^\beta + B_4(p,q)q_\delta\varepsilon_{\mu\nu\alpha\beta}p^\alpha q^\beta\end{aligned} \tag{12.9}$$

where As and Bs are scalar functions. The terms $p_\mu\varepsilon_{\nu\delta\alpha\beta}p^\alpha q^\beta$ and $q_\mu\varepsilon_{\nu\delta\alpha\beta}p^\alpha q^\beta$ are not linearly independent. The requirement of Bose symmetry under interchange of the two vector currents $\Gamma^5_{\mu\nu\delta}(p,q) = \Gamma^5_{\mu\delta\nu}(q,p)$ implies that $A_1(p,q) = -A_2(q,p)$, $B_1(p,q) = -B_4(q,p)$, $B_2(p,q) = -B_3(q,p)$. Using again power counting arguments it is clear that only the functions A_1 and A_2 are divergent, in fact only logarithmically divergent, whereas the functions B_i are finite. Subtractions are, in principle, necessary for the functions A_1 and A_2. However, these functions can be expressed in terms of Bs when definite renormalization conditions are imposed. Imagine, for instance, that we insist on the vector current conservation

$$q^\nu \Gamma^5_{\mu\nu\delta}(p,q) = p^\delta \Gamma^5_{\mu\nu\delta}(p,q) = 0 \tag{12.10}$$

for the renormalized vertex $\Gamma^5_{\mu\nu\delta}$. Then we must have

$$\left.\begin{aligned} A_1 + p\cdot qB_1 + q^2 B_2 = 0 \\ A_2 + p^2 B_3 + p\cdot qB_4 = 0 \end{aligned}\right\} \tag{12.11}$$

and indeed the conditions (12.11) determine the divergent amplitudes A_i in terms of finite B_is. From Bose symmetry $A_1(p,q) = -A_2(q,p)$ follows so that only one of the equations (12.11) is independent.

There are many ways to calculate anomalies. One is to regularize the amplitude (12.5) by some regularization prescription and to calculate Ward identities for the regularized $\Gamma^5_{\mu\nu\delta}$. If the regularization procedure satisfies the constraints which we want to impose as our renormalization constraints then in the limit of no regulator we obtain the anomalous Ward identity we are looking for. Given this identity, by additional finite subtractions one can always transform it into another anomalous Ward identity, satisfying different renormalization constraints. Let us follow this approach choosing to work with the Pauli–Villars regularization (Itzykson & Zuber 1980) of the amplitude (12.5). Dimensional regularization has in this case some problems with ambiguities in the extension of γ_5 to n dimensions (for a review of this matter see, for instance, Ovrut (1983) and references therein). Let us choose as our renormalization conditions the relations (12.10). Thus we insist on vector current conservation for the renormalized vertex $\Gamma^5_{\mu\nu\delta}$.

The Pauli–Villars regularization of the amplitude $\Gamma^5_{\mu\nu\delta}(p,q)$ consists of considering it to be a function of the mass of the fermion circulating in the loop: $\Gamma^5_{\mu\nu\delta}(p,q) = \Gamma^5_{\mu\nu\delta}(p,q,m)$. A regulated amplitude is defined as the difference between the given amplitude and the same amplitude taken at some other value M of the mass (M is the regulator)

$$\Gamma^5_{\mu\nu\delta}(p,q,m,M) = \Gamma^5_{\mu\nu\delta}(p,q,m) - \Gamma^5_{\mu\nu\delta}(p,q,M) \tag{12.12}$$

The regularized amplitude $\Gamma^5_{\mu\nu\delta}(p,q,m,M)$ is finite. This can easily be checked by expanding the integrand in the integrals over d^4k in powers of $1/k^2$: the linear divergence of $\Gamma^5_{\mu\nu\delta}(p,q,m)$ is cancelled by $\Gamma^5_{\mu\nu\delta}(p,q,M)$. At the end one lets the mass M go to infinity and the final answer is finite with no additional subtractions, because it satisfies our renormalization conditions fixing A_i in terms of B_i.

First we show that the regularized amplitude satisfies the 'normal' Ward identities

$$-(p+q)^\mu \Gamma^5_{\mu\nu\delta}(p,q,m,M) = 2m\Gamma^5_{\nu\delta}(p,q,m) - 2M\Gamma^5_{\nu\delta}(p,q,M) \tag{12.13}$$

$$p^\delta \Gamma^5_{\mu\nu\delta}(p,q,m,M) = q^\nu \Gamma^5_{\mu\nu\delta}(p,q,m,M) = 0 \tag{12.14}$$

Indeed, using the relation $\not{p} + \not{q} = (\not{q} + \not{k} - m) - (\not{k} - \not{p} - m)$ we can write down the following equation

$$-(p+q)^\mu\Gamma^5_{\mu\nu\delta}(p,q,m,M)=2\mathrm{i}\int\frac{\mathrm{d}^4k}{(2\pi)^4}\mathrm{Tr}\left[\frac{1}{\not k+\not q-m}\gamma_\nu\frac{1}{\not k-m}\gamma_\delta\gamma_5-(m\to M)\right]$$

$$+2\mathrm{i}\int\frac{\mathrm{d}^4k}{(2\pi)^4}\mathrm{Tr}\left[\gamma_\nu\frac{1}{\not k-m}\gamma_\delta\frac{1}{\not k-\not p-m}\gamma_5-(m\to M)\right]$$

$$+2\mathrm{i}\int\frac{\mathrm{d}^4k}{(2\pi)^4}\mathrm{Tr}\left[\frac{1}{\not k+\not q-m}\gamma_\nu\frac{1}{\not k-m}\gamma_\delta\frac{1}{\not k-\not p-m}2m\gamma_5-(m\to M)\right] \tag{12.15}$$

All the integrals are finite. The first two terms are second rank pseudotensors depending on only one momentum variable and hence vanish. The last term gives just the Ward identity (12.13). Consider now $p^\delta\Gamma^5_{\mu\nu\delta}(p,q,m,M)$. Using $\not p=(\not k-m)-(\not k-\not p-m)$ this becomes

$$p^\delta\Gamma^5_{\mu\nu\delta}(p,q,m,M)=2\mathrm{i}\int\frac{\mathrm{d}^4k}{(2\pi)^4}\mathrm{Tr}\left[\frac{1}{\not k+\not q-m}\gamma_\nu\frac{1}{\not k-\not p-m}\gamma_\mu\gamma_5-(m\to M)\right]$$

$$-2\mathrm{i}\int\frac{\mathrm{d}^4k}{(2\pi)^4}\mathrm{Tr}\left[\frac{1}{\not k+\not q-m}\gamma_\nu\frac{1}{\not k-m}\gamma_\mu\gamma_5-(m\to M)\right] \tag{12.16}$$

The last term vanishes for the same reason as above. The first term also vanishes because we can shift the momentum integration variable to $k'=k-p$ and again apply the same argument. This shift of the integration variable is legitimate because the integral is finite. If it were linearly divergent the translation of the integration variable might involve a non-vanishing surface term which would spoil the argument† (see Problem 12.1). In a quite analogous way we can show that

$$q^\nu\Gamma^5_{\mu\nu\delta}(p,q,m,M)=0$$

Thus, our regularization procedure is consistent with our renormalization requirements (12.10). Our last step is to take the limit $M\to\infty$ in the Ward identity (12.13). The anomaly results from the non-vanishing of the regulator term in (12.13) in that limit. The $\Gamma^5_{\nu\delta}(p,q,m)$ vertex can be calculated explicitly since the Feynman integral is convergent

$$\Gamma^{5\nu\delta}(p,q,m)=2\mathrm{i}\int\frac{\mathrm{d}^4k}{(2\pi)^4}\mathrm{Tr}\left[\frac{1}{\not k+\not q-m}\gamma^\nu\frac{1}{\not k-m}\gamma^\delta\frac{1}{\not k-\not p-m}\gamma_5\right]$$

$$=2\times 4m\mathrm{i}\varepsilon^{\nu\delta\beta\alpha}p_\beta q_\alpha F(p,q,m) \tag{12.17}$$

where

$$F(p,q,m)=\mathrm{i}\int\frac{\mathrm{d}^4k}{(2\pi)^4}\frac{1}{(k+q)^2-m^2}\frac{1}{k^2-m^2}\frac{1}{(p-k)^2-m^2} \tag{12.18}$$

† Note the difference between a shift of the integration variable by a constant and a change of variables $k\to -k$ used in proving (12.7).

and the factor 2 in front of the integral in (12.17) takes account of the crossed diagram.

Using the Feynman parametrization (A.16) and integrating over d^4k one gets

$$F(p,q,m) = -\frac{1}{16\pi^2}\int_0^1 dx \int_0^{1-x} dy[q^2y(1-y) + p^2(1-x)x + 2pqxy - m^2]^{-1} \tag{12.19}$$

It is now straightforward to calculate the limit

$$\lim_{M\to\infty} 2M\Gamma^{5\nu\delta}(p,q,M) = +(i/2\pi^2)\varepsilon^{\nu\delta\beta\alpha}p_\beta q_\alpha \tag{12.20}$$

and the final form of the axial Ward identity consistent with our renormalization constraints reads

$$\left.\begin{aligned} -(p+q)^\mu\Gamma^5_{\mu\nu\delta}(p,q,m) &= 2m\Gamma^5_{\nu\delta}(p,q,m) - (i/2\pi^2)\varepsilon_{\nu\delta\beta\alpha}p^\beta q^\alpha \\ p^\delta\Gamma^5_{\mu\nu\delta} &= q^\nu\Gamma^5_{\mu\nu\delta} = 0 \end{aligned}\right\} \tag{12.21}$$

The anomalous term is m independent.

Another way to obtain the anomalous Ward identity (12.21) is to calculate explicitly the finite functions B_i starting with the unregularized amplitude (12.5) and then to use renormalization constraints (12.11). This we leave as an exercise for the reader.

Different renormalization constraints for the triangle amplitude

The renormalization conditions (12.10) lead to the result (12.21). Whether we should impose them or not, depends on the physics of the problem. This will be illustrated in the next Section. It may happen that we must, instead, insist on the relation

$$-(p+q)^\mu\Gamma^5_{\mu\nu\delta}(p,q,m) = 2m\Gamma^5_{\nu\delta}(p,q,m) \tag{12.22}$$

as our renormalization constraint. The divergent amplitudes A_1 and A_2 are again uniquely specified by the condition (12.22). Indeed, with an additional finite subtraction for the A_1 and A_2

$$a_1\varepsilon_{\mu\nu\delta\alpha}p^\alpha + a_2\varepsilon_{\mu\nu\delta\alpha}q^\alpha \tag{12.23}$$

we see, by comparing (12.21) and (12.22), that the latter is satisfied for

$$a_1 - a_2 = i/2\pi^2$$

Since $a_1(p,q) = -a_2(q,p)$ from the Bose symmetry under interchange of the two vector currents discussed earlier, we conclude that

$$a_1 = -a_2 = i/4\pi^2 \tag{12.24}$$

Thus, with (12.22) as the renormalization condition one gets

$$\left.\begin{aligned} p^\delta\Gamma^5_{\mu\nu\delta}(p,q,m) &= -(\mathrm{i}/4\pi^2)\varepsilon_{\mu\nu\alpha\beta}p^\alpha q^\beta \\ q^\nu\Gamma^5_{\mu\nu\delta}(p,q,m) &= (\mathrm{i}/4\pi^2)\varepsilon_{\mu\delta\alpha\beta}p^\alpha q^\beta \end{aligned}\right\} \quad (12.25)$$

One cannot satisfy simultaneously all three Ward identities (12.2) and (12.3).

Finally let us consider the triangle on the l.h.s. of Fig. 12.1 with the left-handed currents at each vertex and in the limit $m = 0$. At the level of the divergent Feynman integral the bare amplitude $\Gamma^{\mathrm{L}}_{\mu\nu\delta}(p,q)$ formally satisfies the following relation

$$\Gamma^{\mathrm{L}}_{\mu\nu\delta}(p,q) = \tfrac{1}{2}\Gamma_{\mu\nu\delta}(p,q) - \tfrac{1}{2}\Gamma^5_{\mu\nu\delta}(p,q) \quad (12.26)$$

where $\Gamma_{\mu\nu\delta}(p,q)$ is the three-vector current amplitude which has no anomaly. Thus formally the anomaly of $\Gamma^{\mathrm{L}}_{\mu\nu\delta}(p,q)$ can also be calculated from the triangle on the r.h.s. of Fig. 12.1. However, our previous renormalization conditions are inappropriate since now for the Green's function of three identical currents we must require Bose symmetry of the renormalized $\Gamma^5_{\mu\nu\delta}(p,q)$ (which we denote by $\tilde{\Gamma}^5_{\mu\nu\delta}(p,q)$ to distinguish from the AVV amplitude considered earlier) under interchange of any pair of vertices. We can get the right answer by applying a finite renormalization to the previous result (12.21). Adding a finite polynomial term, (12.23), to the amplitudes A_1 and A_2 we get (for $m = 0$)

$$\left.\begin{aligned} -(p+q)^\mu\tilde{\Gamma}^5_{\mu\nu\delta}(p,q) &= (-\mathrm{i}/2\pi^2 + a_1 - a_2)\varepsilon_{\nu\delta\alpha\beta}p^\alpha q^\beta \\ p^\delta\tilde{\Gamma}^5_{\mu\nu\delta}(p,q) &= a_2\varepsilon_{\mu\nu\alpha\beta}p^\alpha q^\beta \\ q^\nu\tilde{\Gamma}^5_{\mu\nu\delta}(p,q) &= a_1\varepsilon_{\mu\delta\alpha\beta}p^\alpha q^\beta \end{aligned}\right\} \quad (12.27)$$

Insisting on symmetry under the interchange $-(p+q)\leftrightarrow p$, $\mu\leftrightarrow\delta$ and $p\leftrightarrow q$, $\delta\leftrightarrow\nu$ one gets the following equations

$$\left.\begin{aligned} -\mathrm{i}/2\pi^2 + a_1 - a_2 &= a_2 \\ a_1 &= -a_2 \end{aligned}\right\} \quad (12.28)$$

Therefore

$$a_1 = -a_2 = \mathrm{i}/6\pi^2 \quad (12.29)$$

and

$$\left.\begin{aligned} -(p+q)^\mu\tilde{\Gamma}^5_{\mu\nu\delta}(p,q) &= -(\mathrm{i}/6\pi^2)\varepsilon_{\nu\delta\alpha\beta}p^\alpha q^\beta \\ p^\delta\tilde{\Gamma}^5_{\mu\nu\delta}(p,q) &= -(\mathrm{i}/6\pi^2)\varepsilon_{\mu\nu\alpha\beta}p^\alpha q^\beta \\ q^\nu\tilde{\Gamma}^5_{\mu\nu\delta}(p,q) &= +(\mathrm{i}/6\pi^2)\varepsilon_{\mu\delta\alpha\beta}p^\alpha q^\beta \end{aligned}\right\} \quad (12.30)$$

Thus, the same triangle amplitude leads to three different sets of anomalous Ward identities depending on the imposed renormalization conditions. The latter must be chosen according to the physical requirements of the problem under consideration. We recall that our present interest is in the amplitude $\Gamma^{\mathrm{L}}_{\mu\nu\delta}(p,q)$ so we combine (12.30) with (12.26).

Important comments

In the next Section we will discuss in more detail the anomaly problem in the context of specific physical theories. Here, to illustrate the meaning of arbitrariness in renormalization conditions for the triangle diagram, we consider two models given by the lagrangians

$$\mathscr{L} = Z_2\bar{\Psi}\mathrm{i}\not{D}\Psi - \tfrac{1}{4}Z_3F_{\mu\nu}F^{\mu\nu} \tag{12.31}$$

and

$$\mathscr{L} = Z_2^{\mathrm{R}}\bar{\Psi}_{\mathrm{R}}\mathrm{i}\not{\partial}\Psi_{\mathrm{R}} + Z_2^{\mathrm{L}}\bar{\Psi}_{\mathrm{L}}\mathrm{i}\not{D}\Psi_{\mathrm{L}} - \tfrac{1}{4}Z_3F_{\mu\nu}F^{\mu\nu} \tag{12.32}$$

respectively, where

$$D_\mu = \partial_\mu + \mathrm{i}gA_\mu, \quad F_{\mu\nu} = \partial_\mu A_\nu - \partial_\nu A_\mu$$

and the lagrangians are written in terms of renormalized quantities. We define the bare currents, for instance vector and axial-vector currents in the theory (12.31), which in terms of the renormalized fields are as follows

$$j_\mu^5(x) = Z_2\bar{\Psi}(x)\gamma_\mu\gamma_5\Psi(x) \tag{12.33}$$

and

$$j_\mu(x) = Z_2\bar{\Psi}(x)\gamma_\mu\Psi(x) \tag{12.34}$$

A comment is in order here on the renormalization of currents in presence of the axial anomaly. In Section 10.1 we have proved that a conserved or partially conserved current satisfying normal Ward identities is not renormalized. This conclusion is no longer valid in presence of the anomaly. The current is still multiplicatively renormalized ${}^{\mathrm{R}}j_\mu^5 = Z_5 j_\mu^5$ but Z_5 is not finite (we will come back to this point shortly). However, as we know from the previous Section, at the one-loop level the current matrix elements are finite after imposing certain renormalization conditions, without additional counterterms. Thus, at this level $Z_5 = 1$.

The first theory is just QED of a single fermion field. The vector current is coupled to the electromagnetic field and the axial-vector current, conserved in the Noether sense, is not coupled to any gauge field. Gauge invariance of the theory requires the vector current conservation. Thus for the triangle diagram involving j_μ^5 and two vector currents we must impose the renormalization conditions (12.10) which lead to the anomalous Ward identity (12.21). In operator language it corresponds to an anomaly in the divergence of the axial-vector current

$$\partial^\mu j_\mu^5(x) = (g^2/16\pi^2)\varepsilon^{\mu\nu\rho\delta}F_{\mu\nu}(x)F_{\rho\delta}(x) \tag{12.35}$$

Equation (12.35) can be verified using the Feynman rules for the operator insertion

$$\varepsilon^{\mu\nu\rho\delta}F_{\mu\nu}(x)F_{\rho\delta}(x) = 4\varepsilon^{\mu\nu\rho\delta}\partial_\mu A_\nu(x)\partial_\rho A_\delta(x) \tag{12.36}$$

(see Section 7.6). Considering e.g. the Green's function $\langle 0|T\partial_\mu A_\nu(z)\partial_\rho A_\delta(z)A_\alpha(y)$ $A_\beta(x)|0\rangle$ in momentum space one gets the Feynman diagram shown in Fig. 12.2

$$\otimes \equiv (\mathrm{i}/16\pi^2)\, g^2 \epsilon^{\mu\nu\rho\delta} F_{\mu\nu} F_{\rho\delta} \equiv (\mathrm{i}/16\pi^2)\, 2g^2 F_{\mu\nu} \tilde{F}^{\mu\nu}$$

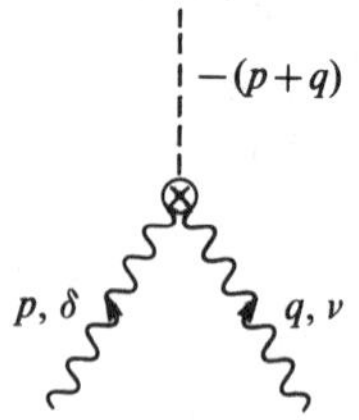

$$\longrightarrow -(\mathrm{i}/2\pi^2)\, g^2 \epsilon^{\nu\delta\mu\rho} p_\mu q_\rho$$

Fig. 12.2

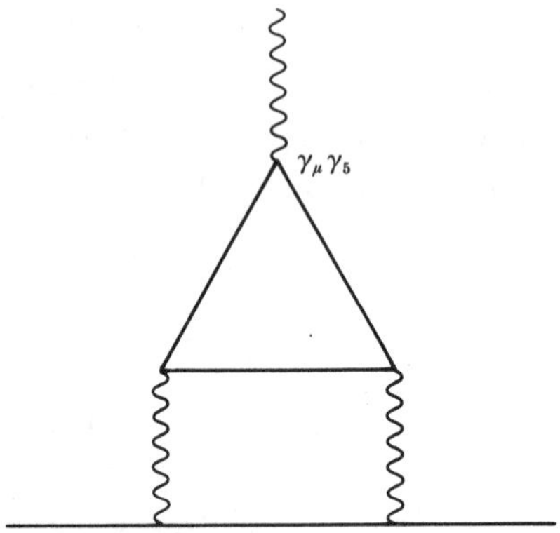

Fig. 12.3

(propagators are, as usual, removed from the external lines). Therefore, in order g^2 the Fourier transform (12.21)† (with $m = 0$) of the divergence $\partial^\mu \langle 0 | T j_\mu^5 j_\nu j_\delta | 0 \rangle$ is given by the Feynman diagram in Fig. 12.2 and the relation (12.35) follows if we recall the general form (10.11) of Ward identities. Using (12.35) we can calculate perturbatively other anomalous Ward identities involving axial-vector current $j_\mu^5(x)$. For instance, the derivative of the Green's function $\Gamma_{\Psi\Psi}^{5\mu} = \langle 0 | T\Psi(x) j^{5\mu}(0)$ $\bar{\Psi}(y) | 0 \rangle$ is anomalous because the latter gets a contribution from the diagram in Fig. 12.3. and in the lowest non-vanishing order in g^2 the anomaly is given by the diagram in Fig. 12.4. The anomalous Ward identity for $\Gamma_{\Psi\Psi}^5$ (derive it using the path integral technique of Sections 10.2 and 12.3) can be used to study the renormalization properties of the axial-vector current. Closely following the procedure of Section 10.1 we conclude that Z_5 is not finite as the operator $\tilde{F}_{\mu\nu} F^{\mu\nu}$ does not have finite matrix elements: e.g. the diagram in Fig. 12.4 is logarithmically divergent.

Our second example, the lagrangian (12.32), is a theory in which only the left-handed chiral current couples to the gauge fields. The triangle diagram with gauge source currents at each vertex should be evaluated imposing symmetry under

† Now with $(-\mathrm{i}g)$ at each vector vertex included.

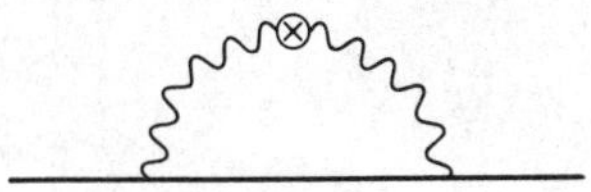

Fig. 12.4

interchange of any pair of indices. Using relations (12.26) and (12.30) we now get the anomaly in the chiral current

$$\partial^\mu j^L_\mu(x) = -(g^2/48\pi^2)F_{\mu\nu}\tilde{F}^{\mu\nu} \tag{12.37}$$

Because of the anomaly the gauge source current is no longer conserved and the theory is not gauge invariant.

Writing the operator equation (12.35) for the bare current we have used the result obtained in the lowest order in g^2 in perturbation theory. Equivalently one can say that the gauge field has been treated as a classical background field. One believes that the anomalies are not modified by higher order corrections, i.e. that (12.35) is exact to any order in g^2 for the bare current in the regularized theory or for the renormalized operators (see also the next Section) with g and $F_{\mu\nu}$ being the renormalized quantities defined by the lagrangian (12.31). For the vector abelian theory considered here this has been proved to all orders in perturbation theory (Adler & Bardeen 1969). The proof is technical but its main point can be summarized as follows: the necessity of renormalization of the higher order corrections in g^2 i.e. of photon insertions in the triangle diagram does not interfere with chiral symmetry, so the chiral anomaly should arise only from the lowest order fermion loops. (On the contrary the scale invariance anomaly discussed in Chapter 7 does obtain corrections because renormalization violates scale invariance.) The Adler–Bardeen theorem also holds in non-abelian gauge theory but it may be regularization scheme dependent (Bardeen 1972), particularly in supersymmetric theories. In gauge theories with chiral gauge symmetry the anomalies must be absent for a consistent theory. According to the Adler–Bardeen theorem generalized to the non-abelian case a theory is completely free of anomalies if, and only if, all the triangle diagram anomalies are absent.

Our last remark in this Section is concerned with generalizing the triangle diagram calculation to the case where several species of fermion fields having some internal degrees of freedom are present. Let the currents in the vertices have coupling matrices $\lambda_1^a, \lambda_2^b, \lambda_3^c$. The currents may be coupled to gauge fields or not and the matrices $\lambda_1^a, \lambda_2^b, \lambda_3^c$ may transform non-trivially under different internal symmetry groups and be block-diagonal in the remaining internal symmetry spaces. For the triangle diagram we get then an extra factor $\frac{1}{2}\mathrm{Tr}[\lambda_1^a\{\lambda_2^b, \lambda_3^c\}]$ where the trace corresponds to summing over different fermions in the loop and the anticommutator is due to the symmetrization over the two diagrams in Fig. 12.5. The necessary modifications of operator equations, like (12.35), will be discussed later on.

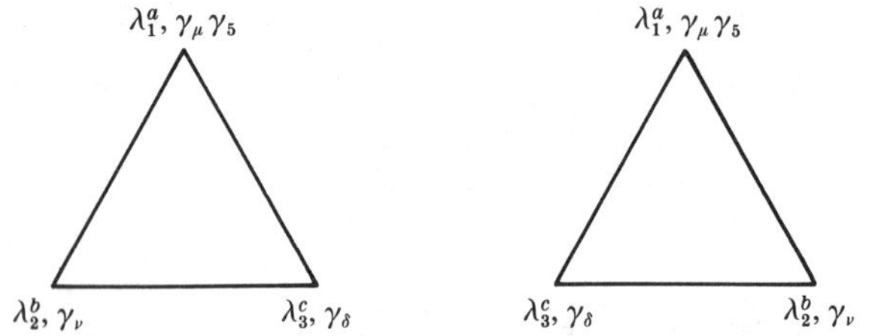

Fig. 12.5

12.2 Some physical consequences of the chiral anomalies

Chiral invariance in spinor electrodynamics

The lagrangian (12.31) of spinor electrodynamics is invariant both under $U(1)$ and axial $U_A(1)$ transformations. However, in quantum theory if we want to preserve vector gauge invariance an anomaly appears in the divergence of the axial-vector current. If we introduce a cut-off which preserves photon gauge invariance then to any order in g^2 the divergence of the unrenormalized axial-vector current is given by (12.35). As discussed in the previous Section, the triangle diagram amplitude turns out to be finite even in the limit of no cut-off but the higher order corrections induce infinities in the axial-current matrix elements. Not all of them can be absorbed in the renormalization constants present in the lagrangian (12.31). Therefore due to the anomaly the axial-vector current is renormalized:

$$^{R}j_\mu^5 = Z_5 j_\mu^5$$

and we must reexpress (12.35) in terms of the renormalized operators. Defining

$$(g^2/8\pi^2)[F_{\mu\nu}(x)\tilde{F}^{\mu\nu}(x)]^R = (g^2/8\pi^2)F_{\mu\nu}(x)\tilde{F}^{\mu\nu}(x) + (Z_5 - 1)\partial^\mu j_\mu^5(x) \qquad (12.38)$$

we may write (12.35) in terms of the renormalized operators

$$\partial^{\mu\,R}j_\mu^5(x) = (g^2/8\pi^2)[F_{\mu\nu}(x)\tilde{F}^{\mu\nu}(x)]^R \qquad (12.39)$$

However this renormalized axial-vector current does not generate chiral transformations. The axial charge is not conserved, it does not commute with the photon field and does not have the correct equal-time commutation relations with the renormalized spinor field, due to the presence of a dimension four operator $(F_{\mu\nu}\tilde{F}^{\mu\nu})^R$ in the divergence of the current (see Section 10.1).

Nevertheless, it turns out that chiral invariance remains a symmetry of the quantum theory based on the lagrangian (12.31) and it is possible to construct the corresponding charge Q^5. Following Adler (1970) we first define a new bare current: symmetry current

$$j_{5\mu}^S(x) = j_\mu^5(x) - (g^2/4\pi^2)A^\nu(x)\tilde{F}_{\mu\nu}(x) \qquad (12.40)$$

which, from (12.35), is conserved

$$\partial^\mu j_{5\mu}^S = 0 \qquad (12.41)$$

Its existence is directly related to our discussion of different renormalization conditions of the triangle diagram; compare (12.21) and (12.22). This current is a finite operator (Bardeen 1974) and does not need renormalization. The current $j^S_{5\mu}$ is conserved but it is explicitly gauge dependent due to the presence of the gauge field in its definition and therefore is not an observable operator. But the associate charge

$$Q_5(t) = \int d^3x j^S_{50}(\mathbf{x}, t) \tag{12.42}$$

is from (12.41) time independent, its commutator with $\Psi(x)$, calculated by use of the canonical commutation relations, is

$$[Q_5(t), \Psi(\mathbf{x}, t)] = -\gamma_5\Psi(x) \tag{12.43}$$

it commutes with the photon field and it is gauge invariant. Let us check the last property. Under the gauge transformation

$$\left.\begin{aligned} \delta\Psi(x) &= -ig\Theta(x)\Psi(x) \\ \delta A_\mu(x) &= -\partial_\mu\Theta(x) \end{aligned}\right\} \tag{12.44}$$

the symmetry current transforms as follows:

$$\delta j^S_{5\mu} = (g^2/4\pi^2)\partial^\nu\Theta(x)\tilde{F}_{\mu\nu}(x) = (g^2/4\pi^2)\partial^\nu(\Theta(x)\tilde{F}_{\mu\nu}(x)) \tag{12.45}$$

Therefore

$$\delta Q_5(t) = \int d^3x\delta j^S_{50}(x) \sim \int d^3x\partial^\nu(\Theta(x)\tilde{F}_{0\nu}(x)) = \int d^3x\partial^k(\Theta(x)\tilde{F}_{0k}(x)) = 0 \tag{12.46}$$

Gauge invariance of Q_5 in spinor electrodynamics follows from the vanishing of $\Theta(x)\tilde{F}_{0k}(x)$ at spatial infinity. The situation is different in non-abelian gauge theories (see Section 8.3) where not all finite gauge transformations can be reached by iterating infinitesimal transformations. We conclude that in spite of the presence of the axial anomaly chiral symmetry remains a good symmetry in massless spinor electrodynamics.

$\pi^0 \to 2\gamma$

The relevance of the chiral anomaly to the decay $\pi^0 \to 2\gamma$ relies on our identification of pions as the Goldstone bosons of the spontaneously broken chiral symmetry of QCD (see Chapter 9). Thus, the axial-vector flavour currents of the $SU(2) \times SU(2)$ symmetric QCD lagrangian (9.1) acquire dynamical sense through their coupling to pions, see (9.65)

$$\langle 0|j^{5a}_\mu(x)|\pi^b\rangle = if_\pi p_\mu\delta^{ab}\exp(-ipx)$$

where p_μ is the pion four-momentum. To calculate the $\pi^0 \to 2\gamma$ decay rate we

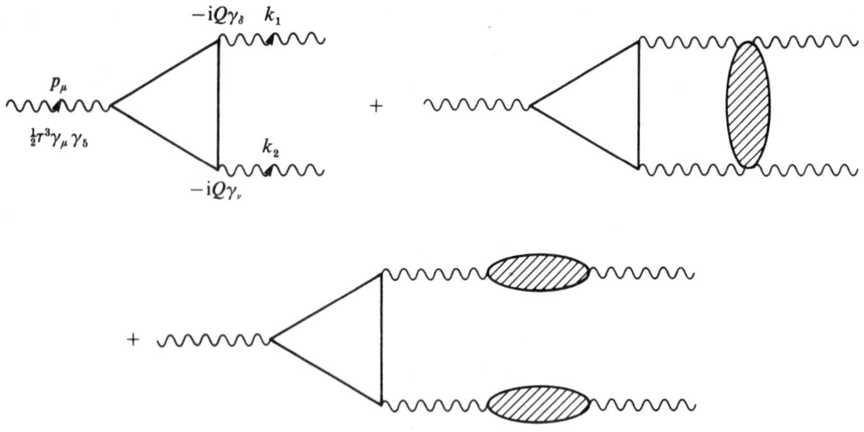

Fig. 12.6

consider the amplitude ($j_\mu^{5a}(x)$ is the renormalized axial isospin current)

$$\int d^4x \exp(-iqx)\langle\gamma_1(k_1)\gamma_2(k_2)|j_\mu^{53}(x)|0\rangle = \int d^4x\, d^4y\, d^4z \exp(+ik_1y)\exp(+ik_2z)$$
$$*\exp(-iqx)\langle 0|Tj_\mu^{53}(x)j_\nu(y)j_\delta(z)|0\rangle\varepsilon_1^\nu(k_1)\varepsilon_2^\delta(k_2) \tag{12.47}$$

and use the PCAC (partial conservation of the axial-vector current) approximation relating physical quantities to the results obtained in the exact symmetry limit $p^2 = m_\pi^2 = 0$. In this limit there is a pion pole contribution to the l.h.s. of (12.47) (see e.g. (2.194), (2.195) and (2.196))

$$\lim_{q^2\to p^2}\int d^4x \exp(-iqx)\langle\gamma_1(k_1)\gamma_2(k_2)|j_\mu^{53}(x)|0\rangle = f_\pi(p_\mu/p^2)\langle\gamma_1\gamma_2|\pi^0\rangle + O(k_1, k_2) \tag{12.48}$$

where $O(k_1, k_2)$ has no pole in p^2.

We now take the divergence of (12.47). The divergence of the r.h.s. of (12.47) is given by the results of the previous Section with the triangle diagram specified as shown in Fig. 12.6. The fermion lines are quark lines, the Pauli matrix τ^3 is the neutral axial-vector coupling matrix and Q is the diagonal quark electric charge matrix. In this case, for $\Gamma_{\mu\nu\delta}^5$ defined by (12.4) we get the following

$$p^\mu\Gamma_{\mu\nu\delta}^5(k_1, k_2) = (i/2\pi^2)\mathrm{Tr}[\tfrac{1}{2}\tau^3Q^2]\varepsilon_{\nu\delta\beta\alpha}k_1^\beta k_2^\alpha \tag{12.49}$$

plus terms of at least third order in k_i from the second diagram: the factor k_1k_2 comes from the gauge-invariant coupling (through the field strength tensor $F_{\mu\nu}$) of external photons and an extra factor $(k_1 + k_2)$ comes from the anomalous divergence term. The effect of the last diagram is only the coupling-constant renormalization. With u and d quarks of three colours we get

$$\mathrm{Tr}[\tfrac{1}{2}\tau^3Q^2] = 3\times\tfrac{1}{2}(\tfrac{4}{9} - \tfrac{1}{9})e^2 = \tfrac{1}{2}e^2 \tag{12.50}$$

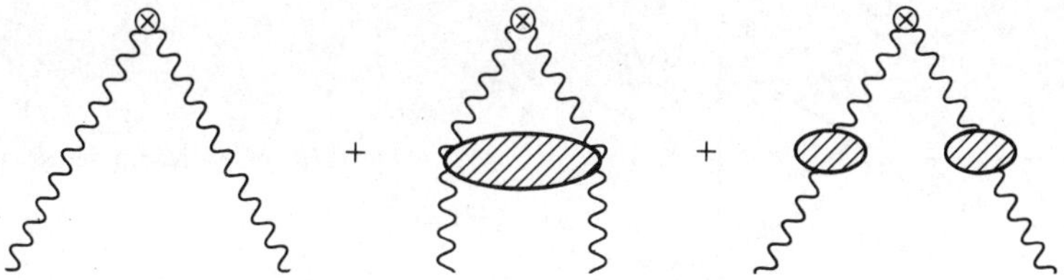

Fig. 12.7

Therefore, from (12.47) and (12.48)

$$\langle\gamma_1\gamma_2|\pi^0\rangle = \frac{1}{f_\pi}\frac{e^2}{4\pi^2}\mathrm{i}\varepsilon_{\nu\delta\beta\alpha}k_1^\beta k_2^\alpha\varepsilon_1^\nu\varepsilon_2^\delta(2\pi)^4\delta(p-k_1-k_2)+O(k_1,k_2,p) \quad (12.51)$$

Our calculation of the divergence of (12.47) is, of course, equivalent to calculating perturbatively the matrix element $\langle\gamma_1\gamma_2|\partial^\mu j_\mu^{5a}(x)|0\rangle$ with the divergence given by

$$\partial^\mu j_\mu^{5a}(x) = (e^2/8\pi^2)\tfrac{1}{2}\mathrm{Tr}\left[\tfrac{1}{2}\tau^a\{Q,Q\}\right]F_{\mu\nu}\tilde{F}^{\mu\nu} \quad (12.52)$$

The contributing diagrams are those shown in Fig. 12.7. Terms $O(k_1,k_2)$ in (12.48) and $O(k_1,k_2,p)$ in (12.51) are at least second and third order in momenta, respectively.

From Lorentz and parity invariance

$$\langle\gamma_1\gamma_2|\pi^0\rangle = f(p^2)\mathrm{i}\varepsilon_{\nu\delta\beta\alpha}k_1^\beta k_2^\alpha\varepsilon_1^\nu\varepsilon_2^\delta(2\pi)^4\delta(p-k_1-k_2) \quad (12.53)$$

Comparing (12.51) and the last equation we conclude that

$$f(0) = \frac{1}{f_\pi}\left(\frac{e^2}{4\pi^2}\right) = \frac{\alpha}{f_\pi\pi} \quad (12.54)$$

and the rate for $\pi^0\to2\gamma$ is

$$\Gamma = \frac{1}{2m_\pi}\sum_{\varepsilon_1,\varepsilon_2}\int\frac{\mathrm{d}^3k_1\mathrm{d}^3k_2}{(2\pi)^6 4k_{10}k_{20}}|f(0)\varepsilon_{\nu\delta\beta\alpha}k_1^\beta k_2^\alpha\varepsilon_1^\nu\varepsilon_2^\delta|^2(2\pi)^4\delta(p-k_1-k_2)$$

$$= \frac{\alpha^2 m_\pi^3}{64\pi^3 f_\pi^2} = 7.63\,\mathrm{eV} \quad (12.55)$$

(Bose statistics implies integration over 2π solid angle only). The experimental value is $\Gamma^{\mathrm{Exp}} = (7.3\pm0.2\,\mathrm{eV})$. We make two important observations. The result (12.55) is entirely given by the axial anomaly. In the absence of the anomaly one would get $f(0)=0$ and PCAC could not be reconciled with the $\pi^0\to2\gamma$ decay rate. Secondly, the agreement with the experimental value relies on the existence of three colour states of quarks.

Chiral anomaly for the axial $U(1)$ current in QCD; $U_A(1)$ problem

Global symmetries of the QCD lagrangian with massless quarks are (see Section 9.1): chiral $SU(n)\times SU(n)$ invariance and invariance under abelian vector $U(1)$ and axial-vector $U_A(1)$ transformation. We study now the anomaly of the

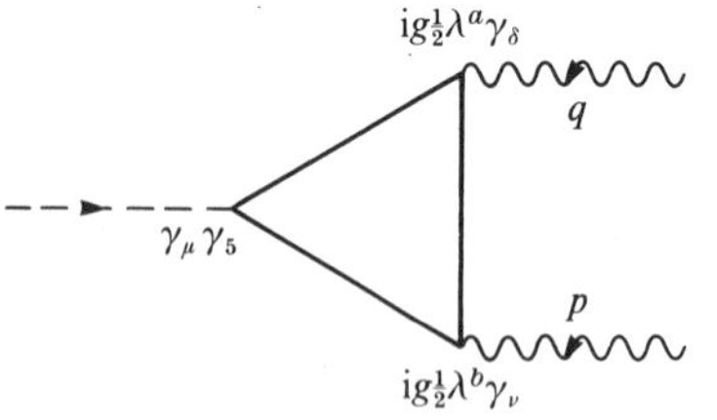

Fig. 12.8

$U_A(1)$ current. The $U_A(1)$ current is a singlet under the colour group and the relevant triangle diagram is shown in Fig. 12.8. We must have $D^\mu j^a_\mu = 0$ to maintain gauge invariance. A derivation of the equation $D^\mu j^a_\mu = 0$ in a *quantized* gauge-invariant theory is given in the next Section. For the triangle diagram we effectively get the condition $\partial^\nu \langle 0|Tj^5_\mu j^a_\nu j^b_\delta|0\rangle = 0$ and therefore we must impose the renormalization constraints (12.10). Thus, by analogy with (12.21), the divergence $-(p+q)^\mu \Gamma^5_{\mu\nu\delta}(p,q)$ of the Fourier transform of the matrix element $\langle 0|Tj^5_\mu j^a_\nu j^b_\delta|0\rangle$ reads

$$-(p+q)^\mu \Gamma^5_{\mu\nu\delta}(p,q) = (ig^2/2\pi^2)n \,\mathrm{Tr}\,[\tfrac{1}{2}\lambda^a \tfrac{1}{2}\lambda^b]\varepsilon_{\nu\delta\alpha\beta}p^\alpha q^\beta \tag{12.56}$$

where n is the number of quark flavours. The operator equation (12.35) can also be easily generalized to the present case if we remember that $U_A(1)$ current is a colour singlet. The only colour-invariant operator which for the triangle diagram gives the result (12.56) is

$$\partial^\mu j^5_\mu(x) = -(n/8\pi^2)\,\mathrm{Tr}\,[G_{\mu\nu}\tilde{G}^{\mu\nu}] \tag{12.57}$$

where Tr is in the internal quantum number space and $G_{\mu\nu} = -igG^a_{\mu\nu}T^a$, $T^a = \frac{1}{2}\lambda^a$. A new feature of (12.57) is that, due to the self-coupling of the non-abelian gauge fields, see (1.43), the anomalous divergence of the $U_A(1)$ current involves not only terms quadratic in gauge fields but also terms of the third order in gauge fields

$$\begin{aligned}\partial^\mu j^5_\mu(x) &= -(n/16\pi^2)\mathrm{Tr}\,[4\partial_\mu A_\nu \partial_\rho A_\sigma + 4\partial_\mu A_\nu [A_\rho, A_\sigma]]\varepsilon^{\mu\nu\rho\sigma}\\ &= -(n/4\pi^2)\varepsilon^{\mu\nu\rho\sigma}\partial_\mu \mathrm{Tr}\,[A_\nu \partial_\rho A_\sigma + \tfrac{2}{3}A_\nu A_\rho A_\sigma]\end{aligned} \tag{12.58}$$

Thus, square diagrams also contribute to the anomaly for the divergence of the axial current. However, this contribution is not independent of the triangle anomaly: given the latter, it is determined by the gauge invariance of the axial-vector current. All the anomalous Ward identities follow from the operator equation (12.57). For an explicit discussion of the square diagrams in the spirit of the previous Section see, for instance, Aviv & Zee 1972.

The anomaly (12.57) of the axial-vector current is crucial for resolving the so-called $U_A(1)$ problem in QCD. As we know from Chapter 9 the massless QCD lagrangian has the chiral $(U(1) \mp U_A(1))$ symmetry. As with the $SU(n) \times SU(n)$ it does not seem to be the symmetry of the hadron spectra so $U_A(1)$ must be broken ($U(1)$ manifests itself in baryon number conservation). However, contrary to the case of the spontaneously broken $SU(n) \times SU(n)$, there is no good candidate for the $U_A(1)$ Goldstone boson in the particle spectrum. The resolution of the problem is in the existence of the anomaly (12.57) and of field configurations with a non-zero topological charge (see Section 8.3). These properties offer a mechanism of breaking of the $U_A(1)$ which does not imply the existence of the massless Goldstone boson coupled to the gauge-invariant $U_A(1)$ current.

As in the first subsection of this Section we observe that the very existence of the anomaly does not destroy the $U_A(1)$ symmetry since the conserved current

$$j^S_{5\mu} = j_{5\mu} - 2nK_\mu, \quad \partial^\mu K_\mu = -(1/16\pi^2)\,\mathrm{Tr}\,[G_{\mu\nu}\tilde{G}^{\mu\nu}]$$

and the conserved charge

$$Q_5 = \int \mathrm{d}^3x j^S_{50}$$

can be defined. Now, however, Q_5 is not gauge invariant: it is not gauge invariant under large gauge transformations $\mathscr{G}_N$ with the winding number N. We have (see Section 8.3 and in particular (8.65) and (8.67) and Problem 8.2)

$$\mathscr{G}_N Q_5 \mathscr{G}_N^{-1} = Q_5 - 2nN$$

From Section 8.3 we also remember that

$$\mathscr{G}_N|\Theta\rangle = \exp(-\mathrm{i}N\Theta)|\Theta\rangle \tag{8.58}$$

Thus we get

$$\mathscr{G}_N \exp(\mathrm{i}\Theta' Q_5)|\Theta\rangle = \exp[-\mathrm{i}N(\Theta + 2n\Theta')]\exp(\mathrm{i}\Theta' Q_5)|\Theta\rangle$$

or

$$\exp(\mathrm{i}\Theta' Q_5)|\Theta\rangle = |\Theta + 2n\Theta'\rangle$$

The conclusion is that the $U_A(1)$ rotation by an angle Θ' changes the $|\Theta\rangle$ vacuum into the $|\Theta + 2n\Theta'\rangle$ vacuum. $U_A(1)$ symmetry is spontaneously broken because the vacuum is not $U_A(1)$ invariant. Different Θ-vacua are degenerate states since for massless fermions H commutes with Q_5. However, it can be seen from the chiral Ward identities that the spontaneous breakdown occurring because of the axial-vector anomaly and the existence of field configurations with non-zero topological charge (hence the set of the Θ-vacua) is consistent with the assumption of no Goldstone boson coupled to the gauge invariant current j^5_μ.

Anomaly cancellation in the $SU(2) \times U(1)$ electroweak theory

We discuss this problem at the level of the triangle diagrams relegating to the next Section the construction of the full anomalous divergence of a non-abelian current coupled to gauge fields. This does not limit the generality of our discussion: a theory is anomaly free if, and only if, the triangle diagram anomalies are absent. For consistency of a quantum gauge theory the gauge source currents must be

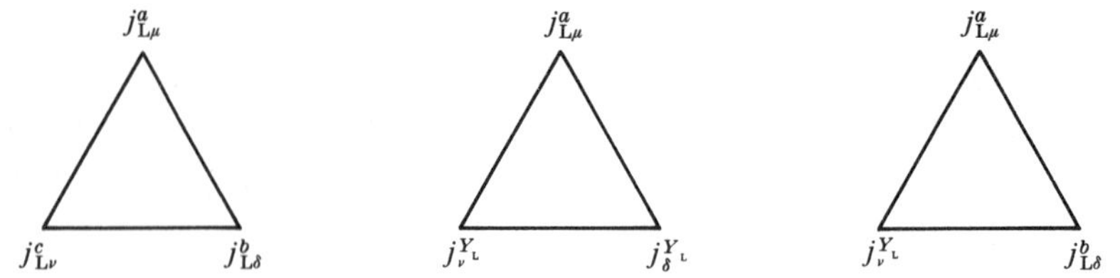

(*a*) (*b*) (*c*)

Fig. 12.9

covariantly conserved, that is anomaly free (Gross & Jackiw 1972; see also Jackiw 1984 and our discussion in the next Section).

In the electroweak theory the $SU(2)$ gauge bosons couple only to the left-handed currents and the $U(1)$ gauge boson couples both to the left- and right-handed currents but with different $U(1)$ charges Y. It will also be convenient to remember that by construction the fermion electric charge operator is given by $Q = T^3 + \frac{1}{2}Y$. Thus, $Y_R = 2Q$ and $Y_L = 2(Q - T^3)$. We can now systematically list possible anomalous triangle diagram contributions to the divergences of the gauge source currents. Take first the $SU(2)$ current $j^a_{L\mu}$: we can have the triangle diagrams shown in Fig. 12.9 with the $j^a_{L\mu}$ in one of the vertices and the corresponding symmetrized diagrams ($j^{Y_L}_\mu$ is the left-handed part of the $U(1)$ current). It is clear from our discussion in the previous Section that the eventual anomalous contribution of these diagrams after summing over all fermions in the loops is proportional to

$$\begin{aligned} &\tfrac{1}{2}\mathrm{Tr}\,[T^a\{T^b, T^c\}] \qquad &&\text{for diagram } (a)\\ &\tfrac{1}{2}\mathrm{Tr}\,[T^a\{Y, Y\}] &&\text{for diagram } (b)\\ &\tfrac{1}{2}\mathrm{Tr}\,[T^a\{Y, T^b\}] &&\text{for diagram } (c)\end{aligned}$$

($T^a = \frac{1}{2}\tau^a$ are $SU(2)$ generators). We immediately see that diagram (*b*) does not contribute because $[Y, T^a] = 0$ and $\mathrm{Tr}\, T^a = 0$. If we express Y in terms of Q the remaining contributions are

$$\mathrm{Tr}[T^a\{Q, T^b\}] = \mathrm{Tr}\,[Q\{T^b, T^a\}] = \delta^{ab}2\,\mathrm{Tr}\,[Q(T^a)^2] = \delta^{ab}\sum_i Q_i \qquad (12.59)$$

where the sum is taken over the charges of all fermions in the loop, and

$$\mathrm{Tr}\,[T^a\{T^b, T^c\}] = 0$$

for the group $SU(2)$.

The anomalous contribution to the divergence of the $U(1)$ current $j^Y_\mu = \frac{1}{2}\sum_i(\bar{\Psi}^i_L\gamma_\mu Y\Psi^i_L + \bar{\Psi}^i_R\gamma_\mu Y\Psi^i_R)$ where the sum is over all fermions in the theory, can be studied in a similar way. We recall that the $U(1)$ coupling is not vector-like because of different Y quantum number assignments to the left- and right-handed fields. In this case all possible triangle diagrams are proportional to one of the following factors

$$\tfrac{1}{2}\mathrm{Tr}[Y\{Y, T^a\}],\quad \tfrac{1}{2}\mathrm{Tr}[Y\{T^a, T^b\}] \quad\text{and}\quad \tfrac{1}{2}\mathrm{Tr}[Y\{Y, Y\}]$$

The first two traces have been already calculated. The last one appears in two types of loops, with left- and right-handed fermions, which contribute with opposite signs because of the $(1 \pm \gamma_5)$ factors (see (12.26)). We calculate

$$\tfrac{1}{2}\mathrm{Tr}[Y\{Y, Y\}] = \mathrm{Tr}\, Y^3 \sim \mathrm{Tr}(Q - T_3)^3 = \mathrm{Tr}[Q^3 - 3Q^2 T_3 + 3Q(T_3)^2 - (T_3)^3] \tag{12.60}$$

Since Q is the same for fermions with either of the chiralities, the Q^3 term is cancelled separately for each fermion when the sum over the two types of loops is taken. The remaining terms give again

$$\sum_i Q_i = 0 \tag{12.61}$$

since $\mathrm{Tr}[Q^2 T^3] = \mathrm{Tr}[(T_3 + \frac{1}{2}Y)^2 T_3] = \mathrm{Tr}[(T_3)^3 + Y(T_3)^2 + \frac{1}{4}Y^2 T_3] = \frac{1}{2}\mathrm{Tr}\, Y = \mathrm{Tr}\, Q$. Thus (12.61) is the only condition for the anomaly cancellation. We conclude that the $SU(2) \times U(1)$ theory is anomaly free if the sum of charges of all fermions in the theory vanishes. Quarks and leptons discovered so far strikingly satisfy this condition: to each $SU(2)$ doublet of leptons corresponds a doublet of quarks, fractionally charged but coming in three colours.

An interesting point to mention in the context of the $SU(2) \times U(1)$ theory is that once we insist on conservation of the chiral source currents for the electroweak gauge fields (one needs suitable sets of fermion multiplets to ensure this property) the baryon number current, $U(1)$ vector current, acquires an anomaly. This is clear if we consider the triangle diagram with the baryon current and two gauge source currents, for instance,

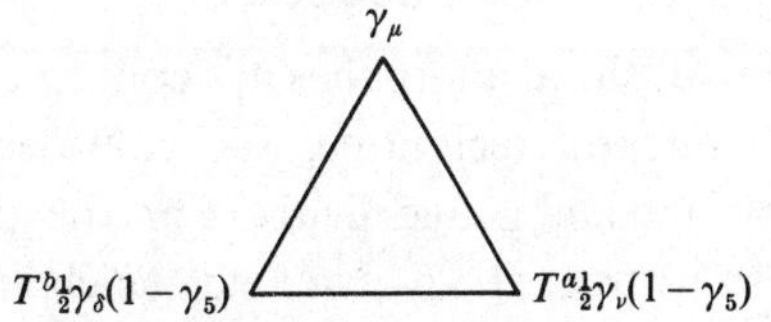

Insisting on chiral gauge current conservation we are formally back to the renormalization conditions (12.10) and the anomaly is placed in the vector current. Notice that the condition of vanishing divergence of two currents can always be imposed on triangle diagrams. Restrictions on the fermion representations appear when we consider triangle diagrams with three gauge currents.

Since $U(1)$ current is $SU(2)$ singlet, in the operator notation we have, analogously to (12.35)

$$\partial^\mu j_\mu = -(1/16\pi^2)\,\mathrm{Tr}[\varepsilon^{\mu\nu\rho\sigma} G_{\mu\nu} G_{\rho\sigma}] \tag{12.62}$$

where $G_{\mu\nu} = -igG^a_{\mu\nu}T^a$ is the $SU(2)$ field strength tensor. An interesting possibility is that the vacuum expectation value of the r.h.s. of (12.62) does not vanish due to some non-perturbative effects like tunnelling, with instantons as the dominant field configurations, or monopoles. One would then expect 'topological' baryon number non-conservation.

Anomaly free models

We conclude with more general remarks. As we know the triangle diagram anomaly is given by a multiple of

$$D_{abc} = \tfrac{1}{2}\mathrm{Tr}[T^a\{T^b, T^c\}] \tag{12.63}$$

where T^a, T^b and T^c are generators of the group of transformations of currents attached to the three vertices. We take all three currents to be gauge source currents. One can distinguish three cases of anomaly-free models (Georgi & Glashow 1972). Firstly, the right-handed fermion loop and the left-handed fermion loop anomalies may cancel. This happens when fermions with either of the chiralities transform according to the same representation of the gauge group, i.e. the theory is vector-like (Appendix B). If this is not the case, for an anomaly-free theory the quantity D_{abc} must vanish. Two cases may be distinguished here. It may be that, for all representations of the group, $D_{abc} = 0$. These are called 'safe' groups. They include $SU(2)$, all orthogonal groups except $SO(6) \approx SU(4)$ and all symplectic groups. The other case is when the group is not safe, like $SU(N)$ with $N > 2$ or $SU(2) \times U(1)$, but there exist some 'safe' representations for which D_{abc} vanishes. This then gives a limitation on the allowed fermion representations.

12.3 Anomalies and the path integral

Introduction

For a systematic discussion of chiral anomalies it is convenient to use the generating functional and the path integral technique. We consider a theory of massless fermions coupled to non-abelian gauge fields. For the purpose of the present discussion it is convenient to treat the gauge fields as classical background fields. Depending on the physical problem at hand we may need to use vector and axial-vector couplings or left- and right-handed couplings. In the first case the gauge-invariant lagrangian density reads†

$$\mathscr{L} = \bar{\Psi}\mathrm{i}\gamma^\mu(\partial_\mu + A^{\mathrm{V}}_\mu + \gamma_5 A^{\mathrm{A}}_\mu)\Psi = \bar{\Psi}\mathrm{i}\gamma^\mu\partial_\mu\Psi + j^{\alpha\mu}_{\mathrm{V}} A^{\mathrm{V}}_{\alpha\mu} + j^{\alpha\mu}_{\mathrm{A}} A^{\mathrm{A}}_{\alpha\mu} \tag{12.64}$$

where

$$A^{\mathrm{V,A}}_\mu = -\mathrm{i}A^{\mathrm{V,A}}_{\alpha\mu} T^\alpha$$

† Since in this section fields A_μ are treated as background fields the coupling constant g is included in the definition of A_μ.

and the T^αs are the generators of the gauge group in fermion space. The transformation properties of the fermion and the gauge fields under infinitesimal vector and axial-vector gauge transformations are

$$\left.\begin{aligned}\Psi' &= (1+\Theta_V)\Psi\\ \delta_{\Theta_V}A_\mu^V &= -\partial_\mu\Theta_V + [\Theta_V, A_\mu^V] = -D_\mu\Theta_V\\ \delta_{\Theta_V}A_\mu^A &= [\Theta_V, A_\mu^A]\end{aligned}\right\} \tag{12.65}$$

$$\left.\begin{aligned}\Psi' &= (1+\gamma_5\Theta_A)\Psi\\ \delta_{\Theta_A}A_\mu^V &= [\Theta_A, A_\mu^A]\\ \delta_{\Theta_A}A_\mu^A &= -\partial_\mu\Theta_A + [\Theta_A, A_\mu^V]\end{aligned}\right\} \tag{12.66}$$

respectively, where as usual $\Theta = -\mathrm{i}\Theta^\alpha T^\alpha$. We see, in particular, that in a non-abelian theory a pure axial-vector interaction with Dirac fermions would not be gauge invariant. The lagrangian (12.64) is also invariant under global $U(1)$ and $U_A(1)$ transformations. However, no gauge bosons are coupled to the corresponding abelian currents.

It is often more convenient to use the left- and right-handed fields and then

$$\left.\begin{aligned}\mathscr{L} &= \bar\Psi_L \mathrm{i}\gamma^\mu(\partial_\mu - \mathrm{i}A_{L\mu}^\alpha T^\alpha)\Psi_L + \bar\Psi_R\mathrm{i}\gamma^\mu(\partial_\mu - \mathrm{i}A_{R\mu}^\alpha T^\alpha)\Psi_R\\ &= \sum_{L,R}\bar\Psi_{L,R}\mathrm{i}\gamma^\mu\partial_\mu\Psi_{L,R} + j_\mu^{L\alpha}A_{L\alpha}^\mu + j_\mu^{R\alpha}A_{R\alpha}^\mu\end{aligned}\right\} \tag{12.67}$$

where

$$j_{L,R\mu}^\alpha = \bar\Psi_{L,R}\gamma_\mu T_{L,R}^\alpha\Psi_{L,R}$$

Invariance under gauge transformations on the chiral fields requires standard transformation properties for the gauge fields $A_{L,R}^\mu$ (see Section 1.3). At the classical level the $U(1)$ currents are conserved: $\partial^\mu j_\mu^V = \partial^\mu j_\mu^A = 0$ and the non-abelian currents are covariantly conserved: $D_\mu j_{L,R}^{\alpha\mu} = 0$ (see (1.61)).

Depending on the physical problem considered the gauge fields in the lagrangian (12.64) or (12.67) can be regarded as dynamical gauge fields or as auxiliary background fields, introduced to write down the generating functional for the anomalous Ward identities involving currents which in the framework of the considered theory are not coupled to any dynamical gauge fields. This is, for instance, the case with the flavour symmetry currents in QCD. Of course, the physical interpretation of the non-abelian structure in the lagrangians (12.64) and (12.67) also depends on the problem. In addition the same fermions may couple to other gauge fields, too, corresponding to a gauge group acting in a different space.

Since, by assumption, the gauge fields are classical background fields, only two types of Feynman diagrams appear in our theory: diagrams where the external fields couple to a free spinor line and diagrams where the external fields couple to a spinor loop. This, however, has no impact on the generality of the discussion of chiral anomalies due to the Adler–Bardeen non-renormalization theorem quoted in Section 12.1.

Abelian anomaly

Let us first consider the generating functional for Ward identities involving the $U(1)$ axial-vector current. We use the lagrangian (12.64) and specify $A_\mu^{\mathrm{A}} = 0$. Thus we consider the axial abelian anomaly in a QCD-like theory. Applying the general formalism developed in Section 10.1 to derive Ward identities we take the functional

$$\exp\{\mathrm{i}Z[A_\mu, \eta, \bar{\eta}]\} = \int \mathscr{D}\Psi\mathscr{D}\bar{\Psi}\exp\left[\mathrm{i}\int \mathrm{d}^4x(\mathscr{L}_{\mathrm{kin}} + j_\alpha^\mu A_\mu^\alpha + \bar{\eta}\Psi + \bar{\Psi}\eta)\right] \quad (12.68)$$

and use its independence of the choice of the integration variables changed according to the transformation

$$\Psi(x) \to \exp[-\mathrm{i}\gamma_5\Theta(x)]\Psi(x), \quad \bar{\Psi}(x) \to \bar{\Psi}(x)\exp[-\mathrm{i}\gamma_5\Theta(x)] \quad (12.69)$$

(the gauge fields A_μ^α are not altered). Naively we then get (10.9) which generates Ward identities corresponding to $\partial^\mu j_\mu^5 = 0$ and is wrong in view of our perturbative calculations and of (12.57) in particular. Using the latter we can anticipate the correct modification of (10.9). It reads

$$\begin{aligned} 0 &= \delta Z[A, \eta, \bar{\eta}]/\delta\Theta|_{\Theta=0} \\ &= \frac{1}{\exp\{\mathrm{i}Z[A, \eta, \bar{\eta}]\}}\int \mathrm{d}^4z \int \mathscr{D}\Psi\mathscr{D}\bar{\Psi}\exp\left[\mathrm{i}\int \mathrm{d}^4x(\bar{\Psi}\mathrm{i}\partial\!\!\!/\Psi + j_\alpha^\mu A_\mu^\alpha + \bar{\eta}\Psi + \bar{\Psi}\eta)\right] \\ &\quad \times \{-(n/8\pi^2)\,\mathrm{Tr}[G_{\mu\nu}(z)\tilde{G}^{\mu\nu}(z)] - \partial^\mu j_\mu^5(z) + \bar{\eta}(-\mathrm{i}\gamma_5\Psi) + (-\mathrm{i}\bar{\Psi}\gamma_5)\eta\} \end{aligned} \quad (12.70)$$

One may ask where the anomalous term comes from in the path integral formalism. The answer has been given by Fujikawa (1980) who noticed that the anomaly is due to the non-invariance of the fermionic path integral measure under the transformation (12.69). We shall come back to this point in Section 12.4.

Equation (12.70) generates the desired Ward identities when we differentiate it functionally with respect to the A_μ^α, η and $\bar{\eta}$ and set $\bar{\eta} = \eta = 0$. For instance the anomalous Ward identity for the matrix element $\langle 0|Tj_\mu^5(z)j_\nu^\alpha(y)j_\delta^\beta(x)|0\rangle$ is obtained from (12.70) by taking the derivative

$$\frac{\delta}{\delta A_\nu^\alpha(y)}\frac{\delta}{\delta A_\delta^\beta(x)}$$

and setting $\bar{\eta} = \eta = 0$. One gets

$$\partial^\mu\langle 0|Tj_\mu^5(z)j_\nu^\alpha(y)j_\delta^\beta(x)|0\rangle = -(n/8\pi^2)\langle 0|T\,\mathrm{Tr}[G_{\mu\rho}\tilde{G}^{\mu\rho}]j_\nu^\alpha j_\delta^\beta|0\rangle$$

and in momentum space this is just equivalent to (12.56). Differentiating with respect to $\bar{\eta}$ and η and setting $\bar{\eta} = \eta = 0$ one gets the anomalous Ward identity for the matrix element $\langle 0|T\Psi j_\mu^5\bar{\Psi}|0\rangle$.

Non-abelian anomaly and gauge invariance

Our next task is to study the generating functional for the anomalous Ward identities involving non-abelian currents. We again consider the functional

$$\exp\{iZ[A_\mu]\} = \int \mathscr{D}\Psi\mathscr{D}\bar{\Psi}\exp\left(i\int d^4x\bar{\Psi}i\not{D}\Psi\right) \tag{12.71}$$

corresponding to the lagrangian (12.64) or (12.67). The Ψ stands for Dirac or chiral fermion fields and A_μ for vector and axial-vector gauge fields or left- and right-handed gauge fields, respectively; other fields are suppressed. First, we study the gauge dependence of the functional $Z[A_\mu]$. The gauge transformation of the fermion fields is again just a change of integration variables but under the accompanying transformation of the gauge fields $A_\mu(x) \to A_\mu(x) + \delta A_\mu(x)$ the functional $Z[A_\mu]$ changes as follows

$$Z[A_\mu] \to Z[A_\mu] - 2\int d^4x\,\mathrm{Tr}\left[\frac{\delta Z[A_\mu]}{\delta A_\mu(x)}\delta A_\mu(x)\right] \tag{12.72}$$

where

$$\frac{\delta}{\delta A_\mu} = \frac{\delta}{\delta A^\alpha_\mu}(-iT^\alpha)$$

and

$$\frac{\delta Z[A_\mu]}{\delta A_\mu(x)} = \frac{1}{\exp\{iZ[A_\mu]\}}\int \mathscr{D}\Psi\mathscr{D}\bar{\Psi}j^\mu(x)\exp\left(i\int d^4x\bar{\Psi}i\not{D}\Psi\right) \equiv \hat{J}^\mu \tag{12.73}$$

and $j_\mu(x)$ is the current coupled to the field $A_\mu(x)$ whereas $\hat{J}^\mu(x)$ is defined by (12.73). The explicit formulae (12.71)–(12.73) and that for δA_μ depend, of course, on whether we use vector and axial-vector fields or chiral fields. If we work with vector and axial-vector gauge fields, then $A_\mu(x)$ in (12.72) includes both A^{V}_μ and A^{A}_μ which transform under vector or axial-vector gauge transformation according to (12.65) and (12.66). On the other hand, if we use chiral fields, the gauge transformations on the left- and right-handed fields separately form a group and we have

$$\delta_\Theta A^{\mathrm{L,R}}_\mu = -\partial_\mu\Theta^{\mathrm{L,R}} + [\Theta^{\mathrm{L,R}}, A^{\mathrm{L,R}}_\mu] = -D_\mu\Theta^{\mathrm{L,R}} \tag{12.74}$$

In each case we may rewrite (12.72) and (12.73) using the following relation

$$\int d^4x\,\mathrm{Tr}[\hat{J}^\mu(x)D_\mu\Theta(x)] = -\int d^4x\,\mathrm{Tr}[D^\mu\hat{J}_\mu(x)\Theta(x)] \tag{12.75}$$

which is easy to verify. With help of (12.75) we get

$$\delta_\Theta Z = 2\int d^4x\,\mathrm{Tr}[X(x)Z[A_\mu]\Theta(x)] \tag{12.76}$$

or

$$\delta_\Theta Z[A_\mu]/\delta\Theta_\alpha(x)|_{\Theta(x)=0} = -X_\alpha(x)Z[A_\mu] \tag{12.77}$$

where

$$-X_{\mathrm{V}}(x) = \partial_\mu\frac{\delta}{\delta A^{\mathrm{V}}_\mu(x)} + \left[A^{\mathrm{V}}_\mu(x), \frac{\delta}{\delta A^{\mathrm{V}}_\mu(x)}\right] + \left[A^{\mathrm{A}}_\mu(x), \frac{\delta}{\delta A^{\mathrm{A}}_\mu(x)}\right] \tag{12.78}$$

(where $X = -\mathrm{i}X^\alpha T^\alpha$) for the vector gauge transformations on the vector and axial-vector fields,

$$-X_{\mathrm{A}}(x) = \partial_\mu \frac{\delta}{\delta A_\mu^{\mathrm{A}}(x)} + \left[A_\mu^{\mathrm{V}}(x), \frac{\delta}{\delta A_\mu^{\mathrm{A}}(x)}\right] + \left[A_\mu^{\mathrm{A}}(x), \frac{\delta}{\delta A_\mu^{\mathrm{V}}(x)}\right] \tag{12.79}$$

for the axial-vector gauge transformations on the vector and axial-vector fields, and

$$-X_{\mathrm{L,R}}(x) = \partial_\mu \frac{\delta}{\delta A_\mu^{\mathrm{L,R}}(x)} + \left[A_\mu^{\mathrm{L,R}}(x), \frac{\delta}{\delta A_\mu^{\mathrm{L,R}}}\right] = D_\mu \frac{\delta}{\delta A_\mu^{\mathrm{L,R}}} \tag{12.80}$$

for chiral gauge transformations on the chiral gauge fields (all commutators are for matrices T^α and not for fields which are c-numbers). By explicit calculation† one verifies that the Xs satisfy the gauge group commutation relations

$$\left.\begin{aligned} [X_{\mathrm{V}}^\alpha(x), X_{\mathrm{V}}^\beta(y)] &= c^{\alpha\beta\gamma} X_{\mathrm{V}}^\gamma \delta(x-y) \\ [X_{\mathrm{V}}^\alpha(x), X_{\mathrm{A}}^\beta(y)] &= c^{\alpha\beta\gamma} X_{\mathrm{A}}^\gamma \delta(x-y) \\ [X_{\mathrm{A}}^\alpha(x), X_{\mathrm{A}}^\beta(y)] &= c^{\alpha\beta\gamma} X_{\mathrm{V}}^\gamma \delta(x-y) \\ [X_{\mathrm{L,R}}^\alpha(x), X_{\mathrm{L,R}}^\beta(y)] &= c^{\alpha\beta\gamma} X_{\mathrm{L,R}}^\gamma \delta(x-y) \end{aligned}\right\} \tag{12.81}$$

which, applied to $Z[A_\mu]$, are equivalent to the relation

$$\delta_{\Theta_1}\delta_{\Theta_2} Z[A_\mu] - \delta_{\Theta_2}\delta_{\Theta_1} Z[A_\mu] = \delta_{[\Theta_1,\Theta_2]} Z[A_\mu] \tag{12.82}$$

expressing the group property of the gauge transformations.

Equations (12.76), (12.77) and (12.81) or (12.82) have several implications. Firstly, for invariance of our quantum theory under one of the previously defined gauge transformations we must have

$$\delta_\Theta Z[A_\mu] = 0 \Rightarrow X(x) Z[A_\mu] = 0 \tag{12.83}$$

where the X is one of the Xs given by (12.78)–(12.80) depending on the considered gauge transformation. In a theory with only vector or only chiral gauge fields, using (12.73) we recover the equation

$$D^\mu j_\mu^{\mathrm{V}} = 0 \quad \text{or} \quad D^\mu j_\mu^{\mathrm{L,R}} = 0 \tag{12.84}$$

obtained in Section 1.3 on the classical level. The non-abelian source currents for gauge fields must be covariantly conserved at any order of perturbation theory for gauge invariance to be maintained in quantum theory. With dynamical gauge fields the generating functional $Z[A_\mu]$ differs from the $Z[J, \eta, \bar{\eta}]$ of the complete theory, where $J, \eta, \bar{\eta}$ are the gauge and fermion field sources, by the absence of the functional integration over $\mathscr{D}A_\mu$ and of the gauge field kinetic terms which are gauge invariant. To study gauge invariance of the complete quantum theory it is sufficient to study the gauge invariance of the $Z[A_\mu]$.

The anomaly means that the variation of $Z[A_\mu]$ under gauge transformation does not vanish

$$D^\mu j_\mu^{\mathrm{L,R}}(x) = G^{\mathrm{L,R}}(x) \tag{12.85}$$

† It is convenient to use a test functional.

or equivalently

$$-X^{\alpha}_{L,R}(x)Z[A_\mu] = G^{\alpha}_{L,R}(x) \tag{12.86}$$

Similarly, if we work with vector and axial-vector currents and insist on the vector current conservation $D^\mu j^V_\mu(x) = 0$ then the divergence of the axial current is anomalous and

$$D^\mu j^A_\mu = \partial^\mu j^A_\mu + [A^\mu_V, j^A_\mu] = G^A(x) \tag{12.87}$$

or

$$-X^{\alpha}_{A}(x)Z[A_\mu] = G^{\alpha}_{A}(x) \tag{12.88}$$

Indeed, from the triangle diagram calculation in Section 12.1 we know partial contributions to the $G^{L,R}_\alpha(x)$ in (12.86) and the $G^A_\alpha(x)$ in (12.88) which are (at the triangle diagram level $D^\mu = \partial^\mu$)

$$\pm(1/12\pi^2)\varepsilon^{\mu\nu\rho\sigma}\mathrm{Tr}[T^\alpha\partial_\mu A^{L,R}_\nu \partial_\rho A^{L,R}_\sigma] \tag{12.89}$$

and

$$-(1/2\pi^2)\varepsilon^{\mu\nu\rho\sigma}\mathrm{Tr}[T^\alpha(\partial_\mu A^V_\nu \partial_\rho A^V_\sigma + \tfrac{1}{3}\partial_\mu A^A_\nu \partial_\rho A^A_\sigma)] \tag{12.90}$$

respectively. In the second expression the first term comes from the AVV diagram and the second one from the AAA diagram. Different numerical coefficients reflect different renormalization conditions for the two diagrams: conservation of the vector currents for the first one and symmetry between all three vertices for the second one. Observe that as long as for the anomaly the minimal form depending only upon the gauge fields and not other external fields is taken (12.86) and (12.88) follow from (12.85) and (12.87), respectively, since the $G^\alpha(A)$s factorize out of the path integral which cancels with the $\exp\{-iZ[A]\}$.

One way to get the full anomalous terms $G^\alpha(A)$ is to use the so-called Wess–Zumino consistency condition (Wess & Zumino 1971). Using (12.81) or (12.82) and (12.85) we get for the anomaly $G^{L,R}_\alpha(x)$ the equation

$$X^\beta_{L,R}(x)G^\alpha_{L,R}(y) - X^\alpha_{L,R}(y)G^\beta_{L,R}(x) = -c^{\alpha\beta\gamma}G^\gamma_{L,R}(x)\delta(x-y) \tag{12.91}$$

The importance of this condition follows from the fact that the operator X is non-linear in the gauge potential A_μ. Therefore the Wess–Zumino condition completely determines $G^\alpha[A_\mu]$ once the term (12.89) is given. After some calculation we can get

$$G^\alpha_{L,R}(x) = \pm(1/12\pi^2)\,\mathrm{Tr}[T^\alpha\partial^\mu\varepsilon_{\mu\nu\rho\sigma}(A^\nu_{L,R}\partial^\rho A^\sigma_{L,R} + \tfrac{1}{2}A^\nu_{L,R}A^\rho_{L,R}A^\sigma_{L,R})] \tag{12.92}$$

The expression for the axial anomaly must satisfy the following consistency conditions (again, see (12.81))

$$\left.\begin{aligned} X^V_\alpha(x)G^A_\beta(y) &= c_{\alpha\beta\gamma}G^A_\gamma(x)\delta(x-y) \\ X^A_\alpha(x)G^A_\beta(y) - X^A_\beta(x)G^A_\alpha(y) &= 0 \end{aligned}\right\} \tag{12.93}$$

and it is much more complicated (Bardeen 1969). All the anomalous Ward identities follow from the relations (12.86) or (12.88) (we may need to introduce additional sources to the functional Z).

Consistent and covariant anomaly

Finally we study the transformation properties of the anomalous current. Naively, this current would be expected to transform covariantly under gauge transformations. To determine the effect of the anomalies, in addition to the gauge variation

$$\left.\begin{aligned} \delta_\Theta A_\mu &= -D_\mu\Theta \\ \delta_\Theta Z[A] &= -2\int \mathrm{d}^4x\,\mathrm{Tr}[(\delta Z/\delta A_\mu)\delta_\Theta A_\mu(x)] = -2\int \mathrm{d}^4x\,\mathrm{Tr}[D_\mu \hat{J}^\mu \Theta(x)] \\ &= \int \mathrm{d}^4x\,\Theta^\alpha(x)G^\alpha(A) \end{aligned}\right\} \quad (12.94)$$

we introduce the variation which defines the current. By definition

$$\tilde{\delta}_B Z[A] = -2\int \mathrm{d}^4x\,\mathrm{Tr}[(\delta Z/\delta A_\mu)\delta A_\mu] = -2\int \mathrm{d}^4x\,\mathrm{Tr}[\hat{J}^\mu B_\mu(x)] \quad (12.95)$$

and

$$\tilde{\delta}_B A_\mu(x) = \delta A_\mu \equiv B_\mu(x)$$

The following operator equation holds

$$\tilde{\delta}_B\delta_\Theta - \delta_\Theta\tilde{\delta}_B = \tilde{\delta}_{[B,\Theta]} \quad (12.96)$$

Applying this operator to the functional $Z[A]$ we obtain

$$\left.\begin{aligned} (\tilde{\delta}_B\delta_\Theta - \delta_\Theta\tilde{\delta}_B)Z[A] &= \int \mathrm{d}^4x\{[\tilde{\delta}_B G_\alpha(A)]\Theta^\alpha(x) - (\delta_\Theta \hat{J}^\mu_\alpha)B^\alpha_\mu(x)\} \\ &= \int \mathrm{d}^4x\,\hat{J}^\mu_\alpha\{[B_\mu(x),\Theta(x)]\}^\alpha \end{aligned}\right\} \quad (12.97)$$

The gauge transformation properties of the non-abelian current follow from (12.97)

$$\int \mathrm{d}^4x(\delta_\Theta \hat{J}^\mu_\alpha)B^\alpha_\mu(x) = -\int \mathrm{d}^4x\{[\Theta,\hat{J}^\mu]_\alpha B^\alpha_\mu - [\tilde{\delta}_B G_\alpha(A)]\Theta^\alpha\} \quad (12.98)$$

The first term on the r.h.s. gives the usual transformation property of the current; the second term is due to the anomaly. One can show (Bardeen & Zumino 1984) that in the case of the anomaly (12.92) there exists the covariant non-abelian current

$$\tilde{J}^{\mathrm{L,R}\mu}_\alpha = \hat{J}^{\mathrm{L,R}\mu}_\alpha + P^{\mathrm{L,R}\mu}_\alpha(A) \quad (12.99)$$

Its explicit form can be found by searching for a polynomial $P^{\mathrm{L,R}\mu}_\alpha$ in gauge fields with an anomalous gauge transformation rule opposite to that of the current $\hat{J}^\mu_\alpha$. Bardeen & Zumino (1984) show that

$$P^{\mathrm{L,R}\mu}_\alpha = \mp(1/24\pi^2)\varepsilon^{\mu\nu\rho\sigma}\mathrm{Tr}[T_\alpha(A_\nu G_{\rho\sigma} + G_{\rho\sigma}A_\nu - A_\nu A_\rho A_\sigma)] \quad (12.100)$$

The current $\tilde{J}^\alpha_\mu$ is also anomalous

$$(D_\mu \tilde{J}^{\mathrm{L,R}\mu})_\alpha = \tilde{G}_\alpha(A) = \pm(1/16\pi^2)\varepsilon^{\mu\nu\rho\sigma}\mathrm{Tr}[T_\alpha G_{\mu\nu}G_{\rho\sigma}] \quad (12.101)$$

The anomalies (12.92) and (12.101) are called consistent and covariant anomalies, respectively. Each can be used to study anomaly cancellation. The physical sense of the original non-abelian current has been discussed earlier. The covariant current may have some physical significance when coupled to other external non-gauged fields.

12.4 Anomalies from the path integral in Euclidean space

Introduction

In this Section we discuss the recently developed techniques of anomaly calculation using the path integrals in Euclidean space (Fujikawa 1980). At the beginning let us summarize the rules of continuation from Minkowski to Euclidean space. For the spatial coordinates we have (see Section 2.3)

$$\left.\begin{array}{ll} x_0 = -\mathrm{i}\hat{x}_4 & \hat{x}_i = (-x, -y, -z) \\ x_i = \hat{x}_i & \hat{x}^\mu = -\hat{x}_\mu \end{array}\right\} \tag{12.102}$$

where $\mu = 1, 2, 3, 4$ and correspondingly

$$\left.\begin{array}{ll} p_0 = -\mathrm{i}\hat{p}_4 & \hat{p}_i = (-p_x, -p_y, -p_z \\ p_i = \hat{p}_i & \hat{p}_\mu = \mathrm{i}(\partial/\partial\hat{x}^\mu) \end{array}\right\} \tag{12.103}$$

where again $\mu = 1, 2, 3, 4$. We define the Euclidean vector potential $\hat{A}_\mu$ as

$$A_0 = \mathrm{i}\hat{A}_4, \quad A_i = -\hat{A}_i \tag{12.104}$$

so that $\hat{A}_\mu$ form a Euclidean four-vector. Thus for the covariant derivative $D_\mu = \partial_\mu + A_\mu = \partial_\mu - \mathrm{i}gA^a_\mu T^a$ we get $(\partial_\mu = (\partial/\partial x_0, -\partial/\partial x_i))$:

$$D_0 = \mathrm{i}\hat{D}_4, \quad D_i = -\hat{D}_i, \quad \hat{D}_\mu = -\partial/\partial\hat{x}^\mu - \mathrm{i}g\hat{A}^a_\mu T^a \tag{12.105}$$

The field strength tensor is

$$G^a_{ij} = \hat{G}^a_{ij} \qquad (i, j = 1, 2, 3), \qquad G^a_{0i} = -\mathrm{i}\hat{G}^a_{0i} \tag{12.106}$$

where

$$\hat{G}^a_{\mu\nu} = -\frac{\partial}{\partial\hat{x}^\mu}\hat{A}^a_\nu + \frac{\partial}{\partial\hat{x}^\nu}\hat{A}^a_\mu + gc^{abc}\hat{A}^b_\mu\hat{A}^c_\nu$$

We define the Euclidean γ-matrices to be hermitean matrices $\hat{\gamma}_n$ obeying

$$\{\hat{\gamma}_\mu, \hat{\gamma}_\nu\} = 2\delta_{\mu\nu} \qquad \mu, \nu = 1, \ldots, 4 \tag{12.107}$$

and explicitly

$$\hat{\gamma}_4 = \gamma_0, \quad \hat{\gamma}_i = -\mathrm{i}\gamma_i, \quad i = 1, 2, 3 \tag{12.108}$$

The matrix $\hat{\gamma}_5$ is taken as

$$\hat{\gamma}_5 = -\hat{\gamma}_1\hat{\gamma}_2\hat{\gamma}_3\hat{\gamma}_4 \tag{12.109}$$

and it is also hermitean. The Dirac operator $i\hat{\not{D}}$ is hermitean. The fermion fields Ψ and $\bar{\Psi}$ are independent variables in the path integral and in Euclidean space we define

$$\hat{\Psi} = \Psi, \qquad \hat{\bar{\Psi}} = i\bar{\Psi} \tag{12.110}$$

Under infinitesimal rotations in Euclidean space $\hat{\Psi}$ and $\hat{\Psi}^\dagger$ transform as follows

$$\delta\hat{\Psi} = \tfrac{1}{8}(\hat{\gamma}_\mu\hat{\gamma}_\nu - \hat{\gamma}_\nu\hat{\gamma}_\mu)\hat{\omega}_{\mu\nu}\hat{\Psi} \tag{12.111}$$

where

$$\omega_{ij} = \hat{\omega}_{ij}, \qquad \omega_{0i} = i\hat{\omega}_{4i}$$

and

$$\delta\hat{\Psi}^\dagger = -\tfrac{1}{8}\hat{\Psi}^\dagger(\hat{\gamma}_\mu\hat{\gamma}_\nu - \hat{\gamma}_\nu\hat{\gamma}_\mu)\hat{\omega}_{\mu\nu} \tag{12.112}$$

so that $\hat{\Psi}_1^\dagger\hat{\Psi}_2$ is a scalar and we identify the transformation properties of $\hat{\bar{\Psi}}$ as those of $\hat{\Psi}^\dagger$ (in Minkowski space $\Psi_1^\dagger\gamma_0\Psi_2$ is a scalar). The Euclidean action is

$$iS = -\hat{S} \tag{12.113}$$

where

$$\hat{S} = \int d^4\hat{x}\,[\tfrac{1}{4}\hat{G}^a_{\mu\nu}\hat{G}^a_{\mu\nu} + \hat{\bar{\Psi}}(i\hat{\not{D}} + iM)\hat{\Psi}]$$

To complete this introduction let us discuss the eigenvalue problem for the hermitean Dirac operator $i\hat{\not{D}}$ in Euclidean space. For simplicity we assume that $i\hat{\not{D}}$ has a purely discrete spectrum. This corresponds to stereographically projecting Euclidean four-space onto a four-sphere and changes the determinant but only by a factor which is independent of the gauge field. Since $i\hat{\not{D}}$ is hermitean it has real eigenvalues λ_n

$$i\hat{\not{D}}\hat{\Psi}_n = \lambda_n\hat{\Psi}_n \tag{12.114}$$

and because

$$\{\hat{\gamma}_5, i\hat{\not{D}}\} = 0 \tag{12.115}$$

we get

$$i\hat{\not{D}}\hat{\gamma}_5\hat{\Psi}_n = -\lambda_n\hat{\gamma}_5\hat{\Psi}_n \tag{12.116}$$

Thus non-vanishing eigenvalues always occur in pairs of opposite sign. We observe also that for $\lambda_n \neq 0$ chiral fields are not eigenvectors of $i\hat{\not{D}}$

$$i\hat{\not{D}}(1 \pm \hat{\gamma}_5)\hat{\Psi}_n = \lambda_n(1 \mp \hat{\gamma}_5)\hat{\Psi}_n \tag{12.117}$$

On the other hand eigenfunctions of the vanishing eigenvalue can always be chosen to be eigenfunctions of $\hat{\gamma}_5$ and thus to be chiral fields. Since $\hat{\gamma}_5^2 = 1$ its eigenvalues are ± 1 and

$$\hat{\gamma}_5(1 \pm \hat{\gamma}_5)\hat{\Psi}_n = \pm(1 \pm \hat{\gamma}_5)\hat{\Psi}_n \tag{12.118}$$

In the rest of this Section we work in Euclidean space and the caret is omitted.

Abelian anomaly

As mentioned in the previous Section the anomaly of the ungauged axial $U(1)$ current $j^5_\mu = \bar{\Psi}\gamma^\mu\gamma_5\Psi$ can be understood as a non-invariance of the path integral measure under local transformations on fermion fields

$$\left.\begin{aligned} \Psi' &= \exp[\mathrm{i}\Theta(x)\gamma_5]\Psi \approx (1 + \mathrm{i}\Theta(x)\gamma_5)\Psi \\ \bar{\Psi}' &= \bar{\Psi}\exp[\mathrm{i}\Theta(x)\gamma_5] \approx \bar{\Psi}(1 + \mathrm{i}\Theta(x)\gamma_5) \end{aligned}\right\} \tag{12.119}$$

The generating functional in the Euclidean space

$$\exp\{-Z[A]\} = \int \mathscr{D}\Psi\mathscr{D}\bar{\Psi}\exp\left(-\int \mathrm{d}^4x\bar{\Psi}\mathrm{i}\not{D}\Psi\right) \tag{12.120}$$

where Ψ and $\bar{\Psi}$ are two independent variables describing Dirac fermions coupled to gauge potentials transforming under a vector-like gauge group, can be defined precisely by expanding Ψ and $\bar{\Psi}$ in terms of the eigenfunctions φ_n of $\mathrm{i}\not{D}$

$$\left.\begin{aligned} &\Psi(x) = \sum_n a_n\varphi_n(x), \quad \bar{\Psi}(x) = \sum_n \varphi_n^\dagger(x)\bar{b}_n \\ &\mathrm{i}\not{D}\varphi_n(x) = \lambda_n\varphi_n(x), \quad \int \mathrm{d}^4x\varphi_n^\dagger(x)\varphi_m(x) = \delta_{mn} \end{aligned}\right\} \tag{12.121}$$

The coefficients a_n and $\bar{b}_n$ are Grassmann variables. The measure becomes

$$\mathscr{D}\Psi\mathscr{D}\bar{\Psi} = \prod_n \mathrm{d}a_n \prod_m \mathrm{d}\bar{b}_m \tag{12.122}$$

up to an irrelevant arbitrary normalization factor. Under local infinitesimal axial transformations (12.119) the lagrangian is transformed into

$$\mathscr{L}'(x) = \mathscr{L}(x) + \partial_\mu\Theta j^\mu_5(x)$$

and therefore

$$\int \mathrm{d}^4x\bar{\Psi}\mathrm{i}\not{D}\Psi \to \int \mathrm{d}^4x\bar{\Psi}\mathrm{i}\not{D}\Psi - \int \mathrm{d}^4x\Theta(x)\partial_\mu j^\mu_5(x) \tag{12.123}$$

If the measure were invariant under axial transformations then using the invariance of the path integral under a change of variables one could derive (10.11) (with $\delta^a\mathscr{L}(x) = 0$) implying the conservation of axial current at the quantum level. However the measure (12.122) is not invariant and following Fujikawa we can explicitly evaluate the Jacobian of the change of variables. We get

$$\Psi'(x) = \sum_n a'_n\varphi_n(x) = \sum_n a_n\exp[\mathrm{i}\Theta(x)\gamma_5]\varphi_n(x) \tag{12.124}$$

or

$$a'_m = \sum_n \int \mathrm{d}^4x\varphi_m^\dagger(x)\exp[\mathrm{i}\Theta(x)\gamma_5]\varphi_n(x)a_n = \sum_n C_{mn}a_n \tag{12.125}$$

Thus (see (2.116))

$$\prod_m \mathrm{d}a'_m = \det{}^{-1}C\prod_n \mathrm{d}a_n \tag{12.126}$$

and the Jacobian for $\mathscr{D}\bar{\Psi}$ gives the identical factor. Furthermore, we have

$$C = \exp[\mathrm{i}\Theta(x)\gamma_5]$$

and

$$\det C = \exp\{\mathrm{i}\,\mathrm{Tr}[\Theta(x)\gamma_5]\} = \exp\left[\mathrm{i}\sum_n \int \mathrm{d}^4x\Theta(x)\varphi_n^\dagger(x)\gamma_5\varphi_n(x)\right] \quad (12.127)$$

where the trace is taken over the whole Hilbert space. Hence the measure is multiplied by

$$\exp\left[-2\mathrm{i}\int \mathrm{d}^4x\Theta(x)\sum_n \varphi_n^\dagger(x)\gamma_5\varphi_n(x)\right] \quad (12.128)$$

which is ill defined as it stands and must be regularized. In our problem we want to choose a regularization procedure which preserves the invariance under the vector-like gauge group with gauge potentials A_μ. Actually, with this fact in mind we have been working in the basis (12.121) of the eigenfunctions of $\mathrm{i}\not{D} = \mathrm{i}(\not{\partial} + \not{A})$ which are gauge invariant i.e. gauge transformed eigenfunctions are eigenfunctions of the gauge transformed operator. Hence we may simply regularize the expression (12.128) with a gaussian cut-off for large eigenvalues λ_n

$$\int \mathrm{d}^4x\Theta(x)\sum_n \varphi_n^\dagger(x)\gamma_5\varphi_n(x)$$

$$\equiv \lim_{M\to\infty}\int \mathrm{d}^4x\Theta(x)\sum_n \varphi_n^\dagger(x)\gamma_5\varphi_n(x)\exp(-\lambda_n^2/M^2)$$

$$= \lim_{M\to\infty}\int \mathrm{d}^4x\Theta(x)\sum_n \tilde{\mathrm{T}}\mathrm{r}[\gamma_5\exp[-(\mathrm{i}\not{D})^2/M^2]\varphi_n(x)\varphi_n^\dagger(x)]$$

$$= \lim_{M\to\infty}\int \mathrm{d}^4x\Theta(x)\lim_{y\to x}\{\gamma_5\exp[-(\mathrm{i}\not{D}_x)^2/M^2]\}_{kl}^{ab}\sum_n[\varphi_n(x)]_l^b[\varphi_n^*(y)]_k^a \quad (12.129)$$

where $\tilde{\mathrm{T}}\mathrm{r}$ means the trace over the group and the Dirac indices. Going over to a plane wave basis and using the completeness relation for the $\varphi_n(x)$ we get

$$\sum_n[\varphi_n(x)]_l^b[\varphi_n^*(y)]_k^a$$

$$= \delta^{ab}\delta^{kl}\sum_n\langle x|n\rangle\langle n|y\rangle = \delta^{ab}\delta^{kl}\langle x|y\rangle$$

$$= \delta^{ab}\delta^{kl}\int\frac{\mathrm{d}^4k}{(2\pi)^4}\langle x|k\rangle\langle k|y\rangle$$

$$= \delta^{ab}\delta^{kl}\int\frac{\mathrm{d}^4k}{(2\pi)^4}\exp(-\mathrm{i}kx)\exp(+\mathrm{i}ky) \quad (12.130)$$

and therefore†

$$\int \mathrm{d}^4x\Theta(x)\sum_n \varphi_n^\dagger(x)\gamma_5\varphi_n(x)$$

$$= \lim_{M\to\infty}\int \mathrm{d}^4x\Theta(x)\int\frac{\mathrm{d}^4k}{(2\pi)^4}\tilde{\mathrm{T}}\mathrm{r}[\gamma_5\exp(+\mathrm{i}kx)\exp[-(\mathrm{i}\not{D})^2/M^2]\exp(-\mathrm{i}kx)]$$

$$= \lim_{M\to\infty}\int \mathrm{d}^4x\Theta(x)\tilde{\mathrm{T}}\mathrm{r}[\gamma_5([\gamma^\mu,\gamma^\nu]G_{\mu\nu})^2]\left(\frac{1}{4M^2}\right)^2\frac{1}{2!}\int\frac{\mathrm{d}^4k}{(2\pi)^4}\exp(+k^\mu k_\mu/M^2)$$

$$= \frac{1}{16\pi^2}\int \mathrm{d}^4x\Theta(x)\tilde{\mathrm{T}}\mathrm{r}[G_{\mu\nu}\tilde{G}^{\mu\nu}] \tag{12.131}$$

where the last trace is over the group indices only. Using (12.123), (12.128) and invariance of the path integral under the change of variables we finally get

$$\partial^\mu j_\mu^5(x) = (\mathrm{i}/8\pi^2)\mathrm{Tr}[G_{\mu\nu}\tilde{G}^{\mu\nu}] \tag{12.132}$$

which corresponds to the previous result (12.57) obtained in Minkowski space by the triangle diagram calculation.

As a by-product of this derivation of the abelian anomaly we can easily understand the content of the so-called index theorem for the Dirac operator $\mathrm{i}\not{D}$. The theorem relates the number of positive chirality minus the number of negative chirality zero modes of the operator $\mathrm{i}\not{D}$ to the topological charge (Pontryagin index) of the background gauge field A_μ. Using the result (12.131) for $\Theta(x)$ independent of x, observing that

$$\int \mathrm{d}^4x\sum_n \varphi_n^\dagger(x)\gamma_5\varphi_n(x) \tag{12.133}$$

vanishes for eigenfunctions with eigenvalue $\lambda_n \neq 0$ since $\gamma_5\varphi_n$ has eigenvalue $-\lambda_n$ and is thus orthogonal to φ_n, and writing

$$\varphi_n^\dagger(x)\gamma_5\varphi_n(x) = \tfrac{1}{4}\varphi_n^\dagger(1+\gamma_5)^2\varphi_n - \tfrac{1}{4}\varphi_n^\dagger(1-\gamma_5)^2\varphi_n$$

for the eigenfunctions with eigenvalue $\lambda_n = 0$, we get

$$n_+ - n_- = Q \tag{12.134}$$

where n_+ (n_-) is the number of positive chirality (negative chirality) zero modes of the operator $\mathrm{i}\not{D}$ and Q is the topological charge defined by (8.62). As in Section 8.3 we assume here that the background gauge fields rapidly approach the pure gauge configuration of $|x_\mu| \to \infty$ so that the Euclidean action is finite and the topological charge Q is well defined. The topological charge density is connected to the anomaly density.

Non-abelian anomaly

We consider now the structure of the non-abelian anomaly in theories of chiral fermions coupled to an external dynamical or auxiliary gauge field A_μ and described

† In Euclidean space we define $\varepsilon^{1234} = -1$ and then $\mathrm{Tr}[\gamma_5\gamma^\mu\gamma^\nu\gamma^\rho\gamma^\delta] = 4\varepsilon^{\mu\nu\rho\delta}$

by the effective action

$$\exp\{-Z[A]\} = \int \mathscr{D}\Psi_L \mathscr{D}\bar{\Psi}_L \exp\left(-\int d^4x \bar{\Psi}_L i\not{D}\Psi_L\right) \qquad (12.135)$$

The left-handed fermions Ψ_L and the right-handed fermions $\bar{\Psi}_L$ are independent variables. In Section 12.3 we have understood the non-abelian anomaly as a breakdown of the gauge invariance of the functional $Z[A]$. This suggests a possible approach to its calculation using the path integral formalism: calculate

$$Z[A] = -\ln \det i\not{D}(A) \qquad (12.136)$$

and its change under gauge transformation. However, as we know from the introduction to this Section the operator $i\not{D}(A)$ maps positive chirality spinors to negative chirality spinors

$$i\not{D}(A)\Psi_L \rightarrow \Psi_R$$

and consequently it does not have a well-defined eigenvalue problem and a well-defined determinant on the space of chiral fields. The only exception is when chiral fermions transform as a real representation of the gauge group i.e. left- and right-handed chiralities transform under the same representation. Then we can formulate our theory in terms of the Dirac fermions $\Psi = \Psi_L + \Psi_R$ and in Euclidean space we are back to the hermitean eigenvalue problem $i\not{D}\Psi_n = \lambda_n \Psi_n$ with real λ_n. In this case $\det i\not{D}$ is real and, as we know from the previous discussion, it can always be regularized in a gauge-invariant anomaly-free manner (in other words positive and negative chirality anomalies cancel each other).

Now consider chiral fermions in a complex representation R. We can always imagine an extended theory with additional fermions in R^* so that

$$Z_{R+R^*}[A] = Z_R[A] + Z_{R^*}[A] = Z_R[A] + Z_R^*[A] = 2\,\mathrm{Re}\, Z_R[A] \qquad (12.137)$$

But $R + R^*$ is a real representation and consequently $Z_{R+R^*}[A]$ is gauge invariant. We arrive at an interesting observation (Alvarez-Gaumé & Witten 1983): in Euclidean space only the imaginary part of the functional $Z_R[A]$ may suffer from the anomalies which give its anomalous variation under gauge transformation. Equivalently, if $\det i\not{D}$ was defined in the space of chiral fermions, we could say that only the phase of $i\not{D}$ may be gauge non-invariant. One can circumvent the problem of defining $\det i\not{D}$ observing that it suffices to calculate the change of $\ln \det i\not{D}$ under a gauge transformation. And

$$\ln \det i\not{D}(A^g) - \ln \det i\not{D}(A) = \ln \det[-i\not{D}^{-1}(A) i\not{D}(A^g)] \qquad (12.138)$$

is meaningful since $\not{D}^{-1}\not{D}$ maps a space of fermion fields with given chirality into itself. The overall phase of the determinant is irrelevant for us and so is the dependence on A; we can put e.g. $A = 0$. Thus we see that the calculation of the non-abelian anomaly from the path integral is a well-defined problem. The main difficulty is to invent an appropriate regularization procedure. For actual calculation and different regularization methods we refer the reader to the original literature (Fujikawa 1984, Balachandran, Marmo, Nair & Trahern 1982, Alvarez-Gaumé & Ginsparg 1984, Leutwyler 1984). Of course the result agrees with (12.93) which in Euclidean space reads

$$G_L^\alpha = D_\mu \frac{\delta Z}{\delta A_{\mu L}^\alpha} = -\frac{\mathrm{i}}{12\pi^2} \varepsilon^{\mu\nu\rho\sigma} \mathrm{Tr}\,[T^\alpha \partial^\mu (A_\nu \partial_\rho A_\sigma - \tfrac{1}{2}\mathrm{i} A_\nu A_\rho A_\sigma)] \qquad (12.139)$$

Problems

12.1 Derive the anomalous Ward identity (12.21) starting with the unregularized integral representation (12.5), performing a shift of the integration variable such that (12.3) is satisfied for this integral representation and studying the surface term induced by this change of the integration variable in the linearly divergent integral.

12.2 Show that Ward identities for the Green's function of the three-vector currents are anomaly free. Discuss this point also in Euclidean space using the material of Section 12.4.

12.3 In Fig. 12.6 replace photons by gluons and check that in QCD there is no anomaly for the axial isospin current.

12.4 Check that

$$\begin{aligned} G_A^\alpha = &-(1/4\pi^2)\varepsilon^{\mu\nu\sigma\tau}\,\mathrm{Tr}\,[\lambda^\alpha[\tfrac{1}{4}F_{\mu\nu}^V F_{\sigma\tau}^V + \tfrac{1}{12}F_{\mu\nu}^A F_{\sigma\tau}^A \\ &+ \tfrac{2}{3}(A_\mu A_\nu F_{\sigma\tau}^V - A_\mu F_{\nu\sigma}^V A_\tau + F_{\mu\nu}^V A_\sigma A_\tau) - \tfrac{8}{3} A_\mu A_\nu A_\sigma A_\tau]] \end{aligned}$$

where

$$V_\mu = -\mathrm{i}\lambda^i V_\mu^i, \quad A_\mu = -\mathrm{i}\lambda^i A_\mu^i$$

$$F_V^{\mu\nu} = \partial^\mu V^\nu - \partial^\nu V^\mu + [V^\mu, V^\nu] + [A^\mu, A^\nu]$$
$$F_A^{\mu\nu} = \partial^\mu A^\nu - \partial^\nu A^\mu + [V^\mu, A^\nu] + [A^\mu; V^\nu]$$

satisfies the consistency conditions (12.93).

12.5 The determinant of an operator A with real positive discrete and increasing without bound eigenvalues λ_n: $Af_n = \lambda_n f_n$ can be defined as

$$\det A = \prod_n \lambda_n = \exp[-(\mathrm{d}\zeta_A/\mathrm{d}t)(0)]$$

where the so-called ζ-function defined as

$$\zeta_A(t) = \sum_n 1/\lambda_n^t$$

converges for $\mathrm{Re}\,t > 2$ and can be analytically extended to a meromophic function of t with poles only at $t=1$ and $t=2$ and is regular at $t=0$. Using the integral representation for the Γ-function

$$\Gamma(t) = \int_0^\infty x^{t-1} \mathrm{e}^{-x}\,\mathrm{d}x$$

check that the ζ_A-function can be calculated as

$$\zeta_A(t) = \frac{1}{\Gamma(t)} \int_0^\infty \mathrm{d}\tau\, \tau^{t-1} \int \mathrm{d}x\, h(x, x, \tau)$$

where

$$h(x, y, \tau) = \sum_n \exp(-\lambda_n \tau) f_n(x) f_n^*(y) = \langle x | \exp(-A\tau) | y \rangle$$

and $h(x, y, \tau)$ obeys the so-called 'heat equation'

$$A_x h(x, y, \tau) = -(\partial/\partial\tau) h(x, y, \tau), \quad h(x, y, \tau = 0) = \delta(x - y)$$

Note also that

$$\zeta_A(t, x, y) \equiv \sum_n \frac{f_n(x) f_n^*(y)}{\lambda_n^t} = \frac{1}{\Gamma(t)} \int_0^\infty \mathrm{d}\tau\, \tau^{t-1} h(x, y, \tau)$$

The last equation can be used to regularize quantities such as

$$\sum_n \varphi_n^+(x) \gamma_5 \varphi_n(x) = \lim_{\substack{t \to 0 \\ y \to x}} \mathrm{Tr}\,[\gamma_5 \zeta_A(x, y, t)]$$

with $A = (\not{D})^2$. The heat kernel $h(x, y, \tau)$ has then the asymptotic expansion (Nielsen, Grisaru, Römer & van Nieuwenhuizen 1978)

$$h(x, y, \tau) = \frac{1}{16\pi^2 \tau^2} \exp\left[-\frac{(x - y)^2}{4\tau}\right] \sum_{n=0}^\infty a_n(x, y) \tau^n$$

for small τ. Check that

$$\zeta_{\not{D}^2}(x, x, 0) = (1/16\pi^2) a_2(x, x)$$

12.6 Determine the abelian and non-abelian chiral anomalies in $2n$-dimensional space-time (Frampton & Kephart 1983, Zumino, Wu & Zee 1984, Leutwyler 1984)

12.7 Prove (12.134) starting with the anomalous divergence in Euclidean space

$$\partial^\mu j_\mu^5 = 2m \bar{\Psi} \gamma_5 \Psi - 2iQ(x)$$

(Integrate over the Euclidean volume, take the vacuum expectation value and then take the limit $m \to 0$.)

13
Effective lagrangians

13.1 Non-linear realization of the symmetry group

Non-linear σ-model

To become familiar with the effective lagrangian technique it is useful to follow the original approach and to begin with the discussion of the $SU(2) \times SU(2)$ invariant σ-model in which the chiral symmetry has been spontaneously broken: $SU_L(2) \times SU_R(2) \rightarrow SU_V(2)$. We consider the σ-model with bosons only. The quadruplet of real fields $\Phi_0 = (\pi^1 \pi^2 \pi^3 \sigma_0)^T$ transforming as $(2, 2)$ under the chiral group splits then into massless Goldstone bosons and a massive $\sigma = \sigma_0 - v$ meson which transform under the unbroken subgroup $SU(2)$ as a triplet and a singlet, respectively. The σ mass is proportional to the vacuum expectation value v and can be made arbitrarily heavy. Imagine we are interested in the physics at low energy, $E \ll m_\sigma$, only. It is natural to expect that the low energy region can be effectively described in terms of the Goldstone boson fields, with the σ meson decoupled. However, an inspection of the lagrangian (9.38) shows that this does not happen: the $\pi\pi\sigma$ coupling is proportional to v and therefore, e.g. σ-exchange contribution to the low energy $\pi\pi \rightarrow \pi\pi$ scattering cannot be neglected when $m_\sigma \rightarrow \infty$. The reason is that a naive decoupling of the σ field by its elimination from the lagrangian (9.38) explicitly breaks the $SU(2) \times SU(2)$ invariance of the lagrangian. Both the $\pi\pi\pi\pi$ and the $\pi\pi\sigma$ couplings originate from the $SU(2) \times SU(2)$ invariant term $(\sigma_0^2 + \boldsymbol{\pi}^2)^2$ in the original lagrangian. Equivalently, decoupling of the σ means setting $\sigma_0 = v$ and this breaks the invariance since $(\boldsymbol{\pi}, \sigma_0)$ does not transform any more as a quadruplet under $SU(2) \times SU(2)$. Can we find an effective low energy description in terms of the Goldstone bosons only, which would be $SU(2) \times SU(2)$ invariant?

Our goal is to implement the $SU(2) \times SU(2)$ symmetry in such a way that the two irreducible multiplets, the triplet and the singlet of the unbroken $SU(2)$ which are contained in the quadruplet Φ_0 do not mix under general $SU(2) \times SU(2)$ transformations. This can be achieved if we use the freedom of choice of the broken symmetry vacuum (Section 9.5) among the degenerate set $|\alpha\rangle$

$$|\alpha\rangle = \exp(+\mathrm{i}\alpha_a X_a)|0\rangle \tag{13.1}$$

The corresponding transformation of the field operators reads

$$\Phi' = \exp(\mathrm{i}\alpha_a X_a)\Phi\exp(-\mathrm{i}\alpha_a X_a) \tag{13.2}$$

or in the matrix form

$$\Phi' = \exp(-\mathrm{i}\alpha_a X_a)\Phi \tag{13.3}$$

where the X_as are broken generators in (13.1) and (13.2) and their matrix representation in (13.3). Let us take some $\Phi_0 = (\boldsymbol{\pi}(x), \sigma_0(x))^{\mathrm{T}}$ in the originally defined vacuum. First of all we observe that at a given point x the Goldstone boson fields can be always set to zero by a suitable redefinition of the vacuum. In fact using the real representation (B.38) for the generators we can write

$$\Phi_0(x) = \begin{pmatrix} \pi^1(x) \\ \pi^2(x) \\ \pi^3(x) \\ \sigma_0(x) \end{pmatrix} = \exp(-\mathrm{i}\xi_a X_a)\begin{pmatrix} 0 \\ 0 \\ 0 \\ \sigma(x) \end{pmatrix} = \begin{pmatrix} -\sigma(\xi_1/\xi)\sin\xi \\ -\sigma(\xi_2/\xi)\sin\xi \\ -\sigma(\xi_3/\xi)\sin\xi \\ \sigma\cos\xi \end{pmatrix} \tag{13.4}$$

where

$$\xi = (\xi_1^2 + \xi_2^2 + \xi_3^2)^{1/2}$$

and

$$\sigma^2(x) = \boldsymbol{\pi}^2(x) + \sigma_0^2(x)$$

and therefore due to (13.3) a rotation of the vacuum by the angle $\boldsymbol{\xi}(x)$ sets the new Goldstone boson fields to zero locally at x. Any chiral transformation on the vector $\Phi_0(x)$ can then be written as follows

$$\Phi_0'(x) = \exp(-\mathrm{i}\alpha_a G_a)\Phi_0(x) = \begin{pmatrix} \pi_1'(x) \\ \pi_2'(x) \\ \pi_3'(x) \\ \sigma_0'(x) \end{pmatrix} = \exp[-\mathrm{i}\xi_a'(\xi,\alpha)X_a]\begin{pmatrix} 0 \\ 0 \\ 0 \\ \sigma(x) \end{pmatrix} \tag{13.5}$$

where

$$G_a \equiv (X_a, Y_a), \quad \mathrm{Tr}[G_j G_i] = \delta_{ji}$$

and X_a and Y_a are broken and unbroken generators of $SU(2)\times SU(2)$, respectively, with the same $\sigma(x)$ as in (13.4) because $\sigma^2(x) = \boldsymbol{\pi}^2(x) + \sigma_0^2(x)$ is invariant under chiral rotations. Thus, a general chiral transformation on $\Phi_0(x)$ can be represented by the new angles $\xi_a'(\boldsymbol{\xi},\boldsymbol{\alpha})$ specifying the vacuum orientation which sets the transformed Goldstone boson fields locally to zero. The vector $(0\,0\,0\,\sigma(x))^{\mathrm{T}}$ refers now to the new, rotated, vacuum.

Since $\sigma(x)$ is an $SU_{\mathrm{V}}(2)$ singlet we suspect that in general a chiral transformation can be factorized into an unbroken-$SU(2)$-subgroup transformation (which in our case acts on the $SU(2)$ singlet $\sigma(x)$) and into a rotation of the local vacuum orientation. However, before we generalize our discussion, for the purpose of further considerations let us check the transformation properties of the $\xi_a(x)$s under the unbroken $SU(2)$ isospin group

$$\exp(-\mathrm{i}u_a Y_a)\Phi_0 = \exp(-\mathrm{i}u_a Y_a)\exp(-\mathrm{i}\xi_a X_a)\exp(\mathrm{i}u_a Y_a)\exp(-\mathrm{i}u_a Y_a)\begin{pmatrix}0\\0\\0\\\sigma\end{pmatrix}$$

$$= \exp(-\mathrm{i}\xi_a R_{ab} X_b)\begin{pmatrix}0\\0\\0\\\sigma\end{pmatrix} = \exp(-\mathrm{i}\xi'_b X_b)\begin{pmatrix}0\\0\\0\\\sigma\end{pmatrix} \tag{13.6}$$

Hence

$$\xi'_b = R_{ab}\xi_a$$

R is the matrix representation of the unbroken $SU(2)$ under which the matrices X_a transform

$$\exp(-\mathrm{i}u_c Y_c) X_a \exp(\mathrm{i}u_c Y_c) = R_{ab} X_b$$

i.e. $\xi_a(x)$s indeed transform linearly under the $SU_V(2)$, belong to the same representation as X_as and transform according to R^T. These results can be generalized as follows (Coleman, Wess & Zumino 1969, Callan, Coleman, Wess & Zumino 1969). We consider a theory with some symmetry group G which is spontaneously broken to a subgroup H. Let G be a compact, connected, semisimple Lie group. From the properties of the exponentials it follows that, in some neighbourhood of the identity of G, every group element can be decomposed uniquely into a product of the form

$$g = \exp(\mathrm{i}\xi_a X_a)\exp(\mathrm{i}u_a Y_a) \tag{13.7}$$

where X_as and Y_as are broken and unbroken generators of the group G, respectively.† If Φ_0 is a field which transforms according to a linear representation of G, we can recast it as

$$\Phi_0(x) = \exp[-\mathrm{i}\xi_a(x) X_a]\Phi(x) \equiv U\Phi \tag{13.8}$$

(X_a is the matrix representation of the broken generator X_a appropriate for $\Phi_0(x)$) where under a general G transformation $g(\alpha)$

$$\left.\begin{aligned}\boldsymbol{\xi}' &= \boldsymbol{\xi}'(\boldsymbol{\xi}, \boldsymbol{\alpha})\\ \Phi' &= \exp[-\mathrm{i}u_i(\boldsymbol{\xi}, \boldsymbol{\alpha})Y_i]\Phi\end{aligned}\right\} \tag{13.9}$$

† The scalar fields $\xi(x)$ are coordinates of the manifold of left cosets G/H at each point of space-time. The set of group elements $l(\xi) = \exp(\mathrm{i}\xi_a X_a)$ parametrizes this manifold. Equation (13.7) means that once we have a parametrization $l(\xi)$ each group element g can be uniquely decomposed into a product $g = l \cdot h$ where $h \in H$. The l is the representative member of the coset to which g belongs and h connects l to g within the coset.

and $\boldsymbol{\xi}'(\boldsymbol{\xi},\boldsymbol{\alpha})$ and $\mathbf{u}(\boldsymbol{\xi},\boldsymbol{\alpha})$ are non-linear functions of $\boldsymbol{\xi}$ and $\boldsymbol{\alpha}$. Indeed, using (13.7)

$$\begin{aligned}\Phi_0'(x) &= \exp(-\mathrm{i}\alpha_a G_a)\Phi_0(x) = \exp(-\mathrm{i}\alpha_a G_a)\exp[-\mathrm{i}\xi_a(x)X_a]\Phi(x)\\ &= \exp[-\mathrm{i}\xi_a'(\boldsymbol{\xi},\boldsymbol{\alpha})X_a]\exp[-\mathrm{i}u_a(\boldsymbol{\xi},\boldsymbol{\alpha})Y_a]\Phi(x)\end{aligned} \tag{13.10}$$

since $\exp(-\mathrm{i}\alpha\cdot G)\exp(-\mathrm{i}\zeta\cdot X)$ belongs to G and can be decomposed as in (13.7)†. Thus, a general global transformation belonging to the group G is realized as a non-linear transformation on the fields $\xi_a(x)$ and as a gauge (local) transformation, belonging to the unbroken H, on the multiplet Φ. We say that the fields (ξ,Φ) provide a non-linear realization of the group G. Under transformations belonging to the unbroken subgroup H the fields $\xi_i(x)$ transform linearly, according to the generalized (13.6).

Let us go back to the σ-model. Given the transformation rule (13.5) we observe that the field $\sigma(x)$ can be decoupled at $E \ll m_\sigma$ in a chirally invariant manner by setting $\sigma(x) = v$ or equivalently by imposing on the field $\Phi_0(x)$ a chirally invariant constraint $\Phi_0^{\mathrm{T}}\Phi_0 = v^2$. After rescaling $\Phi_0/v \to \Phi_0$ we obtain the so-called non-linear σ-model described by the lagrangian

$$\mathscr{L} = \tfrac{1}{2}f_\pi^2\partial_\mu\Phi_0^{\mathrm{T}}\partial^\mu\Phi_0, \quad \Phi_0^{\mathrm{T}}\Phi_0 = 1, \quad f_\pi^2 = v^2 \tag{13.11}$$

Imagine that using the lagrangian (13.11) we want to calculate matrix elements involving Goldstone bosons. We can work either with Green's functions of currents or with the on-mass-shell matrix elements. The first method is more general (see e.g. Gasser & Leutwyler 1984) and consists in gauging the full chiral group, introducing the external gauge fields by changing the ordinary derivatives acting on the linearly transforming fields Φ_0 into covariant derivatives and expanding the generating functional, given by the vacuum-to-vacuum amplitude in the presence of external gauge fields, in powers of external fields. If we are interested in the on-mass-shell matrix elements only, it is best to interpret the transforming non-linearly $\xi_a(x)$s rather than the $\pi_a(x)$s as the pion fields (Weinberg 1968, Coleman, Wess & Zumino 1969). Such a redefinition of the pion field does not change the S-matrix elements if the redefinition is local and leaves the origin of field space invariant and as long as the new object has the right quantum numbers. Different definitions lead to different matrix elements off the mass-shell but they all give the same results on the mass-shell. We shall briefly discuss this second method. Irrespective of the method, however, one finds that beyond the tree level our theory has UV divergences which require counterterms of higher and higher order in derivatives of fields reflecting the non-renormalizability of the non-linear σ-model. This leads us to the following important comment on the physical significance of the effective lagrangian.

We have arrived at the lagrangian (13.11) by freezing the σ degree of freedom in the linear σ-model. However, one can convince oneself that this lagrangian is

† A product of g with an arbitrary group element and in particular with some $l(\xi)$ defines another $l(\xi')$ and an h according to $g\cdot l(\xi) = l(\xi')\cdot h$ where $\xi' = \xi'(\xi, g)$ and $h = h(\xi, g)$.

the most general one in order p^2 which is Lorentz, parity and chiral invariant and can be constructed in terms of the real $O(4)$ vector Φ_0 of unit length $\Phi_0^T\Phi_0 = 1$.[†] Thus we can think about Φ_0 as representing the degrees of freedom of composite Goldstone bosons in the $SU(2) \times SU(2)$ invariant QCD as well. The real virtue of the effective lagrangians is that they provide the most general systematic low energy expansion in the light particle sector, invariant under full symmetry G of the underlying theory. In this context the non-renormalizability of (13.11) is not a problem. Working to order p^4 or higher we anyway have to supplement the lagrangian (13.11) with additional terms with free parameters. Let us again consider QCD. Its chiral symmetry is spontaneously broken into the isovector subgroup H. There are Goldstone bosons, composites of quark fields which form a multiplet under H, and possibly other states also with well-defined transformation properties under H. Due to spontaneous symmetry breaking chiral multiplets split into H-multiplets with quite different masses in general. Imagine we are interested in the low energy description of the strong interactions. We are not able yet to get it directly from QCD in terms of its fundamental parameter Λ (and quark masses, if explicit breaking of chiral symmetry is taken into account). There is also no reason to expect that the linear σ-model is a good approximation to the low energy QCD. On the other hand the non-linear effective lagrangian is a tool for exploiting QCD as much as possible without solving the confinement problem, namely for exploiting its underlying chiral symmetry. This happens at the expense of a set of new unknown constants at every level of the low energy expansion.

Effective lagrangian in the $\xi_a(x)$ basis

We turn back to the problem mentioned earlier of calculating the on-mass-shell matrix elements from effective lagrangians. Because of the constraint $\Phi_0^T\Phi_0 = 1$ the form (13.11) is not convenient for constructing the Feynman rules. They can be derived easily if we work in the $\xi_a(x)$ basis. We can introduce the covariant derivative of the field $\xi(x)$, defined by its transformation properties under general transformation $G(\alpha)$ as follows

$$[(D_\mu\xi)_a(x)]' \equiv (D_\mu\xi)_b R_{ba}(\exp[-iu_c(\xi, \alpha)Y_c]) \tag{13.12}$$

where the matrix R_{ba} is defined by the relation

$$\exp[-iu_a(\xi, \alpha)Y_a]X_b\exp[iu_a(\xi, \alpha)Y_a] = R_{bc}(\exp[-iu_a(\xi, \alpha)Y_a])X_c \tag{13.13}$$

and Y_a and X_a are representation matrices of unbroken and broken generators of G, respectively, in the representation to which the Φ_0 belongs. The functions $u_a(\xi, \alpha)$ are defined by (13.9). The covariant derivative can be constructed explicitly by considering the derivative $\partial_\mu\Phi_0$ which transforms under G transformations in the same way as Φ_0 does. Using the non-linear notation (13.8) for a vector Φ_0

[†] $\Phi_0^T\partial_\mu\Phi_0$ vanishes and $\Phi_0^T\partial^\mu\partial_\mu\Phi_0 = -\partial_\mu\Phi_0^T\partial^\mu\Phi_0$.

we can, on general grounds, expand as follows

$$\partial_\mu(U\Phi) = \exp[-\mathrm{i}\xi_a(x)X_a]\{\mathrm{i}\partial_\mu\xi_a(x)D_{ab}(\xi)X_b\Phi + [\mathrm{i}\partial_\mu\xi_a(x)E_{ab}(\xi)Y_b + \partial_\mu]\Phi\} \tag{13.14}$$

and correspondingly for $\partial_\mu(U\Phi)'$ with $\xi_a \to \xi'_a$, $D_{ab} \to D'_{ab}$, $E_{ab} \to E'_{ab}$, $\Phi \to \Phi'$ where the functions $D(\xi)$ and $E(\xi)$ are defined by†

$$\exp(\mathrm{i}\xi_a X_a)\partial_\mu \exp(-\mathrm{i}\xi_a X_a) = \mathrm{i}\partial_\mu\xi_a[D_{ab}(\xi)X_b + E_{ab}(\xi)Y_b] \tag{13.15}$$

The transformation properties under G of both sides in (13.14) are the same as of the Φ_0, so the object in the braces must be a multiplet of H and have the local gauge transformation law (13.9). Thus

$$[\mathrm{i}\partial_\mu\xi'_a(\boldsymbol{\xi},\boldsymbol{\alpha})D'_{ab}(\xi')X_b + (\mathrm{i}\partial_\mu\xi'_a E'_{ab}(\xi')Y_b + \partial_\mu)]\Phi'$$

$$= \exp(-\mathrm{i}u_c Y_c)[\mathrm{i}\partial_\mu\xi_a D_{ab}X_b + (\mathrm{i}\partial_\mu\xi_a E_{ab}Y_b + \partial_\mu)]\underbrace{\exp[\mathrm{i}u_c(\boldsymbol{\xi},\boldsymbol{\alpha})Y_c]\Phi'}_{\Phi}$$

Identifying

$$(D_\mu\xi)_b = \mathrm{i}\partial_\mu\xi_a D_{ab}(\xi) \equiv \mathrm{Tr}[U^\dagger\partial_\mu U X_b] \tag{13.16}$$

$$D_\mu\Phi = [\partial_\mu + \mathrm{i}\partial_\mu\xi_a E_{ab}(\xi)Y_b]\Phi \equiv \{\partial_\mu + \mathrm{Tr}[U^\dagger\partial_\mu U Y_b]Y_b\}\Phi \tag{13.17}$$

and equating coefficients of X_a we recover for the object $(D_\mu\xi)_b$ the transformation law (13.12). Similarly we also get

$$(D_\mu\Phi)' = \exp[-\mathrm{i}u_a(\boldsymbol{\xi},\boldsymbol{\alpha})Y_a]D_\mu\Phi \tag{13.18}$$

Thus, under general transformation $G(\alpha)$ functions Φ, $D_\mu\Phi$ and $(D_\mu\xi)_b$ all transform according to some finite matrix representations of the group H. An important conclusion following from these transformation rules is that a function of Φ, $D_\mu\Phi$ and $(D_\mu\xi)_b$ which is invariant under H transformations is also automatically invariant under G transformations.

The explicit form of the covariant derivative can be obtained from the relation (13.15). First, expanding the l.h.s. (13.15) in power series one can check the identity

$$\exp(\mathrm{i}\xi_a X_a)\partial_\mu\exp(-\mathrm{i}\xi_a X_a)$$

$$= \sum_{n=0}^{\infty}\frac{(-\mathrm{i})^{n+1}}{(n+1)!}[[\ldots[[\partial_\mu\xi_a\cdot X_a, \underset{1}{\xi_b\cdot X_b}], \underset{2}{\xi_d\cdot X_d}]_{\underset{\ldots}{\ldots}} \underset{n}{\xi_w\cdot X_w}] \tag{13.19}$$

The commutators give

$$\partial_\mu\xi_a\cdot \mathrm{i}c_{abc}\underset{1}{\xi_b}\cdot \mathrm{i}c_{cde}\underset{2}{\xi_d}\underset{\ldots}{\ldots}\mathrm{i}c_{uwz}\underset{n}{\xi_w}G_z$$

where

$$[G_a, G_b] = \mathrm{i}c_{abc}G_c$$

† Those who are familiar with differential geometry recognize that in the expansion (13.15) $D_{ab}(\xi)$ represents the *vielbein* in G/H and $E_{ab}(\xi)$ is an H-connection.

is the G group algebra and G_z is a broken generator X for n even, and unbroken Y for n odd. Denoting

$$\mathrm{i}c_{abc}\xi_b = (c\cdot\xi)_{ac} \tag{13.20}$$

one gets

$$\exp(\mathrm{i}\xi_a\cdot X_a)\partial_\mu \exp(-\mathrm{i}\xi_a\cdot X_a) = \sum_{n=0}^{\infty}\frac{(-\mathrm{i})^{n+1}}{(n+1)!}\partial_\mu\xi_a(c\cdot\xi)^n_{az}G_z \tag{13.21}$$

and summing the terms even in n

$$D_{ab}(\xi) = \sum_{m=0}^{\infty}\frac{(-1)^{m+1}}{(2m+1)!}(c\cdot\xi)^{2m}_{ab} = -\left(\frac{\sin c\cdot\xi}{c\cdot\xi}\right)_{ab} \tag{13.22}$$

Similarly

$$E_{ab}(\xi) = \sum_{m=0}^{\infty}\frac{\mathrm{i}(-1)^m}{(2m+2)!}(c\cdot\xi)^{2m+1}_{ab} = \mathrm{i}\left(\frac{1-\cos c\cdot\xi}{c\cdot\xi}\right)_{ab} \tag{13.23}$$

Using the transformation properties of the covariant derivative $(D_\mu\xi)_a$ the effective lagrangian of the non-linear σ-model can be written down in the $\xi(x)$ basis. It reads:

$$\mathscr{L} = \tfrac{1}{2}f_\pi^2(D_\mu\xi)_a(D^\mu\xi)_a \tag{13.24}$$ †

As expected, there is no dependence on the ξ field other than that present in the covariant derivative. Any such dependence would break G invariance because one may always set $\xi(x)$ to zero at any given x by rotating the vacuum. Using the explicit form of the covariant derivative we can now easily read-off the Feynman rules.

We can extend the effective lagrangian description to include the interaction of Goldstone bosons with other fields which may be relevant for the effective low energy theory. According to the previously described general approach a multiplet Ψ_0 transforming linearly under G is replaced by the fields $(\xi(x), \Psi)$ where

$$\Psi_0 = \exp[-\mathrm{i}\xi_a(x)X_a]\Psi$$

providing a non-linear realization of the group G which is linear on the unbroken subgroup H. Any H-invariant form built of $\Psi(x)$, $D_\mu\Psi(x)$ and $(D_\mu\xi(x))_a$ is automatically G invariant. For definiteness let us take Ψ to be a fermion field and G to be the chiral symmetry spontaneously broken to the isospin subgroup H. The leading low energy terms are

$$\bar{\Psi}\Psi,\quad \bar{\Psi}\gamma_\mu\gamma_5 Y^a\Psi(D^\mu\xi)^a,\quad \bar{\Psi}\gamma_\mu D^\mu\Psi,\ldots$$

The presence of the first term reflects the fact that the chiral symmetry of the

† Or $\mathscr{L} = \frac{1}{2}f_\pi^2 g_{ab}\partial_\mu\xi^a\partial^\mu\xi^b$ where g_{ab} is given in terms of the *vielbein* $D_{ab}(\xi)$ and is a G-invariant metric on G/H.

lagrangian tells us nothing about the fermion masses as the symmetry may be spontaneously broken. The second term contains the fermion–fermion–pion vertex and the effective gradient coupling is known as the Adler zero for the soft pion emission. If we want to use the effective lagrangian beyond the tree approximation, terms of higher order in derivatives must be systematically included.

Matrix representation for Goldstone boson fields

In QCD-like theories with chiral $U(N) \times U(N)$ symmetry spontaneously broken to $SU_V(N) \times U_V(1)$† it is very convenient to collect the Goldstone boson fields into a unitary matrix U transforming linearly under chiral $U(N) \times U(N)$ transformations. Let us consider a general matrix

$$\tilde{\Sigma} = \bar{\Psi}_{Lj}\Psi_{Ri} \tag{13.25}$$

where $\Psi^i_{L,R}$ are chiral fermion fields with flavour i. Under chiral transformations $\tilde{\Sigma}$ transforms linearly and according to (9.85)

$$\tilde{\Sigma}' = U_R \tilde{\Sigma} U^{\dagger}_L \tag{13.26}$$

Matrices U_R and U_L act in the space of fermions Ψ_R and Ψ_L respectively, and have the form

$$U_{L,R} = \exp(-i\alpha^a_{L,R} T^a) \tag{13.27}$$

where $a = 0, 1, \ldots, N^2 - 1$ and $T^0 = \mathbb{1}$ and T^a, $a = 1, 2, 3, \ldots, N^2 - 1$, are the $SU(N)$ generators in the fermion representation with $\mathrm{Tr}\,[T^a T^b] = \frac{1}{2}\delta^{ab}$. The transformations under the unbroken vector group $U_V(N)$ are represented by $\alpha_L = \alpha_R$ and axial transformations by $\alpha_L = -\alpha_R$. Observe that now we always work with $SU(N)$ generators T^a and do not need to introduce explicitly the broken and unbroken generators of the full chiral $SU(N) \times SU(N)$. Thus a change of the vacuum orientation among the degenerate set corresponds to $\tilde{\Sigma} \to \tilde{\Sigma}'$ where

$$\tilde{\Sigma}' = \exp(i\alpha_a T_a)\tilde{\Sigma}\exp(i\alpha_a T_a) \tag{13.28}$$

where $a = 0, 1, \ldots, N^2 - 1$. Imagine now that we take $\tilde{\Sigma} = \mathbb{1}$ which according to (9.87) is an *Ansatz* for the spontaneous breaking $SU(N) \times SU(N) \to SU_V(N)$. The matrix

$$\Sigma = \exp(2i\alpha_a T_a)\mathbb{1} \equiv U \cdot \mathbb{1} \tag{13.29}$$

corresponds to a rotated vacuum and represents the degrees of freedom necessary to describe the Goldstone boson sector. We check that any matrix $\tilde{\Sigma}$ written as a product of a unitary and hermitean matrix $\Sigma = U \cdot h$ transforms under general chiral transformations as follows

$$\Sigma' = U_R U h U^{\dagger}_L = U_R U U^{\dagger}_L U_L h U^{\dagger}_L = U' h' \tag{13.30}$$

so U also transforms linearly and h undergoes only vector transformations. Thus

† See Section 12.2 for the discussion of the spontaneous breaking of $U_A(1)$.

once Σ is written in this product form any full $U_V(N)$ multiplet can be decoupled from the matrix h in the chirally invariant way. Since (13.29) is in the product form it provides us with the desired chirally invariant decoupling of all but Goldstone degrees of freedom. Actually, the unitary matrix $U(x)$ contains N^2 fields

$$U(x) = \exp[\tfrac{1}{3}\mathrm{i}\alpha_0(x)]\exp[2\mathrm{i}\alpha_a(x)T_a] \tag{13.31}$$

where $a = 1,\ldots,N^2 - 1$ and T^a are hermitean and traceless. For instance in the case of $U(3)\times U(3)$ it includes the η' degree of freedom which due to the abelian anomaly, is not a Goldstone boson (see Section 12.2). The single component field $\alpha_0(x)$ is related to the determinant of U

$$\det U = \exp(\mathrm{i}\alpha_0) \tag{13.32}$$

Under a $U_A(1)$ axial transformation

$$U' = \exp(2\mathrm{i}\beta)U \tag{13.33}$$

Under an arbitrary chiral $SU(N)\times SU(N)$ transformation

$$\exp(-\tfrac{1}{3}\mathrm{i}\alpha_0)U' = \exp(-\mathrm{i}\beta^a_{\mathrm{R}}T^a)\exp(2\mathrm{i}\alpha^c T^c)\exp(\mathrm{i}\beta^a_{\mathrm{L}}T^a) = \exp[2\mathrm{i}\alpha'_c(\beta_{\mathrm{R}},\beta_{\mathrm{L}},\alpha)T_c] \tag{13.34}$$

where $c = 1,\ldots,N^2 - 1$ so the fields $(\boldsymbol{\alpha}, h)$ provide a non-linear realization of this chiral group. One can also easily check that α_as transform linearly under vector transformations $U_{\mathrm{R}} = U_{\mathrm{L}}$.

Because U is an unitary matrix $UU^\dagger = 1$, it is impossible to write a $U(3)\times U(3)$-invariant interaction without derivatives. In order p^2 there are two $U(3)\times U(3)$ invariants

$$I_1 = \mathrm{Tr}[\partial_\mu U\partial^\mu U^\dagger]$$
$$I_2 = -\partial_\mu\alpha_0\partial^\mu\alpha_0 = \{\mathrm{Tr}[U^\dagger\partial_\mu U]\}^2$$

(from (13.32) $\alpha_0 = -\mathrm{i}\ln\det U = -\mathrm{i}\,\mathrm{Tr}\ln U$). On the basis of arguments which are not discussed in this book such as the Okubo–Zweig–Iizuka rule and the large N_c (number of colours) limit of QCD (Veneziano 1979, Witten 1979, Di Vecchia 1980) one expects that the I_1 term is the dominant one. Since

$$U^{-1}\partial_\mu U \equiv \mathrm{i}\partial_\mu\alpha_a D_{ab}(\alpha_c)T_b \equiv (D_\mu\alpha)_b T_b \tag{13.35}$$

where T_b form a group, the effective lagrangian can now be written in one of the following equivalent ways

$$\left.\begin{aligned}\mathscr{L} &= \tfrac{1}{4}f_\pi^2\,\mathrm{Tr}[\partial_\mu U\partial^\mu U^\dagger]\\ &= \tfrac{1}{4}f_\pi^2\,\mathrm{Tr}[U^{-1}\partial_\mu U(U^{-1}\partial^\mu U)^\dagger]\\ &= \tfrac{1}{2}f_\pi^2\,\mathrm{Tr}[U^{-1}\partial_\mu UT_a]\,\mathrm{Tr}[(U^{-1}\partial^\mu U)^\dagger T_a]\\ &= \tfrac{1}{8}f_\pi^2(D_\mu\alpha)_b(D\ \ \alpha)_b^\dagger\end{aligned}\right\} \tag{13.36}$$

Using the definition (13.35) and (13.30) one can easily check that for general chiral transformations the 'covariant derivative' $(D_\mu\alpha)_a$ transforms only under *global* rotations of the unbroken $SU_V(N)$. Thus if we want to include in the effective lagrangian other multiplets of the unbroken H we can work with their ordinary derivatives.

13.2 Effective lagrangians and anomalies

In QCD and in models of composite quarks and leptons one starts from a fundamental lagrangian describing a non-abelian gauge theory with elementary fermions. However, usually we are unable to solve the original theory and find the spectrum and the interaction of the composite states. The virtue of the effective lagrangian approach is that it provides a systematic method of isolating the composite states that are relevant at low energy and of studying their interactions in terms of a finite number of free effective parameters. The basic requirement in the construction of an effective lagrangian is that it possesses the same symmetries as the original theory. If this theory has anomalies the effective lagrangian must also satisfy the same anomalous transformation laws. Then the anomalous Ward identities will be satisfied at least to the tree order.

Abelian anomaly

In QCD the axial current has the $U_A(1)$ anomaly (Section 12.2)

$$\partial_\mu j_5^\mu = -2N(1/32\pi^2)\varepsilon^{\mu\nu\rho\sigma}G_{\mu\nu}G_{\rho\sigma} \equiv 2Nq(x) \tag{12.57}$$

where N is the number of flavours, determined by the response of the theory to the axial $U_A(1)$ transformations on the fermion fields. The effective lagrangian (13.36) describing the interaction of Goldstone bosons is, however, invariant under the $U_A(1)$ transformations. We want to generalize our effective theory so that (12.57) is satisfied by virtue of the equations of motion. Then the anomalous Ward identities will be satisfied to the tree order. The problem can be solved (Rosenzweig, Schechter & Trahern 1980, Di Vecchia & Veneziano 1980, Witten 1980) by introducing the pseudovector 'glueball' field $q(x)$ into the effective lagrangian which is modified to

$$\mathscr{L} = \tfrac{1}{4}f_\pi^2 \operatorname{Tr}[\partial_\mu U \partial^\mu U^\dagger] + \tfrac{1}{2}\mathrm{i}q(x)\operatorname{Tr}[\ln U - \ln U^\dagger] + cq^2(x) \tag{13.37}$$

Every term is invariant under $SU(N) \times SU(N)$ transformation. Under $U_A(1)$ transformation $U' = \exp(\mathrm{i}\Theta)U$ the action transforms as follows

$$S' = S - \Theta N \int \mathrm{d}^4x\, q(x) \tag{13.38}$$

The $U_A(1)$ matter current found by Noether's theorem is

$$j_\mu^5 = \tfrac{1}{4}\mathrm{i}f_\pi^2 \operatorname{Tr}[\partial_\mu U^\dagger U - U^\dagger \partial_\mu U] \tag{13.39}$$

and therefore

$$\partial^\mu j_\mu^5 = \tfrac{1}{4}\mathrm{i}f_\pi^2 \operatorname{Tr}[U \Box U^\dagger - U^\dagger \Box U] \tag{13.40}$$

On the other hand from the equations of motion we get

$$\tfrac{1}{4}f_\pi^2\Box U + \tfrac{1}{2}\mathrm{i}q(x)(U^\dagger)^{-1} = 0 \tag{13.41}$$

Multiplying (13.41) on the left with $U^\dagger$ and subtracting the complex conjugate equation gives

$$\partial^\mu j_\mu^5 = 2Nq(x)$$

Once the effective field $q(x)$ has been introduced into the lagrangian the presence of the chirally invariant term $q^2(x)$ is crucial for the consistency of this approach† and for the successful phenomenology. The additional free parameter c is needed for the phenomenological solution to the $U_A(1)$ problem i.e. for the correct description of the masses of the pseudoscalar nonet (including η') in the framework of the effective lagrangian approach (see Problem 13.1). Of course, to this end one must also introduce a term which accounts for the explicit breaking of the chiral symmetry by the quark masses. The quark mass terms transform under $SU(3) \times SU(3)$ as $(3,\bar{3})\oplus(\bar{3},3)$ so we must construct, from U, terms which transform under $SU(3) \times SU(3)$ in the same way. Since the components of U transform as $(3,\bar{3})$, to lowest order in the number of derivatives the necessary correction is

$$\Delta\mathscr{L} = \tfrac{1}{4}f_\pi^2\{\mathrm{Tr}\,[MU] + \mathrm{Tr}\,[M^\dagger U^\dagger]\}$$

where M is some matrix which must be a multiple of the quark mass matrix.

It is convenient to eliminate the field $q(x)$ from the lagrangian (13.37) by using its equation of motion. We finally get

$$\mathscr{L} = \tfrac{1}{4}f_\pi^2\,\mathrm{Tr}\,[\partial_\mu U\partial^\mu U^\dagger] + \tfrac{1}{4}f_\pi^2\{\mathrm{Tr}\,[MU] + \mathrm{Tr}\,[M^\dagger U^\dagger]\} + \frac{1}{8c}\{\mathrm{Tr}\,[\ln U - \ln U^\dagger]\}^2 \tag{13.42}$$

Effects connected to the existence of a non-vanishing vacuum angle Θ can also be studied by adding to the lagrangian (13.42) a term $\Theta q(x)$.

The Wess–Zumino term

In Chapter 12 we have related the presence of the non-abelian anomaly in theories with chiral fermions to the gauge invariance breaking of the functional $Z[A]$, see (12.76)

$$\delta_\Theta Z[A] = \int \mathrm{d}^4x[-\Theta_\alpha(x)X_\alpha(x)]Z[A]$$

where A denotes dynamical or auxiliary gauge fields coupled to the fermionic

† In general arbitrary chiral-invariant functions of $q(x)$ are possible. The term $q^2(x)$ is the only one which survives in the limit $N_c \to \infty$ where N_c is the number of colours (see, for instance, Di Vecchia 1980).

currents and $Z[A]$ is the effective action of the theory with fermions integrated out

$$\exp\{iZ[A]\} = \int \mathscr{D}\Psi\mathscr{D}\bar{\Psi}\exp\left(i\int d^4x\mathscr{L}\right)$$

In Section 12.3 we have also found certain consistency conditions (integrability conditions) which must be satisfied in the presence of anomalies. Our question now is how to take account of the non-abelian anomaly using the effective lagrangian for a QCD-like theory with $SU(N) \times SU(N)$ symmetry spontaneously broken to $SU_V(N)$. The Goldstone boson fields are collected in the matrix $U = \exp[2i\alpha^a(x)T^a]$, $a = 1, 2, \ldots, N^2 - 1$. The field $\alpha_0(x)$ in (13.31) which is a singlet under $SU(N) \times SU(N)$ is irrelevant for the present discussion. The term (13.36) constructed in the previous Section is chirally invariant and does not satisfy the Wess–Zumino consistency condition. To find the correct generalization of the effective lagrangian we imagine an addition to the original fermionic lagrangian of the effective Goldstone fields coupled to fermions. When fermions are integrated out one now gets the effective action $Z[\alpha^a, A]$ (rather than $Z[A]$) which should satisfy the Wess–Zumino condition.

Let us consider a finite axial gauge transformation

$$Z[\alpha', A'] = \exp\left\{\int d^4x\Theta_a(x)[-X_a^A(x) + V_a(x)]\right\}Z[\alpha, A] \tag{13.43}$$

where $X_a^A(x)$ is given in (12.79) and $V_a(x)$ denotes an infinitesimal axial gauge transformation operating on the Goldstone boson fields α_a. Using the notation

$$\int d^4x\Theta_a(-X_a^A + V_a) = \Theta\cdot(-X^A + V) \tag{13.44}$$

we observe that

$$[\Theta\cdot(-X^A + V)]^n Z[\alpha, A] = [\Theta\cdot(-X^A + V)]^{n-1}(\Theta\cdot G_A) = (-\Theta\cdot X^A)^{n-1}(\Theta\cdot G_A) \tag{13.45}$$

where G_A^a is the axial anomaly, see (12.87), depending only on the gauge fields. Therefore the solution to (13.43) reads

$$Z[\alpha', A'] = Z[\alpha, A] - \frac{\exp(-\Theta\cdot X^A) - 1}{\Theta\cdot X^A}\Theta\cdot G_A \tag{13.46}$$

We know that for each $\boldsymbol{\alpha}(x)$ there exists an axial rotation such that $\boldsymbol{\alpha}'(x) = 0$. Setting $\Theta^a(x) = -\alpha^a(x)$ we get

$$Z[0, A'] = Z[\alpha, A] - \frac{\exp(\alpha\cdot X^A) - 1}{\alpha\cdot X^A}\alpha\cdot G_A \tag{13.47}$$

Since $Z[0, A']$ does not involve the fields $\alpha^a(x)$ we can assume $Z[0, A'] = 0$ and then

$$Z[\alpha, A] = -\frac{1 - \exp(\alpha\cdot X^A)}{\alpha\cdot X^A}\alpha\cdot G_A = \int_0^1 dt\exp(t\alpha\cdot X^A)\alpha\cdot G_A \tag{13.48}$$

The explicit solution for $Z[\alpha, A]$ can be used for reading off the vertices of the effective lagrangian which describe the consequence of the anomalies in the Ward identities. Expanding the exponential and using the explicit expression for X^{A}, see (12.79), and G_{A} (see Problem 12.4) one can work out the effective vertices order-by-order in $\alpha^a(x) = \pi^a(x)/f_\pi$. An interesting point is that due to the derivatives with respect to the gauge fields A present in the X^{A} the effective action does not vanish for $A = 0$. Thus the anomalous effective lagrangian contains vertices describing interactions between Goldstone bosons. Using (12.79) and Problem 12.4 one finds, after a short calculation, that the lowest order term is

$$\frac{1}{6\pi^2 f_\pi^5} \varepsilon^{\mu\nu\sigma\rho} \operatorname{Tr}[\pi\partial_\mu\pi\partial_\nu\pi\partial_\sigma\pi\partial_\rho\pi] \tag{13.49}$$

where $\pi = \pi^a T^a$. In a similar way one can also work out vertices involving pseudoscalars and one vector abelian gauge field describing the photon coupled to the neutral component of the $SU_{\mathrm{V}}(N)$ current (see Wess & Zumino 1971 and Problem 13.2).

Recently further attention has been paid to the construction and analysis of the Wess–Zumino terms in the effective action (see for instance Witten 1983, Adkins, Nappi & Witten 1983, Alvarez-Gaumé & Ginsparg 1984).

Problems

13.1 In terms of the free parameters M and c obtain the mass spectrum of the pseudoscalar mesons from the quadratic in the fields $\alpha^a(x)$ part of the lagrangian (13.42); $U = \exp(\mathrm{i}\alpha^i T^i)$, $i = 0, 1, \ldots, 8$ (Veneziano 1979).

13.2 Using (13.48) and the axial anomaly $G_{\mathrm{A}}(A)$ given in Problem 12.4 find the effective vertices describing the processes $\pi^0 \to 2\gamma$, $\gamma \to \pi^+\pi^-\pi^0$ and $\gamma + \gamma \to 3\pi$ (consider the two-flavour case).

14
Introduction to supersymmetry

14.1 Introduction

This last Chapter is intended as a brief introduction to supersymmetric field theories. In recent years they have become a topic of intense research. The motivation for this research is so far purely theoretical. The standard model $SU(3) \times SU(2) \times U(1)$ describes very accurately the physics at presently accessible energies and in particular there is as yet no single phenomenological indication that supersymmetry might be relevant in nature. On the contrary, present experimental results tell us that if it is it has to be badly broken in the energy range explored so far. Nevertheless, there are several theoretical reasons to expect that the standard model is not a complete theory. The main motivation for considering supersymmetry in the search for such a theory is the improved convergence properties in the UV of supersymmetric field theories. It is this feature which we eventually discuss in this Chapter, after introducing the formalism of superfields.

UV divergences, which are softer than in non-supersymmetric theories, may be crucial for solving the problem of the 'naturalness' of the Higgs sector, considered as one of the most serious defects of standard electroweak theory. Higgs particles are scalar particles and their masses are subject to quadratic divergences in perturbation theory. For example, one-loop radiative corrections to elementary scalar masses are

$$\delta m^2 = g^2 \int^{\Lambda} \frac{\mathrm{d}^4 k}{(2\pi)^4} \frac{1}{k^2 - m^2} = O\left(\frac{\alpha}{\pi}\right)\Lambda^2$$

Since the theory is renormalizable we can take the limit $\Lambda \to \infty$ and as long as we treat these masses as free parameters there is no reason to pay attention to the magnitude of δm^2: bare parameters and corrections to them are not observable quantities. However, ultimately we want to understand the magnitude of these masses and then we must require

$$\delta m^2 \lesssim O(m^2)$$

Thus the quadratic divergence should be physically cut off at mass scale Λ of the order of the scalar particle mass and originating from new physics beyond the standard model. The standard model requires the mass of the scalars to be $O(100\,\mathrm{GeV})$, since otherwise the couplings in the Higgs sector would be too large for perturbation theory to be valid. So we need

$$\Lambda \lesssim 1\,\mathrm{TeV}$$

On the other hand if the standard model is a complete theory of everything but gravity the only cut-off which remains at our disposal is due to the fact that all particles participate in the gravitational interactions. The energy at which gravity and quantum effects become of comparable strength can be estimated from the only expression with the dimension of energy that can be formed from the fundamental constants c, $\hbar$ and G_{Newton}. We have then

$$\Lambda \sim M_{\mathrm{Planck}} = c^2(\hbar c/G_{\mathrm{Newton}})^{1/2} \approx 10^{19}\,\mathrm{GeV}$$

Two strategies have been proposed to resolve the above dilemma. One is to postulate that the scalar Higgs particle is composite with the compositeness scale $\Lambda \lesssim 1\,\mathrm{TeV}$ which provides a natural cut-off in the quadratically divergent loops. This 'technicolour' scenario has been discussed in Chapter 11. Interesting as it is it cannot easily be reconciled with small fermion masses and with the magnitude of the flavour-changing neutral interactions.

The other strategy is to cancel the loops noticing that the boson and fermion loops have opposite signs. This occurs in supersymmetric field theories and is part of the non-renormalization theorem to be discussed at the end of this Chapter. Of course this cancellation cannot be exact since supersymmetry, if at all relevant, must be a broken symmetry. What we need, however, is

$$\delta m^2 = +O(\alpha/\pi)\Lambda^2|_{\mathrm{bosons}} - O(\alpha/\pi)\Lambda^2|_{\mathrm{fermions}} \lesssim 1\,\mathrm{TeV}^2$$

or effectively

$$|m_{\mathrm{B}}^2 - m_{\mathrm{F}}^2| \lesssim 1\,\mathrm{TeV}^2$$

Thus, this strategy leads to the expectation that the supersymmetric partners of known particles have masses lighter than $O(1\,\mathrm{TeV})$.

The reader interested in studying supersymmetry at a level going beyond our introductory discussion is advised to consult e.g. Gates *et al* (1983), Sohnius (1985), Wess & Bagger (1983), Nilles (1984) and references therein.

14.2 The supersymmetry algebra

The continuous global symmetries of quantum field theory are generated by infinitesimal transformations

$$\delta_a \varphi = \mathrm{i}\Theta[Q_a, \varphi] \tag{14.1}$$

where Θ is an infinitesimal parameter. The conserved charges Q_a are expressed in terms of fields in the way given by Noether's procedure. The group of the Poincaré transformations with the Lorentz generators $M^{\mu\nu}$ and the energy–momentum operators P^{μ} satisfying the Poincaré algebra (see Problem 7.1) and various groups of internal symmetry transformations with generators transforming under the Lorentz group as spin zero operators and forming a Lie algebra

$$[Q_a, Q_b] = \mathrm{i}c_{abc}Q_c \tag{14.2}$$

are well known to be important symmetries of physical theories. Actually, Coleman & Mandula (1967) have proved that all generators, other than the Poincaré group generators, of symmetry transformations which form Lie groups with real parameters must have spin zero. Thus, any symmetry group of the S-matrix whose generators obey well-defined *commutation relations* must commute with the Poincaré group. Since the Casimir operators of the Poincaré group are the mass square operator $P^2 = P_\mu P^\mu$ and the generalized spin operator $W^2 = W_\mu W^\mu$ where W^μ is the Pauli–Lubarski vector

$$W^\mu = -\tfrac{1}{2}\varepsilon^{\mu\nu\rho\sigma}P_\nu M_{\rho\sigma}$$

one can conclude that all members of an irreducible multiplet of the internal symmetry group must have the same mass and the same spin.

The Coleman–Mandula theorem, however, leaves open the possibility of including symmetry operations whose generators obey anticommutation relations. Such a generalization of a Lie algebra is called a graded algebra or a superalgebra. The generators of a graded Lie algebra consist of even elements A_i and odd elements B_α and have commutation rules of the form

$$\left.\begin{aligned}[A_i, A_j] &= c_{ijk}A_k\\ [A_i, B_\alpha] &= g_{i\alpha\beta}B_\beta\\ \{B_\alpha, B_\beta\} &= h_{\alpha\beta i}A_i\end{aligned}\right\} \tag{14.3}$$

We are interested in an extension of the Poincaré algebra into a graded Lie algebra which contains the Poincaré algebra as its subalgebra. This is easily accomplished if we introduce spinorial charge Q which is a Majorana spinor or equivalently a left-handed Weyl spinor Q_α $(\alpha = 1, 2)$ where

$$Q = \begin{pmatrix} Q_\alpha \\ \bar{Q}^{\dot{\alpha}} \end{pmatrix}, \qquad \bar{Q} = (Q^\alpha, \bar{Q}_{\dot{\alpha}}) \tag{14.4}$$

where $\bar{Q} = Q^\dagger\gamma^0$ and $Q_\alpha^\dagger = \bar{Q}_{\dot{\alpha}}$ (at this point we advise the reader to consult Appendix C for the two-component notation). Thus we have

$$\left.\begin{aligned}[M_{\mu\nu}, Q_\alpha] &= \tfrac{1}{2}(\sigma_{\mu\nu})_\alpha{}^\beta Q_\beta\\ [M_{\mu\nu}, \bar{Q}_{\dot{\alpha}}] &= -\tfrac{1}{2}(\bar{\sigma}_{\mu\nu})^{\dot{\beta}}{}_{\dot{\alpha}}\bar{Q}_{\dot{\beta}}\end{aligned}\right\} \tag{14.5}$$

The anticommutator $\{Q_\alpha, Q_\beta^\dagger\}$ transforms under the Lorentz transformations as $(\tfrac{1}{2}, \tfrac{1}{2})$ and, since P_μ is the only generator of the Poincaré algebra in such a representation, we must have

$$\{Q_\alpha, \bar{Q}_{\dot\beta}\} = 2(\sigma^\mu)_{\alpha\dot\beta} P_\mu \tag{14.6}$$

The factor 2 is the normalization convention and the sign is determined by the requirement that the energy $E = P_0$ should be a semi-positive definite operator

$$\sum_{\alpha=1}^{2} \{Q_\alpha, Q_\alpha^\dagger\} = 2\,\mathrm{Tr}\,[\sigma^\mu P_\mu] = 4P_0 \tag{14.7}$$

The supersymmetry generators must commute with the momenta

$$[Q_\alpha, P_\mu] = [\bar{Q}_{\dot\alpha}, P_\mu] = 0 \tag{14.8}$$

Indeed, the commutator of the Q with a P_μ could contain the Lorentz representations $(1, \frac{1}{2})$ and $(0, \frac{1}{2})$. Since there are no $(1, \frac{1}{2})$ generators present we get, as the most general possibility,

$$[Q_\alpha, P_\mu] = C(\sigma_\mu)_{\alpha\dot\beta} \bar{Q}^{\dot\beta}$$

and the adjoint equation

$$[\bar{Q}^{\dot\beta}, P_\mu] = C^*(\bar\sigma_\mu)^{\dot\beta\alpha} Q_\alpha$$

Therefore

$$[[Q_\alpha, P_\mu], P_\nu] = CC^*(\sigma_\mu \bar\sigma_\nu)_\alpha{}^\beta Q_\beta$$

and from the Jacobi identity and from $[P_\mu, P_\nu] = 0$ we get $C = 0$ (a graded Lie algebra is defined by (14.3) and by the requirement that the graded Jacobi identities are fulfilled; the graded cyclic sum is defined just as the cyclic sum, except that there is an additional minus sign if two fermionic operators are interchanged). Finally, the anticommutator of Q_α with Q_β, by Lorentz covariance, must be a linear combination of Poincaré generators in the representations (0, 0) and (1, 0) of the Lorentz group. In addition, due to (14.8) and the Jacobi identity they must commute with P_μ. Such operators do not exist and we are left with (14.5), (14.6), (14.8) and

$$\{Q_\alpha, Q_\beta\} = 0 \tag{14.9}$$

which together with the Poincaré algebra form the simplest supersymmetric algebra which has the Poincaré group as space-time symmetry. It is called $N = 1$ supersymmetry, meaning that there is one two-spinor supercharge Q_α. Haag, Łopuszański & Sohnius (1975) have shown that only two-spinor supercharges are acceptable as generators of graded Lie algebras of symmetries of the S-matrix which are consistent with relativistic quantum field theory. Thus the only possible extension of the $N = 1$ supersymmetry is to have $N > 1$ two-spinor supercharges $Q_{\alpha i}$, $i = 1, \ldots, N$, which carry some representation of the internal symmetry†

$$[Q_{\alpha i}, T_j] = c_{ijk} Q_{\alpha k} \tag{14.10}$$

† In the $N = 1$ case the only non-trivially acting internal symmetry is a single $U(1)$, known under the name of R-symmetry: $[Q_\alpha, R] = Q_\alpha$, $[\bar{Q}_\alpha, R] = -\bar{Q}_\alpha$. Since under parity $Q_\alpha \leftrightarrow \bar{Q}^\alpha$ we must have $R \to -R$ i.e. the $U(1)$ symmetry group is chiral. In the four-spinor notation $[Q, R] = \mathrm{i}\gamma_5 Q$.

We shall limit our discussion to $N = 1$ supersymmetry which has been the most extensively studied one. In this case the full symmetry group of physical theory can be a direct product of $N = 1$ supersymmetry and of some internal symmetry group.

To conclude these considerations we note the four-component form of (14.6) (remember that in four-component notation Q is a Majorana spinor)

$$\{Q, \bar{Q}\} = 2\gamma^{\mu} P_{\mu} \tag{14.11}$$

14.3 Simple consequences of the supersymmetry algebra

We notice that

$$\left.\begin{aligned} [Q_{\alpha}, W^2] &\neq 0 \\ [Q_{\alpha}, P^2] &= 0 \end{aligned}\right\} \tag{14.12}$$

i.e. supersymmetry multiplets contain different spins but are always degenerate in mass. Supersymmetry transformations relate fermions to bosons as follows immediately from the fact that the Q_{α}s have spin one-half. Since in nature we do not observe mass degeneracy among particles of different spin, supersymmetry must be broken either explicitly or spontaneously if it is relevant to physics. Spontaneous breaking is theoretically more appealing.

The vacuum structure of supersymmetric theories is of a special character. It follows immediately from (14.7) that if there exists a supersymmetrically invariant state

$$Q_{\alpha}|0\rangle = 0 \quad \text{and} \quad \bar{Q}_{\dot{\alpha}}|0\rangle = 0 \tag{14.13}$$

then obviously

$$E_{\text{vac}} = 0 \tag{14.14}$$

Since from (14.7) the spectrum of H is semi-positive definite, $E_{\text{vac}} = 0$ implies that the supersymmetrically invariant state is always at the absolute minimum of the potential i.e. it is the true vacuum state. Thus, if the supersymmetric state exists, it is the ground state and supersymmetry is not spontaneously broken. In this respect supersymmetry is different from ordinary symmetries with which a symmetric state may exist without being the ground state. Only if there does not exist a state invariant under supersymmetry is supersymmetry spontaneously broken and from

$$Q_{\alpha}|0\rangle \neq 0 \quad \text{and/or} \quad \bar{Q}_{\alpha}|0\rangle \neq 0 \tag{14.15}$$

and from (14.7) it follows that

$$E_{\text{vac}} > 0$$

Thus supersymmetry is unbroken if, and only if, $E_{\text{vac}} = 0$. It is spontaneously

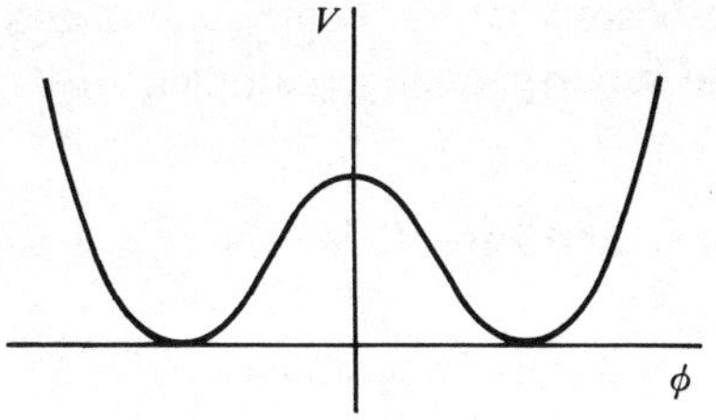

Fig. 14.1

broken if, and only if, the energy of the vacuum is greater than zero. An important fact is that a non-zero vacuum expectation value of a scalar field does not necessarily mean that supersymmetry is spontaneously broken. This is illustrated in Fig. 14.1 which shows the case when the expectation value of a scalar field breaks an internal symmetry but does not break supersymmetry because the vacuum energy is zero.

In the case of spontaneously broken supersymmetry one can prove an analogue of Goldstone's theorem. Consider the state obtained by acting on the vacuum with Q_α (or $\bar{Q}_{\dot{\alpha}}$)

$$|\Psi\rangle = Q_\alpha|0\rangle \quad (\text{or}\,|\Psi\rangle = \bar{Q}_{\dot{\alpha}}|0\rangle) \tag{14.16}$$

By the definition of spontaneously broken symmetry, $|\Psi\rangle$ is not zero. It is a state of odd fermion number and, because $[Q_\alpha, H] = 0$, it is degenerate with the vacuum. A state with such a property is a state of a single massless fermion of momentum $\mathbf{p}$, in the limit $\mathbf{p} \to 0$. So spontaneously broken supersymmetry requires the existence of a massless fermion, the so-called Goldstone fermion or goldstino.

Finally let us construct a hermitean operator

$$Q = (1/\sqrt{2})(Q_\alpha + Q_{\dot{\alpha}}^\dagger) \tag{14.17}$$

for $\alpha = 1$ or 2 and restrict our attention to the subspace of states with zero three-momentum $\mathbf{p} = 0$.† In the space of states with $\mathbf{p} = 0$ we have

$$Q^2 = H \tag{14.18}$$

and all states of non-zero energy are paired by the action of Q. Given any boson state $|b\rangle$ of non-zero energy E one gets a fermion state $|f\rangle$ acting with Q on $|b\rangle$

$$Q|b\rangle = E^{1/2}|f\rangle$$

and vice versa

$$Q|f\rangle = E^{1/2}|b\rangle$$

However, the zero-energy states are annihilated by Q and therefore they are not

† For a detailed discussion of the supersymmetric spectrum, including massless states, see e.g. Sohnius (1985).

paired in this way. Any bosonic or fermionic state of zero energy (vacuum state) forms a one-dimensional supersymmetry multiplet.

14.4 Superspace and superfields for $N = 1$ supersymmetry

Superspace

To formulate a supersymmetric field theory we must work out representations of supersymmetry algebra on field operators. The supersymmetry transformations must act linearly on field multiplets and leave the action invariant. An elegant formalism in which $N = 1$ supersymmetry is manifest is the formalism of superfields in superspace. To construct the superfield multiplets we first represent the supersymmetry algebra in terms of differential operators acting in superspace. To do so we can imitate the construction of differential operators representing the Poincaré generators in Minkowski space which can proceed as follows.

Take a quantum field $\Phi(x)$ which depends on the four coordinates x^μ. We can consider $\Phi(x)$ to have been translated from $x^\mu = 0$

$$\Phi(x) = \exp(\mathrm{i}x\cdot P)\Phi(0)\exp(-\mathrm{i}x\cdot P) \tag{14.19}$$

or, differentially,

$$\mathrm{i}[P_\mu, \Phi] = \partial_\mu\Phi \tag{14.20}$$

The important point is that this transformation is compatible with the multiplication law

$$\exp(\mathrm{i}x\cdot P)\exp(\mathrm{i}y\cdot P) = \exp[\mathrm{i}(x + y)\cdot P] \tag{14.21}$$

which holds because the operators P_μ commute with each other.

The differential version of a Lorentz transformation can be derived writing

$$\begin{aligned}\Phi'(x) &= \exp(-\tfrac{1}{2}\mathrm{i}\omega\cdot M)\Phi(x)\exp(\tfrac{1}{2}\mathrm{i}\omega\cdot M)\\ &= \exp(\mathrm{i}x'\cdot P)\exp(-\tfrac{1}{2}\mathrm{i}\omega\cdot M)\Phi(0)\exp(\tfrac{1}{2}\mathrm{i}\omega\cdot M)\exp(-\mathrm{i}x'\cdot P)\end{aligned} \tag{14.22}$$

where $\omega\cdot M = \omega^{\mu\nu}M_{\mu\nu}$ and x' is determined by the equation

$$\begin{aligned}\exp(-\tfrac{1}{2}\mathrm{i}\omega\cdot M)\exp(\mathrm{i}x\cdot P) &= \exp(\mathrm{i}x'\cdot P)\exp(-\tfrac{1}{2}\mathrm{i}\omega\cdot M)\\ &= \exp[\mathrm{i}x\cdot(P - \tfrac{1}{2}\mathrm{i}\omega\cdot[M, P])]\exp(-\tfrac{1}{2}\mathrm{i}\omega\cdot M) + O(\omega^2)\end{aligned} \tag{14.23}$$

or explicitly, for infinitesimal $\omega^{\mu\nu}$ and using (see Problem 7.1)

$$[M^{\mu\nu}, P^\lambda] = -\mathrm{i}(g^{\mu\lambda}P^\nu - g^{\nu\lambda}P^\mu)$$

we get

$$x'^\lambda = x^\lambda + \omega^{\lambda\nu}x_\nu \tag{14.24}$$

Finally, writing

$$\exp(-\tfrac{1}{2}\mathrm{i}\omega\cdot M)\Phi(0)\exp(\tfrac{1}{2}\mathrm{i}\omega\cdot M) = \exp(-\tfrac{1}{2}\mathrm{i}\omega\cdot\Sigma)\Phi(0) \tag{14.25}$$

where the Σ is some matrix representation of the algebra of the $M_{\mu\nu}$ which depends on the spin of Φ, and using (14.22) one has

$$\Phi'(x) = \exp(-\tfrac{1}{2}\mathrm{i}\omega\cdot\Sigma)\Phi(x') \tag{14.26}$$

In the differential form we get

$$\mathrm{i}[M_{\mu\nu}, \Phi(x)] = (x_\mu\partial_\nu - x_\nu\partial_\mu + \mathrm{i}\Sigma_{\mu\nu})\Phi(x) \tag{14.27}$$

This result follows essentially from the multiplication law (14.23).

We can now repeat a similar construction for the case of the $N = 1$ supersymmetry algebra. The first observation is that the presence of the anticommutator in the supersymmetry algebra calls for an extension of the Minkowski space into a superspace including some Grassmann parameters, in which differential operators will act representing the supersymmetry algebra. This stems from the fact that the algebra must be exponentiated into a group i.e. the product of two group elements like

$$\exp(\mathrm{i}\Theta^\alpha Q_\alpha)\exp(\mathrm{i}\bar{Q}_{\dot\alpha}\bar\Theta^{\dot\alpha}) \tag{14.28}$$

where Θ^α and $\bar\Theta^{\dot\alpha}$ are for the time being arbitrary parameters, must be a group element. This is indeed assured if we take Θ^α and $\bar\Theta^{\dot\beta}$ to be anticommuting components of two-spinors

$$\{\Theta^\alpha, \Theta^\beta\} = \{\bar\Theta^{\dot\alpha}, \bar\Theta^{\dot\beta}\} = \{\Theta^\alpha, \bar\Theta^{\dot\beta}\} = 0 \tag{14.29}$$

Since

$$\left.\begin{aligned} &[\Theta^\alpha Q_\alpha, \bar{Q}_{\dot\beta}\bar\Theta^{\dot\beta}] = 2\Theta^\alpha(\sigma^\mu)_{\alpha\dot\beta}\bar\Theta^{\dot\beta}P_\mu \\ &[\Theta^\alpha Q_\alpha, \Theta^\beta Q_\beta] = [\bar{Q}_{\dot\beta}\bar\Theta^{\dot\beta}, \bar{Q}_{\dot\alpha}\bar\Theta^{\dot\alpha}] = 0 \end{aligned}\right\} \tag{14.30}$$

we can write the product (14.28) as a group element using the Baker–Campbell–Hausdorf formula (see Problem 2.9) in which the multiple commutators vanish for generators of the supersymmetry algebra

$$\exp(\mathrm{i}\Theta^\alpha Q_\alpha)\exp(\mathrm{i}\bar{Q}_{\dot\alpha}\bar\Theta^{\dot\alpha}) = \exp[\mathrm{i}(\Theta^\alpha Q_\alpha + \bar{Q}_{\dot\alpha}\bar\Theta^{\dot\alpha} + \mathrm{i}\Theta^\alpha\sigma^\mu{}_{\alpha\dot\beta}\bar\Theta^{\dot\beta}P_\mu)] \tag{14.31}$$

The commonly used choice for Θ and $\bar\Theta$ is the 'real' superspace $z = (x^\mu, \Theta^\alpha, \bar\Theta^{\dot\alpha})$ with $\bar\Theta^{\dot\alpha} = (\Theta^\alpha)^*$ which in the four-component notation corresponds to choosing Θ_i, $i = 1, \ldots, 4$, as a Majorana spinor.

As for the Poincaré group we can discuss separately the Lorentz transformations L which form a subgroup of the super-Poincaré group and supertranslations which form a coset manifold with respect to L. Any element of the super-Poincaré group can be uniquely decomposed into a product

$$S(x, \Theta, \bar\Theta)L(\omega) \tag{14.32}$$

where a finite supertranslation $S(x, \Theta, \bar\Theta)$ is parametrized as follows

$$S(x, \Theta, \bar\Theta) = \exp(\mathrm{i}x\cdot P + \mathrm{i}\Theta^\alpha Q_\alpha + \mathrm{i}\bar{Q}_{\dot\alpha}\bar\Theta^{\dot\alpha}) \tag{14.33}$$

The multiplication law for two successive supertranslations, analogous to (14.21), can be evaluated with the help of the Baker–Campbell–Hausdorf formula and of (14.30)

$$S(y,\eta,\bar{\eta})S(x,\Theta,\bar{\Theta}) = S(x',\Theta',\bar{\Theta}') \tag{14.34}$$

where

$$\left.\begin{aligned} x'^{\mu} &= x^{\mu} + y^{\mu} - \mathrm{i}\Theta^{\alpha}\sigma^{\mu}{}_{\alpha\beta}\bar{\eta}^{\beta} + \mathrm{i}\eta^{\alpha}\sigma^{\mu}{}_{\alpha\beta}\bar{\Theta}^{\beta} \\ \Theta' &= \eta + \Theta \\ \bar{\Theta}' &= \bar{\eta} + \bar{\Theta} \end{aligned}\right\} \tag{14.35}$$

In complete analogy with (14.23), multiplication with the Lorentz group gives

$$L(\omega)S(x,\Theta,\bar{\Theta}) = S(x',\Theta',\bar{\Theta}')L(\omega) \tag{14.36}$$

where, for infinitesimal transformations, using (14.5) we get

$$\left.\begin{aligned} x'^{\lambda} &= x^{\lambda} + \omega^{\lambda\nu}x_{\nu} \\ \Theta'^{\alpha} &= \Theta^{\alpha} - \tfrac{1}{4}\mathrm{i}\omega^{\mu\nu}(\sigma_{\mu\nu})_{\beta}{}^{\alpha}\Theta^{\beta} \\ \bar{\Theta}'^{\dot{\alpha}} &= \bar{\Theta}^{\dot{\alpha}} + \tfrac{1}{4}\mathrm{i}\omega^{\mu\nu}(\bar{\sigma}_{\mu\nu})^{\dot{\alpha}}{}_{\dot{\beta}}\bar{\Theta}^{\dot{\beta}} \end{aligned}\right\} \tag{14.37}$$

A superfield $\Phi(x,\Theta,\bar{\Theta})$ is now defined as a function of the parameters Θ and $\bar{\Theta}$ in addition to x_{μ} such that (see (14.19))

$$\Phi(x,\Theta,\bar{\Theta}) = S(x,\Theta,\bar{\Theta})\Phi(0,0,0)S^{-1}(x,\Theta,\bar{\Theta}) \tag{14.38}$$

with the coordinate transformations given by (14.35), and

$$L(\omega)\Phi(x,\Theta,\bar{\Theta})L^{-1}(\omega) = \exp(-\tfrac{1}{2}\mathrm{i}\omega\cdot\Sigma)\Phi(x',\Theta',\bar{\Theta}') \tag{14.39}$$

where for infinitesimal ω the parameters $x',\Theta',\bar{\Theta}'$ are given by (14.37). Expanding

$$\begin{aligned} \Phi(x',\Theta',\bar{\Theta}') &= \Phi(x,\Theta,\bar{\Theta}) + (y - \mathrm{i}\Theta\sigma\bar{\eta} + \mathrm{i}\eta\sigma\bar{\Theta})(\partial\Phi/\partial x) + \eta(\partial\Phi/\partial\Theta) \\ &\quad + \bar{\eta}(\partial\Phi/\partial\bar{\Theta}) + \cdots = \Phi(x,\Theta,\bar{\Theta}) + \mathrm{i}y[P,\Phi] + \mathrm{i}[\eta Q,\Phi] + \mathrm{i}[\bar{Q}\bar{\eta},\Phi] + \cdots \end{aligned} \tag{14.40}$$

we get the differential form of the supersymmetry generators acting on a superfield $\Phi(x,\Theta,\bar{\Theta})$

$$\left.\begin{aligned} \mathrm{i}[P_{\mu},\Phi] &= \partial_{\mu}\Phi \\ \mathrm{i}[\eta^{\alpha}Q_{\alpha},\Phi] &= \eta^{\alpha}(\partial/\partial\Theta^{\alpha} + \mathrm{i}\sigma^{\mu}{}_{\alpha\beta}\bar{\Theta}^{\beta}\partial_{\mu})\Phi \\ \mathrm{i}[\bar{Q}_{\dot{\alpha}}\bar{\eta}^{\dot{\alpha}},\Phi] &= \bar{\eta}^{\dot{\alpha}}(\partial/\partial\bar{\Theta}^{\dot{\alpha}} + \mathrm{i}\Theta^{\beta}\sigma^{\mu}{}_{\beta\dot{\alpha}}\partial_{\mu})\Phi \end{aligned}\right\} \tag{14.41}$$

Using (14.37) and (14.39) we also get

$$\mathrm{i}[M_{\mu\nu},\Phi] = x_{\mu}\partial_{\nu} - x_{\nu}\partial_{\mu} + \tfrac{1}{2}\mathrm{i}\Theta^{\alpha}(\sigma_{\mu\nu})_{\alpha}{}^{\beta}(\partial/\partial\Theta^{\beta}) - \tfrac{1}{2}\mathrm{i}\bar{\Theta}^{\dot{\beta}}(\bar{\sigma}_{\mu\nu})^{\dot{\alpha}}{}_{\dot{\beta}}(\partial/\partial\bar{\Theta}^{\dot{\alpha}}) + \mathrm{i}\Sigma_{\mu\nu} \tag{14.42}$$

If Φ does not have overall spinor indices we can factorize out η in the last two equations of (14.41), e.g.

$$\mathrm{i}[Q_{\alpha},\Phi] = (\partial/\partial\Theta^{\alpha} + \mathrm{i}\sigma^{\mu}{}_{\alpha\beta}\bar{\Theta}^{\beta}\partial_{\mu})\Phi \tag{14.43}$$

The important notion in supersymmetry is that of covariant derivatives. As usual one wants to generalize

$$\frac{\partial}{\partial\Theta^\alpha}\Phi\to D_\alpha\Phi, \qquad \frac{\partial}{\partial\bar{\Theta}^{\dot\alpha}}\Phi\to \bar{D}_{\dot\alpha}\Phi$$

so that the quantities $D_\alpha\Phi$ and $\bar{D}_{\dot\alpha}\Phi$ transform under supertranslations as the fields themselves. (Note that this requirement is satisfied by $\partial_\mu\Phi$.) Thus we require

$$\left.\begin{aligned}\{D_\beta, Q_\alpha\} &= 0\\ \{D_\beta, \bar{Q}_{\dot\alpha}\} &= 0\end{aligned}\right\} \tag{14.44}$$

and similarly for $\bar{D}_{\dot\beta}$. This gives

$$\left.\begin{aligned}D_\beta &= \partial/\partial\Theta^\beta - \mathrm{i}\sigma^\mu{}_{\beta\dot\beta}\bar{\Theta}^{\dot\beta}\partial_\mu\\ \bar{D}_{\dot\beta} &= -\partial/\partial\bar{\Theta}^{\dot\beta} + \mathrm{i}\Theta^\beta\sigma^\mu{}_{\beta\dot\beta}\partial_\mu\end{aligned}\right\} \tag{14.45}$$

One can check that

$$\left.\begin{aligned}\{D_\alpha, D_\beta\} &= \{\bar{D}_{\dot\alpha}, \bar{D}_{\dot\beta}\} = 0\\ \{D_\alpha, \bar{D}_{\dot\beta}\} &= 2\mathrm{i}\sigma^\mu{}_{\alpha\dot\beta}\partial_\mu\end{aligned}\right\} \tag{14.46}$$

In the four-component notation we have

$$D = \begin{pmatrix} D_\alpha \\ \bar{D}^{\dot\alpha}\end{pmatrix}, \qquad \Theta = \begin{pmatrix}\Theta_\alpha \\ \bar{\Theta}^{\dot\alpha}\end{pmatrix}, \qquad \bar{\Theta} = (\Theta^\alpha, \bar{\Theta}_{\dot\alpha})$$

where

$$\bar{D}^{\dot\alpha} = \varepsilon^{\dot\alpha\dot\beta}\bar{D}_{\dot\beta} = \partial/\partial\bar{\Theta}_{\dot\alpha} - \mathrm{i}\bar{\sigma}^{\mu\dot\alpha\beta}\Theta_\beta\partial_\mu \tag{14.47}$$

(since $\partial/\partial\bar{\Theta}_{\dot\beta} = -\varepsilon^{\dot\beta\dot\alpha}\partial/\partial\bar{\Theta}^{\dot\alpha}$) and consequently

$$D = \partial/\partial\bar{\Theta} - \mathrm{i}\gamma^\mu\Theta\partial_\mu \tag{14.48}$$

Superfields

We are now prepared to construct superfields as power series expansions in Θ^α and $\bar{\Theta}^{\dot\alpha}$. The most general superfield in the two-component notation reads (we denote it by $F(x, \Theta, \bar{\Theta})$ reserving the Φ for the chiral superfields)

$$\begin{aligned}F(x,\Theta,\bar{\Theta}) = f(x) &+ \Theta\chi(x) + \bar{\Theta}\bar{\chi}(x) + \Theta\Theta m(x) + \bar{\Theta}\bar{\Theta}n(x) + \Theta\sigma^\mu\bar{\Theta}v_\mu(x)\\ &+ \Theta\Theta\bar{\Theta}\bar{\lambda}(x) + \bar{\Theta}\bar{\Theta}\Theta\lambda(x) + \Theta\Theta\bar{\Theta}\bar{\Theta}d(x)\end{aligned} \tag{14.49}$$

This is the most general superfield due to the vanishing of the square of each component of Θ^α and $\bar{\Theta}^{\dot\alpha}$. Remember that

$$\Theta\Theta = \Theta^\alpha\Theta_\alpha = \varepsilon_{\alpha\beta}\Theta^\alpha\Theta^\beta, \qquad \bar{\Theta}\bar{\Theta} = -\bar{\Theta}^{\dot\alpha}\bar{\Theta}_{\dot\alpha} = -\varepsilon_{\dot\alpha\dot\beta}\bar{\Theta}^{\dot\alpha}\bar{\Theta}^{\dot\beta}$$

or in other words

$$\left.\begin{aligned}\Theta^\alpha\Theta^\beta &= -\tfrac{1}{2}\varepsilon^{\alpha\beta}\Theta\Theta\\ \bar{\Theta}^{\dot\alpha}\bar{\Theta}^{\dot\beta} &= \tfrac{1}{2}\varepsilon^{\dot\alpha\dot\beta}\bar{\Theta}\bar{\Theta}\end{aligned}\right\} \tag{14.50}$$

Assuming that the superfield has no overall Lorentz indices, this superfield contains four complex scalar fields $f(x), m(x), n(x)$ and $d(x)$, two spinors $\chi(x)$ and $\lambda(x)$ in the $(\frac{1}{2}, 0)$ representation and two unrelated spinors $\bar{\chi}(x)$ and $\bar{\lambda}(x)$ in the $(0, \frac{1}{2})$ representation of the Lorentz group, and one complex vector field $v_\mu(x)$. Thus we have altogether 16 bosonic and 16 fermionic degrees of freedom (a degree of freedom is an unconstrained single real field).

The transformation laws under supersymmetric transformations for the components of $F(x, \Theta, \bar{\Theta})$ can be obtained by comparing powers of Θ and $\bar{\Theta}$ in the equation

$$\delta F = \delta_\eta F + \delta_{\bar{\eta}} F = \delta f(x) + \Theta\delta\chi(x) + \bar{\Theta}\delta\bar{\chi}(x) + \Theta\Theta\delta m(x) + \cdots \tag{14.51}$$

where the l.h.s. of (14.51) is given by (14.41). Note that it follows from (14.41) that the variation of the highest component of the superfield must always be a space-time derivative.

One can see that the general superfield yields the component multiplet $(f, \chi, \bar{\chi}, m, n, v_\mu, \lambda, \bar{\lambda}, d)$ which is reducible under supertranslations. To get irreducible component multiplets one must impose some conditions on the superfields. These conditions must be covariant with respect to supersymmetric transformations but otherwise there is no general rule concerning their choice. One widely used set of such conditions is that which defines a so-called chiral (left-handed) superfield $\Phi(x, \Theta, \bar{\Theta})$ with

$$\bar{D}_{\dot{\alpha}}\Phi = 0 \tag{14.52}$$

and an antichiral (right-handed) superfield $\bar{\Phi}(x, \Theta, \bar{\Theta})$ with

$$D_\alpha\bar{\Phi} = 0 \tag{14.53}$$

Using (14.45) and solving the first-order partial differential equations we get

$$\Phi(x, \Theta, \bar{\Theta}) = \exp(-\mathrm{i}\Theta^\beta\sigma^\mu{}_{\beta\dot{\alpha}}\bar{\Theta}^{\dot{\alpha}}\partial_\mu)\hat{\Phi}(x, \Theta) \tag{14.54}$$

and

$$\bar{\Phi}(x, \Theta, \bar{\Theta}) = \exp(\mathrm{i}\Theta^\beta\sigma^\mu_{\beta\dot{\alpha}}\bar{\Theta}^{\dot{\alpha}}\partial_\mu)\hat{\bar{\Phi}}(x, \bar{\Theta}) \tag{14.55}$$

The fields $\hat{\Phi}(x, \Theta)$ and $\hat{\bar{\Phi}}(x, \bar{\Theta})$ have the expansion

$$\hat{\Phi}(x, \Theta) = A(x) + 2\Theta\Psi(x) - \Theta\Theta F(x) \tag{14.56}$$

$$\hat{\bar{\Phi}}(x, \bar{\Theta}) = \bar{A}(x) + 2\bar{\Psi}(x)\bar{\Theta} - \bar{\Theta}\bar{\Theta}\bar{F}(x) \tag{14.57}$$

and each contains Weyl spinors of only one chirality. The chiral superfield $\Phi(x, \Theta, \bar{\Theta})$ is formally obtained from $\hat{\Phi}(x, \Theta)$ by the translation (14.54)

$$\begin{aligned}\Phi(x, \Theta, \bar{\Theta}) = \hat{\Phi}(x^\mu - \mathrm{i}\Theta^\beta\sigma^\mu{}_{\beta\dot{\alpha}}\bar{\Theta}^{\dot{\alpha}}, \Theta) &= A(x) - \mathrm{i}\Theta\sigma^\mu\bar{\Theta}\partial_\mu A(x) \\ &- \tfrac{1}{4}\Theta\Theta\bar{\Theta}\bar{\Theta}\partial^2 A(x) + 2\Theta\Psi(x) + \mathrm{i}\Theta\Theta\partial_\mu\Psi(x)\sigma^\mu\bar{\Theta} - \Theta\Theta F(x)\end{aligned} \tag{14.58}$$

and the analogous expression also holds for $\bar{\Phi}(x, \Theta, \bar{\Theta})$. In deriving (14.58) we have used the relation

$$(\Theta\sigma^\mu\bar{\Theta})(\Theta\sigma^\nu\bar{\Theta}) = \tfrac{1}{2}g^{\mu\nu}(\Theta\Theta)(\bar{\Theta}\bar{\Theta}) \tag{14.59}$$

The transformation rules under supertranslation of the component fields A, Ψ and F follow from the general formula (14.51) and read

$$\left.\begin{aligned} \delta A &= \delta_\eta A + \delta_{\bar{\eta}} A = 2\eta^\alpha \Psi_\alpha \\ \delta \Psi_\alpha &= -\eta_\alpha F - \mathrm{i}(\sigma^\mu \bar{\eta})_\alpha \partial_\mu A \\ \delta F &= -2\mathrm{i}\partial_\mu \Psi \sigma^\mu \bar{\eta} \end{aligned}\right\} \tag{14.60}$$

We see explicitly that the component multiplet of the chiral superfield is irreducible under supertranslations. Let us also mention that the chiral superfield is not the only one used in the construction of supersymmetric theories. In particular, supersymmetric gauge theories require vector superfields defined by the condition $V(x, \Theta, \bar{\Theta}) = V^\dagger(x, \Theta, \bar{\Theta})$ that contain spin one bosons. In the following we shall, however, limit ourselves to using chiral superfields in constructing a supersymmetric lagrangian.

14.5 Supersymmetric lagrangian; Wess–Zumino model

Let us begin our considerations with the observation that the product of two superfields is again a superfield

$$F_1(x, \Theta, \bar{\Theta}) F_2(x, \Theta, \bar{\Theta}) = F_3(x, \Theta, \bar{\Theta}) \tag{14.61}$$

This follows from the representation (14.41) of the superalgebra which is of first order in differential operators. From two superfields Φ_1 and Φ_2 of the same chirality one can construct another chiral superfield

$$\Phi_3 = \Phi_1 \Phi_2 \tag{14.62}$$

and from two superfields of different chirality – two vector superfields

$$\left.\begin{aligned} V_\mathrm{S} &= \tfrac{1}{2}(\Phi_1 \bar{\Phi}_2 + \bar{\Phi}_1 \Phi_2) \\ V_\mathrm{A} &= \tfrac{1}{2}\mathrm{i}(\Phi_1 \bar{\Phi}_2 - \bar{\Phi}_1 \Phi_2) \end{aligned}\right\} \tag{14.63}$$

corresponding to the symmetrized and antisymmetrized product, respectively. As we mentioned earlier, we are going to use chiral fields only. The Φ_3 in (14.62) is chiral due to the fact that $\bar{D}$ is a first-order differential operator which obeys the Leibniz rule

$$\bar{D}(\Phi_1 \Phi_2) = (\bar{D}\Phi_1)\Phi_2 + \Phi_1 \bar{D}\Phi_2 \tag{14.64}$$

A very important chiral superfield is the one which, as we shall see, can be identified with the kinetic part of the supersymmetric lagrangian. It reads

$$\Phi_\mathrm{K} = \tfrac{1}{4}\Phi \bar{D}^2 \bar{\Phi} \tag{14.65}$$

where $\bar{\Phi} \equiv \Phi^\dagger$ and it is a left-handed chiral superfield since $\bar{D}^2\bar{\Phi}$ is a left-handed chiral superfield (see (14.46))

$$\bar{D}\bar{D}^2\bar{\Phi} = 0 \tag{14.66}$$

and the product of two left-handed superfields is again a left-handed superfield.

We will need the notion of integration over the Grassmann variables Θ and $\bar{\Theta}$. This has been introduced in Section 2.5 and we just add the following definitions

$$\int d^2\Theta = \int d\Theta^2\, d\Theta^1, \qquad \int d^2\bar{\Theta} = \int d\bar{\Theta}^1\, d\bar{\Theta}^2 \tag{14.67}$$

Therefore

$$\int d^2\Theta\Theta\Theta = \int d^2\bar{\Theta}\bar{\Theta}\bar{\Theta} = -2 \tag{14.68}$$

We also note that the Dirac δ-functions are

$$\left.\begin{array}{ll} \delta(\Theta_\alpha) = \Theta_\alpha, & \delta(-\Theta_\alpha) = -\delta(\Theta_\alpha) \\ \delta^{(2)}(\Theta) = -\frac{1}{2}\Theta\Theta, & \delta^{(2)}(\bar{\Theta}) = -\frac{1}{2}\bar{\Theta}\bar{\Theta} \end{array}\right\} \tag{14.69}$$

The crucial observation for constructing supersymmetric lagrangians is the fact that for the general scalar superfield (14.49) the integral

$$S_D = \frac{1}{4}\int d^4x\, d^2\Theta\, d^2\bar{\Theta} F(x, \Theta, \bar{\Theta}) \tag{14.70}$$

and for the chiral superfield the integral

$$S_F = \frac{1}{4}\int d^4x\, d^2\Theta\Phi(x, \Theta, \bar{\Theta}) + \text{h.c.} \tag{14.71}$$

are invariant under supertranslations. Indeed we see from (14.49) that

$$\int d^2\Theta\, d^2\bar{\Theta} F(x, \Theta, \bar{\Theta}) = 4d(x) \tag{14.72}$$

and we remember that the variation of $d(x)$ under supertranslation is a four-derivative. Thus S_D is an invariant. For chiral fields S_D vanishes but the $\int d^4x\, d^2\Theta$ alone gives an invariant result. Under $\int d^4x$ we can use the expansion (14.56) since the transformation (14.54) in its infinitesimal form gives $\delta\Phi$ which is a four-derivative. Thus

$$S_F = \frac{1}{2}\int d^4x F(x) + \text{h.c.} \tag{14.73}$$

where $F(x)$ is the highest component of the chiral superfield, and S_F is an invariant of supertranslations since $\delta F(x)$ is also a four-derivative. (We always assume that the fields fall off fast enough at infinity.) Of course the superfields in S_D and S_F must in general arise as products of 'single-particle' superfields since the lagrangian has to contain terms at least bilinear in these elementary fields.

As the last step towards constructing the supersymmetric lagrangian let us do some dimensional analysis. Clearly, from (14.7) and (14.28) each Θ and $\bar{\Theta}$ carries a mass dimension $-\frac{1}{2}$. Thus taking the canonical dimensions 1 and $\frac{3}{2}$ for the boson and fermion fields we conclude that the whole 'single-particle' superfield has dimension 1. The derivatives D_α and $\bar{D}_{\dot{\alpha}}$ have dimension $\frac{1}{2}$ each and from (14.68) the dimension of the measure $\mathrm{d}^4x\,\mathrm{d}^2\Theta\,\mathrm{d}^2\bar{\Theta} \equiv \mathrm{d}z$ is -2 and the dimension of the chiral measure $\mathrm{d}^4x\,\mathrm{d}^2\Theta$ is -3. In order to have a renormalizable theory we have to construct a dimensionless action built of single-particle superfields so that no negative-dimension coupling constants are present.

We are now ready to write down the supersymmetric lagrangian for a chiral superfield Φ with complex field components A, Ψ^α and F describing four bosonic and four fermionic degrees of freedom. In four dimensions this is the smallest possible number of degrees of freedom which can be contained in a superfield since any multiplet must contain a spinor which has at least two complex (Weyl) or four real (Majorana) components. It is clear from our considerations that the most general action for Φ (the so-called Wess–Zumino action) with the superkinetic term quadratic in the field reads

$$S_{\mathrm{WZ}} = \frac{1}{4}\int \mathrm{d}^4x\,\mathrm{d}^2\Theta(\tfrac{1}{8}\Phi\bar{D}^2\bar{\Phi} - \tfrac{1}{2}m\Phi^2 - \tfrac{1}{3}g\Phi^3) + \text{h.c.} \tag{14.74}$$

(the term $\lambda\Phi$ can be eliminated by a redefinition of fields).

In terms of the field components denoted as follows

$$\operatorname{Re} A \equiv A, \operatorname{Im} A \equiv B, \operatorname{Re} F \equiv F, \operatorname{Im} F \equiv -G$$

and

$$\Psi = \begin{pmatrix} \Psi_\alpha \\ \bar{\Psi}^{\dot{\alpha}} \end{pmatrix}$$

where Ψ is the four-component Majorana spinor, one can derive from (14.74) the following lagrangian density

$$\begin{aligned}\mathscr{L} = {} & \tfrac{1}{2}(\partial_\mu A\partial^\mu A + \partial_\mu B\partial^\mu B + \mathrm{i}\bar{\Psi}\not{\partial}\Psi + F^2 + G^2) - m(AF + BG + \tfrac{1}{2}\bar{\Psi}\Psi) \\ & - g[(A^2 - B^2)F + 2ABG + \bar{\Psi}(A - \mathrm{i}\gamma_5 B)\Psi] + 4\text{-div}\end{aligned} \tag{14.75}$$

This is the lagrangian of the original Wess–Zumino supersymmetric model. It gives the following equations of motion

$$\left.\begin{aligned} \partial^2 A &= -mF - 2g(AF + BG + \tfrac{1}{2}\bar{\Psi}\Psi) \\ \partial^2 B &= -mG - 2g(AG - BF - \tfrac{1}{2}\bar{\Psi}\mathrm{i}\gamma_5\Psi) \\ \mathrm{i}\not{\partial}\Psi &= m\Psi + 2g(A - \mathrm{i}\gamma_5 B)\Psi \\ F &= mA + g(A^2 - B^2) \\ G &= mB + 2gAB \end{aligned}\right\} \tag{14.76}$$

An important observation is that the equations for the fields F and G are purely algebraic and consequently the F and G can be eliminated from the lagrangian and from the equations of motion. We then get

$$\begin{aligned}\mathscr{L} = {} & \tfrac{1}{2}(\partial_\mu A\partial^\mu A - m^2A^2) + \tfrac{1}{2}(\partial_\mu B\partial^\mu B - m^2B^2) + \tfrac{1}{2}\bar{\Psi}(\mathrm{i}\not{\partial} - m)\Psi - mgA(A^2 + B^2) \\ & - g\bar{\Psi}(A - \mathrm{i}\gamma_5 B)\Psi - \tfrac{1}{2}g^2(A^2 + B^2)^2\end{aligned} \tag{14.77}$$

and correspondingly

$$\left.\begin{aligned}(\partial^2+m^2)A &= -mg(3A^2+B^2)-2g^2A(A^2+B^2)-g\bar{\Psi}\Psi\\(\partial^2+m^2)B &= -2mgAB-2g^2B(A^2+B^2)+g\bar{\Psi}\mathrm{i}\gamma_5\Psi\end{aligned}\right\} \tag{14.78}$$
$$(\mathrm{i}\not{\partial}-m)\Psi=2g(A-\mathrm{i}\gamma_5B)\Psi$$

The elimination of the auxiliary fields F and G gives, however, a theory which is supersymmetric only 'on-shell'. The lagrangian (14.77) is invariant under transformations obtained from the supersymmetry transformations by elimination of the fields F and G, using their own equations of motion, and the on-shell algebra closes only if the equations of motion of the dynamical fields hold. This is understandable by counting the off-shell and on-shell number of degrees of freedom for scalar and fermion fields.

14.6 Supergraphs and the non-renormalization theorem

Perturbation theory in superspace may be developed as an extension of ordinary perturbation theory. Our aim is to calculate superfield Green's functions ($z\equiv(x,\Theta,\bar{\Theta})$)

$$\langle 0|T\Phi(z_1)\ldots\Phi(z_k)\ldots\bar{\Phi}(z_{k+1})\ldots\bar{\Phi}(z_n)|0\rangle \tag{14.79}$$

from which one can obtain the component-field Green's functions by power series expansion in Θ_1, $\bar{\Theta}_1\ldots\Theta_n$, $\bar{\Theta}_n$. We use the generating functional technique with generating functionals defined in Section 2.6. For discussing the necessary counterterms in particular, it is very convenient (see Section 4.2) to calculate the effective action

$$\begin{aligned}\Gamma[\Phi\ldots\bar{\Phi}\ldots] &= \sum_n\int \mathrm{d}^4x_1\ldots\mathrm{d}^4x_n\mathrm{d}^2\Theta_1\ldots\mathrm{d}^2\bar{\Theta}_n\Gamma^{(n)}(z_1,\ldots,z_n)\Phi(z_1)\ldots\bar{\Phi}(z_n)\\ &= \sum_n\int\prod_{i=1}^n\left[\frac{\mathrm{d}^4p_i}{(2\pi)^4}\right]\mathrm{d}^2\Theta_1\cdots\mathrm{d}^2\bar{\Theta}_n(2\pi)^4\delta\left(\sum_{i=1}^n p_i\right)\tilde{\Gamma}^{(n)}(p_1,\ldots,p_n;\Theta_1,\ldots,\bar{\Theta}_n)\\ &\quad\times\Phi(p_1,\Theta_1)\ldots\Phi(p_n,\bar{\Theta}_n)\end{aligned} \tag{14.80}$$

where the $\Gamma^{(n)}$s are the 1PI Green's functions.

Let us begin with several auxiliary remarks. The difference between $\bar{D}_{\dot{\alpha}}$ and $-\partial/\partial\bar{\Theta}^{\dot{\alpha}}$ is a derivative term which does not contribute under the d^4x integral. Thus, using the equivalence of integration over and differentiation with respect to Grassmann variables and our conventions (14.50) and (14.67) we can convert

$$\bar{D}^2\to 2\mathrm{d}^2\bar{\Theta} \tag{14.81}$$

under the d^4x integral. Similarly, we have

$$D^2\to 2\mathrm{d}^2\Theta \tag{14.82}$$

For the operators $\bar{D}^2D^2$ and $D^2\bar{D}^2$ acting on chiral and antichiral fields, respectively, we can write

$$\bar{D}^2 D^2 \Phi = [\bar{D}^2, D^2]\Phi = (-16\partial^2 - 8\mathrm{i}D\sigma^\mu \partial_\mu \bar{D})\Phi = -16\partial^2 \Phi \tag{14.83}$$

and similarly

$$D^2 \bar{D}^2 \bar{\Phi} = -16\partial^2 \bar{\Phi} \tag{14.84}$$

Thus, for example

$$\int \mathrm{d}^2\Theta\Phi = \int \mathrm{d}^2\Theta(-\bar{D}^2 D^2/16\partial^2)\Phi = \int \mathrm{d}^2\Theta\, \mathrm{d}^2\bar{\Theta}(-D^2/8\partial^2)\Phi \tag{14.85}$$

and
$$\int \mathrm{d}^2\Theta\Phi^2 = \int \mathrm{d}^2\Theta\Phi(-\bar{D}^2 D^2/16\partial^2)\Phi = \int \mathrm{d}^2\Theta(\bar{D}^2/2)[\Phi(-D^2/8\partial^2)\Phi]$$
$$= \int \mathrm{d}^2\Theta\, \mathrm{d}^2\bar{\Theta}\Phi(-D^2/8\partial^2)\Phi \tag{14.86}$$

where we have used the fact that $\bar{D}\Phi = 0$.

The free-field part of the action (14.74) can now be written as follows

$$S_0 = \frac{1}{4}\int \mathrm{d}^4x\, \mathrm{d}^2\Theta(\tfrac{1}{8}\Phi\bar{D}^2\bar{\Phi} - \tfrac{1}{2}m\Phi^2) + \text{h.c.}$$
$$= \int \mathrm{d}^4x\, \mathrm{d}^2\Theta\, \mathrm{d}^2\bar{\Theta}[\tfrac{1}{8}\bar{\Phi}\Phi - \tfrac{1}{8}m\Phi(-D^2/8\partial^2)\Phi$$
$$-\tfrac{1}{8}m\bar{\Phi}(-\bar{D}^2/8\partial^2)\bar{\Phi}] \quad = \int \mathrm{d}^4x\, \mathrm{d}^2\Theta\, \mathrm{d}^2\bar{\Theta}\tfrac{1}{16}(\Phi, \bar{\Phi})A\begin{pmatrix}\Phi\\ \bar{\Phi}\end{pmatrix} \tag{14.87}$$

where
$$A = \begin{pmatrix} mD^2/4\partial^2 & 1 \\ 1 & m\bar{D}^2/4\partial^2 \end{pmatrix} \tag{14.88}$$

To derive the Feynman rules we shall use the generating functional technique. We define the functional derivative for the chiral fields by the equation

$$\frac{\delta}{\delta\Phi(x, \Theta, \bar{\Theta})}\int \mathrm{d}^4x'\, \mathrm{d}^2\Theta' \mathscr{F}[\Phi(x', \Theta', \bar{\Theta}')] = \mathscr{F}'[\Phi(x, \Theta, \bar{\Theta})] \tag{14.89}$$

This definition implies

$$\frac{\delta\Phi(x', \Theta', \bar{\Theta}')}{\delta\Phi(x, \Theta, \bar{\Theta})} = \tfrac{1}{2}\bar{D}^2[\delta^{(2)}(\Theta' - \Theta)\delta^{(2)}(\bar{\Theta}' - \bar{\Theta})\delta^{(4)}(x' - x)] \tag{14.90}$$

(to get (14.89) integrate by parts using (14.90)) and analogously for $\bar{\Phi}$.

The generating functional of the theory is

$$W[J, \bar{J}] = \exp(\mathrm{i}S_{\text{int}}[\delta/\mathrm{i}\delta J, \delta/\mathrm{i}\delta\bar{J}])W_0[J, \bar{J}] \tag{14.91}$$

where $W_0[J, \bar{J}]$ is the generating functional for the superfield Green's functions:

$$W_0[J, \bar{J}] = \left\langle 0 \middle| T\exp\left[\mathrm{i}\int \mathrm{d}^4x\mathrm{d}^2\Theta\mathrm{d}^2\bar{\Theta}\tfrac{1}{2}(\Phi, \bar{\Phi})\begin{pmatrix} -(D^2/4\partial^2)J \\ -(\bar{D}^2/4\partial^2)\bar{J}\end{pmatrix}\right]\middle| 0 \right\rangle \tag{14.92}$$

and J and $\bar{J}$ are chiral sources. Thus

$$G^{(n),0}_{\Phi\cdots\Phi}(z_1,\ldots,z_n)=(-\mathrm{i})^n\frac{\delta}{\delta J(z_1)}\cdots\frac{\delta}{\delta \bar{J}(z_n)}W_0[J,\bar{J}]|_{J=\bar{J}=0} \tag{14.93}$$

Using the path integral representation for the $W_0[J,\bar{J}]$

$$W_0[J,\bar{J}]=\int\mathscr{D}\Phi\mathscr{D}\bar{\Phi}\exp\left\{\mathrm{i}\int\mathrm{d}z\left[\tfrac{1}{16}(\Phi,\bar{\Phi})A\begin{pmatrix}\Phi\\ \bar{\Phi}\end{pmatrix}+\tfrac{1}{2}(\Phi,\bar{\Phi})B\right]\right\} \tag{14.94}$$

where $\mathrm{d}z=\mathrm{d}^4x\mathrm{d}^2\Theta\mathrm{d}^2\bar{\Theta}$ and

$$B=\begin{pmatrix}-(D^2/4\partial^2)J\\ -(\bar{D}^2/4\partial^2)\bar{J}\end{pmatrix}$$

and performing the Gaussian integration over Φ and $\bar{\Phi}$ one gets (see (2.81))

$$W_0[J,\bar{J}]=\exp\left(-\mathrm{i}\int\mathrm{d}zB^{\mathrm{T}}A^{-1}B\right) \tag{14.95}$$

where

$$A^{-1}=\begin{pmatrix}-\dfrac{m\bar{D}^2}{4(\partial^2+m^2)} & 1+\dfrac{m^2\bar{D}^2D^2}{16\partial^2(\partial^2+m^2)}\\ 1+\dfrac{m^2D^2\bar{D}^2}{16\partial^2(\partial^2+m^2)} & -\dfrac{mD^2}{4(\partial^2+m^2)}\end{pmatrix} \tag{14.96}$$

Finally we have

$$W_0[J,\bar{J}]=\exp\left[-2\mathrm{i}\int\mathrm{d}z\left(-\bar{J}\frac{1}{\partial^2+m^2}J+\tfrac{1}{2}J\frac{\frac{1}{4}mD^2}{\partial^2(\partial^2+m^2)}J+\tfrac{1}{2}\bar{J}\frac{\frac{1}{4}m\bar{D}^2}{\partial^2(\partial^2+m^2)}\bar{J}\right)\right] \tag{14.97}$$

Using (14.90), (14.93), (14.97) and integrating by parts when necessary we obtain the free superpropagators. For instance

$$\begin{aligned}G^{(2),0}_{\Phi\bar{\Phi}}(z_1,z_2)&=-2\mathrm{i}(\delta/\delta J(z_1))(\delta/\delta\bar{J}(z_2))\int\mathrm{d}z_3\,\mathrm{d}z_4\bar{J}(z_3)\delta^{(4)}(\Theta_{34})G(x_3-x_4)J(z_4)\\&=-2\mathrm{i}\int\mathrm{d}z_3\,\mathrm{d}z_4[\tfrac{1}{2}\bar{D}^2_1\delta(z_{14})]\delta^{(4)}(\Theta_{34})G(x_3-x_4)[\tfrac{1}{2}D^2_2\delta(z_{23})]\\&=-2\mathrm{i}\tfrac{1}{2}\bar{D}^2_1\tfrac{1}{2}D^2_2[\delta^{(4)}(\Theta_{12})G(x_1-x_2)]\end{aligned} \tag{14.98}$$

where

$$\delta(z_{ij})\equiv\delta^{(4)}(\Theta_i-\Theta_j)\delta^{(4)}(x_i-x_j)\equiv\delta^{(4)}(\Theta_{ij})\delta^{(4)}(x_i-x_j)$$

and

$$(\partial^2_1+m^2)G(x_1-x_2)=\delta^{(4)}(x_1-x_2)$$

and

$$\left.\begin{aligned}D^2_1\delta^{(2)}(\Theta_1-\Theta_2)&=-D^2_1\tfrac{1}{2}(\Theta_1-\Theta_2)^2=2\exp[\mathrm{i}(\Theta_1-\Theta_2)\sigma^\mu\bar{\Theta}_1\partial^{(1)}_\mu]\\ \bar{D}^2_1\delta^{(2)}(\bar{\Theta}_1-\bar{\Theta}_2)&=-\bar{D}^2_1\tfrac{1}{2}(\bar{\Theta}_1-\bar{\Theta}_2)^2=2\exp[-\mathrm{i}\Theta_1\sigma^\mu(\bar{\Theta}_1-\bar{\Theta}_2)\partial^{(1)}_\mu]\end{aligned}\right\} \tag{14.99}$$

(remember that $\partial/\partial\Theta^\alpha = -\varepsilon_{\alpha\beta}\partial/\partial\Theta_\beta$). To obtain the component-field propagators we can use (14.99), (14.54) and (14.55) to write

$$\hat{G}^{(2),0}_{\Phi\bar{\Phi}}(z_1,z_2) = \langle 0|T\hat{\Phi}(x_1,\Theta_1)\hat{\bar{\Phi}}(x_2,\bar{\Theta}_2)|0\rangle = -2\mathrm{i}\exp(\mathrm{i}2\Theta_1\sigma^\mu\bar{\Theta}_2\partial^{(1)}_\mu)G(x_1-x_2) \tag{14.100}$$

and then use (14.56) and (14.57) to expand in Θ_1 and $\bar{\Theta}_2$. Analogously, we get†

$$\hat{G}^{(2),0}_{\Phi\Phi}(z_1,z_2) = 2\mathrm{i}m\frac{\bar{D}_2^2 D_2^2}{16\partial^2}\bar{D}_1^2[G(x_1-x_2)\delta^{(4)}(\Theta_{12})] \tag{14.101}$$

$$= -2\mathrm{i}m\bar{D}_1^2[\delta^{(4)}(\Theta_{12})G(x_1-x_2)] \tag{14.101a}$$

In momentum space we thus get, e.g.

$$G(x_1-x_2) = -\frac{1}{(2\pi)^4}\int \mathrm{d}^4p\frac{\exp[-\mathrm{i}p(x_1-x_2)]}{p^2-m^2}$$

$$\hat{G}^{(2),0}_{\Phi\bar{\Phi}} = \frac{2\mathrm{i}}{p^2-m^2}\exp[(2\Theta_1\bar{\Theta}_2 - \Theta_1\bar{\Theta}_1 + \Theta_2\bar{\Theta}_2)^{\alpha\dot{\beta}}\sigma^\mu_{\alpha\dot{\beta}}p_\mu] \tag{14.102}$$

It is, however, more convenient to take as free superpropagators the following

$$\left.\begin{aligned} G^{(2),0}_{\Phi\bar{\Phi}} &= \frac{2\mathrm{i}}{p^2-m^2}\delta^{(4)}(\Theta_{12}) \\ G^{(2),0}_{\Phi\Phi} &= \frac{2\mathrm{i}}{p^2(p^2-m^2)}\tfrac{1}{4}mD^2\delta^{(4)}(\Theta_{12}) \\ G^{(2),0}_{\bar{\Phi}\bar{\Phi}} &= \frac{2\mathrm{i}}{p^2(p^2-m^2)}\tfrac{1}{4}m\bar{D}^2\delta^{(4)}(\Theta_{12}) \end{aligned}\right\} \tag{14.103‡}$$

where

$$D^2 = D^2(p,\Theta_1)$$
$$D_\alpha(p,\Theta) = \partial/\partial\Theta^\alpha - (\sigma^\mu\bar{\Theta})_\alpha p_\mu$$

Using this convention we associate the factors $\frac{1}{2}\bar{D}_1^2$ and $\frac{1}{2}D_2^2$ present in (14.98) and $\frac{1}{2}\bar{D}_2^2$ and $\frac{1}{2}\bar{D}_1^2$ present in (14.101) with the chiral vertices rather than with the propagators. Notice that these factors cancel the p^{-2} factors in the $\Phi\Phi$ and $\bar{\Phi}\bar{\Phi}$ propagators (14.103).

The vertices can be obtained from the general formula (14.91) where

$$S_{\mathrm{int}}\left[\frac{\delta}{\mathrm{i}\delta J},\frac{\delta}{\mathrm{i}\delta\bar{J}}\right] = -\tfrac{1}{12}g\int \mathrm{d}^4x\left[\mathrm{d}^2\Theta\left(\frac{1}{\mathrm{i}\delta J}\right)^3 + \mathrm{d}^2\bar{\Theta}\left(\frac{1}{\mathrm{i}\delta\bar{J}}\right)^3\right] \tag{14.104}$$

Let us consider, for instance, the three-point Green's function $G^{(3)}_{\Phi\Phi\Phi}(z_1,z_2,z_3)$ in the first order in g. The first term in (14.104) gives, up to the overall combinatorial

† $\bar{D}_1^2[\delta^{(4)}(\Theta_{12})f(x_1-x_2)] = \bar{D}_2^2[\delta^{(4)}(\Theta_{12})f(x_1-x_2)]$

‡ Factors 2 in the propagators correspond to our normalization (14.74) appropriate for the real component fields, see (14.75).

Fig. 14.2

factor,

$$G^{(3)}_{\Phi\Phi\Phi}(z_1,z_2,z_3) = -S_{\text{int}}\left[\frac{\delta}{\text{i}\delta J}\right]\frac{\delta}{\delta J(z_1)}\frac{\delta}{\delta J(z_2)}\frac{\delta}{\delta J(z_3)}W_0[J,\bar{J}]$$

$$\sim -\text{i}g\int \text{d}^4x_4\text{d}^2\Theta_4 G^{(2),0}_{\Phi\Phi}(z_1,z_4)G^{(2),0}_{\Phi\Phi}(z_2,z_4)G^{(2),0}_{\Phi\Phi}(z_3,z_4) \quad (14.105)$$

where $G^{(2),0}_{\Phi\Phi}$ are given by (14.101). It is now convenient to regroup the factors $\frac{1}{2}\bar{D}^2$ present in each $G^{(2),0}_{\Phi\Phi}$ (due to the functional differentiation over δJ) as follows: use one $\frac{1}{2}\bar{D}^2$ to replace $\int\text{d}^2\Theta_4$ by $\text{d}^4\Theta_4$, associate the other two $\frac{1}{2}\bar{D}^2$s with the Φ^3 vertex and leave the remaining $\bar{D}^2$s for use in the adjacent vertices. Thus a Φ^3 vertex has two factors $\frac{1}{2}\bar{D}^2$ acting on two of three propagators entering it. We then get the following rules:

(i) free propagators are given by (14.103)
(ii) vertices are read from S_{int} with an extra $\frac{1}{2}\bar{D}^2$ (or $\frac{1}{2}D^2$) for each chiral (or antichiral) superfield, but omitting one $\frac{1}{2}\bar{D}^2$ (or $\frac{1}{2}D^2$) for converting $\text{d}^2\Theta$ (or $\text{d}^2\bar{\Theta}$) into $\text{d}^4\Theta$
(iii) for each vertex there is an integration over $\text{d}^4\Theta$ and for each loop the usual integration over $\text{d}^4k/(2\pi)^4$
(iv) to obtain the effective action we compute amputated 1PI diagrams i.e. we replace the external propagators by the appropriate superfield with a $\frac{1}{2}\bar{D}^2$ (or $\frac{1}{2}D^2$) omitted at a vertex for each external chiral (or antichiral) superfield

The counting of the $\frac{1}{2}\bar{D}^2$ factors becomes even more evident when we consider the one-loop and two-loop diagrams in Fig. 14.2. In the first case after amputating the external lines we are left with two propagators i.e. with four factors $\frac{1}{2}\bar{D}^2$ or $\frac{1}{2}D^2$. Two of them are used to convert two $\text{d}^2\Theta$ integrals into $\text{d}^4\Theta$ integrals and two are left as vertex factors, in agreement with our general rules. In the second case the internal vertices have two $\frac{1}{2}\bar{D}^2$ or $\frac{1}{2}D^2$ factors each.

Let us use these rules to calculate the amplitudes corresponding to the 1PI diagrams given in Fig. 14.3. Diagram (*a*), after integrating by parts, gives

$$\Gamma(\Phi,\bar{\Phi}) \sim g^2\int\frac{\text{d}^4p}{(2\pi)^4}\int\frac{\text{d}^4k}{(2\pi)^4}\frac{1}{k^2(p+k)^2}\int\text{d}^4\Theta_1\text{d}^4\Theta_2\Phi(-p,\Theta_1)$$
$$*\bar{\Phi}(p,\Theta_2)\delta^{(4)}(\Theta_{12})\bar{D}^2_1 D^2_2\delta^{(4)}(\Theta_{12})$$
$$\sim g^2\int\frac{\text{d}^4p}{(2\pi)^4}\int\frac{\text{d}^4k}{(2\pi)^4}\frac{1}{k^2(p+k)^2}\int\text{d}^4\Theta\Phi(-p,\Theta)\bar{\Phi}(p,\Theta) \quad (14.106)$$

We have used the fact that (see (14.99))

$$\bar{D}^2_1 D^2_2\delta^{(4)}(\Theta_{12})_{\Theta_1=\Theta_2} = 4$$

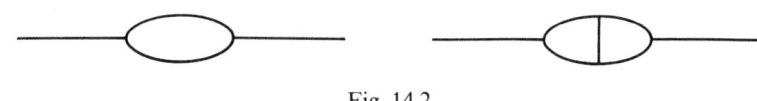

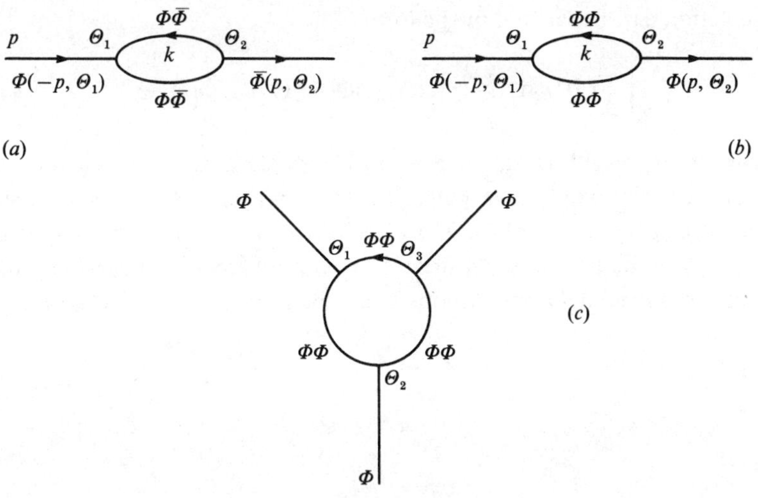

Fig. 14.3

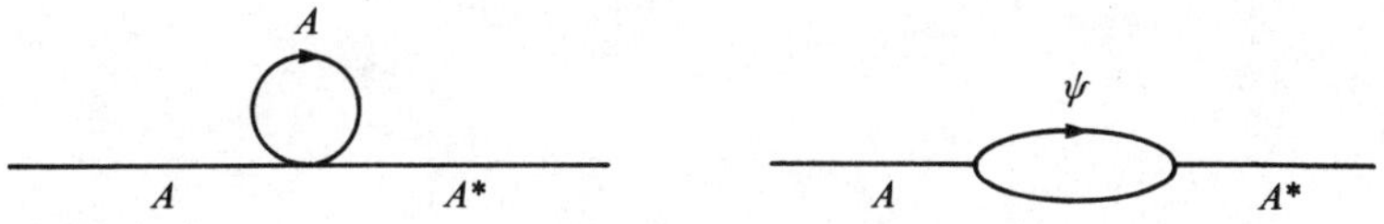

Fig. 14.4

This diagram is logarithmically divergent and contributes to the wave-function renormalization constant for the superfield ($\int d^4\Theta\Phi\bar{\Phi}$ counterterm).

Writing down the amplitudes for the diagrams (*b*) and (*c*) in Fig. 14.3 we immediately see that they vanish. Indeed we end up with integrals $\int d^4\Theta\Phi\Phi = 0$ because the integrand is chiral. Thus, we see that at the one-loop level there is no mass and coupling-constant renormalization other than that induced by the wave-function renormalization, neither infinite nor finite. Expressed in terms of the component fields this result means that the contributions to the mass renormalization constant of e.g. the diagrams in Fig. 14.4 generated by the lagrangian (14.77) (we now use the complex field notation $A \equiv A + iB$) cancel each other. Actually there is no *infinite* renormalization of the mass and the coupling-constant to any order of perturbation theory. This is the non-renormalization theorem for chiral superfields. (In general, there is finite renormalization of these parameters in higher orders. It vanishes when renormalizing at vanishing external momenta.)

The above-mentioned validity of the non-renormalization theorem to any order of perturbation theory follows from the fact that, by integrating by parts the factors $\frac{1}{2}\bar{D}^2$ or $\frac{1}{2}D^2$ one-by-one, the contribution from any Feynman diagram to the effective action can be written in the form

$$\int \prod_{i=1}^{n} \mathrm{d}^4 p_i \mathrm{d}^4 \Theta f(p_1, \ldots, p_n) \Phi(p_1, \Theta_1) \ldots \bar{\Phi}(p_n, \Theta_n) \tag{14.107}$$

i.e. it can be expressed as an integral over a single $\mathrm{d}^4\Theta$ and with the function $f(p_1, \ldots, p_n)$ translationally invariant. Of course, our result (14.106) is the simplest example of the general rule (14.107). From (14.107) it follows in particular that all vacuum-to-vacuum diagrams vanish because without any superfields the expression is annihilated by the $\mathrm{d}^4\Theta$ integration. Thus, the normalization factor in (14.97) is one.

Appendix A
Feynman rules and Feynman integrals

We use the conventions of Appendix A in Bjorken & Drell (1965).

Feynman rules for the $\lambda\Phi^4$ theory:

propagator

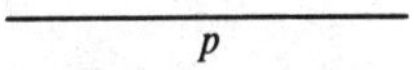

$$\frac{\mathrm{i}}{p^2 - m^2 + \mathrm{i}\varepsilon}$$

loop integration

$$\int \frac{\mathrm{d}^4 k}{(2\pi)^4}$$

vertex

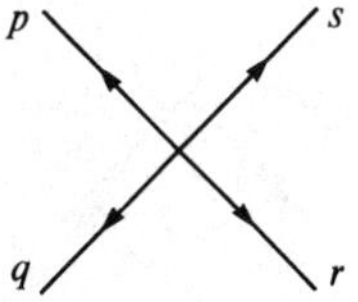

$$-\mathrm{i}\lambda, \qquad p + q + r + s = 0$$

symmetry factors

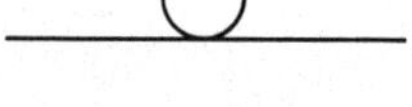

$S = \frac{1}{2}$

$S = \frac{1}{6}$

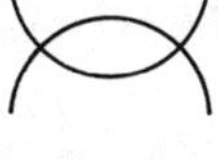

$S = \frac{1}{2}$

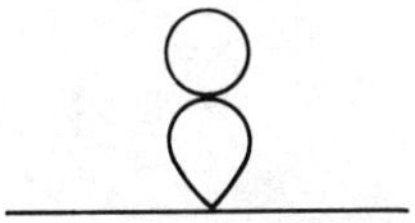

$S = \frac{1}{4}$

vertex counterterm

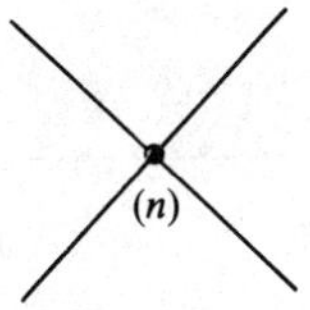

$$-\mathrm{i}\lambda(Z_1 - 1)^{(n)}$$

mass counterterm

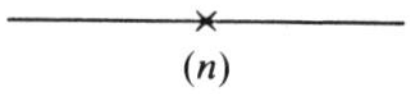

$-\mathrm{i}(Z_0 - 1)^{(n)} m^2$

wave-function renormalization counterterm

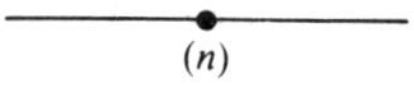

$+\mathrm{i}(Z_3 - 1)^{(n)} p^2$

Feynman rules for QED

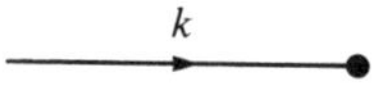

incoming electron with momentum k: $u(k)$

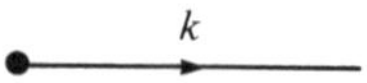

outgoing electron with momentum k: $\bar{u}(k)$

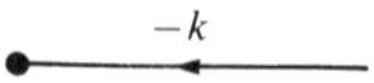

outgoing positron with momentum k: $v(k)$

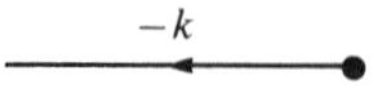

incoming positron with momentum k: $\bar{v}(k)$

fermion propagator

$$\frac{\mathrm{i}}{\not{p} - m + \mathrm{i}\varepsilon}$$

photon propagator

$$\frac{-\mathrm{i}}{k^2 + \mathrm{i}\varepsilon}\left[g_{\mu\nu} - (1 - a)\frac{k_\mu k_\nu}{k^2}\right]$$

fermion–fermion–photon vertex

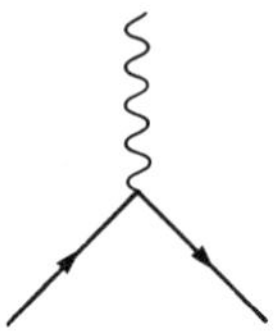

$-\mathrm{i}e\gamma_\mu$

fermion wave-function renormalization counterterm

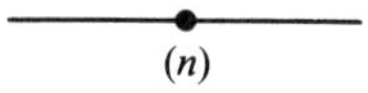

$\mathrm{i}(Z_2 - 1)^{(n)}\not{p}$

fermion mass counterterm

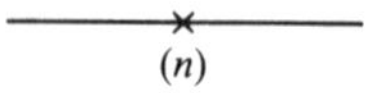

$\mathrm{i}m(Z_0 - 1)^{(n)}$
$\equiv \mathrm{i}\delta_m^{(n)}$

photon wave-function renormalization counterterm

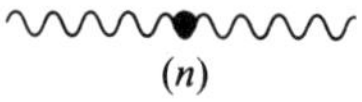

$-\mathrm{i}(Z_3 - 1)^{(n)}$
$(k^2 g_{\mu\nu} - k_\mu k_\nu)$

vertex counterterm

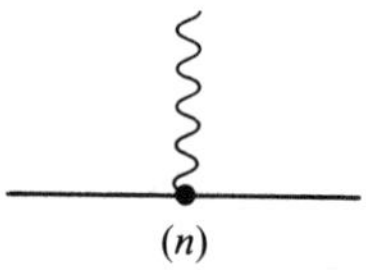

$-\mathrm{i}e(Z_1 - 1)^{(n)}\gamma_\mu$

Fermion loop parametrization

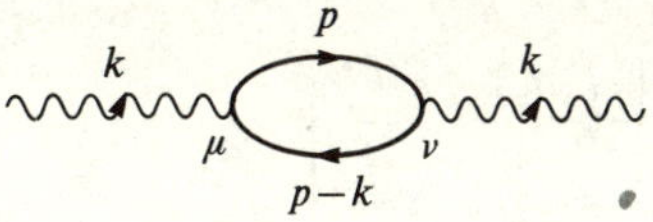

$$= (-1)\,\mathrm{Tr}\left[\frac{\mathrm{i}}{\not{p}-m}\gamma_\mu \frac{\mathrm{i}}{\not{p}-\not{k}-m}\gamma_\nu\right](-e^2)$$

Feynman rules for QCD

three-gluon coupling

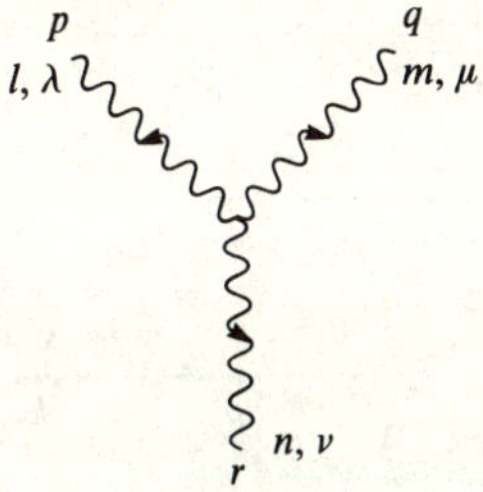

$$gc_{lmn}[(p-r)_\mu g_{\lambda\nu} + (r-q)_\lambda g_{\mu\nu} + (q-p)_\nu g_{\lambda\mu}]$$

four-gluon coupling

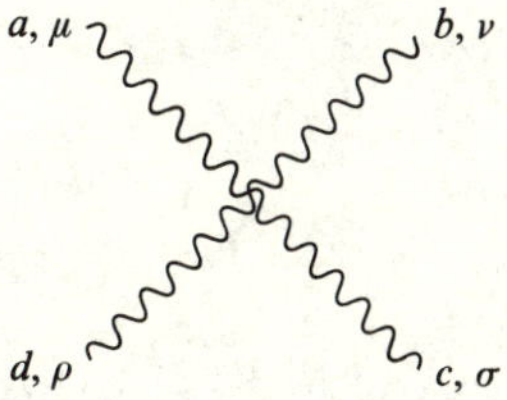

$$(-\mathrm{i}g^2)[c_{abe}c_{cde}(g_{\mu\sigma}g_{\nu\rho} - g_{\mu\rho}g_{\nu\sigma}) + c_{ace}c_{bde}(g_{\mu\nu}g_{\rho\sigma} - g_{\mu\rho}g_{\nu\sigma}) + c_{ade}c_{cbe}(g_{\mu\sigma}g_{\nu\rho} - g_{\mu\nu}g_{\rho\sigma})]$$

quark–quark–gluon vertex

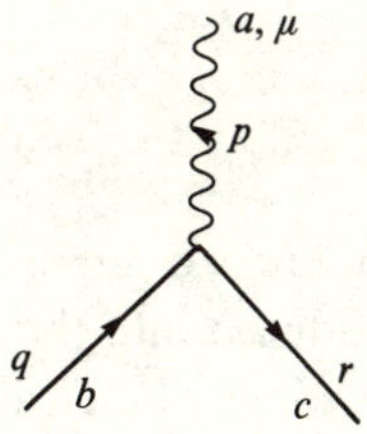

$\mathrm{i}g\gamma_\mu(T^a)_{bc}$
$\mathrm{Tr}\,[T^aT^b] = \frac{1}{2}\delta^{ab}$

ghost–ghost–gluon vertex

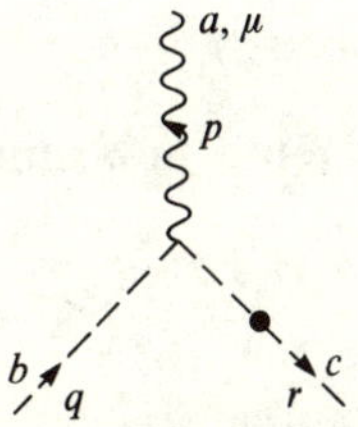

$-gc_{abc}r_\mu$

gluon propagator

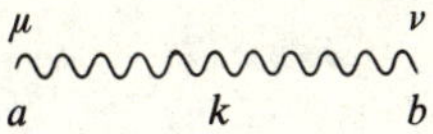

covariant gauge:

$$-\frac{1}{2a}(\partial_\mu A^\mu_a)^2$$

$$\frac{-\mathrm{i}}{k^2+\mathrm{i}\varepsilon}\delta_{ab}\left[g_{\mu\nu} - (1-a)\frac{k_\mu k_\nu}{k^2}\right]$$

Coulomb gauge: $\partial_\mu A_a^\mu - (n_\mu \partial^\mu)(n_\mu A_a^\mu) = 0$, $n_\mu = (1, 0, 0, 0)$

$$\frac{-\mathrm{i}}{k^2 + \mathrm{i}\varepsilon}\delta_{ab}\left[g_{\mu\nu} - \frac{k\cdot n(k_\mu n_\nu + k_\nu n_\mu) - k_\mu k_\nu}{(k\cdot n)^2 - k^2} \right]$$

axial gauge: $n_\mu A_a^\mu = 0$, $\quad n^2 = 0$ or $n^2 < 0$

$$\frac{-\mathrm{i}}{k^2 + \mathrm{i}\varepsilon}\delta_{ab}\left[g_{\mu\nu} - \frac{k_\mu n_\nu + k_\nu n_\mu}{k\cdot n} + \frac{n^2}{(n\cdot k)^2} k_\mu k_\nu \right]$$

ghost propagator

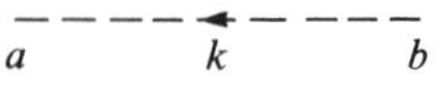

$$-\frac{\mathrm{i}}{k^2 + \mathrm{i}\varepsilon}\delta_{ab}$$

quark propagator

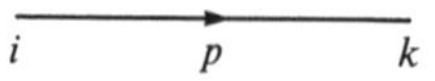

$$\delta_{ik}\frac{\mathrm{i}}{\not p - m + \mathrm{i}\varepsilon}$$

symmetry factors

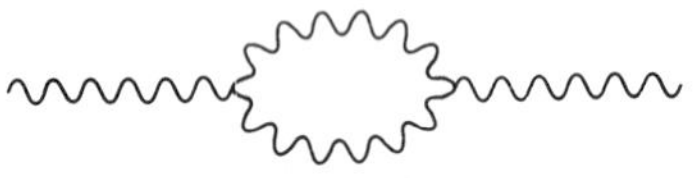

$$S = \frac{1}{2!}$$

$$S = \frac{1}{2!}$$

$$S = \frac{1}{3!}$$

For each fermion and ghost loop there is a minus sign. As for $\lambda\Phi^4$ and for QED, counterterms can be explicitly introduced into the Feynman rules. Feynman rules for counterterms are trivial modifications of the rules given above and they follow immediately from (8.8).

Dirac algebra in *n* dimensions

$$\{\gamma^\mu, \gamma^\nu\} = 2g^{\mu\nu} \tag{A.1}$$

$$g^\mu{}_\mu = n \tag{A.2}$$

$$\gamma^\mu \gamma_\mu = n \tag{A.3}$$

$$\gamma_\mu \not a \gamma^\mu = (2 - n)\not a \tag{A.4}$$

$$\gamma^\mu \not a \not b \gamma_\mu = 4a\cdot b + (n - 4)\not a \not b \tag{A.5}$$

$$\gamma^\mu \not a \not b \not c \gamma_\mu = -2\not c \not b \not a - (n - 4)\not a \not b \not c \tag{A.6}$$

$$\mathrm{Tr}\,\mathbb{1} = 4 \quad \mathrm{Tr}\,(\mathrm{odd}\,\gamma) = 0 \quad \mathrm{Tr}\,[\gamma^\mu \gamma^\nu] = 4g^{\mu\nu} \tag{A.7}$$

In four dimensions ($\varepsilon_{0123} = -\varepsilon^{0123} = 1$)

$$\gamma^5 = \gamma_5 = \mathrm{i}\gamma^0\gamma^1\gamma^2\gamma^3 = \mathrm{i}\varepsilon_{\alpha\beta\gamma\rho}\gamma^\alpha\gamma^\beta\gamma^\gamma\gamma^\rho/4! \tag{A.8}$$

$$\mathrm{Tr}\,[\gamma^5\gamma^\alpha\gamma^\beta\gamma^\gamma\gamma^\rho] = +\mathrm{i}\varepsilon^{\alpha\beta\gamma\rho}\,\mathrm{Tr}\,\mathbb{1} \tag{A.9}$$

$$\gamma_\mu\gamma_\rho\gamma_\nu = (S_{\mu\rho\nu\sigma} + \mathrm{i}\varepsilon_{\mu\rho\nu\sigma}\gamma_5)\gamma^\sigma \tag{A.10}$$

$$S_{\mu\rho\nu\sigma} = g_{\mu\rho}g_{\nu\sigma} + g_{\rho\nu}g_{\mu\sigma} - g_{\mu\nu}g_{\rho\sigma} \tag{A.11}$$

The Dirac representation for γs is

$$\gamma^0 = \begin{pmatrix} 1 & 0 \\ 0 & -1 \end{pmatrix} \quad \gamma^i = \begin{pmatrix} 0 & \sigma^i \\ -\sigma^i & 0 \end{pmatrix} \quad \gamma_5 = \begin{pmatrix} 0 & 1 \\ 1 & 0 \end{pmatrix} \tag{A.12}$$

where σ^is are 2×2 Pauli matrices.

$$(\gamma^5)^\dagger = \gamma^5, \quad (\gamma^\mu)^\dagger = \gamma_\mu = \begin{cases} \gamma^\mu & \mu = 0 \\ -\gamma^\mu & \mu \geqslant 1 \end{cases} \tag{A.13}$$

Feynman parameters

$$\frac{1}{ab} = \int_0^1 \mathrm{d}x \frac{1}{[(1-x)b + xa]^2} \tag{A.14}$$

$$\frac{1}{a^n b} = n\int_0^1 \mathrm{d}x \frac{x^{n-1}}{[(1-x)b + xa]^{n+1}} \tag{A.15}$$

$$\begin{aligned}\frac{1}{abc} &= 2\int_0^1 \mathrm{d}x \int_0^1 \mathrm{d}y \frac{x}{[axy + bx(1-y) + c(1-x)]^3} \\ &= 2\int_0^1 \mathrm{d}u \int_0^{1-u} \mathrm{d}w \frac{1}{[aw + b(1-u-w) + cu]^3}\end{aligned} \tag{A.16}$$

$$\frac{1}{a_1 a_2 \ldots a_n} = (n-1)! \int_0^1 \mathrm{d}x_n\,\mathrm{d}x_{n-1} \ldots \mathrm{d}x_2$$

$$* \frac{x_n^{n-2} x_{n-1}^{n-3} \ldots x_3^1 x_2^0}{[(1-x_n)a_n + x_n[(1-x_{n-1})a_{n-1} + x_{n-1}[\cdots + x_3[(1-x_2)a_2 + x_2 a_1]]\ldots]^n} \tag{A.17}$$

Feynman integrals in *n* dimensions

$$I_0 = \int \frac{\mathrm{d}^n p}{(2\pi)^n} \frac{1}{(p^2 + 2k\cdot p + M^2 + \mathrm{i}\varepsilon)^\alpha} = \frac{\mathrm{i}(-\pi)^{n/2}}{(2\pi)^n} \frac{\Gamma(\alpha - \frac{1}{2}n)}{\Gamma(\alpha)} \frac{1}{(M^2 - k^2 + \mathrm{i}\varepsilon)^{\alpha - n/2}} \tag{A.18}$$

$$I_\mu = \int \frac{\mathrm{d}^n p}{(2\pi)^n} \frac{p_\mu}{(p^2 + 2k\cdot p + M^2 + \mathrm{i}\varepsilon)^\alpha} = -k_\mu I_0 \tag{A.19}$$

$$I_{\mu\nu} = \int \frac{d^n p}{(2\pi)^n} \frac{p_\mu p_\nu}{(p^2 + 2k\cdot p + M^2 + i\varepsilon)^\alpha} = I_0 \left[k_\mu k_\nu + \tfrac{1}{2} g_{\mu\nu}(M^2 - k^2) \frac{1}{\alpha - \frac{1}{2}n - 1} \right] \tag{A.20}$$

$$I_{\mu\nu\rho} = \int \frac{d^n p}{(2\pi)^n} \frac{p_\mu p_\nu p_\rho}{(p^2 + 2k\cdot p + M^2 + i\varepsilon)^2}$$

$$= -I_0 \left[k_\mu k_\nu k_\rho + \tfrac{1}{2}(g_{\mu\nu}k_\rho + g_{\mu\rho}k_\nu + g_{\nu\rho}k_\mu)(M^2 - k^2) \frac{1}{\alpha - \frac{1}{2}n - 1} \right] \tag{A.21}$$

Also, by convention

$$\int d^n p (p^2)^{-1} = 0$$

α-representation

$$\frac{1}{p^2 - m^2 + i\varepsilon} = \frac{1}{i} \int_0^\infty d\alpha \exp[i\alpha(p^2 - m^2 + i\varepsilon)] \tag{A.22}$$

Gaussian integrals

$$\int d^4k \exp[i(ak^2 + 2b\cdot k)] = \frac{1}{i}\left(\frac{\pi}{a}\right)^2 \exp\left(-i\frac{b^2}{a}\right) \tag{A.23}$$

$$\int d^4k\, k_\mu \exp[i(ak^2 + 2b\cdot k)] = \frac{1}{i}\left(\frac{\pi}{a}\right)^2 \exp\left(-i\frac{b^2}{a}\right)\left(-\frac{b_\mu}{a}\right) \tag{A.24}$$

$$\int d^4k\, k_\mu k_\nu \exp[i(ak^2 + 2b\cdot k)] = \frac{1}{i}\left(\frac{\pi}{a}\right)^2 \exp\left(-i\frac{b^2}{a}\right)\left(\frac{iag_{\mu\nu} + 2b_\mu b_\nu}{2a^2}\right) \tag{A.25}$$

λ-parameter integrals

$$\int_0^\infty \frac{d\lambda}{\lambda} \exp[i(A + i\varepsilon)\lambda] = -\ln(A + i\varepsilon) + \infty \tag{A.26}$$

$$\int_0^\infty \frac{d\lambda}{\lambda} [\exp(iA\lambda) - \exp(iB\lambda)] \exp(-\varepsilon\lambda) = \ln\frac{B + i\varepsilon}{A + i\varepsilon} \tag{A.27}$$

$$\int_0^\infty \frac{d\lambda}{\lambda^2} \exp(-\varepsilon\lambda) f(\lambda) \underset{\varepsilon\to 0}{=} \int_0^\infty \frac{d\lambda}{\lambda} \exp(-\varepsilon\lambda) \frac{df(\lambda)}{d\lambda} \tag{A.28}$$

Feynman integrals in light-like gauge $n\cdot A = 0$, $n^2 = 0$

Using

$$\frac{1}{a^\alpha b^\beta} = \frac{\Gamma(\alpha+\beta)}{\Gamma(\alpha)\Gamma(\beta)} \int_0^1 dx \frac{x^{\alpha-1}(1-x)^{\beta-1}}{[ax + b(1-x)]^{\alpha+\beta}} \tag{A.29}$$

one can show that

$$I(\alpha) = \int \frac{d^n k}{(2\pi)^n} \frac{1}{(k^2 - 2p\cdot k + M^2)^\alpha (k\cdot n)^\beta} = \frac{1}{(p\cdot n)^\beta} \int \frac{d^n k}{(2\pi)^n} \frac{1}{(k^2 - 2p\cdot k + M^2)^\alpha} \tag{A.30}$$

$$I_\mu(\alpha) = \int \frac{d^n k}{(2\pi)^n} \frac{k_\mu}{(k^2 - 2p\cdot k + M^2)^\alpha (k\cdot n)^\beta} = I(\alpha)\left[p_\mu - \frac{\beta}{2(\alpha - \frac{1}{2}n - 1)}(M^2 - p^2)\frac{n_\mu}{p\cdot n}\right] \tag{A.31}$$

$$\begin{aligned} I_{\mu\nu}(\alpha) = \int \frac{d^n k}{(2\pi)^n} \frac{k_\mu k_\nu}{(k^2 - 2p\cdot k + M^2)^\alpha (k\cdot n)^\beta} &= I(\alpha)\left[\tfrac{1}{2}(M^2 - p^2)\frac{g_{\mu\nu}}{\alpha - 1 - \frac{1}{2}n} \right. \\ &+ p_\mu p_\nu - \frac{p_\mu n_\nu + p_\nu n_\mu}{p\cdot n}(M^2 - p^2)\frac{\beta}{2(\alpha - 1 - \frac{1}{2}n)} \\ &\left. + \frac{n_\mu n_\nu}{(p\cdot n)^2}(M^2 - p^2)^2 \frac{\beta(\beta+1)}{4(\alpha - 1 - \frac{1}{2}n)(\alpha - 2 - \frac{1}{2}n)} \right] \end{aligned} \tag{A.32}$$

Convention for the logarithm

The logarithm has a cut along the negative real axis ($\ln z = \ln|z| + \mathrm{i}\arg z$, $-\pi < \arg z < \pi$). With this convention the rule for the logarithm of a product is

$$\ln(ab) = \ln a + \ln b + \eta(a, b) \tag{A.33}$$

where

$$\eta(a, b) = 2\pi\mathrm{i}[\Theta(-\mathrm{Im}\, a)\Theta(-\mathrm{Im}\, b)\Theta(+\mathrm{Im}\, ab) - \Theta(\mathrm{Im}\, a)\Theta(\mathrm{Im}\, b)\Theta(-\mathrm{Im}\, ab)] \tag{A.34}$$

Important consequences are:

(i) $\ln(ab) = \ln a + \ln b$ if $\mathrm{Im}\, a$ and $\mathrm{Im}\, b$ have different signs
(ii) $\ln(a/b) = \ln a - \ln b$ if $\mathrm{Im}\, a$ and $\mathrm{Im}\, b$ have the same sign
(iii) If a and b are real then

$$\ln(ab + \mathrm{i}\varepsilon) = \ln(a + \mathrm{i}(\varepsilon/b)) + \ln(b + \mathrm{i}(\varepsilon/a))$$

The integration $\int_0^1 (\mathrm{d}x/(ax+b))$ for arbitrary complex a and b requires some care. Naively we have $a^{-1}\ln(ax+b)$ but if for $0 < x < 1$ the argument can be real and negative, one must actually split the integral into two intervals. It is easier to first divide out a:

$$\frac{1}{a}\int_0^1 \frac{\mathrm{d}x}{x + b/a} = \frac{1}{a}\ln\left(x + \frac{b}{a}\right)\Bigg|_0^1 \tag{A.35}$$

No matter what the sign is of $\mathrm{Im}\, b/a$ the argument of ln never crosses the cut for real x. Thus the answer is $(1/a)\ln[(a+b)/b]$ since the imaginary parts of $(1 + b/a)$ and b/a have the same sign.

Spence functions:

$$F(\xi) = \int_0^\xi \frac{\ln(1+x)}{x}\,\mathrm{d}x \tag{A.36}$$

$$F(\xi) + F(1/\xi) = \tfrac{1}{6}\pi^2 + \tfrac{1}{2}\ln^2\xi \tag{A.37}$$

$$F(-\xi) + F(\xi - 1) = -\tfrac{1}{6}\pi^2 + \ln\xi\ln(1-\xi) \tag{A.38}$$

$$F(1) = \tfrac{1}{12}\pi^2, \quad F(-1) = -\tfrac{1}{6}\pi^2 \tag{A.39}$$

$$F(\xi) = \xi - \tfrac{1}{4}\xi^2 + \tfrac{1}{9}\xi^3 + \cdots \tag{A.40}$$

$$L(x) = \int_0^x \frac{\ln(1-u)}{u}\,\mathrm{d}u = F(-x) \tag{A.41}$$

$$L(x) = \int_0^x \frac{\ln|1-u|}{u}\,\mathrm{d}u \pm \mathrm{i}\pi\ln x \tag{A.42}$$

Indefinite integrals

$$\int(\mathrm{d}z/z)\ln(z - z_0) = \tfrac{1}{2}\ln^2 z - L(z_0/z) \tag{A.43}$$

$$\int(\mathrm{d}z/z)\ln(z + z_0) = \ln z\ln z_0 + F(z/z_0) \tag{A.44}$$

Appendix B

Elements of group theory

For a concise introduction to group theory and its application in particle physics see, for instance, Werle (1966). Here we recall only those definitions and properties most useful in reading this book, with no attempt for a complete and self-consistent presentation.

Definitions

Let H be a proper (i.e. different from the group G itself and from the unit element alone) subgroup of G and g, an arbitrary element of G. The sets gH and Hg with fixed $g \in G$, are called left and right cosets of H. Two left cosets $g_i H$ and $g_k H$ (or two right cosets Hg_i and Hg_k) of the same subgroup H contain exactly the same elements of G or have no common elements at all. Taking all different left (or right) cosets of H we can decompose the group G into the sum

$$G = H + g_1 H + \cdots + g_{\nu-1} H \tag{B.1}$$

where $g_1 \in G$, $g_1 \notin H$, $g_2 \in G$, $g_2 \notin H$, $g_2 \notin g_1 H$ etc. The number ν is called the index of H in G. We can define equivalence classes of elements of G; two elements g and g' are considered equivalent if they belong to the same coset with respect to H. The set of all cosets is a manifold denoted by G/H.

Two elements $g, h \in G$ are said to be mutually conjugate if there exists such an element $a \in G$ that

$$h = aga^{-1}$$

One can divide all the group elements into separate classes of conjugate elements.

If a set H with elements h is a subgroup of G then the set $H' = gHg^{-1}$ is another subgroup of G which is called conjugate to H. If for arbitrary $g \in G$ all the conjugate subgroups gHg^{-1} are identical (i.e. contain the same elements) $gHg^{-1} = H$, we call H a normal subgroup. A normal subgroup H of G consists of whole undivided classes of G. Groups which contain no proper normal subgroups are said to be *simple*. Groups which contain no proper normal abelian subgroups are said to be *semisimple*.

Consider the products of an element of the coset gH by an element of the coset fH where H is a normal subgroup of G. All such products belong to the coset gfH. Hence we can define a new type of multiplication of cosets

$$(gH)(fH) = (gfH) \tag{B.2}$$

The set of all different cosets of a normal subgroup H is a group with respect to this multiplication law. This so-called quotient group F is not a subgroup of G as its elements are whole sets (cosets). We write symbolically $F = G/H$.

A function φ is said to be defined on a group G if to each element $g \in G$ a unique complex number $\varphi(g)$ is assigned. In the case of a continuous k-parameter group $G_{(k)}$ a function $\varphi(g)$ can be regarded as an ordinary function $\varphi(\alpha_1, \ldots, \alpha_k)$ of k real variables that are necessary to specify the group elements. A group is called *compact* if the group manifold M is compact i.e. if every infinite subset of M contains a sequence converging to a limit in M, or equivalently, if any function $\varphi(\alpha_i)$ which is continuous in all points of M is also bounded. For compact groups we can define the average of a group function by the expression

$$\mathrm{Av}[\varphi] = (1/V) \int \varphi(\alpha)\rho(\alpha)\, \mathrm{d}\alpha \tag{B.3}$$

where

$$V = \int \rho(\alpha)\, \mathrm{d}\alpha \tag{B.4}$$

and $\rho(\alpha)$ is a properly chosen weight function such that

(i) $\mathrm{Av}[\varphi(g)] = c$ if $\varphi(g) = \text{const}$ for all $g \in G$
(ii) $\mathrm{Av}[\varphi(g)] \geqslant 0$ if $\varphi(g) \geqslant 0$
(iii) $\mathrm{Av}[\varphi(hg)] = \mathrm{Av}[\varphi(gh)] = \mathrm{Av}[\varphi(g)]$ for any fixed $h \in G$.

All representations of a semisimple group are fully reducible. Unitary representations of any group are fully reducible. Every representation of a finite group or of a compact Lie group is equivalent to a unitary representation. Lie groups which are not compact may have representations which are not equivalent to any unitary representation. Any finite group has a finite number and any compact Lie group has a countable infinite number of inequivalent, unitary, irreducible representations which are all finite dimensional. Non-compact Lie groups have uncountable numbers of inequivalent, unitary irreducible representations which are all infinite dimensional.

Transformation of operators

Consider the set of all possible linear operators $\hat{O}$ which transform the vectors $|v\rangle$ of some Hilbert space $\mathscr{H}$ into vectors $|w\rangle$ of the same space $|w\rangle = \hat{O}|v\rangle$. Suppose that under a group of transformations G the vectors of $\mathscr{H}$ are transformed as follows

$$|w_g\rangle = U(g)|w\rangle, \qquad |v_g\rangle = U(g)|v\rangle \tag{B.5}$$

where $U(g) = \exp(\mathrm{i}\alpha^a Q^a)$ are unitary operators and Q^a are generators of G. The operator $\hat{O}_g$ which transforms $|v_g\rangle$ into $|w_g\rangle$ is given by

$$\hat{O}_g = U(g)\hat{O}U^{-1}(g) \tag{B.6}$$

Let us define the linear transformation $T(g)$ which transforms $\hat{O} \to \hat{O}_g$

$$\hat{O}_g = T(g)\hat{O} = U(g)\hat{O}U^{-1}(g) \tag{B.7}$$

The transformation $T(g) = \exp(-\mathrm{i}\alpha^a T^a)$ form a unitary representation of G. For infinitesimal transformations we have

$$\hat{O}_g \cong (1 - \mathrm{i}\delta\alpha^a T^a)\hat{O} = \hat{O} + \mathrm{i}\delta\alpha^a[Q^a, \hat{O}] \tag{B.8}$$

Complex and real representations

Let T_i, $i = 1, \ldots, N$ form a representation of generators of a group G

$$[T_i, T_j] = \mathrm{i}c_{ijk}T^k \tag{B.9}$$

The structure constants c_{ijk} are real if T_i are hermitean. Taking the complex conjugate of these commutation relations we see that $(-T_i^*)$ also form a representation of the group

$$[(-T_i^*), (-T_j^*)] = \mathrm{i}c_{ijk}(-T^{k*}) \tag{B.10}$$

If T_i and $(-T_i^*)$ are equivalent i.e.

$$-T_i^* = UT_iU^\dagger \tag{B.11}$$

where U is a unitary transformation, the representation is said to be real (the term real is motivated by an alternative convention in which $\mathrm{i}T_i$ is the representation matrix). For instance for 2 and 2* of $SU(2)$ we have

$$-T_i^* = -T_i^{\mathrm{T}} = -\tfrac{1}{2}\tau_i^{\mathrm{T}} \tag{B.12}$$

and

$$\mathrm{i}\tau_2(-\tfrac{1}{2}\tau_i^{\mathrm{T}})(\mathrm{i}\tau_2)^\dagger = \tfrac{1}{2}\tau_i \tag{B.13}$$

But this is not true for $SU(N)$, $N > 2$.

Take a set of n fields $\Phi_a(x)$ which transform according to $n \times n$-dimensional representation matrices T_i:

$$[Q^i, \Phi_a] = -(T^i)_{ab}\Phi_b \tag{B.14}$$

Then the fields $\Phi_a^\dagger$ transform according to the complex conjugate representation (take the hermitean conjugate of (B.14)):

$$[Q^i, \Phi_a^\dagger(x)] = +T_{ab}^{i*}\Phi_b^\dagger(x) = -(-T_{ab}^{i*})\Phi_b^\dagger(x) \tag{B.15}$$

Let all the left-handed fermion fields Ψ_L and $(\Psi^C)_L$ (C means charge conjugation, see Appendix C) in a theory form the representation f_L under some symmetry group. From (C.3)

$$\Psi_R = C\overline{(\Psi^C)_L}^T \quad \text{and} \quad (\Psi^C)_R = (\Psi_L)^C = C\bar{\Psi}_L{}^T \tag{B.16}$$

so Ψ_R and $(\Psi^C)_R$ transform under the complex conjugate representation f_L^*. If f_L is real then the left-handed and the right-handed fermions transform according to the same representation of the symmetry group. The theory is said to be vector-like. If f_L is complex ($f_R = f_L^* \neq f_L$) the theory is said to be chiral.

QCD is vector-like because

$$\begin{aligned} f_L &= 3 + 3^* \\ &\ (u_L)(u^C)_L \end{aligned} \tag{B.17}$$

while

$$f_R = f_L^* = 3^* + 3 = f_L \tag{B.18}$$

The standard model $SU(3)\times SU(2)\times U(1)$ is chiral. For one family we have:

$$\begin{matrix} f_L = (3, 2, \tfrac{1}{6}) & + (3^*, 1, -\tfrac{2}{3}) & + (3^*, 1, \tfrac{1}{3}) & + (1, 2, -\tfrac{1}{2}) & + (1, 1, 1) \\ \begin{pmatrix} u \\ d \end{pmatrix}_L & (u^C)_L & (d^C)_L & \begin{pmatrix} \nu \\ e^- \end{pmatrix}_L & (e^+)_L \end{matrix} \tag{B.19}$$

and

$$\begin{matrix} f_R = f_L^* = (3^*, 2, -\tfrac{1}{6}) & + (3, 1, \tfrac{2}{3}) & + (3, 1, -\tfrac{1}{3}) & + (1, 2, \tfrac{1}{2}) & + (1, 1, -1) \neq f_L \\ \begin{pmatrix} d^C \\ -u^C \end{pmatrix}_R & u_R & d_R & \begin{pmatrix} e^+ \\ -\nu^C \end{pmatrix}_R & e_R^- \end{matrix} \tag{B.20}$$

(We have used the equivalence $2^* = 2$ and the transformation rule (B.13)).

Traces

Let

$$[T^a, T^b] = \mathrm{i}c_{abc}T^c, \; c_{abc} \equiv c^{abc} \tag{B.21}$$

Then

$$\mathbf{T}^2 = \sum_{a,k} T^a_{ik} T^a_{kj} = \delta_{ij} C_R \tag{B.22}$$

where R means representation R.

$$\mathrm{Tr}\,[T^a T^b] = T_R \delta^{ab} \tag{B.23}$$

$$T_R d(G) = C_R d(R) \tag{B.24}$$

where $d(G)$ is the dimension of the group and $d(R)$ is the dimension of the representation R.

For $SU(N)$:

$$d(R) = N \quad \text{for the fundamental representation } (F)$$
$$d(R) = N^2 - 1 = d(G) \quad \text{for the adjoint representation } (A)$$

$$T_F = \tfrac{1}{2}, \quad C_F = \frac{N^2-1}{2N}$$
$$T_A = C_A = N$$

$\sum_{i,k} c_{aik} c_{bik} = \delta_{ab} C_R = N\delta_{ab}$ because in the adjoint representation $(T^a)_{ik} = -\mathrm{i}c_{aik}$.

For $SU(3)$ in the fundamental representation $T^a = \frac{1}{2}\lambda^a$ with

$$\left.\begin{aligned}
&\lambda^i = \begin{pmatrix} \sigma^i & 0 \\ 0 & 0 \end{pmatrix}, i = 1,2,3; \sigma^1 = \begin{pmatrix} 0 & 1 \\ 1 & 0 \end{pmatrix}, \sigma^2 = \begin{pmatrix} 0 & -i \\ i & 0 \end{pmatrix}, \sigma^3 = \begin{pmatrix} 1 & 0 \\ 0 & -1 \end{pmatrix} \\
&\lambda^4 = \begin{pmatrix} & & 1 \\ & 0 & \\ 1 & & \end{pmatrix} \quad \lambda^5 = \begin{pmatrix} & & -\mathrm{i} \\ & 0 & \\ \mathrm{i} & & \end{pmatrix} \quad \lambda^6 = \begin{pmatrix} 0 & 0 \\ 0 & \sigma^1 \end{pmatrix} \\
&\lambda^7 = \begin{pmatrix} 0 & 0 \\ 0 & \sigma^2 \end{pmatrix} \quad \lambda^8 = \frac{1}{\sqrt{3}} \begin{pmatrix} 1 & & \\ & 1 & \\ & & -2 \end{pmatrix}
\end{aligned}\right\} \tag{B.25}$$

$$\{T^a, T^b\} = \tfrac{1}{3}\delta^{ab}\mathbb{1} + d^{abc}T^c, \; d^{abc} = d_{abc} \tag{B.26}$$

is totally symmetric. One has

$$\begin{aligned}
1 &= c_{123} = 2c_{147} = 2c_{246} = 2c_{257} = 2c_{345} = -2c_{156} \\
&= -2c_{367} = 2c_{458}/\sqrt{3} = 2c_{678}/\sqrt{3} \\
1/\sqrt{3} &= d_{118} = d_{228} = d_{338} = -d_{888} \\
-1/2\sqrt{3} &= d_{448} = d_{558} = d_{668} = d_{778} \\
\tfrac{1}{2} &= d_{146} = d_{157} = d_{247} = d_{256} = d_{344} = d_{355} = -d_{366} = -d_{377}
\end{aligned}$$

$$T^aT^b = \tfrac{1}{2}(\mathrm{i}c^{abn} + d^{abn})T^n + \tfrac{1}{6}\delta^{ab}, \quad \mathrm{Tr}\,[T^aT^b] = \tfrac{1}{2}\delta^{ab} \tag{B.27}$$

$$T^aT^bT^c = \tfrac{1}{2}(d^{abn} + \mathrm{i}c^{abn})T^nT^c + \tfrac{1}{6}\delta^{ab}T^c, \quad \mathrm{Tr}\,[T^aT^bT^c] = \tfrac{1}{4}(d^{abc} + \mathrm{i}c^{abc}) \tag{B.28}$$

$$\mathrm{Tr}\,[T^aT^bT^cT^d] = \tfrac{1}{8}(d^{abn} + \mathrm{i}c^{abn})(d^{cdn} + \mathrm{i}c^{cdn}) + \tfrac{1}{12}\delta^{ab}\delta^{cd} \tag{B.29}$$

$$\mathrm{Tr}\,[T^aT^bT^aT^b] = \tfrac{1}{8}(d^{abn}d_{abn} - c^{abn}c_{abn}) + \tfrac{8}{12} = \tfrac{1}{8}(d^2 - f^2) + \tfrac{8}{12}, \quad d^2 = \tfrac{40}{3}, \quad f^2 = 24 \tag{B.30}$$

σ-model

We derive the transformation properties of the π and σ fields in the σ-model. We assume the nucleon doublets N_R and N_L transforming under the $SU_R(2) \times SU_L(2)$ as $(\frac{1}{2}, 0)$ and $(0, \frac{1}{2})$ respectively. Therefore

under $SU_R(2)$

$$\begin{aligned}
N_R &\to N_R - \mathrm{i}\delta\boldsymbol{\alpha}_R \cdot (\tfrac{1}{2}\boldsymbol{\tau})N_R \\
N_L &\to N_L
\end{aligned} \tag{B.31}$$

under $SU_L(2)$

$$\begin{aligned} N_L &\to N_L - i\delta\boldsymbol{\alpha}_L\cdot(\tfrac{1}{2}\boldsymbol{\tau})N_L \\ N_R &\to N_R \end{aligned} \tag{B.32}$$

(remember that $\tau_R^a N_R = \tau^a \otimes P_R N_R = \tau^a \otimes (P_R + P_L) N_R = \tau^a N_R$). For invariance of the lagrangian (9.27) under each of these groups of transformations one needs that the change of $\sigma' + i\boldsymbol{\pi}\cdot\boldsymbol{\tau}$ compensates the change of the fermion fields:

$$\sigma' + i\boldsymbol{\pi}\cdot\boldsymbol{\tau} \to \sigma' + i\boldsymbol{\pi}\cdot\boldsymbol{\tau} - i\delta(\sigma' + i\boldsymbol{\pi}\cdot\boldsymbol{\tau}) \tag{B.33}$$

where for $SU_L(2)$

$$\begin{aligned} -i\delta(\sigma' + i\boldsymbol{\pi}\cdot\boldsymbol{\tau}) &= -i\delta\boldsymbol{\alpha}_L(\tfrac{1}{2}\boldsymbol{\tau})(\sigma' + i\boldsymbol{\pi}\cdot\boldsymbol{\tau}) \\ &= (-i\tfrac{1}{2}\boldsymbol{\tau})[\sigma'\delta\boldsymbol{\alpha}_L + (\boldsymbol{\pi}\times\delta\boldsymbol{\alpha}_L)] + \tfrac{1}{2}\boldsymbol{\pi}\cdot\delta\boldsymbol{\alpha}_L \end{aligned} \tag{B.34}$$

$$(\boldsymbol{\tau}\cdot\mathbf{a}\,\boldsymbol{\tau}\cdot\mathbf{b} = \mathbf{a}\cdot\mathbf{b} + i\boldsymbol{\tau}\cdot\mathbf{a}\times\mathbf{b})$$

and for $SU_R(2)$

$$\begin{aligned} -i\delta(\sigma' + i\boldsymbol{\pi}\cdot\boldsymbol{\tau}) &= (\sigma' + i\boldsymbol{\pi}\cdot\boldsymbol{\tau})(i\tfrac{1}{2}\boldsymbol{\tau})\delta\boldsymbol{\alpha}_R \\ &= (-i\tfrac{1}{2}\boldsymbol{\tau})[-\sigma'\delta\boldsymbol{\alpha}_R + (\boldsymbol{\pi}\times\delta\boldsymbol{\alpha}_R)] - \tfrac{1}{2}\boldsymbol{\pi}\cdot\delta\boldsymbol{\alpha}_R \end{aligned} \tag{B.35}$$

Therefore we get the following transformation of fields:

$SU_L(2)$

$$\left.\begin{aligned} \sigma' &\to \sigma' + \tfrac{1}{2}\boldsymbol{\pi}\cdot\delta\boldsymbol{\alpha}_L \\ \boldsymbol{\pi} &\to \boldsymbol{\pi} - \tfrac{1}{2}\boldsymbol{\pi}\times\delta\boldsymbol{\alpha}_L - \tfrac{1}{2}\sigma'\delta\boldsymbol{\alpha}_L \end{aligned}\right\} \tag{B.36}$$

$SU_R(2)$

$$\left.\begin{aligned} \sigma' &\to \sigma' - \tfrac{1}{2}\boldsymbol{\pi}\cdot\delta\boldsymbol{\alpha}_R \\ \boldsymbol{\pi} &\to \boldsymbol{\pi} - \tfrac{1}{2}\boldsymbol{\pi}\times\delta\boldsymbol{\alpha}_R + \tfrac{1}{2}\sigma'\delta\boldsymbol{\alpha}_R \end{aligned}\right\} \tag{B.37}$$

The above results correspond to the commutation relations (9.30) (as can be immediately seen using (9.32)). By taking appropriate linear combinations we also get the relations (9.28).

Sometimes it is convenient to work in the real four-dimensional representation $\Phi \equiv (\pi^1, \pi^2, \pi^3, \sigma')^T$. It is easy to check that

$$[Q^a, \Phi_b] = -(T^a)_{bi}\Phi_i$$

where

$$T^1 = \left(\begin{array}{cc|cc} 0&0&0&0\\ 0&0&-i&0\\ \hline 0&i&0&0\\ 0&0&0&0 \end{array}\right) \quad T^2 = \left(\begin{array}{cc|cc} 0&0&i&0\\ 0&0&0&0\\ \hline -i&0&0&0\\ 0&0&0&0 \end{array}\right) \quad T^3 = \left(\begin{array}{cc|cc} 0&-i&0&0\\ i&0&0&0\\ \hline 0&0&0&0\\ 0&0&0&0 \end{array}\right)$$

$${}_5T^1 = \left(\begin{array}{cc|cc} 0&0&0&i\\ 0&0&0&0\\ \hline 0&0&0&0\\ -i&0&0&0 \end{array}\right) \quad {}_5T^2 = \left(\begin{array}{cc|cc} 0&0&0&0\\ 0&0&0&i\\ \hline 0&0&0&0\\ 0&-i&0&0 \end{array}\right) \quad {}_5T^3 = \left(\begin{array}{cc|cc} 0&0&0&0\\ 0&0&0&0\\ \hline 0&0&0&i\\ 0&0&-i&0 \end{array}\right) \tag{B.38}$$

Appendix C

Chiral, Weyl and Majorana spinors

Definitions

It is often convenient to work with chiral fields Ψ_R and Ψ_L defined by the decomposition of the Dirac spinor

$$\Psi = \tfrac{1}{2}(1-\gamma_5)\Psi + \tfrac{1}{2}(1+\gamma_5)\Psi \equiv \Psi_L + \Psi_R$$

We recall here some of their useful properties. For instance, in terms of Ψ_R and Ψ_L one gets:

$$\left.\begin{aligned} \bar{\Psi}\Psi &= \bar{\Psi}_R\Psi_L + \bar{\Psi}_L\Psi_R \\ \bar{\Psi}\gamma_\mu\Psi &= \bar{\Psi}_L\gamma_\mu\Psi_L + \bar{\Psi}_R\gamma_\mu\Psi_R \\ \bar{\Psi}\gamma_\mu\gamma_5\Psi &= \bar{\Psi}_R\gamma_\mu\Psi_R - \bar{\Psi}_L\gamma_\mu\Psi_L \end{aligned}\right\} \tag{C.1}$$

Furthermore, in many problems it is better to work with one type of chiral field Ψ_R or Ψ_L, only. The description in terms of e.g. the left-handed fields Ψ_L can be obtained if we replace Ψ_R by their charge conjugate partners. In general the charge conjugation is defined as follows:†

$$\Psi^C = C\bar{\Psi}^T \tag{C.2}$$

where $C^T = C^\dagger = -C$, $C^2 = -1$, $CC^\dagger = 1$, $C^{-1}\gamma_\mu C = -\gamma_\mu^T$ and $C^{-1}\gamma_5 C = \gamma_5^T$, and in the Dirac representation for γ-matrices (see Appendix A)

$$C = i\gamma^2\gamma^0 = \begin{pmatrix} 0 & -i\sigma^2 \\ -i\sigma^2 & 0 \end{pmatrix}$$

We then easily get

$$(\Psi_R)^C = (\Psi^C)_L \underset{df}{\equiv} \tilde{\Psi}_L \tag{C.3}$$

† In quantum field theory we seek a unitary operator $\hat{C}$ which generates the transformation

$$\hat{C}\Psi_\alpha(x)\hat{C}^{-1} = C_{\alpha\beta}\bar{\Psi}_\beta(x) = (C\gamma_0^T)_{\alpha\beta}\Psi_\beta^\dagger(x)$$

and we can formulate our theory in the terms of the above-defined left-handed fields $\tilde{\Psi}_L$ and fields Ψ_L instead of fields Ψ_R and Ψ_L. In particular, after a short calculation, one gets (for anticommuting spinors)

$$\left.\begin{aligned} \bar{\Psi}\gamma_\mu\Psi &= \bar{\Psi}_L\gamma_\mu\Psi_L - \overline{\tilde{\Psi}}_L\gamma_\mu\tilde{\Psi}_L \\ \bar{\Psi}\gamma_\mu\gamma_5\Psi &= -\overline{\tilde{\Psi}}_L\gamma_\mu\tilde{\Psi}_L - \bar{\Psi}_L\gamma_\mu\Psi_L \\ \bar{\Psi}\Psi &= \overline{(\tilde{\Psi}_L)^C}\Psi_L + \text{h.c.} \end{aligned}\right\} \tag{C.1a}$$

To get the lagrangian (9.11) in terms of fields Ψ_L and $\tilde{\Psi}_L$ we must remember about replacing $t^a{}_r$ in D_μ in (9.1) by the complex conjugate representations $t^a{}_{\bar{r}}$ (see Appendix B).

Another notation worth mentioning is the one using the two-component Weyl spinors. Writing

$$\Psi(x) = \begin{pmatrix} u(x) \\ v(x) \end{pmatrix}$$

and choosing the 'chiral' representation for the γ-matrices

$$\gamma_5 = \begin{pmatrix} -1 & 0 \\ 0 & 1 \end{pmatrix}, \quad \gamma^i = \begin{pmatrix} 0 & \sigma^i \\ -\sigma^i & 0 \end{pmatrix}, \quad \gamma^0 = \begin{pmatrix} 0 & 1 \\ 1 & 0 \end{pmatrix} \tag{C.4}$$

we get

$$\Psi_L = \begin{pmatrix} u \\ 0 \end{pmatrix}, \quad \Psi_R = \begin{pmatrix} 0 \\ v \end{pmatrix}$$

and, for instance,

$$\left.\begin{aligned} \bar{\Psi}\gamma^i\Psi &= -u^\dagger\sigma^i u + v^\dagger\sigma^i v \\ \bar{\Psi}\gamma^i\gamma_5\Psi &= u^\dagger\sigma^i u + v^\dagger\sigma^i v \\ \bar{\Psi}\Psi &= u^\dagger v + \text{h.c.} \end{aligned}\right\} \tag{C.5}$$

Also the lagrangian (9.11) can easily be rewritten in terms of Weyl spinors $u(x)$ and $v(x)$.

Next we can use charge conjugation to eliminate e.g. $u(x)$. In chiral representation, see (V.4)

$$C = \begin{pmatrix} -i\sigma^2 & \\ & +i\sigma^2 \end{pmatrix} \equiv -i\gamma^2\gamma^0 \tag{C.6}$$

and

$$(\Psi(x))^C = \begin{pmatrix} -i\sigma^2 v^*(x) \\ +i\sigma^2 u^*(x) \end{pmatrix}$$

Defining

$$\tilde{v}(x) = +i\sigma^2 u^*(x)$$

or

$$u(x) = -i\sigma^2 \tilde{v}^*(x) \tag{C.7}$$

we get ($\sigma^2\sigma^i\sigma^2 = -(\sigma^i)^{\mathrm{T}}$ and spinors anticommute)

$$\left.\begin{aligned} \bar{\Psi}\gamma^i\Psi &= -\tilde{v}^\dagger\sigma^i\tilde{v} + v^\dagger\sigma^i v \\ \bar{\Psi}\gamma^i\gamma_5\Psi &= \tilde{v}^\dagger\sigma^i\tilde{v} + v^\dagger\sigma^i v \\ \bar{\Psi}\Psi &= \tilde{v}^{\mathrm{T}}(\mathrm{i}\sigma^2)v - \mathrm{i}v^\dagger\sigma^2\tilde{v}^* \end{aligned}\right\} \tag{C.8}$$

The two-component complex Weyl spinors $u(x)$ and $v(x)$ describe massless fermions in one helicity state each, together with their antiparticles in the opposite helicity state (see also the end of this Appendix). The Dirac spinor describes a fermion with two degrees of freedom related by the parity operation, and its corresponding antiparticle. Under

$$\text{CP:} \quad \begin{pmatrix} u \\ v \end{pmatrix} \to \begin{pmatrix} -\mathrm{i}\sigma^2 u^* \\ +\mathrm{i}\sigma^2 v^* \end{pmatrix} \tag{C.9}$$

$$\text{P:} \quad \begin{pmatrix} u \\ v \end{pmatrix} \to \begin{pmatrix} v \\ u \end{pmatrix} \tag{C.10}$$

The Majorana spinor is defined by the relation

$$\Psi^{\mathrm{C}} = \Psi \tag{C.11}$$

i.e.

$$\left.\begin{aligned} u(x) &= -\mathrm{i}\sigma^2 v^*(x) \\ v(x) &= \mathrm{i}\sigma^2 u^*(x) \end{aligned}\right\} \tag{C.12}$$

for the associated Weyl spinors. Thus

$$\Psi_{\mathrm{M}} = \begin{pmatrix} -\mathrm{i}\sigma^2 v^*(x) \\ v(x) \end{pmatrix} = \begin{pmatrix} u(x) \\ \mathrm{i}\sigma^2 u^*(x) \end{pmatrix} \tag{C.13}$$

is represented by only one Weyl spinor. It can describe a massive self-conjugate fermion with two spin degrees of freedom and with the Lorentz invariant mass term $-\mathrm{i}(v^\dagger\sigma^2 v^* - v^{\mathrm{T}}\sigma^2 v)$. Four components of a Majorana spinor may be also viewed as the real and imaginary parts of a two-component Weyl spinor. This can be seen in the Majorana representation:

$$\left.\begin{aligned} \gamma^0 &= \begin{pmatrix} 0 & \sigma^2 \\ \sigma^2 & 0 \end{pmatrix} \quad & \gamma^1_{\mathrm{M}} &= \begin{pmatrix} \mathrm{i}\sigma^3 & 0 \\ 0 & \mathrm{i}\sigma^3 \end{pmatrix} \\ \gamma^2_{\mathrm{M}} &= \begin{pmatrix} 0 & -\sigma^2 \\ \sigma^2 & 0 \end{pmatrix} \quad & \gamma^3_{\mathrm{M}} &= \begin{pmatrix} -\mathrm{i}\sigma^1 & 0 \\ 0 & -\mathrm{i}\sigma^1 \end{pmatrix} \end{aligned}\right\} \tag{C.14}$$

Lorentz transformation properties of Weyl spinors

Consider the group $SL(2, C)$ of 2×2 complex matrices M of determinant one. Its connection to the Lorentz group is analogous to the connection of the group $SU(2)$ of two-dimensional unitary unimodular matrices to the rotation group (e.g.

Werle 1966) and can be established through the Pauli matrices. To every space-time point x we can assign a matrix

$$\sigma\cdot x=\sum_{\mu=0}^{3}\sigma^{\mu}\cdot x_{\mu}=\begin{pmatrix} x_0+x_3 & x_1-ix_2 \\ x_1+ix_2 & x_0-x_3 \end{pmatrix} \tag{C.15}$$

where

$$\sigma^0=\begin{pmatrix} 1 & 0 \\ 0 & 1 \end{pmatrix}$$

and σ^i, $i=1,2,3$, are Pauli matrices such that

$$\det(\sigma\cdot x)=x^2$$

The transformation $x\to x'$ is defined by the equation

$$\sigma x'=M\sigma xM^{\dagger},\quad \det M=1 \tag{C.16}$$

Since

$$\det(\sigma x')=x'^2=\det(\sigma x)=x^2 \tag{C.17}$$

the transformation $x'^{\mu}=\Lambda^{\mu}{}_{\nu}(M)x^{\nu}$ is a Lorentz transformation. The group $SL(2,C)$ is homomorphic to the restricted Lorentz group $\det\Lambda=1$, $\Lambda^0{}_0\geqslant 1$: $L_+^{\uparrow}=SL(2,C)/Z_2$.

The group $SL(2,C)$ has two inequivalent spinor representations

$$\left.\begin{aligned} \varphi'_{\alpha}&=M_{\alpha}{}^{\beta}\varphi_{\beta} \\ \bar{\chi}'_{\dot{\alpha}}&=(M^*)_{\dot{\alpha}}{}^{\dot{\beta}}\bar{\chi}_{\dot{\beta}} \end{aligned}\right\} \tag{C.18}$$

where φ and χ are two-component fields (the bar and the dotted indices are just the notation). From (C.16) and (C.18) it follows that σ^{μ} has the following index structure: $\sigma^{\mu}_{\alpha\dot{\alpha}}$. The antisymmetric tensors $\varepsilon^{\alpha\beta}$ and $\varepsilon_{\alpha\beta}$

$$\left.\begin{aligned} \varepsilon^{12}&=-\varepsilon^{21}=1 \\ \varepsilon_{21}&=-\varepsilon_{12}=1 \end{aligned}\right\} \tag{C.19}$$

(and the same for dotted indices) satisfy

$$\varepsilon^{\alpha\beta}\varepsilon_{\beta\gamma}=\delta^{\alpha}{}_{\gamma}=\delta_{\gamma}{}^{\alpha} \tag{C.20}$$

and, since M is unimodular, they are invariant under $SL(2,C)$ transformations

$$\left.\begin{aligned} \varepsilon_{\alpha\beta}&=M_{\alpha}{}^{\gamma}M_{\beta}{}^{\delta}\varepsilon_{\gamma\delta} \\ \varepsilon^{\alpha\beta}&=\varepsilon^{\gamma\delta}M_{\gamma}{}^{\alpha}M_{\delta}{}^{\beta} \end{aligned}\right\} \tag{C.21}$$

In compact notation the first equation (C.21) reads

$$\varepsilon=M\varepsilon M^{\mathrm{T}} \tag{C.22}$$

or

$$\varepsilon^{-1}M\varepsilon=(M^{-1})^{\mathrm{T}} \tag{C.23}$$

so that $(M^{-1})^{\mathrm{T}}$ is equivalent to M. The spinor

$$\varphi^\alpha = \varepsilon^{\alpha\beta}\varphi_\beta \tag{C.24}$$

transforms as follows

$$\varphi'^\alpha = (M^{-1})_\beta{}^\alpha \varphi^\beta \tag{C.25}$$

Analogously $\varepsilon^{-1}M^*\varepsilon = (M^{*-1})^{\mathrm{T}}$ and

$$\bar{\chi}^{\dot\alpha} = (M^{*-1})_{\dot\beta}{}^{\dot\alpha}\bar{\chi}^{\dot\beta} \tag{C.26}$$

We see that

$$\varphi_1 \cdot \varphi_2 = \varphi_1{}^\alpha \varphi_{2\alpha} = -\varphi_{1\alpha}\varphi_2{}^\alpha = \varphi_2{}^\alpha \varphi_{1\alpha} = \varphi_2 \cdot \varphi_1 \tag{C.27}$$

and

$$\bar{\chi}_1\bar{\chi}_2 = \bar{\chi}_{1\dot\alpha}\bar{\chi}_2^{\dot\alpha} = -\bar{\chi}_1{}^{\dot\alpha}\bar{\chi}_{2\dot\alpha} = \bar{\chi}_{2\dot\alpha}\bar{\chi}_1^{\dot\alpha} = \bar{\chi}_2\bar{\chi}_1 \tag{C.28}$$

are invariant under $SL(2, C)$ transformations.

The spinor representations (C.18) of $SL(2, C)$ transform as $(\frac{1}{2}, 0)$ and $(0, \frac{1}{2})$ representations of the Lorentz group, respectively. We recall that Lorentz transformations (see Problem 7.1)

$$U(\alpha) = \exp(-\tfrac{1}{2}\mathrm{i}\omega_{\mu\nu}M^{\mu\nu})$$

can also be parametrized in terms of

$$N_i^\pm = \tfrac{1}{2}(J_i \pm \mathrm{i}K_i) \tag{C.29}$$

where

$$\left.\begin{aligned} J_i &= \tfrac{1}{2}\varepsilon_{ijk}M_{jk} \\ K_i &= M_{0i} \\ [J_i, J_j] &= \mathrm{i}\varepsilon_{ijk}J_k \\ [K_i, K_j] &= -\mathrm{i}\varepsilon_{ijk}J_k \\ [J_i, K_j] &= \mathrm{i}\varepsilon_{ijk}K_k \end{aligned}\right\} \tag{C.30}$$

and

$$\left.\begin{aligned} [N_i^+, N_j^-] &= 0 \\ [N_i^\pm, N_j^\pm] &= \mathrm{i}\varepsilon_{ijk}N_k^\pm \end{aligned}\right\} \tag{C.31}$$

We see that the N_i^+ and N_i^- obey the Lie algebra of $SU(2)$. The finite-dimensional (non-unitary) representations of the Lorentz group are labelled by the pair (m, n) where $m(m+1)$ is the eigenvalue of the $N_i^+N_i^+$, and $n(n+1)$ of the $N_i^-N_i^-$. However the two $SU(2)$ groups are not independent (it is not a direct product) as they are interchanged under the parity operation

$$J_i \to J_i, \qquad K_i \to -K_i \tag{C.32}$$

and also under hermitean conjugation. Taking into account the relation between the $SL(2, C)$ and the Lorentz group we see that the two-component spinor representations (C.18) transform as $(\frac{1}{2}, 0)$ and $(0, \frac{1}{2})$, respectively.

When parity is relevant we need the Dirac spinor representation $(\frac{1}{2},0)\oplus(0,\frac{1}{2})$. Using the general form of the Lorentz transformation for the Dirac spinor

$$S=\exp(\tfrac{1}{4}\mathrm{i}\tilde{\sigma}^{\mu\nu}\omega_{\mu\nu}),\quad \tilde{\sigma}^{\mu\nu}=\tfrac{1}{2}\mathrm{i}[\gamma^{\mu},\gamma^{\nu}]$$

and the chiral representation for the γ-matrices one can see that a Dirac spinor can be written as

$$\left.\begin{aligned}\Psi&=\begin{pmatrix}\varphi_{\alpha}\\ \bar{\chi}^{\dot{\alpha}}\end{pmatrix}\\ \bar{\Psi}&=(\bar{\varphi}_{\dot{\alpha}},\chi^{\alpha})\begin{pmatrix}0&1\\1&0\end{pmatrix}=(\chi^{\alpha},\bar{\varphi}_{\dot{\alpha}})\end{aligned}\right\}\tag{C.33}$$

where the transformation properties of φ and χ are defined by (C.18) and (C.26). Indeed, introducing the notation

$$\sigma^{\mu}=(1,\sigma^{i}),\quad \bar{\sigma}^{\mu}=(1,-\sigma^{i})\tag{C.34}$$

we have

$$\gamma^{\mu}=\begin{pmatrix}0&\sigma^{\mu}\\ \bar{\sigma}^{\mu}&0\end{pmatrix}\tag{C.35}$$

and

$$\tilde{\sigma}^{\mu\nu}=\tfrac{1}{2}\mathrm{i}[\gamma^{\mu},\gamma^{\nu}]\equiv\begin{pmatrix}\sigma^{\mu\nu}&0\\0&\bar{\sigma}^{\mu\nu}\end{pmatrix}\tag{C.36}$$

where

$$\left.\begin{aligned}\sigma^{\mu\nu}&=\tfrac{1}{2}\mathrm{i}(\sigma^{\mu}\bar{\sigma}^{\nu}-\sigma^{\nu}\bar{\sigma}^{\mu})\\ \bar{\sigma}^{\mu\nu}&=\tfrac{1}{2}\mathrm{i}(\bar{\sigma}^{\mu}\sigma^{\nu}-\bar{\sigma}^{\nu}\sigma^{\mu})\end{aligned}\right\}\tag{C.37}$$

It is then straightforward to see that the Lorentz transformation matrix S can be written as follows

$$S=\exp(\tfrac{1}{4}\mathrm{i}\tilde{\sigma}^{\mu\nu}\omega_{\mu\nu})=\begin{pmatrix}M&0\\0&(M^{-1})^{\dagger}\end{pmatrix}\tag{C.38}$$

where

$$M=\exp(\tfrac{1}{4}\mathrm{i}\sigma^{\mu\nu}\omega_{\mu\nu})=\exp(\tfrac{1}{4}\sigma^{i}\sigma^{j}\omega_{ij})\tag{C.39}$$

Using $(M^{-1})^{\dagger}=(M^{*-1})^{\mathrm{T}}$, (C.18) and (C.26) we get (C.33). We also note the following transformation rules

$$\left.\begin{aligned}\sigma^{\mu}&=(\sigma^{\mu})_{\alpha\dot{\beta}},\quad \bar{\sigma}^{\mu}=(\bar{\sigma}^{\mu})^{\dot{\alpha}\beta}\\ \sigma^{\mu\nu}&=(\sigma^{\mu\nu})_{\alpha}{}^{\beta},\quad \bar{\sigma}^{\mu\nu}=(\bar{\sigma}^{\mu\nu})^{\dot{\alpha}}{}_{\dot{\beta}}\\ (\bar{\sigma}^{\mu})^{\dot{\alpha}\beta}&=(\sigma^{\mu})^{\beta\dot{\alpha}}=\varepsilon^{\beta\rho}\varepsilon^{\dot{\alpha}\dot{\sigma}}(\sigma^{\mu})_{\rho\dot{\sigma}}\end{aligned}\right\}\tag{C.40}$$

Using the relation

$$\sigma^{\mu}_{\alpha\dot{\alpha}}\bar{\sigma}_{\mu}{}^{\dot{\beta}\beta}=2\delta_{\alpha}{}^{\beta}\delta_{\dot{\alpha}}{}^{\dot{\beta}}\tag{C.41}$$

one derives the Fierz transformation

$$(\Theta\sigma^{\mu}\bar{\Theta})(\bar{\chi}\bar{\sigma}_{\mu}\chi)=-2(\Theta\chi)(\bar{\chi}\bar{\Theta})\tag{C.42}$$

or

$$(\Theta\sigma^{\mu}\bar{\Theta})(\chi\sigma_{\mu}\bar{\chi})=2(\Theta\chi)(\bar{\chi}\bar{\Theta})\tag{C.43}$$

In the present notation we identify

$$\left.\begin{array}{c} u=\{\varphi_\alpha\}, \quad v=\{\bar{\chi}^{\dot{\alpha}}\}, \quad u^\dagger=\{\bar{\varphi}_{\dot{\alpha}}\}, \quad v^\dagger=\{\chi^\alpha\} \\ \mathrm{i}\sigma^2=\varepsilon^{\alpha\beta}, \quad -\mathrm{i}\sigma^2=\varepsilon_{\alpha\beta} \end{array}\right\} \tag{C.44}$$

and

$$\left.\begin{array}{r} \bar{\Psi}\Psi=\chi^\alpha\varphi_\alpha+\bar{\varphi}_{\dot{\alpha}}\bar{\chi}^{\dot{\alpha}} \\ \bar{\Psi}\gamma^m\Psi=\chi\sigma^m\bar{\chi}+\bar{\varphi}\bar{\sigma}^m\varphi \end{array}\right\} \tag{C.45}$$

where

$$\bar{\sigma}^{m\dot{\alpha}\alpha}=\varepsilon^{\alpha\beta}\varepsilon^{\dot{\alpha}\dot{\beta}}\sigma^m_{\beta\dot{\beta}}$$

which coincide with those in (C.5). C may be represented as

$$C=\begin{pmatrix} \varepsilon_{\alpha\beta} & 0 \\ 0 & \varepsilon^{\dot{\alpha}\dot{\beta}} \end{pmatrix} \tag{C.46}$$

and

$$\begin{pmatrix} \varphi_\alpha \\ \bar{\chi}^{\dot{\alpha}} \end{pmatrix}^{\mathrm{C}}=\begin{pmatrix} \chi_\alpha \\ \bar{\varphi}^{\dot{\alpha}} \end{pmatrix} \tag{C.47}$$

For Majorana spinors $\varphi^\alpha=\chi^\alpha$.

Free particle solutions of the massless Dirac equation

Finally we recall the properties of the free particle solutions of the massless Dirac equation written as chiral states and Weyl spinors. One can start with the four-component notation and use the 'chiral' representation (C.4) for the γ-matrices. Then the Dirac equation may be written in a split two-component form, with

$$\left.\begin{array}{l} \Psi=\begin{pmatrix} u \\ v \end{pmatrix} \\ \Psi_{\mathrm{L}}=P_{\mathrm{L}}\Psi=\begin{pmatrix} u \\ 0 \end{pmatrix} \\ \Psi_{\mathrm{R}}=P_{\mathrm{R}}\Psi=\begin{pmatrix} 0 \\ v \end{pmatrix} \end{array}\right\} \tag{C.48}$$

as

$$\mathrm{i}\frac{\partial}{\partial t}\begin{pmatrix} u \\ v \end{pmatrix}=\boldsymbol{\sigma}\cdot\mathbf{p}\begin{pmatrix} -u \\ +v \end{pmatrix} \tag{C.49}$$

The positive energy solutions with momentum $\mathbf{p}$ are

$$\begin{pmatrix} u^{(+)}_{\mathbf{p}}(x) \\ v^{(+)}_{\mathbf{p}}(x) \end{pmatrix}=E^{1/2}\begin{pmatrix} u^{(+)}(p) \\ v^{(+)}(p) \end{pmatrix}\exp[-\mathrm{i}(Et-\mathbf{px})] \tag{C.50}$$

where $u^{(+)}(p)$ and $v^{(+)}(p)$ must satisfy the equation

$$\boldsymbol{\sigma}\cdot\mathbf{p}\begin{pmatrix} u^{(+)}(p) \\ v^{(+)}(p) \end{pmatrix}=E\begin{pmatrix} -u^{(+)}(p) \\ +v^{(+)}(p) \end{pmatrix} \tag{C.51}$$

i.e. $v^{(+)}(p)$ describes a right-handed (positive helicity) particle, and $u^{(+)}(p)$ a left-handed (negative helicity) one. For $\mathbf{p}$ in the z direction we get

$$v^{(+)}(p) = \begin{pmatrix} 1 \\ 0 \end{pmatrix} \quad u^{(+)}(p) = \begin{pmatrix} 0 \\ 1 \end{pmatrix} \tag{C.52}$$

The negative energy solutions (states with the same $\mathbf{p}$ as above) are ($E = +|\mathbf{p}|$)

$$\begin{pmatrix} u_{\mathbf{p}}^{(-)}(x) \\ v_{\mathbf{p}}^{(-)}(x) \end{pmatrix} = E^{1/2} \begin{pmatrix} u^{(-)}(-p) \\ v^{(-)}(-p) \end{pmatrix} \exp[\mathrm{i}(Et - (-\mathbf{p})\mathbf{x})] \tag{C.53}$$

where

$$\boldsymbol{\sigma}\cdot(-\mathbf{p}) \begin{pmatrix} u^{(-)}(-p) \\ v^{(-)}(-p) \end{pmatrix} = E \begin{pmatrix} -u^{(-)}(-p) \\ +v^{(-)}(-p) \end{pmatrix} \tag{C.54}$$

i.e. $v^{(-)}(-p)$ and $u^{(-)}(-p)$ are respectively right-handed and left-handed (in agreement with our definition (C.48) of R and L by the projection operators P_{R} and P_{L}) but with respect to $(-\mathbf{p})$. We get now

$$v^{(-)}(-p) = \begin{pmatrix} 0 \\ 1 \end{pmatrix} \quad u^{(-)}(-p) = \begin{pmatrix} 1 \\ 0 \end{pmatrix} \tag{C.55}$$

Under the charge conjugation operation

$$\Psi^{\mathrm{C}} = C\gamma_0 \Psi^* = \begin{pmatrix} & -\mathrm{i}\sigma^2 \\ +\mathrm{i}\sigma^2 & \end{pmatrix} \Psi^*$$

and therefore

$$\begin{aligned}(\Psi_{\mathrm{L}}^{(-)})^{\mathrm{C}} &= E^{1/2} \exp[-\mathrm{i}(Et - (-\mathbf{p})\mathbf{x}] \begin{pmatrix} 0 \\ +\mathrm{i}\sigma^2 u^{(-)}(-p) \end{pmatrix} \\ &= E^{1/2} \exp[-\mathrm{i}(Et - (-\mathbf{p})\mathbf{x}] \begin{pmatrix} 0 \\ v^{(-)}(-p) \end{pmatrix}\end{aligned} \tag{C.56}$$

i.e. it describes a particle with $E > 0$, momentum $(-\mathbf{p})$ and positive helicity (with respect to $(-\mathbf{p})$). Similarly, $(\Psi_{\mathrm{R}}^{(-)})^{\mathrm{C}}$ describes a left-handed particle with $E > 0$ and with momentum $(-\mathbf{p})$. We can also write (since $v^+(-p) = v^{(-)}(-p)$):

$$\begin{aligned}(u_{\mathbf{p}}^{(-)}(x))^{\mathrm{C}} &= v_{-\mathbf{p}}^{+}(x) \\ (v_{\mathbf{p}}^{(-)}(x))^{\mathrm{C}} &= u_{-\mathbf{p}}^{(+)}(x)\end{aligned} \tag{C.57}$$

Normalization of boson states:

$$\int \mathrm{d}^3x\, f_{\mathbf{p}}^*(x)\mathrm{i}\overleftrightarrow{\partial}_0 f_{\mathbf{p}'}(x) = \int \mathrm{d}^3x \exp(\mathrm{i}px)\mathrm{i}\overleftrightarrow{\partial}_0 \exp(-\mathrm{i}p'x) = (2\pi)^3\delta(\mathbf{p} - \mathbf{p}')2p_0$$

$$a\overleftrightarrow{\partial}_0 b = a\left(\frac{\partial b}{\partial t}\right) - \left(\frac{\partial a}{\partial t}\right)b, \quad f_{\mathbf{p}} = \exp(-\mathrm{i}px)$$

Massless fermion states also are normalized to $(2\pi)^3\delta(\mathbf{p} - \mathbf{p}')2p_0$.

References

Abbott, L.F. (1982). *Acta Phys. Pol.* **B13**, 33

Abers, E.S. & Lee, B.W. (1973). *Phys. Rep.* **9C**, 1

Adkins, H., Nappi, C. & Witten, E. (1983). *Nucl. Phys.* **B228**, 552

Adler, S.L. (1969). *Phys. Rev.* **177**, 2426

Adler, S.L. (1970). In *Lectures on elementary particles and quantum field theory*, eds. Deser, S., Grisaru, M. & Pendleton, H., MIT Press, Cambridge, Massachusetts

Adler, S. L. & Bardeen, W.A. (1969). *Phys. Rev.* **182**, 1517

Adler, S.L., Collins, J.C. & Duncan, A. (1977). *Phys. Rev.* **D15**, 1712

Aitchison, I.J.R. & Hey, A.J.G. (1982). *Gauge theories in particle physics*, Hilger, Bristol

Altarelli, G. & Parisi, G. (1977). *Nucl. Phys.* **B126**, 298

Alvarez-Gaumé, L. & Baulieu, L. (1983). *Nucl. Phys.* **B212**, 255

Alvarez-Gaumé, L. & Ginsparg, P. (1984). *Nucl. Phys.* **B243**, 449

Alvarez-Gaumé, L. & Witten, E. (1983). *Nucl. Phys.* **B234**, 269

Appelquist, T. & Carazzone, J. (1975). *Phys. Rev.* **D11**, 2856

Aubert, J.J. *et al* (1974). *Phys. Rev. Lett.* **33**, 1404

Augustin, J.E. *et al* (1974). *Phys. Rev. Lett.* **33**, 1406

Aviv, R. & Zee, A. (1972). *Phys. Rev.* **D5**, 2372

Balachandran, A.P., Marmo, G., Nair, V.P. & Trahern, C.G. (1982). *Phys. Rev.* **D25**, 2713

Bardeen, W.A. (1969). *Phys. Rev.* **184**, 1848

Bardeen, W.A. (1972). In *Proc. of the XVI Intern. Conf. on High Energy Physics, NAL.*, eds. Jackson, J.D. & Roberts, A, Fermilab, Batavia

Bardeen, W.A. (1974). *Nucl. Phys.* **B75**, 246

Bardeen, W.A., Fritzsch, H. & Gell-Mann, M. (1973). In *Scale and Conformal Symmetry in Hadron Physics*, ed. Gatto, R., Wiley, New York

Bardeen, W.A. & Zumino, B. (1984). *Nucl. Phys.* **B244**, 421

Bassetto, A., Ciafaloni, M. & Marchesini, G. (1980). *Nucl. Phys.* **B163**, 477

Bassetto, A., Ciafaloni, M., Marchesini, G. & Mueller, A.H. (1982). *Nucl. Phys.* **B207**, 189

Bassetto, A., Dalbosco, M., Lazzizzera, I. & Soldati, R. (1985). *Phys. Rev.* **D31**, 2012

Bassetto, A., Lazzizzera, I. & Soldati, R. (1985). *Nucl. Phys.* **B236**, 319

Baulieu, L. & Thierry Mieg, J. (1982). *Nucl. Phys.* **B197**, 477

Becchi, C., Rouet, A. & Stora, R. (1974). *Phys. Lett.* **52B**, 344

Becchi, C., Rouet, A. & Stora, R. (1976). *Ann. Phys.* **98**, 287

Belavin, A.A., Polyakov, A.M. & Zamolodchikov, A.B. (1984). *Nucl. Phys.* **B241**, 333

Bell, J. & Jackiw, R. (1969). *Nuovo Cimento* **60A**, 47

Bernard, C., Duncan, A., Lo Secco, J. & Weinberg, S. (1975) *Phys. Rev.* **D12**, 792

Bernard, C.W. & Weinberg, E.J. (1977). *Phys. Rev.* **D15**, 3656

Bernstein, J. (1974). *Rev. Mod. Phys.* **46**, 7

Bjorken, J.D. (1969). *Phys. Rev.* **179**, 1547

Bjorken, J.D. & Drell, S.D. (1964). *Relativistic Quantum Mechanics*, McGraw–Hill Inc., New York

Bjorken, J.D. & Drell, S.D. (1965). *Relativistic Quantum Fields*, McGraw–Hill Inc., New York

Bogoliubov, N.N. & Shirkov, D.V. (1959). *Introduction to the Theory of Quantized Fields*, Interscience Publishers Inc., New York

Bollini, C.G. & Giambiagi, J.J. (1972). *Phys. Letters* **40B**, 566

Bonora, L. & Tonin, M. (1981). *Phys. Lett.* **98B**, 48

Bott, R. (1956). *Bull. Soc. Math. France* **84**, 251

Boulware, D.G. (1970). *Ann. Phys. N.Y.* **56**, 140

Buras, A.J. (1980). *Rev. Mod. Phys.* **52**, 199

Callan, C.G., Coleman, S. & Jackiw, R. (1970). *Ann. Phys. N.Y.* **59**, 42

Callan, C.G., Dashen, R.F. & Gross, D.J. (1976). *Phys. Lett.* **63B**, 334

Callan, C.G., Jr., Coleman, S., Wess, J. & Zumino, B. (1969). *Phys. Rev.* **177**, 2247

Caswell, W. (1974). *Phys. Rev. Lett.* **33**, 224

Chanowitz, M.S. & Ellis, J. (1973). *Phys. Rev.* **D7**, 2490

Cheng, R.S. (1972). *J. Math. Phys.* **13**, 1723

Coleman, S. (1974). In *Laws of hadronic matter. 1973 International School of Subnuclear Physics, Erice*, ed. Zichichi, A. Academic Press, New York

Coleman, S. (1979). In *Proceedings of the 1977 International School of Subnuclear Physics, Erice*, ed. Zichichi, A. Plenum Press, New York

Coleman, S. & Gross, D.J. (1973). *Phys. Rev. Lett.* **31**, 851

Coleman, S. & Mandula, J. (1967). *Phys. Rev.* **159**, 1251

Coleman, S. & Weinberg, E. (1973). *Phys. Rev.* **D7**, 1888

Coleman, S., Wess, J. & Zumino, B. (1969). *Phys. Rev.* **177**, 2239

Collins, J.C. (1974). *Phys. Rev.* **D10**, 1213

Collins, J.C. (1984). *Renormalization*, Cambridge University Press, Cambridge

Collins, J.C., Duncan, A. & Joglekar, S.D. (1977). *Phys. Rev.* **D16**, 438

Crewther, R. (1979). In *Field theoretical methods in particle physics*, ed. Rühl, W., Plenum Publishing Corporation, New York

Curci, G. & Ferrari, R. (1976). *Phys. Lett.* **63B**, 91

Das, T., Guralnik, G.S., Mathur, U.S., Low, F.E. & Young, J.E. (1967). *Phys. Rev. Lett.* **18**, 759

Dashen, R. (1969). *Phys. Rev.* **183**, 1245

Dashen, R. & Neuberger, H. (1983). *Phys. Rev. Lett.* **50**, 1897

Dashen, R. & Weinstein, M. (1969). *Phys. Rev.* **183**, 1291

de Rafael, E. (1979). *Lectures on Quantum Electrodynamics*, University of Barcelona UAB–FT–D–1, Barcelona

de Rafael, E. & Rosner, J. (1974). *Ann. Phys. N.Y.* **82**, 369

Dine, M., Fischler, W. & Srednicki, M. (1981). *Phys. Lett.* **104B**, 199

Di Vecchia, P. (1980). In *Field Theory and Strong Interactions*, ed. Urban, P., Springer-Verlag, Vienna

Di Vecchia, P. & Veneziano, G. (1980). *Nucl. Phys.* **B171**, 253

Dokshitzer, Y.L., Dyakonov, D.I. & Troyan, S.I. (1980). *Phys. Rep.* **58C**, 269

Eden, R.J., Landshoff, P.V., Olive, D.I. & Polkinghorne, J.C. (1966). *The analytic S-Matrix*, Cambridge University Press, Cambridge

Ellis, J. (1977). In *Weak and electromagnetic interactions at high energy, Proc. of 1976 Les Houches Summer School*, eds. Balian, R. & Llewellyn Smith, C.H., North–Holland, Amsterdam
Ellis, R.K. *et al.* (1979). *Nucl. Phys.* **B152**, 285
Fahri, E. & Susskind, L. (1981). *Phys. Rep* **74C**, 277
Falck, N.K., Hirshfeld, A.C. & Kubo, J. (1983). *Phys. Lett.* **125B**, 175
Feynman, R.P. (1972). *Photon–Hadron Interaction*. Benjamin, Reading, Massachusetts
Feynman, R.P. (1977). In *Weak and electromagnetic interactions at high energy, Proc. of 1976 Les Houches Summer School*, eds. Balian, R. & Llewellyn Smith, C.H., North–Holland, Amsterdam
Fock, V. (1926). *Zeit. f. Phys.* **39**, 226
Frampton, P.H. & Kephart, T.W. (1983). *Phys. Rev.* **D28**, 1010
Fritzsch, H., Gell-Mann, M. & Leutwyler, H. (1973). *Phys. Lett.* **47B**, 365
Fujikawa, K. (1980). *Phys. Rev.* **D21**, 2848
Fujikawa, K. (1984). *Phys. Rev.* **D29**, 285
Furmański, W., Petronzio, R. & Pokorski, S. (1979). *Nucl. Phys.* **B155**, 253
Gasiorowicz, S. (1966). *Elementary Particle Physics*, Wiley, New York
Gasser, J. & Leutwyler, H. (1982). *Phys. Rep.* **87C**, 77
Gasser, J. & Leutwyler, H. (1984). *Ann. Phys.* **158**, 142
Gasser, J. & Leutwyler, H. (1985*a*). *Nucl. Phys.* **B250**, 465
Gasser, J. & Leutwyler, H. (1985*b*). *Nucl. Phys.* **B250**, 517
Gates, S.J., Jr., Grisaru, M.T., M. Roček, M. & Siegel, W. (1983). *Superspace, Frontiers in Physics*, Benjamin–Cummings, Reading, Massachusetts
Gell-Mann, M. (1964). *Phys. Lett.* **8**, 214
Georgi, H. & Glashow, S.L. (1972). *Phys. Rev.* **D6**, 429
Georgi, H. & Pais, A. (1977). *Phys. Rev.* **D16**, 3520
Gildener, E. (1976). *Phys. Rev.* **D13**, 1025
Gildener, E. & Weinberg, S. (1976). *Phys. Rev.* **D13**, 3333
Glashow, S.L. (1961). *Nucl. Phys.* **22**, 579
Glashow, S.L., Iliopoulos, J. & Maiani, L. (1970). *Phys. Rev.* **D2**, 185
Goldstone, J. (1961). *Nuovo Cimento* **19**, 154
Goldstone, J., Salam, A. & Weinberg, S. (1962). *Phys. Rev.* **127**, 965
Greenberg, O.W. (1964). *Phys. Rev. Lett.* **13**, 598
Gribov, V.N. & Lipatov, I.N. (1972). *Sov. J. Nucl. Phys.* **15**, 438
Gross, D.J. (1976). In *Methods in field theory, Les Houches 1975*, eds. Balian, R. & Zinn-Justin, J., North–Holland, Amsterdam
Gross, D.J. & Jackiw, R. (1972). *Phys. Rev.* **D6**, 477
Gross, D. & Neveu, A. (1974). *Phys. Rev.* **D10**, 3235
Gross, D.J. & Wilczek, F. (1973). *Phys. Rev. Lett.* **30**, 1343
Haag, R., Łopuszański, J.T. & Sohnius, M.F. (1975). *Nucl. Phys.* **B88**, 257
Hasert, F.J. *et al.* (1973). *Phys. Lett.* **46B**, 138
Han, M-Y. & Nambu, Y. (1965). *Phys. Rev.* **139B**, 1006
Hirshfeld, A.C. & Leschke, H. (1981). *Phys. Lett.* 101B, 48
Itzykson, C. & Zuber, J.B. (1980). *Quantum field theory,* McGraw–Hill, New York
Jackiw, R. (1972). In *Lectures on current algebra and its applications* by Treiman, S.B., Jackiw, R. & Gross, D.J., Princeton University Press, Princeton, New Jersey
Jackiw, R. (1980). *Rev. Mod. Phys.* **52**, 661
Jackiw, R. (1984). In *Relativity, groups and topology II*; *Les Houches, Session XL*, 1983, eds. De Witt, B.S. & Stora, R., Elsevier Science Publishers B.V., Amsterdam
Jones, D.R.T. (1974). *Nucl. Phys.* **B75**, 531

Kallosch, R.E. (1978). *Nucl. Phys.* **B141**, 141
Kinoshita, T. (1962). *J. Math. Phys.* **3**, 650
Klein, O. (1939). In *New Theories in Physics, Proceedings of the Conference organized by the International Union of Physics and the Polish Intellectual Co-operation Committee,* Warsaw, May 30–June 3. International Institute of Intellectual Co-operation (Scientific Collection), Paris
Lane, K. (1974). *Phys. Rev.* **D10**, 1353
Lee, B.W. (1976). In *Methods in Field Theory, Les Houches 1975*, eds. Balian, R. & Zinn-Justin. J., North-Holland, Amsterdam
Lee, B.W., Quigg, C. & Thacker, H.B. (1977). *Phys. Rev.* **D16**, 1519
Lee, T.D. & Nauenberg, M. (1964). *Phys. Rev.* **133**, 1549
Leibbrandt, G. (1984). *Phys. Rev.* **D29**, 1699
Leutwyler, H. (1984). *Acta Phys. Pol.* **B15**, 383
Llewellyn Smith, C.H. (1978). *Acta Phys. Austr.* **Supp.19**, 331
Llewellyn Smith, C.H. (1980). In *Quantum flavourdynamics, quantum chromodynamics and unified theories*, eds. Mahanthappa, K.T. & Randa, J., Plenum Press, New York
London, F. (1927). *Zeit. f. Phys.* **42**, 375
Mandelstam, S. (1983). *Nucl. Phys.* **B213**, 149
Marciano, W. & Pagels, H. (1978). *Phys. Rep.* **36C**, 137
Matthews, P.T. (1949). *Phys. Rev.* **76**, 1254
Nambu, Y. (1960). *Phys. Rev. Lett.* **4**, 380
Nambu, Y. (1966). In *Preludes in theoretical physics*, eds. de Shalit. A., Feshbach, H. & Van Hove, L., North–Holland, Amsterdam
Nielsen, N.K. (1977). *Nucl. Phys.* **B120**, 212
Nielsen, N.K. (1978). *Nucl. Phys.* **B140**, 499
Nielsen, N.K., Grisaru, M.T., Römer, H. & van Nieuwenhuizen, P. (1978). *Nucl. Phys.* **B140**, 477
Nilles, H.P. (1984). *Phys. Rep.* **110**, 1
Novikov, V.A., Shifman, M.A., Vainshtein, A.I. & Zakharov, V.I. (1984). *Phys. Rep.* **116**, 103
Ojima, I. (1980). *Prog. Theor. Phys.* **64**, 625
Ovrut, B.A. (1983). *Nucl. Phys.* **B213**, 241
Ovrut, B.A. & Schnitzer, H.J. (1981). *Nucl. Phys.* **B179**, 381
Pagels, H. (1975). *Phys. Rep.* **16C**, 219
Parisi, G. & Petronzio, R. (1979). *Nucl. Phys.* **B154**, 427
Pauli, W. (1933). *Handbuch der Physik*, 2 Aufl., Vol. 24, Teil 1, p. 83, Geiger & Scheel, Berlin
Peccei, R.D. & Quinn, H.R. (1977). *Phys. Rev.* **D16**, 1791
Peskin, M.E. (1981). *Nucl. Phys.* **B185**, 197
Politzer, H.D. (1973). *Phys. Rev. Lett.* **30**, 1346
Preskill, J. (1981). *Nucl. Phys.* **B177**, 21
Pritchard, D.J. & Stirling, W.J. (1979). *Nucl. Phys.* **B165**, 237
Review of Particle Properties (1984). *Rev. Mod. Phys.* **56**, S1
Reya, E. (1981). *Phys. Rep.* **69C**, 195
Rosenzweig, C., Schechter, J. & Trahern, C.G. (1980). *Phys. Rev.* **D21**, 3388
Sachrajda, C.T.C. (1983*a*). In *Gauge theories in high energy physics, Les Houches, Session XXXVII, 1981*, eds. Gaillard, M.K. & Stora, R. North-Holland Amsterdam
Sachrajda, C.T.C. (1983*b*). In *Proc. of the XIVth International Symposium on Multiparticle Dynamics, Lake Tahoe, June 1983*, ed. Gunion, J.F. University of Davis, Davis
Salam, A. (1968). In *Elementary particle physics*, Nobel Symp. No 8, ed. Svartholm, N. Almqvist Wilsell, Stockholm

Salam, A. & Strathdee, J. (1970). *Phys. Rev.* **D2**, 2869
Salam, A. & Ward, J.C. (1964). *Phys. Lett.* **13**, 168
Schwinger, J. (1949). *Phys. Rev.* **76**, 790
Schwinger, J. (1957). *Ann. Phys. N.Y.* **2**, 407
Shaw, R. (1955). Ph.D. Thesis, Cambridge University
Shifman, M.A., Vainshtein, A.I. & Zakharov, V.I. (1979). *Nucl. Phys.* **B147**, 385
Siegel, W. (1979). *Phys. Lett.* **84B**, 193
Slavnov, A.A. & Faddeev, L.D. (1980). *Gauge fields: introduction to quantum theory*, Benjamin–Cummings, Reading, Massachusetts
Sohnius, M.F. (1985). *Phys. Rep.* **128**, 39
Sterman, G., Townsend, P.K. & van Nieuwenhuizen, P. (1978). *Phys. Rev.* **D17**, 1501.
Tarasov, O.V., Vladimirov, A.A. & Zharkov, A.Y. (1980). *Phys. Lett.* **93B**, 429
Taylor, J.C. (1976). *Gauge theories of weak interactions*, Cambridge University Press, Cambridge
't Hooft, G. (1971*a*). *Nucl. Phys.* **B33**, 173
't Hooft, G. (1971*b*). *Nucl. Phys.* **B35**, 167
't Hooft, G. & Veltman, M. (1972). *Nucl. Phys.* **B44**, 189
't Hooft, G. & Veltman, M. (1973). *Diagrammer*, CERN yellow report
Veneziano, G. (1979). *Nucl. Phys.* **B159**, 213
Weinberg, S. (1967*a*). *Phys. Rev. Lett.* **18**, 507
Weinberg, S. (1967*b*). *Phys. Rev. Lett.* **19**, 1264
Weinberg, S. (1968). *Phys. Rev.* **166**, 1568
Weinberg, S. (1977). In *Festschrift for I.I. Rabi*, ed. Motz, L. New York Acad. Sci., New York
Weinberg, S. (1978). *Phys. Rev. Lett.* **40**, 223
Werle, J. (1966). *Relativistic theory of reactions*, Polish Scientific Publishers, Warsaw
Wess, J. & Bagger, J. (1983). *Supersymmetry and Supergravity*, Princeton University Press, Princeton
Wess, J. & Zumino, B. (1971). *Phys. Lett.* **37B**, 95
Weyl, H. (1919). *Ann. Physik* **59**, 101
Weyl, H. (1929). *Zeit. f. Phys.* **56**, 330
Wilczek, F. (1978). *Phys. Rev. Lett.* **40**, 279
Wilson, K.G. (1969*a*). *Phys. Rev.* **179**, 1499
Wilson, K.G. (1969*b*). *Phys. Rev.* **181**, 1909
Witten, E. (1979). *Nucl. Phys.* **B156**, 269
Witten, E. (1980). *Ann. Phys.* **128**, 363
Witten, E. (1983). *Nucl. Phys.* **B223**, 422
Yang, C.N. & Mills, R.L. (1954). *Phys. Rev.* **96**, 191
Yennie, D.R., Frautschi, S.C. & Suura, H. (1961). *Ann. Phys.* **13**, 379
Zimmermann, W. (1970). In *Lectures on elementary particles and quantum field theory, 1970 Brandeis Summer Institute in Theoretical Physics*, eds. Deser, S., Grisaru, M. & Pendleton, H., MIT Press, Cambridge, Massachusetts
Zumino, B., Wu, Y.S. & Zee, A. (1984). *Nucl. Phys.* **B239**, 477
Zweig, G. (1964). CERN preprints 8182/TH.401 and 8419/TH.412, unpublished

Index